HETEROGENEOUS CATALYSIS IN PRACTICE

McGraw-Hill Chemical Engineering Series

BUILDING THE LITERATURE OF A PROFESSION

Fifteen prominent chemical engineers first met in New York more than 50 years ago to plan a continuing literature for their rapidly growing profession. From industry came such pioneer practitioners as Leo H. Baekeland, Arthur D. Little, Charles L. Reese, John V. N. Dorr, M. C. Whitaker, and R. S. McBride. From the universities came such eminent educators as William H. Walker, Alfred H. White, D. D. Jackson, J. H. James, Warren K. Lewis, and Harry A. Curtis. H. C. Parmelee, then editor of *Chemical and Metallurgical Engineering*, served as chairman and was joined subsequently by S. D. Kirkpatrick as consulting editor.

After several meetings, this committee submitted its report to the McGraw-Hill Book Company in September 1925. In the report were detailed specifications for a correlated series of more than a dozen texts and reference books which have since become the McGraw-Hill Series in Chemical Engineering and which became the cornerstone of the chemical engineering curriculum.

From this beginning there has evolved a series of texts surpassing by far the scope and longevity envisioned by the founding Editorial Board. The McGraw-Hill Series in Chemical Engineering stands as a unique historical record of the development of chemical engineering education and practice. In the series one finds the milestones of the subject's evolution: industrial chemistry, stoichiometry, unit operations and processes, thermodynamics, kinetics, and transfer operations.

Chemical engineering is a dynamic profession, and its literature continues to evolve. McGraw-Hill and its consulting editors remain committed to a publishing policy that will serve, and indeed lead, the needs of the chemical engineering profession during the years to come.

THE SERIES

Bailey and Ollis: *Biochemical Engineering Fundamentals*
Bennett and Myers: *Momentum, Heat, and Mass Transfer*
Beveridge and Schechter: *Optimization: Theory and Practice*
Carberry: *Chemical and Catalytic Reaction Engineering*
Churchill: *The Interpretation and Use of Rate Data—The Rate Concept*
Clarke and Davidson: *Manual for Process Engineering Calculations*
Coughanowr and Koppel: *Process Systems Analysis and Control*
Danckwerts: *Gas Liquid Reactions*
Gates, Katzer, and Schuit: *Chemistry of Catalytic Processes*
Harriott: *Process Control*
Johnson: *Automatic Process Control*
Johnstone and Thring: *Pilot Plants, Models, and Scale-up Methods in Chemical Engineering*
Katz, Cornell, Kobayashi, Poettmann, Vary, Elenbaas, and Weinaug: *Handbook of Natural Gas Engineering*
King: *Separation Processes*
Knudsen and Katz: *Fluid Dynamics and Heat Transfer*
Lapidus: *Digital Computation for Chemical Engineers*
Luyben: *Process Modeling, Simulation, and Control for Chemical Engineers*
McCabe and Smith, J. C.: *Unit Operations of Chemical Engineering*
Mickley, Sherwood, and Reed: *Applied Mathematics in Chemical Engineering*
Nelson: *Petroleum Refinery Engineering*
Perry and Chilton (Editors): *Chemical Engineers' Handbook*
Peters: *Elementary Chemical Engineering*
Peters and Timmerhaus: *Plant Design and Economics for Chemical Engineers*
Ray: *Advanced Process Control*
Reed and Gubbins: *Applied Statistical Mechanics*
Reid, Prausnitz, and Sherwood: *The Properties of Gases and Liquids*
Resnick: *Process Analysis and Design for Chemical Engineers*
Satterfield: *Heterogeneous Catalysis in Practice*
Sherwood, Pigford, and Wilke: *Mass Transfer*
Slattery: *Momentum, Energy, and Mass Transfer in Continua*
Smith, B. D.: *Design of Equilibrium Stage Processes*
Smith, J. M.: *Chemical Engineering Kinetics*
Smith, J. M., and Van Ness: *Introduction to Chemical Engineering Thermodynamics*
Thompson and Ceckler: *Introduction to Chemical Engineering*
Treybal: *Mass Transfer Operations*
Van Winkle: *Distillation*
Volk: *Applied Statistics for Engineers*
Walas: *Reaction Kinetics for Chemical Engineers*
Wei, Russell, and Swartzlander: *The Structure of the Chemical Processing Industries*
Whitwell and Toner: *Conservation of Mass and Energy*

HETEROGENEOUS CATALYSIS IN PRACTICE

Charles N. Satterfield

Professor of Chemical Engineering
Massachusetts Institute of Technology

McGraw-Hill Book Company

New York St. Louis San Francisco Auckland Bogotá Hamburg
Johannesburg London Madrid Mexico Montreal New Delhi
Panama Paris São Paulo Singapore Sydney Tokyo Toronto

HETEROGENEOUS CATALYSIS IN PRACTICE

5 6 7 8 9 0 DODO 8 9 8 7

This book was set in Times Roman by Automated Composition Service, Inc.
The editors were Julienne V. Brown, Sibyl Golden, and Scott Amerman;
the production supervisor was John Mancia.
The drawings were done by ANCO/Boston.
R. R. Donnelley & Sons Company was printer and binder.

Library of Congress Cataloging in Publication Data

Satterfield, Charles N.
Heterogeneous catalysis in practice.

(McGraw-Hill chemical engineering series)
Includes index.
1. Heterogeneous catalysis. I. Title.
TP156.C35S27 660.2′995 80-10513
ISBN 0-07-054875-7

To Anne
Mark and Karen
Joye and Thomas

CONTENTS

PREFACE

Studies of solid catalysts and reactions catalyzed by solids have burgeoned in recent years, stimulated by an increasing number and variety of applications in industry. Significant contributions have come from individuals or groups whose formal academic education was in one or more of a wide variety of disciplines: these have ranged over the entire field of chemistry (including organic chemistry, inorganic chemistry, physical chemistry, chemical kinetics, and surface chemistry) to solid-state and surface physics, ceramics, physical metallurgy, and chemical reaction engineering.

One's first impression, which may be reinforced by further study, is apt to be that it is a vast and confusing field replete with an enormous quantity of perhaps significant but empirical facts intermixed with perhaps useful theories. The situation is not surprising when one reflects that heterogeneous catalysis in practice is concerned with controlling the rate and direction of a chemical reaction, whose basic mechanism is frequently understood only in broad outline; by means of a complex solid substance, typically rather poorly characterized, selected from one or more of the many elements of the periodic table.

In such a situation there is the need for an overview of the landscape to identify features that provide orientation. This book has been written for chemists, chemical engineers, and others who seek such an overview, and especially for those who have had little previous exposure to heterogeneous catalysis and would like an introduction to the subject. The term "practice" in the title is to warn the reader that attention is devoted primarily to catalysts and reactions that are of industrial significance for large-scale operations and utilized under practicable conditions of pressure, temperature, and contact time, often processing impure reactants or mixtures. At present, theoretical concepts are the most successful in interpretation of the reactions of small molecules such as hydrogen, carbon monoxide, oxygen, and nitrogen on well-characterized surfaces. In the present volume, theory has not been neglected, but the attention devoted to various theoretical concepts is in some proportion to those that have stood the test of time and also seem to be of present practical importance or of some value for prediction for those reactions of industrial interest. Properly applied and with appreciation for their limitations, these various correlations, hypotheses, and theories can be a useful guide for effectively employing knowledge, past experience, and intuition.

This book is intended to provide a comprehensive introduction to the kinds of information that one needs to know in order to work with solid catalysts in the laboratory, pilot plant, or commercial installations. For those concerned with chemical reaction engineering it may provide some perspective on the chemical aspects that must be considered in reactor design in addition to the mathematical aspects treated in numerous texts. In this respect the present volume can be useful as a text or reference.

To some degree, the value of an introductory treatment such as this may be inversely proportional to its size, and therefore it has been attempted, at the risk of oversimplification, to reduce each topic to its essentials. It is hoped that readers may find the balance appropriate and useful for their needs. Appendix A provides an annotated guide to the vast literature. Because of their rather specialized nature, polymerization reactions, photocatalysis, and electrocatalysis have not been considered. Experience has shown that many laboratory studies of practical catalysts are vitiated by inadequate experimental procedures, and especially by lack of recognition of the possible effects of mass- and heat-transfer gradients. Some suggestions on design of experiments and warning signals to look for in analyzing data are made in Chapter 11.

A number of problems are given in Appendix B with which the reader, whether a student or practicing scientist or engineer, can test his or her mastery of the material. Most of these deal with experimental data or are drawn from situations that have occurred in practice. Most real problems do not arrive conveniently packaged to fit within the neat framework of a textbook, and some of those included here likewise do not correspond specifically to the subject matter taken up in any one chapter. Therefore the problems are deliberately grouped together, but the order in which they appear corresponds approximately, where appropriate, to the order in which topics are taken up in the book. As a further challenge to the reader, a few problems are included that may or may not involve catalysis as such, and they require for their solution more of a general knowledge of chemistry.

An author is indebted to countless individuals in ways that are frequently difficult to recognize and to acknowledge explicitly. This book is the outgrowth of class notes that I first began to write more than 15 years ago for use in a course in catalysis, directed primarily to seniors and first-year graduate students in chemical engineering. Expanded and revised versions have also been further tested by recent use in intensive courses in industry, and I have profited from innumerable comments and suggestions from neophyte students and experienced practitioners alike. Among the many who have contributed valuable advice and criticisms are: George A. Huff, Jr., William J. Linn, John P. Longwell, William H. Manogue, Michael A. Serio, John H. Sinfelt, Preetinder S. Virk, James Wei, and James F. Weiher. I am also indebted to Craig Abernethy for his careful typing of the manuscript and I wish to express my especial gratitude to my wife, Anne, for help in a variety of ways.

Charles N. Satterfield

CHAPTER

ONE

INTRODUCTION AND BASIC CONCEPTS

1.1 INTRODUCTION

The concept of catalysis as a method of controlling the rate and direction of a chemical reaction has captured the imagination of scientists and technologists since Berzelius in 1835 coordinated a number of disparate observations on chemical transformations by attributing them to a "catalytic force" and coined the term *catalysis* to refer to the "decomposition of bodies" by this force. At about the same time Mitscherlich introduced the term *contact action* for a similar group of phenomena. Ideas of what constitutes a catalyst and the mechanism of catalytic activity have undergone continuous refinement since, spurred by the enormous industrial importance of catalysts as illustrated by the variety of catalytic processes characteristic of modern petroleum refineries and of the chemical industries. Most of these processes involve solid catalysts, and an understanding of catalysis from both the theoretical and practical point of view is essential to chemists and chemical engineers.

In practice catalysis is primarily a technology which draws on many fields such as organic chemistry, surface chemistry, chemical kinetics, thermodynamics, solid-state physics, ceramics, and physical metallurgy. No unified theory of catalysis exists, and there are frequently several alternative, and not necessarily mutually exclusive, theoretical "explanations" for any given set of facts.

A basic concept is that a catalyzed reaction involves the transitory adsorption (almost always chemisorption) of one or more of the reactants onto the surface of the catalyst, rearrangement of the bonding, and desorption of the products. This leads to three groups of theories of catalysis:

1. The *geometrical theories* emphasize the importance of the correspondence between the geometrical configuration of the active atoms at the surface of the catalyst

and the arrangement of the atoms in the portion of the reacting molecule that adsorbs on the catalyst, this portion sometimes being called the *index group*. In one sense the usefulness of this approach is limited in that seldom can one change the geometrical arrangement of atoms in the catalyst surface without changing something else. Studies of reaction rates on different crystal faces of a metal have shown that the rates indeed may change with geometry, and it is found that the introduction of defects by cold rolling of a sheet of metal, by grinding, or by radioactive bombardment may substantially change the rate of a reaction if the reaction temperature is sufficiently low that the defects do not rapidly anneal or that the structure does not assume a more stable configuration.

An aspect of the geometrical approach of great usefulness is the observation that reaction selectivity may be markedly altered by the number and arrangement of sites required for competing reactions, which leads to the concepts of the importance of "ensembles" or specific grouping of atoms at the catalyst surface and *structure sensitivity* as affected by particle size, alloying, and other variables (Chap. 6).

2. The *electronic theories* proceed from the fact that chemisorption involves the distortion or displacement of electron clouds, and they attempt to relate activity to the electronic properties of the catalyst. This may be in terms of the electronic structure of the solid as a whole, or in terms of the orbitals around individual atoms. In the charge-transfer theory of catalysis (see, e.g., Volkenstein, 1963) it is postulated that the reaction rate is controlled by the availability of charge carriers–electrons or holes–in the catalyst. These are visualized as being nonlocalized; i.e., a sea of electrons or holes is available. Chemisorption is then related to the electronic properties of the catalyst–for example, the ease or difficulty of removal or donation of an electron to or from the lattice–as predicted by applying band theory as developed for metals and semiconductors.

   This approach, of considerable interest in the 1950s, is now seen to be too broad and is inadequate or inapplicable for most cases. More recently attention has been directed to the properties of atoms as individual entities and to the electronic effects caused by the nearest neighbors in the solid rather than by the solid as a whole. In many cases it is difficult to separate geometrical effects from localized electronic effects; the relative importance of the two probably varies greatly from case to case.

3. The above two theories represent primarily a physical approach in that the catalyst is regarded as essentially a static material having the property of converting reactant to product. The *chemical approach* on the other hand regards the catalyst as a chemical intermediate that forms an unstable, surface, transitory complex with the reactants. This decomposes into the final products, returning the catalyst to its initial state. The rates of these processes and the structures formed are assumed to obey chemical principles. If the energy of formation of the unstable intermediate is low, the affinity between catalyst and reactants will be weak and the overall rate is limited by the rate of formation of the intermediate. If the energy of formation is high, the intermediate compound will be stable and the rate is limited by the rate of breakup of this intermediate.

This leads to the concept that the maximum rate is obtained when the bonds between the adsorbed complex and the catalyst surface are neither too strong nor too weak. This is a useful concept but limited in that the energetics are generally unknown, more than one intermediate is frequently involved, and one is more generally concerned with selectivity rather than activity as such.

With time, each of these groups of theories has evolved, and the relative emphasis upon each has changed. The pioneering work of Sabatier, summarized in his book, *Catalysis in Organic Chemistry*, in 1918, emphasized the chemical approach. In the following decades geometrical factors received much attention under the impact of Baladin's *multiplet hypothesis* (Balandin, 1969), although it is apparent that geometrical factors alone cannot explain most variations of catalytic activity. In the 1950s the solid-state properties of catalysts received much attention subsequent to the vigorous development of solid-state electronic devices such as the transistor and the availability of ultrapure materials into which controlled trace amounts of known additives had been incorporated. However, the interpretation of catalytic effects by electronic theories has usually been ambiguous, and recent years have seen a reemphasis upon the chemical viewpoint, incorporating a more sophisticated understanding of the nature and behavior of chemisorbed species and bonding, an approach which stems particularly from rapid advances in various instrumental methods.

Both physical and chemical viewpoints may provide insight. It is desirable to be able to relate catalytic activity to certain specific properties of the catalyst surface, yet an understanding of the mechanism of action and a successful search for new and more effective catalysts may proceed predominantly through the chemical approach, which relates catalytic behavior to the vast body of knowledge concerning chemical reactions. The approaches are, of course, interrelated. The fundamental question about which little is yet known is how the surface structure of a solid catalyst causes the reactants to be adsorbed, the chemical bonds to be rearranged, and the products to be desorbed. There is no such thing as a "good" catalyst per se. A substance is or is not a good catalyst only with respect to a specific reaction.

The temperature range of practical interest in catalysis is primarily (although not exclusively) between about 20 and 500°C.[1] Below ambient temperatures, most catalytic reactions of practical interest go too slowly and indeed would become more costly because of the necessity to provide refrigeration or cryogenic cooling. Above approximately 500°C or so, it becomes increasingly difficult to achieve selectivity unless the desired product is unusually stable.

It is important for the technologist to understand the method of thinking and framework of theory within which fundamental investigators view their studies, so

[1]Some major exceptions are ortho-to-para hydrogen conversion, which is carried out industrially at cryogenic temperatures; ammonia oxidation to form nitrogen oxides, hydrogen cyanide synthesis by the Andrussow process, partial oxidation of methanol to formaldehyde on a silver catalyst; and steam reforming of natural gas and naphthas, all of which are high-temperature reactions. Catalysts for the oxidation of pollutants such as carbon monoxide and hydrocarbons and reduction of nitrogen oxides from automobile-engine exhaust must also be stable and effective at high temperatures.

as to be able to utilize theories and advances in fundamental understanding and yet not be sidetracked by trying to apply them under the wrong conditions. The practicing technologist is primarily concerned with the *effect* of the catalyst: how the rate and direction of the reaction are altered by changes in catalyst composition and by changes in feed composition, pressure, temperature, degree of recycle, and reaction time. He or she is concerned with the incorporation of the catalyst into a process, how poisons may inadvertently be introduced into the catalyst system by the other portions of the process, and how this can be guarded against. These catalysts are usually highly active and of complex composition; they must be mechanically rugged, show good stability over long periods of time, and have the requisite activity and selectivity. The reactions with which the technologist is concerned are determined by the economic utility of the process. Frequently a variety of catalysts will give nearly the same performance, and the final selection involves an economic balance on the entire process, which will include such factors as catalyst cost and frequency and difficulty of replacement and/or regeneration.

Scientific investigators, on the other hand, are concerned primarily with *mechanism*. In trying to simplify their systems for more fundamental interpretation, they frequently use catalysts of as simple a composition as possible, for example, pure metal films or single pure metals or compounds, even if they are relatively inactive and would not be used in practice. The problem of mechanical strength and stability over long periods of time is of lesser importance. Their reactants are usually highly pure and are chosen for experimental convenience or because of some unusual feature of the reaction. Usually they consist of small and simple molecules. Many of the valuable fundamental studies in providing insight into the causes of catalyst behavior have not been made with reactions at all, but rather have been studies of the structure of catalyst surfaces or of the nature and properties of adsorbed species.

Every theory is based on some model, and in catalysis the model will usually depart to a significant extent from reality. Although the theoretical framework of the model may provide a structure for organization and correlation of facts or suggest a direction for profitable future investigation, theories of catalysis at present must be used with caution in attempting to predict behavior under new conditions. These theoretical approaches can be of value to the technologist provided he or she understands their limitations and does not confuse theoretical, enthusiastic assertions for verified fact.

1.2 INDUSTRIAL HETEROGENEOUS CATALYSTS

The first heterogenous catalytic process of industrial significance, introduced in about 1875, utilized platinum to oxidize SO_2 to SO_3, which was then converted to sulfuric acid by absorption in an aqueous solution of the acid. This came to replace the lead chamber process for manufacture of H_2SO_4, in which the same series of reactions were catalyzed by a homogeneous catalyst, nitrogen oxides, probably via nitrosylsulfuric acid, $HNOSO_4$. In the contact process platinum in turn was superseded by a catalyst comprising vanadium oxide and potassium sulfate on a silica carrier, which was less susceptible to poisoning, and essentially this same composition is used in present-day SO_2 reactors.

Other inorganic chemical catalytic processes followed, notably the development by Ostwald in about 1903 of the oxidation of ammonia on a platinum gauze to form nitrogen oxides for conversion to nitric acid and the synthesis of ammonia from the elements during the period of about 1908 to 1914 by a process developed by Bosch utilizing a catalyst developed by Mittasch. The industrial synthesis of methanol from carbon monoxide and hydrogen appeared in about 1923, and the synthesis of hydrocarbons from carbon monoxide and hydrogen by the Fischer-Tropsch process in the 1930s. The catalytic partial oxidation of methanol to formaldehyde as an industrial process started in Germany about 1890, that of naphthalene to phthalic anhydride was commercialized in the 1920s, and that of benzene to maleic anhydride in 1928. The partial oxidation of ethylene to ethylene oxide was commercialized by Union Carbide in 1937.

In the processing of petroleum for fuels the first catalytic process, catalytic cracking, appeared in about 1937. This first used an acid-treated clay, then later a synthetic silica-alumina catalyst, and more recently zeolite crystals incorporated in a silica-alumina matrix. Reforming, using a molybdena/alumina catalyst to increase the octane number of gasoline by cyclization of paraffins and dehydrogenation to aromatics, was introduced in the United States and Germany just prior to World War II. In the early 1950s this was superseded by a reforming process using a platinum/alumina catalyst. Catalytic hydrocracking first came into use in England and Germany prior to World War II; subsequently it became relatively uneconomic, but more recently, with the advent of new types of catalysts, it has been revived. Hydrodesulfurization and hydrotreating processes have grown rapidly during the past two decades and are now of major importance in petroleum processing. Table 1.1 lists these and a representative

Table 1.1 Some heterogeneous catalysts of industrial importance

Reaction	Catalyst and reactor type (continuous operation unless otherwise noted)
	Dehydrogenation
C_4H_{10} (butane) $\longrightarrow$ butenes and C_4H_6 (butadiene)	Cr_2O_3-Al_2O_3 (fixed bed, cyclic)
Butenes $\longrightarrow$ C_4H_6 (butadiene)	Fe_2O_3 promoted with Cr_2O_3 and K_2CO_3, or $Ca_8Ni(PO_4)_6$ (fixed bed, continuous, in presence of steam)
$C_6H_5C_2H_5 \longrightarrow C_6H_5CH{=}CH_2$ (ethyl benzene $\longrightarrow$ styrene)	Fe_2O_3 promoted with Cr_2O_3 and K_2CO_3 (fixed bed)
CH_4 or other hydrocarbons + $H_2O \longrightarrow$ $CO + H_2$ (steam reforming)	Supported Ni (fixed bed)
$(CH_3)_2CHOH \longrightarrow CH_3COCH_3 + H_2$ (isopropanol $\longrightarrow$ acetone + hydrogen); $CH_3CH(OH)C_2H_5 \longrightarrow CH_3COC_2H_5 + H_2$	ZnO

Table 1.1 Some heterogeneous catalysts of industrial importance (*Continued*)

Reaction	Catalyst and reactor type (continuous operation unless otherwise noted)
	Hydrogenation
Of edible fats and oils	Raney Ni, or Ni on a support (slurry reactor, batch or continuous)
Various hydrogenations of fine organic chemicals	Pd on C or other support (slurry reactor, batch or continuous) (other supported metals may also be used)
$C_6H_6 + 3H_2 \longrightarrow C_6H_{12}$	Ni or noble metal on support (fixed bed)
$N_2 + 3H_2 \longrightarrow 2NH_3$	Fe promoted with Al_2O_3, K_2O, CaO, and MgO (adiabatic fixed beds)
$C_2H_2 \longrightarrow C_2H_6$ (selective hydrogenation of C_2H_2 impurity in C_2H_4 from thermal-cracking plant)	Pd on Al_2O_3 or sulfided Ni on support (adiabatic fixed bed)
	Oxidation
$SO_2 + \frac{1}{2}O_2 \longrightarrow SO_3$	V_2O_5 plus K_2SO_4 on SiO_2 (adiabatic, fixed beds)
$2NH_3 + \frac{5}{2}O_2 \longrightarrow 2NO + 3H_2O$	90% Pt–10% Rh wire gauze, oxidizing conditions
$NH_3 + CH_4 + \text{air} \longrightarrow HCN$ (Andrussow process)	90% Pt–10% Rh wire gauze, under net reducing conditions
$C_{10}H_8$ or 1,2-$C_6H_4(CH_3)_2 + O_2 \longrightarrow C_6H_4(CO)_2O$ (naphthalene or *o*-xylene + air $\longrightarrow$ phthalic anhydride)	Supported V_2O_5 (multitube fixed bed)
n-C_4H_8 or $C_6H_6 + O_2 \longrightarrow C_4H_2O_3$ (butene or benzene + air $\longrightarrow$ maleic anhydride)	Supported V_2O_5 (multitube fixed bed)
$C_2H_4 + \frac{1}{2}O_2 \longrightarrow (CH_2)_2O$ (ethylene oxide)	Supported Ag
$CH_3OH + O_2 \longrightarrow CH_2O + H_2$ and/or H_2O	Ag or Fe_2O_3-MoO_3
$C_3H_6 + O_2 \longrightarrow CH_2{=}CHCHO$ (acrolein) and/or $CH_2{=}CHCOOH$ (acrylic acid)	Cu_2O or multimetallic oxide compositions
$C_3H_6 + NH_3 + \frac{3}{2}O_2 \longrightarrow CH_2{=}CHCN + 3H_2O$	Complex metal molybdates or multimetallic oxide compositions (fluid bed)
Complete oxidation of CO, hydrocarbons, in pollution control, as of auto exhaust	Pt or Pt-Pd, pellet or monolith support
$C_2H_4 + \frac{1}{2}O_2 + CH_3COOH \longrightarrow CH_3COOCH{=}CH_2$ (vinyl acetate)	Pd on acid-resistant support (vapor phase, multitube fixed bed)
$C_4H_8 + \frac{1}{2}O_2 \longrightarrow C_4H_6 + H_2O$	Promoted ferrite spinels
	Acid-catalyzed reactions
Catalytic cracking	Zeolite (molecular sieve) in SiO_2-Al_2O_3 matrix (fluid bed)

Table 1.1 Some heterogeneous catalysts of industrial importance (*Continued*)

Reaction	Catalyst and reactor type (continuous operation unless otherwise noted)
Acid-catalyzed reactions (*Continued*)	
Hydrocracking	Metal (e.g., Pd) on zeolite (adiabatic fixed beds)
Isomerization	Metal on acidified Al_2O_3 (fixed bed), zeolites
Catalytic reforming	Pt, Pt-Re, or Pt-Ir on acidified Al_2O_3 (adiabatic, fixed or moving bed)
Polymerization	H_3PO_4 on clay (fixed bed)
Hydration, e.g., propylene to isopropyl alcohol	Mineral acid or acid-type ion-exchange resin (fixed bed)
Reactions of synthesis gas	
$CO + 2H_2 \longrightarrow CH_3OH$	ZnO promoted with Cr_2O_3, or Cu^I-ZnO promoted with Cr_2O_3 or Al_2O_3 (adiabatic, fixed beds or multitube fixed bed)
$CO + 3H_2 \longrightarrow CH_4 + H_2O$ (methanation)	Supported Ni (fixed bed)
$CO + H_2 \longrightarrow$ paraffins, etc. (Fischer-Tropsch synthesis)	Fe with promoters (fluid bed)
Other	
Oxychlorination (e.g., $C_2H_4 + 2HCl + \frac{1}{2}O_2 \longrightarrow C_2H_4Cl_2 + H_2O$)	$CuCl_2/Al_2O_3$ with KCl promoter
Hydrodesulfurization	Co-Mo/Al_2O_3 or Ni-Mo/Al_2O_3 (adiabatic, fixed beds)
$SO_2 + 2H_2S \longrightarrow 3S + 2H_2O$ (Claus process)	Al_2O_3 (fixed beds)
$H_2O + CO \longrightarrow CO_2 + H_2$ (water-gas shift)	Fe_3O_4 promoted with Cr_2O_3 (adiabatic fixed bed); for a second, lower temperature stage, Cu-ZnO supported on Al_2O_3 or SiO_2

selection of the catalysts used in the principal industrial heterogeneous catalytic processes of present or recent importance together with the type of reactor commonly utilized.

1.3 DEFINITIONS

1.3.1 Catalyst

The basic concept of a catalyst is that of a *substance that in small amount causes a large change*. More precise definitions of catalysis and of what constitutes a catalyst have gradually evolved as understanding of the causes of catalytic phenomena has grown. Even today there is no universal agreement on definitions, the point of view varying somewhat depending upon the investigator; for example, as between the fundamental investigator and the practitioner, and among researchers concerned with heterogeneous catalysis, homogeneous catalysis, polymerization reactions, and enzymes. For present purposes however our definition is: *A catalyst is a substance that increases the rate of reaction without being appreciably consumed in the process.*

This basic concept, stemming from the chemical approach to catalysis, is that a reaction involves a cyclic process in which a site on a catalyst forms a complex with reactants, from which products are then desorbed, thereby restoring the original site. This leads to the idea that a catalyst is unaltered by the reaction it catalyzes, but this is misleading. A catalyst may undergo major changes in its structure and composition as part of the mechanism of its participation in the reaction. A pure metal catalyst will frequently change in surface roughness or crystal structure on use. The ratio of oxygen to metal in a metal oxide catalyst will frequently change with temperature and composition of the contacting fluid. In both cases, however, there is no stoichiometric relationship between such changes and the overall stoichoimetry of the catalyzed reaction. Many so-called polymerization catalysts or initiators are not termed catalysts within the above definition. Thus, in the use of an organic peroxide to initiate a polymerization reaction, the ratio of peroxide consumed to quantity of monomer reacted is indeed nonstoichiometric, but the peroxide becomes completely consumed in the process; hence it cannot be regarded as a true catalyst.

A catalyst is defined as a *substance*; the acceleration of a rate by an energy-transfer process is not regarded as catalysis by this definition. Excluded cases include excitation by thermal energy (increased temperature), by bombardment of reactants by charged or high-energy particles, by electric discharge, or by photochemical irradiation. For example, the reaction of hydrogen and oxygen is increased by irradiation by ultraviolet light and even more so if a small amount of mercury vapor is present and illumination is by a mercury vapor lamp. Here the reaction is accelerated by energy transfer to the reacting gases from mercury atoms, which in turn are activated by irradiation. As a second example, the rate of thermal decomposition of a vapor at low pressure is usually increased by the addition of a second "inert" gas which furnishes activation energy by collisions. These various methods of accelerating the rate of a reaction are more clearly understood as aspects of the mechanisms of homogeneous gas-phase reactions rather than as catalysis.

The same substance may act as a catalyst in one set of circumstances but as a reagent in another. Thus the oxidation of a mixture of *o*-xylene and air to phthalic anhydride is catalyzed by V_2O_5. *o*-Xylene by itself may also be oxidized by contacting it alone with V_2O_5, in which case the V_2O_5 acts as a reagent and becomes stoichiometrically reduced to a lower oxide. Indeed the ability to alternate easily between two or more oxidation states is characteristic of many oxidation catalysts. Thus insight into the reasons for their catalytic activities may come from studies using the same substance as a reagent.

A catalyst *cannot* change the ultimate equilibrium determined by thermodynamics; its role is restricted to accelerating the rate of approach to equilibrium. This point is developed later.

1.3.2 Catalyst Activity

The *activity* of a catalyst refers to the rate at which it causes the reaction to proceed to chemical equilibrium. The rate may be expressed in any of several ways (Sec. 3.2). The performance of an industrial reactor is frequently given in terms of a space-time yield (STY), which is the quantity of product formed per unit time per unit volume of reactor.

1.3.3 Catalyst Selectivity

There are usually many chemical compositions of free energy that are intermediate between that of the reactants and that of the state of complete chemical equilibrium. The *selectivity* of a catalyst is a measure of the extent to which it accelerates the reaction to form one or more of the desired products that are usually intermediates, instead of those formed by reaction to the overall state of lowest free energy. The selectivity will usually vary with pressure, temperature, reactant composition, and extent of conversion as well as with the nature of the catalyst. For precision one should refer to the selectivity of a catalyzed reaction under specified conditions. The selectivity is determined in the first instance by the *functionality* of the catalyst, but also in part by thermodynamic equilibrium considerations. Thus, a certain undesired product may be largely avoided if it is possible to operate under conditions in which the equilbrium concentration of the product is negligible (see Sec. 1.4).

A frequently cited example of catalyst selectivity caused by function is ethanol decomposition. Over copper the reaction proceeds as

$$C_2H_5OH \longrightarrow CH_3CHO + H_2 \tag{1.1}$$

Over alumina it proceeds as

$$C_2H_5OH \longrightarrow C_2H_4 + H_2O \tag{1.2}$$

or

$$2C_2H_5OH \longrightarrow C_2H_5OC_2H_5 + H_2O \tag{1.3}$$

Selectivity here is associated with the fact that copper adsorbs hydrogen whereas alumina adsorbs water.

A catalyst may be useful for either its activity or its selectivity, or both. If a variety of products are possible, selectivity is usually the more important. Activity can usually be increased by increasing the temperature, although frequently increased temperature shortens the life of the catalyst or increases undesirable thermal reactions. It may also decrease the maximum conversion obtainable if the reaction is exothermic and is limited in extent by thermodynamic equilibrium. If a variety of products are thermodynamically possible, increased temperature may either increase or decrease selectivity, depending upon the overall kinetics and the desired product. Thus, for the general case of $A \rightarrow B \rightarrow C$, if B is the desired product, some intermediate temperature is usually optimum; if C is desired, increased temperature improves selectivity since it helps drive the reaction to completion. If B is desired, the maximum selectivity for this type of kinetics occurs at the lowest conversion.

Selectivity is usually defined as the percentage of the consumed reactant that forms the desired product. It is usually a function of degree of conversion and reaction conditions. *Yield* is an engineering or industrially used term which refers to the quantity of product formed per quantity of feedstock (reactant) consumed in the overall reactor operation. Within this overall operation there may be recycle of various reactants, as after separation. Yield is frequently reported on a weight basis, and hence a yield exceeding 100 percent (w/w) may be obtained, for example, in a partial oxidation process in which oxygen is introduced into the product molecule with high selectivity. In the fuels industry products are conventionally sold on a volume rather than a weight basis; hence a yield exceeding 100 percent (v/v) may be obtained when the products are of lower density than the reactants.

1.3.4 Negative Catalyst

A *negative catalyst* is a substance which decreases the rate of reaction. This is usually found only with a reaction that proceeds by the formation and disappearance of free radicals.

The negative catalyst acts by interfering with these free-radical processes, converting radicals into less active forms, or removing them from reaction. An example is the use of lead alkyls such as tetraethyllead to improve the antiknock properties of gasoline in an internal-combustion engine. After compression, a part of the gasoline-air mixture inside the engine cylinder may spontaneously ignite before the combustion wave initiated by the spark plug reaches it, thus producing a sudden and uncontrolled pressure increase or "knock." The degradation products from the lead compound, most probably some kind of finely dispersed lead oxide and lead oxyhalides, interfere with the preignition reactions. Organic peroxides are probably formed in the early stages of hydrocarbon oxidation reactions. The role of the lead compounds may be to destroy these peroxides, which otherwise would split into free radicals and thus initiate a rapidly propagating reaction.

The mechanism of action of a negative catalyst is different from that of most positive catalysts, and a negative catalyst should be described more meaningfully as a "reaction inhibitor." As another example, oxidation inhibitors such as phenolic compounds and amines contain one or more labile hydrogen atoms and act by transferring them to an active free radical. This exchange of an active radical for a less active one causes a slowing down of the overall reaction.

1.3.5 Heterohomogeneous Catalysis

A catalyst may sometimes act by generating free radicals, which then propagate a chain reaction in the bulk of the reacting fluid. Examples are well documented for various liquid-phase reactions in which free radicals have been trapped or otherwise identified. For gas-phase reaction the phenomenon is sometimes postulated to occur, especially for vapor-phase oxidation reactions at elevated temperatures. There is little solid experimental evidence to indicate the extent, if any, to which it is actually significant for conditions and systems of practical interest. Good evidence exists for heterohomogeneous catalysis in the reaction of hydrogen and oxygen gases at low pressures, where the mean free path is high.

At pressures of 1 atm and higher, heterohomogeneous catalysis probably is of significance only for combinations of pressure, temperature, and gas composition close to those under which the homogeneous reaction rate of itself would be significant. That is, the catalytic reaction triggers off an incipient homogeneous reaction.

The above description is the classical form of heterohomogeneous catalysis, but a catalytic reaction and a homogeneous reaction may interact with each other in various subtle ways such that the net effect is not simply the sum of the two. Instead of free radicals being desorbed into the gas phase, a molecular intermediate may be formed and released instead. Its fate depends upon the relative probability of being adsorbed and reacted on another site of the catalyst in contrast to reacting in a different manner in the gas phase. The overall behavior of the system may depend not only on the void fraction of a packed bed of catalyst and the amount of open volume downstream, but also on geometrical and other factors. In many catalytic oxidations the exit gas must be rapidly quenched to avoid over-oxidation or decomposition of the desired product, but there is little evidence one way or the other as to whether heterohomogeneous catalysis contributes to these processes under practical conditions.

In some reactions of hydrocarbons, e.g., steam reforming of methane and other hydrocarbons to form synthesis gas, the extent to which carbon may deposit on the catalyst and gradually inactivate it is determined in part by the balance between homogeneous reactions leading to carbonaceous deposit precursors and the desired heterogeneous reaction. In various hydrotreating processes in the fuels industry, thermal reactions occur simultaneously with catalytic reactions, causing interactions that may be only dimly perceived.

1.3.6 Sites

Under reaction conditions all solid catalysts are nonuniform or heterogeneous in the sense that chemical and physical properties will vary with location on the surface. Even in a pure metal the atoms at specific locations, such as at lattice defects and at edges and corners of crystallites, are different from atoms in a surface plane. The heterogeneity of catalyst surfaces can be demonstrated and to some extent characterized by a variety of methods. The variation of the differential heat of adsorption with coverage or the change in activation energy of adsorption with coverage may be measured, or temperature-programmed desorption studies may be used. More than one maximum in chemisorption isotherms may be observed (Fig. 2.2), showing that more than one kind of chemisorption may occur. Some catalysts may be effectively poisoned by adsorption of an amount of material comprising much less than a monolayer, demonstrating that only a fraction of such a surface is effective for reaction.

These facts and others led to the concept, introduced by H. S. Taylor (1948), that reaction takes place only on specific locations on the catalyst, termed *sites*. Those that are active for one reaction may not be so for a second reaction, but it is usually difficult to identify their identity and structure precisely. In some cases a site may be a group or cluster of neighboring atoms on the catalyst surface; sometimes it may actually be a species adsorbed onto the catalyst. The term *active center* is frequently used as a synonym for *site*, or to refer to a group of sites. A catalyst will frequently undergo reconstruction during reaction, causing a change in total area and nature of the surface and possibly a change in number and nature of the sites. For some reactions on metals the rate is independent of the size, shape, or other physical characteristics of the metal crystallite and is proportional only to the total number of metal atoms exposed to the reactant. (This is typically about 10^{15} atoms per square centimeter.) Such reactions are termed *structure-insensitive* in contrast to *structure-sensitive* reactions whose rates do vary with the detailed structure of the surface. The terms *facile* for *structure-insensitive* and *demanding* for *structure-sensitive* are also used. The concentration of active sites on acid catalysts is usually much less than that on metals and is typically of the order of magnitude of about 10^{11} sites per square centimeter.

1.3.7 Turnover Number

The *turnover number*, or *turnover frequency*, is the number of molecules that react per site per unit time. As a basic measure of true catalytic activity, this is a useful concept, but it is limited by the difficulty of determining the true number of active sites. In general it is easier to do this for metals than for nonmetal catalysts since techniques such as selective chemisorption are available to measure the exposed surface area of metals. For acid catalysts the measurement of site concentration by poisoning or adsorption of bases may be ambiguous and may lead to erroneously high values since sites may be active for sorption but not for reaction. As with rates of reaction in general, the turnover number is a function of pressure, temperature, and composition of the reacting fluid.

1.3.8 Functionality

Some reactions involve the formation and subsequent reaction of various intermediates. Some of the reaction steps may be catalyzed by one kind of site, and others by a second kind of site. When these steps occur in series, both kinds of sites must be in proximity to one another in order for the overall reaction to occur, and usually they are both on the same catalyst particle. In some cases an intermediate can be desorbed from one kind of site into the bulk fluid and adsorbed on a second site. Then an intimate mechanical mixture of two kinds of particles, each possessing only one kind of site, can effectively catalyze the overall reaction, although either one by itself is relatively ineffective. An example is the isomerization of an n-paraffin to an isoparaffin on a platinum catalyst supported on an acidic base. The n-paraffin is first dehydrogenated to an n-olefin. This isomerizes to an isoolefin, which is hydrogenated to an isoparaffin. The hydrogenation and dehydrogenation steps occur on platinum, the isomerization steps on an acid site (Sec. 9.5).

A bifunctional catalyst provides two kinds of sites. Some single compounds also behave in a "multifunctional" manner. For example, substances such as Cr_2O_3, MoO_2, and WS_2 exhibit acidic as well as hydrogenation-dehydrogenation activity.

1.3.9 Naming of Catalysts and Catalyst Structures

Most catalysts are complex, and often the terms by which they are described only list the active elements present and the support, without specifying the form in which the element may exist either in the catalyst as manufactured or under reaction conditions. In large part this stems from uncertainties concerning the actual composition under reaction conditions. For example, a so-called $CoMo/Al_2O_3$ catalyst is commonly used for hydrodesulfurization. The solidus (slash) separates the active elements, Co and Mo, from the support, Al_2O_3. The catalyst is usually supplied in the form of the metal oxide, which is converted to a sulfide before use. Its actual structure is highly complex.

Sometimes a catalyst is described as a compound, but that compound as such may actually not be effective. For example, a common industrial catalyst for methanol synthesis is zinc oxide with which chromia is incorporated. This is sometimes described as a "zinc chromite" catalyst, but this may be misleading since true zinc chromite as a crystalline spinel is relatively inactive. Zinc chromite functions primarily as a textural promoter that minimizes sintering (Sec. 4.6.1). Again the structure of the active catalyst is quite complex.

The council of the International Union of Pure and Applied Chemistry (IUPAC) has adopted a recommended set of symbols and terminology for heterogeneous catalysis, which is published in Volume 26 of *Advances in Catalysis* (Burwell, 1977). The development of definitions is also being considered by the Committee on Catalysts (D32) of the American Society for Testing Materials (ASTM).

The fact that catalysts obey all the normal principles of chemistry is to be borne in mind. The catalyst structure, both physical and chemical, can be markedly affected by the environment, and indeed this principle is frequently utilized to maintain a catalyst in a desired state. In catalytic reforming reactions an acidified support is required, but the degree of acidity may be gradually nullified by formation of carbonaceous deposits (coke) or adsorption of impurities from the feed stream. The acidity may be restored by adding small concentrations of an organohalogen to the feed. In the reactor this decomposes to the corresponding acid, which adsorbs onto the catalyst support and is incorporated into it, thus producing acidic sites. A CoMo catalyst used for hydrogenation is frequently more active in the sulfide than in the metallic form. In the absence of sulfur compounds in a feed stream, a small concentration of H_2S may be added to the reactant to maintain the catalyst in the desired structure. The selectivity of a metal oxide catalyst for a partial oxidation reaction is frequently caused by a specific crystallographic form or a specific compound.

This may become converted irreversibly to a different and inactive form if the reaction mixture becomes too highly oxidizing or too reducing in character.

1.3.10 Catalyst Deactivation

A catalyst may lose its activity or its selectivity for a wide variety of reasons. The causes may be grouped loosely into

1. Poisoning
2. Fouling
3. Reduction of active area by sintering or migration
4. Loss of active species

A catalyst *poison* is an *impurity* present in the *feed stream* that reduces catalyst activity. In a complex reaction it may affect one reaction step more than another; hence the selectivity towards a desired reaction may be improved by deliberately adding a poison. It adsorbs on active sites of the catalyst, and if not adsorbed too strongly, it is gradually desorbed when the poison is eliminated from the feed stream. The phenomenon is then temporary. If adsorption is strong, the effect is permanent. The desorption may be enhanced by reaction with the fluid. Thus in a hydrogenation reaction a metallic catalyst may be poisoned by adsorption of a sulfur compound, but desorption may be enhanced by its conversion to H_2S by reaction with H_2. If a reaction product is strongly adsorbed, the reaction may be termed *self-poisoned* or *self-inhibited* (see also Sec. 3.6).

The term *fouling* is generally used to describe a physical blockage such as the deposit of dust or fine powder or carbonaceous deposits (coke). In the latter case activity can usually be restored by removal of the coke by burning.

Sintering is an irreversible physical process leading to a reduction of effective catalytic area. It may consist of growth of metal crystallites on a support (Sec. 6.4) or of a decrease in area of a nonsupported catalyst.

The particular active species may also be converted to another form less active or selective, as is the case with certain complex metal oxides used in partial oxidation reactions. A complex metal oxide crystal may also decompose into other compounds, sometimes caused by loss of a particular element via volatilization of a compound. A somewhat amorphous catalyst may crystallize, or a compound active in one crystal habit may be converted into a less active crystalline form. A supported metal catalyst may be reduced in activity or selectivity by becoming alloyed with a metallic impurity or by reaction with the support; for example, a nickel/alumina catalyst may be converted to a nickel aluminate.

Various examples will be discussed in conjunction with specific cases or groups of cases. Sintering of supported metal catalysts is discussed in Sec. 6.4. The formation of carbonaceous deposits, which may be regarded primarily as a fouling mechanism, is discussed in Sec. 6.5 and Chap. 7. An example of deliberate poisoning to enhance selectivity at the cost of reduced activity is the addition of an organochlorine compound to ethylene in the commercial process for making ethylene oxide (Sec. 8.3).

In addition to the above, catalysts may also deteriorate because of slow crumbling of a support by chemical attack, by physical grinding in an agitated vessel, or for reasons that are poorly understood. Sometimes a poison lowers selectivity because it itself is a catalyst for an undesired side reaction; thus a trace of metal compound in a feed stream may lead to the deposit of elemental metal, which may catalyze hydrogenation-dehydrogenation reactions where such are not desired.

1.4 THERMODYNAMICS AND ENERGETICS

A true heterogeneous catalyst accelerates the rate of approach to equilibrium but cannot alter that equilibrium. This is readily seen by considering a simple reversible reaction $A \rightleftharpoons B$. The standard free-energy change is expressed as $\Delta G^\circ = -RT \ln K =$

$-RT \ln (a_B/a_A)$, where a_B and a_A are the activities of product and reactant respectively. The presence of the solid catalyst cannot change $\Delta G°$, and hence does not change the ratio a_B/a_A. (If, however, reactants and products were dissolved in a homogeneous catalyst, e.g., a mineral acid which in effect altered their structure, the activity ratio and the equilibrium composition could likewise change.)

In many practical cases a large number of products are possible under equilibrium conditions, but only one or a related group are desired. The selectivity of a catalyst may be related to its ability to direct one reaction essentially to equilibrium while having little or no effect on alternate pathways, so the most stable products are not necessarily formed. Selectivity effects are intimately related to the selective chemisorption characteristics of the catalyst. In the conversion of CO and H_2 to CH_3OH, products such as paraffins, olefins, and higher alcohols are more stable thermodynamically under synthesis conditions but are scarcely formed on the usual ZnO/Cr_2O_3 catalyst. Paraffins and olefins, undesired in CH_3OH synthesis, are readily formed from CO and H_2 on Fe or Co as in the Fischer-Tropsch synthesis (Sec. 10.2). Because these products are more favored thermodynamically and the Fe or Co catalysts are of sufficient activity, reaction can be carried out at lower pressures and temperatures (for example, 1 to 20 atm, 150 to 300°C) than those required for CH_3OH synthesis. The quantities of CH_3OH then formed are insignificant because under these milder reaction conditions the amount that can exist under equilibrium conditions is small.

Consideration of alternate reaction pathways in order of increasing difficulty based on structural or kinetic insights may be a fruitful way of searching for an appropriate catalyst within the overall constraints set by thermodynamics. As an example, consider the isomerization of an olefin such as 2-butene on an acid catalyst. A wide variety of transformations are possible. In general, cis-trans isomerization occurs most readily, and then double-bond migration. Carbon skeletal rearrangement is the most difficult. Cis-trans equilibrium without other changes can probably be achieved on a mild acid catalyst, but strong acidity is required for skeletal rearrangement. Correspondingly, if skeletal rearrangement is achieved, the easier reactions will probably all proceed essentially to equilibrium (Sec. 7.5).

Under actual process conditions it is sometimes possible to improve the performance over that allowed by thermodynamics by removing a product either chemically or physically. In a dehydrogenation reaction, the concentration of hydrogen in the products may be lowered by adding oxygen to convert it to water; in other cases a product may be removed by *ab*sorption or *ad*sorption into another phase under dynamic conditions.

According to the principle of microscopic reversibility, if a catalytic reaction proceeds by a single step, then a catalyst that accelerates the rate of a forward reaction should also accelerate the rate of the reverse reaction. This can be illustrated by the reaction $A \rightleftharpoons B$. The equilibrium constant, K, equals k_1/k_2. Since the value of K is independent of the presence or absence of the catalyst, a catalyst which increases k_1 should also increase k_2. The same argument can be extended to a reaction occurring in a series of steps one of which is rate-limiting, all others being in equilibrium with one another. Again a catalyst which accelerates the rate-limiting step in the forward direction should accelerate the rate in the reverse direction.

The complications in applying this seemingly powerful generalization are twofold. First, operating conditions (pressure, temperature, and/or composition) must perforce be different when carrying out the forward reaction than the reverse, since they must lie on opposite sides of the equilibrium condition. This change in operating conditions between the two sides of the equilibrium may have a significant effect on the catalyst. Increasing the temperature may cause a rapid decrease in activity with time, such as may be caused by crystal growth. Increasing the pressure may cause a shift in the relative amounts of adsorbed species on a catalyst, thus altering the catalyst activity and/or selectivity. Secondly, in practice a multiplicity of reactions will frequently occur, and side reactions may be much more significant when approached from one side of equilibrium than from the other. In hydrogenation-dehydrogenation reactions involving organic compounds, a nickel catalyst is usually highly active for hydrogenation, but is is frequently ineffective for dehydrogenation because carbon is formed on the catalyst surface by side reactions and the catalyst rapidly loses activity.

This generalization is probably of greatest use in a preliminary search for catalysts, and its greatest applicability is when few or no side products can be formed in the reaction direction of interest. Thus active catalysts for the synthesis of ammonia from the elements are also active for its decomposition.

1.4.1 Reaction Pathways

Consider a gas-phase reaction which may occur either homogeneously or may be catalyzed heterogeneously. The two reaction pathways occur simultaneously, but in order for the catalyzed reaction to be observed it must occur measurably faster than the homogeneous reaction. We now inquire into how the catalyst brings this about.

The rate of a single elementary step of a homogeneous reaction is proportional to a rate constant k that varies with temperature according to the Arrhenius relationship, $k = Ae^{-E/RT}$, where E is the activation energy and R the gas constant. The preexponential factor A is a constant that in collision theory is identified as a collision number for bimolecular processes and, for unimolecular processes, as a frequency factor or the probability of reaction of an activated molecule.

The ability of a catalyst to increase a reaction rate can be ascribed in a general way to its causing a reduction in the activation energy of the reaction. But even for the simplest kind of reaction, the single-step conversion of A to B, the situation is considerably more complicated than this simple statement may imply. The catalyzed reaction involves three rate processes: adsorption, the formation and breakup of an activated complex, and desorption of products. Each of these has its own activation energy. The rate of each is also determined by the total surface area of the catalyst present (or, more precisely, the number of active sites) and by the concentration on the catalyst surface of various adsorbed species. The idealized homogeneous reaction has a single activation energy, and its rate is a function of the gas-phase concentration.

In order for a reaction to be noticeably catalyzed, the various factors that determine the overall rate of the heterogeneous reaction must in the entirety of their interactions outweigh that of the different group of factors that determine the rate of the homogeneous reaction. Generally the most important effect of the catalyst is to provide a pathway whereby the activation energy for the formation of the intermediate surface complex is considerably less than for the homogeneous reaction. Because the activation energy appears in the rate equation as an exponent, a slight change in activation energy has a marked effect on the rate.

The rate of the catalyzed reaction is proportional to the active surface area, and the rate of a homogeneous reaction is proportional to the volume of fluid. Hence the maximum ratio of catalyzed rate to the homogeneous rate occurs with the use of high-area (porous) catalyst pellets packed in a reactor. Catalysts such as those frequently utilized in practice have areas of the order of 100 m^2/g or more. Arguments based on the absolute theory of reaction rates show that, although it is not usually observed, it is entirely possible for the observed rate of a catalyzed reaction to proceed faster than the corresponding homogeneous reaction even when the activation energy for the rate-limiting step on the catalyst is no less than that for the homogeneous reaction (Schlosser, 1972, pp. 57-65).

The changes in energy associated with the different steps in a simple exothermic reaction can be depicted as shown in Fig. 1.1. E_{hom} is the activation energy for the homogeneous reaction, E_{ads} for adsorption of reactants onto the catalyst, E_{cat} for the formation of the activated complex, and E_{des} for the desorption of products. λ_{ads} is the heat of adsorption of reactants, taken to be exothermic, and λ_{des} is the heat of desorption of products, taken to be endothermic. The overall energy change upon reaction is ΔH and is, of course, the same for the two pathways.

From experimental rate data, an *apparent activation energy* can be calculated from the slope of an Arrhenius plot of the log of an observed rate constant as a function of the reciprocal of the absolute temperature. To proceed from this to calculation

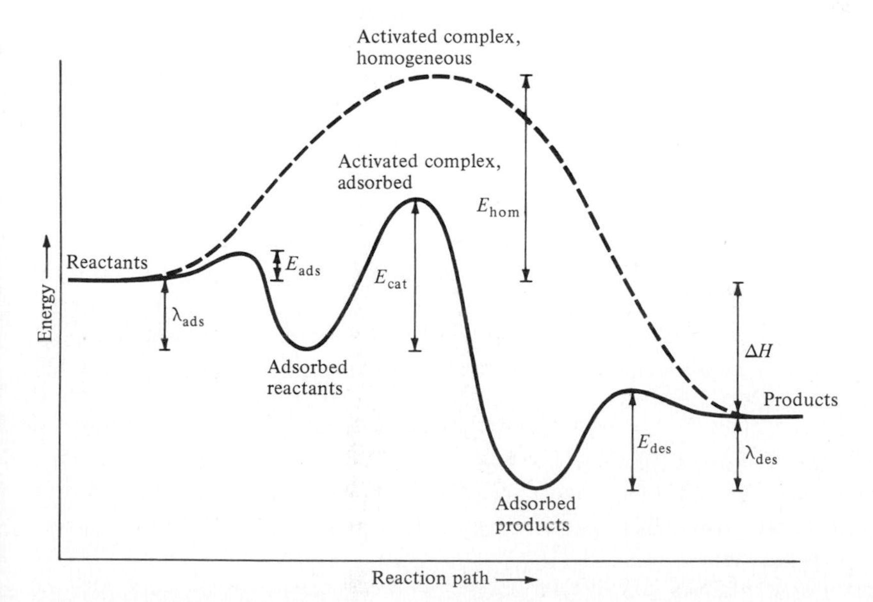

Figure 1.1 Energy changes associated with individual steps of a reaction.

of the activation energy of a surface process in the general case requires a knowledge or assumption of the mechanism of the surface reaction, identification of the rate-controlling step, and heats of adsorption and desorption, as developed in Chap. 3. This treatment also shows that there is no reason why a straight-line relationship between log k and $1/T$ should be expected to be encountered over a substantial range of temperature. It will usually be found that the apparent or effective activation energy for a catalyzed reaction is less than that for the same reaction proceeding homogeneously, both being determined from the slope of an Arrhenius plot. It is to be emphasized, however, that there is no fundamental reason why this should always be the case. Nevertheless, a consequence of this common behavior is that with increased temperature a point will usually be reached beyond which the rate of a catalyzed reaction will be exceeded by the rate of the homogeneous reaction.

With some endothermic reactions high temperatures are required in order for a substantial amount of the product to be present at equilibrium, and it may be found that these temperatures are so high that no significant increase in rate is achieved by any catalyst. An example is the dehydrogenation of ethane to form ethylene and hydrogen, in which a temperature of about 725°C is required for 50 percent conversion to equilibrium at atmospheric pressure. Little increase in rate is obtained with a heterogeneous catalyst, and the process is carried out industrially as a homogeneous thermal reaction. For conversion of a higher paraffin such as butane to butene, however, a specified degree of conversion to equilibrium can be obtained at considerably lower temperatures than with ethane, and a substantial increase in rate is observed with use of a catalyst such as chromia-alumina.

1.5 CLASSIFICATION AND SELECTION OF CATALYSTS

Catalysts effective in practice range from minerals used with little or no further processing, to simple massive metals, to substances of precise and complex composition. The latter may have to be carefully prepared under closely controlled conditions, and their effectiveness in use may also require careful control of the environment in the reactor. The desired catalytic action may range from the acceleration of a simple inorganic reaction which can go in only one direction to a highly selective organic reaction which may involve complicated interactions among many intermediate species.

In the choosing or developing of a catalyst the difficulty of the problem may vary greatly as indicated by a scheme of the order of increasing complexity, suggested by S. Z. Roginskii (1968):

- I. Selection among known catalysts
  - A. For known reactions
  - B. For reactions analogous to known catalytic reactions
  - C. For new reactions
- II. Search for new catalysts
  - A. For well-known catalytic reactions

B. For reactions analogous to those well known
C. For reactions of new types, having no analogues among well-known reactions

Only a few broad generalizations can be offered at this point about correlations between the nature of a catalyst and the reactions it catalyzes. More specific correlations within groups of catalysts or groups of reactions will be discussed later.

Solids exhibiting catalytic activity possess, in general, strong interatomic fields, such as that involved in ionic or metallic bonding. Organic covalent compounds in general are noncatalytic. A fundamental requirement is that the desired catalytic structure be stable under reaction conditions; e.g., a desired metal must remain metallic and not be converted to an inactive compound. Table 1.2 lists some of the types of reactions that are catalyzed by metals, metal oxides, and acids, and some that require both metal and acid functionality.

The metals that catalyze hydrogenation reactions do so usually because they adsorb hydrogen with dissociation and the bonding is not too strong. These are essentially the elements in group VIII (Fe, Co, Ni, and the platinum-group metals) plus possibly copper in group 1B (see Table 1.3). However hydrogen adsorption is also affected by impurities in the metal and by various species adsorbed on the surface. Thus under a particular set of circumstances hydrogen adsorption can be low on pure copper but may be considerably enhanced by the presence of impurities, e.g., carbonaceous residues. Almost all of these reactions are of the homolytic type (those involving noncharged intermediates).

Some metals also catalyze oxidations by chemisorbing oxygen, but the base metals in general cannot be used in the metallic form since they usually are converted to the oxide throughout their bulk. Only the platinum-group metals (Ru, Rh, Pd, Os, Ir, Pt) and silver and gold are sufficiently resistant. Of these, gold has little adsorptivity in general and hence has little catalytic activity, and within the platinum group osmium and iridium are very scarce. Of the others, Pt and Pd are the most important. Note however that the oxide form of many metallic elements, both base and noble, may be active for oxidation reactions.

Oxygen is more strongly adsorbed by metals than is hydrogen; apparently the bonds are generally too strong to be easily rearranged, so higher temperatures are generally required for oxidations to occur on metals than for hydrogenations.

The oxide catalysts may be divided into two groups by structure. If the structure is ionic and oxygen atoms can be readily transferred to or from the lattice, the substance may be a good catalyst for partial oxidation reactions. The easy movement of oxygen atoms (it is not clear in some cases whether these are radicals or ions) causes the compounds to be generally nonstoichiometric, and the ratio of oxygen to metal may vary significantly with the composition of the reacting mixture. Examples are the complex molybdates containing several metals and various multimetallic oxide compositions. The second group consists of dehydrogenation catalysts in which the oxygen is more tightly bound and the oxide must not be reducible to the metal by hydrogen at the reaction temperatures. Thus the oxides suitable for partial oxidation reactions are generally not suitable for dehydrogenations. Some oxides such as chromia may be regarded as being intrinsically dual-function in nature, catalyzing acid-type reac-

Table 1.2 Classification of heterogeneous catalysts according to principle functions*

Metals		Metal oxides		Acids		Metal plus acid	
Function	Examples	Function	Examples	Function	Examples	Function	Examples
Hydrogenation Hydrogenolysis†	Ni, Pd, Pt (Cu)	Partial oxidation	Complex metal molybdates	Hydration	Acid-type ion-exchange resin	Paraffin isomerization	Pt/acidified support
Oxidation	Ag, Pt		Multimetallic oxide compositions	Polymerization	H_3PO_4 on carrier	Hydrogenolysis†	Pd/zeolite
		Dehydrogenation	Fe_2O_3, ZnO, Cr_2O_3/Al_2O_3	Cracking Hydrogen transfer‡ Disproportionation§	SiO_2–Al_2O_3, zeolites in acid form		

*Some reactions are too complex to be simply classified. These include reactions of synthesis gas, the introduction of amine or nitrile groups, hydrodesulfurization and hydrodenitrogenation, and others.

†*Hydrogenolysis* is addition of hydrogen across a single bond to cause splitting into two molecules, for example, $C_2H_6 + H_2 \longrightarrow 2CH_4$, also termed *hydrocracking*. It can occur on a metal by itself or, more rapidly, on a dual-function catalyst (Secs. 6.1, 9.5, and 9.7).

‡Hydrogen transfer involves protons or hydride ions formed by rupture of a C—H bond and transferred to another molecule, for example, $C_2H_4 + C_4H_{10} \rightleftharpoons C_2H_6 + C_4H_8$.

§An example of disproportionation is

$$2\,C_6H_5CH_3 \rightleftharpoons C_6H_6 + C_6H_4(CH_3)_2$$

Table 1.3 Periodic table of the elements

Period	Group IA	Group IIA	Group IIIB	Group IVB	Group VB	Group VIB	Group VIIB	Group VIII			Group IB	Group IIB	Group IIIA	Group IVA	Group VA	Group VIA	Group VIIA	Group 0
1 1*s*	1 H																1 H	2 He
2 2*s*2*p*	3 Li	4 Be											5 B	6 C	7 N	8 O	9 F	10 Ne
3 3*s*3*p*	11 Na	12 Mg											13 Al	14 Si	15 P	16 S	17 Cl	18 Ar
4 4*s*3*d*4*p*	19 K	20 Ca	21 Sc	22 Ti	23 V	24 Cr	25 Mn	26 Fe	27 Co	28 Ni	29 Cu	30 Zn	31 Ga	32 Ge	33 As	34 Se	35 Br	36 Kr
5 5*s*4*d*5*p*	37 Rb	38 Sr	39 Y	40 Zr	41 Nb	42 Mo	43 Tc	44 Ru	45 Rh	46 Pd	47 Ag	48 Cd	49 In	50 Sn	51 Sb	52 Te	53 I	54 Xe
6 6*s*(4*f*)5*d*6*p*	55 Ca	56 Ba	57* La	72 Hf	73 Ta	74 W	75 Re	76 Os	77 Ir	78 Pt	79 Au	80 Hg	81 Tl	82 Pb	83 Bi	84 Po	85 At	86 Rn
7 7*s*(5*f*)6*d*	87 Fr	88 Ra	89† Ac															

*Lanthanide series 4*f*	58 Ce	59 Pr	60 Nd	61 Pm	62 Sm	63 Eu	64 Gd	65 Tb	66 Dy	67 Ho	68 Er	69 Tm	70 Yb	71 Lu
†Actinide series 5*f*	90 Th	91 Pa	92 U	93 Np	94 Pu	95 Am	96 Cm	97 Bk	98 Cf	99 Es	100 Fm	101 Md	102 No	103 Lw

tions as well as dehydrogenation reactions. Metals as such usually are not suitable for dehydrogenations because under representative reaction conditions they become rapidly deactivated by carbonaceous deposits.

A number of solids are acidic (Chap. 7) and can catalyze a wide variety of reactions similar to those catalyzed by strong mineral acids. These include materials in which two or more elements are tightly linked together in the structure via oxygen atoms, such as in silica-alumina and in various zeolites (crystalline aluminosilicates and related materials). Solid acids may also be formed by treating alumina so as to incorporate a halogen into its structure. Many salts such as metal sulfates and phosphates exhibit little or no acidity as prepared in the hydrated form but acquire moderate acid strength after heat treatment. The development of acidity here is associated with the gradual removal of water. In the case of acid-catalyzed reactions the strength and nature of the acidity (Lewis or Brönsted acids or both) are of central importance rather than the particular elements present as such.

Paraffin isomerization and hydrocracking are examples of reactions requiring that the catalyst incorporate a metallic and an acidic function in order to accelerate each of two different intermediate steps (or groups of steps) in the overall reaction.

A number of industrially significant reactions are catalyzed by metal sulfides, but the mechanisms of action are perhaps not as well understood. A metal sulfide such as nickel sulfide is an effective catalyst for hydrogenation reactions, and this is of particular importance in dealing with feed streams containing sulfur compounds. These can adsorb on other metal catalysts, such as platinum, thereby making them inactive.

Some catalysts, such as a mixture of cobalt and molybdenum on a support, actually are more active in the sulfide form than in the metallic form for at least some hydrogenations. Metal sulfides such as those of tungsten, of a mixture of cobalt and molybdenum, or of a mixture of nickel and molybdenum are active for hydrodesulfurization and hydrodenitrogenation (Secs. 9.8 and 9.9). Some sulfides are probably significantly acidic, thereby being bifunctional in themselves and capable of catalyzing both hydrogenation-dehydrogenation reactions and acid-catalyzed reactions.

Many industrial catalysts are more complex than Table 1.1 implies. Promoters and carriers play an important role, and the trend is towards more precise tailoring of the catalyst structure, in which each ingredient and each step in preparation contributes towards better activity, selectivity, or stability.

1.6 HOMOGENEOUS CATALYSTS

In its widest sense homogeneous catalysis occurs when the catalyst and the reactants are both in the same phase, either gas or liquid. In more recent years the term has come to be applied more specifically to the use of a solution of certain organometallic compounds in which a central metal atom is surrounded by a regular pattern of atoms or molecules, known as *ligands*, with which it is coordinated. Depending upon the nature of the ligands, the metal atom may be in a low-positive, zero, or low-negative state. Several different structures may exist in equilibrium in solution simultaneously, with different reactivities, but since the catalyst is dissolved in the reacting medium, each molecule of a particular structure acts like any other. In many cases the structures, although complicated, have been well characterized, and there are systematic correlations of the structure, the nature of the ligands, and the catalytic activity of the catalyst complex.

The reactions of industrial interest are primarily hydroformylation ("oxosynthesis"), carbonylation, addition of HCN, and olefin polymerization. Homogeneous catalysts can also be effective for hydrogenations and isomerizations, but they do not appear to be used industrially for this purpose, nor do they appear to have been useful for direct oxidations with oxygen.

A recently developed and useful process synthesizes CH_3COOH from CH_3OH and CO by use of a rhodium complex activated with HI, which catalyzes an insertion reaction:

$$CH_3OH + CO \longrightarrow CH_3COOH \tag{1.4}$$

Operating conditions are 150 to 200°C, and pressures are 1 to 4 MPa. This succeeds an earlier process using a cobalt catalyst that required pressures of 20 to 30 MPa and a temperature of about 230°C. Closely related to the above is hydroformylation in which an olefin, CO, and H_2 react to form an aldehyde, for example:

$$RCH{=}CH_2 + CO + H_2 \longrightarrow RCH_2CH_2CHO \tag{1.5}$$

As with acetic acid synthesis the first commercial catalysts used cobalt, but very high pressures were also required. More recent processes use a rhodium catalyst instead, which is more active, and thus much lower pressures can be utilized.

Another class of homogeneously catalyzed reactions involve a redox (*red*uction and *ox*idation) system in which intermediates alternate cyclically between two oxidation states. An example is the Wacker process for oxidizing ethylene to acetaldehyde. Reaction is carried out in an aqueous solution containing palladium(II) and copper(II) chlorides. $PdCl_2$ reacts stoichiometrically with C_2H_4 to form Pd^0, which is reconverted to $PdCl_2$ by reaction with $CuCl_2$. The basic reactions can be represented as follows:

$$C_2H_4 + PdCl_2 + H_2O \longrightarrow CH_3CHO + Pd^0 + 2HCl \tag{1.6}$$

$$2CuCl_2 + Pd^0 \xrightarrow{H_2O} 2CuCl + PdCl_2 \tag{1.7}$$

$$C_2H_4 + 2CuCl_2 + H_2O \xrightarrow{PdCl_2} CH_3CHO + 2CuCl + 2HCl \tag{1.8}$$

$$2CuCl + 2HCl + \tfrac{1}{2}O_2 \longrightarrow 2CuCl_2 + H_2O \tag{1.9}$$

The overall reaction is:

$$C_2H_4 + \tfrac{1}{2}O_2 \longrightarrow CH_3CHO \tag{1.10}$$

Reaction (1.6) involves the intermediate formation of a palladium complex with chlorine and ethylene. The reactions can be carried out in a one-stage process in which all four reactions proceed simultaneously in one vessel, or the first three reactions [(1.6), (1.7), and (1.8)] can be carried out in one vessel and regeneration of $CuCl_2$ (1.9) in a separate oxidation reactor.

There are no a priori guidelines to indicate in advance whether a homogeneous liquid-phase catalytic process will be more or less economical than a heterogeneous vapor-phase process, where both processes produce the same product. Some factors to consider are the relative degree of selectivity in the two processes and the ease of control to avoid runaway reactions or explosions. Liquid-phase operation in a stirred vessel, with either a homogeneous or heterogeneous catalyst, provides high heat capacity, which makes temperature control easier, but corrosion problems are often more severe in the liquid phase than in the vapor phase. A homogeneous catalyst may be poisonous, and one must also consider how to separate it from the product, preferably in a form that can be readily reused.[2] With an expensive catalyst such as rhodium, extremely high recoveries must be achieved for an economically viable process. In many cases purity specifications also set a stringent limit on the allowable catalyst concentration in the product.

Considerable attention has been paid in recent years to possible means for eliminating the separation problem while retaining the reaction characteristics of a particular homogeneous catalyst. A porous carrier may be saturated with a solution of the homogeneous catalyst, to be used in a vapor-phase process, or the homogeneous catalyst may be attached to a solid backbone such as a high polymer. This approach has reached commercialization in the use of immobilized enzymes adsorbed on a porous carrier for conversion of glucose to fructose.

Few gas-catalyzed reactions are used industrially, although a wide variety of such reactions are known. In the now obsolete lead chamber process for manufacture of sulfuric acid, nitrogen oxides were added to a mixture of sulfur dioxide and air. The oxidation and conversion of sulfur dioxide to sulfuric acid was catalyzed by the formation of an intermediate metastable compound, nitrosylsulfuric acid, $HNOSO_4$. Bromine will catalyze the gas-phase oxidation of hydrocarbons, but such a reaction does not appear to have been commercialized. The bromine initiates the reaction by forming a free radical by hydrogen abstraction, the hydrogen bromide thus formed being converted back to bromine by oxidation.

REFERENCES

Balandin, A. A.: *Adv. Catal.*, **19**, 1 (1969).
Burwell, R. L., Jr.: *Adv. Catal.*, **26**, 351 (1977).
Roginskii, S. Z., in A. A. Balandin et al. (eds.): *Scientific Selection of Catalysts*. Keter Pub. House, Kiryat Moshe, P.O. Box 7145, Israel. English translation, 1968.
Schlosser, E.-G.: *Heterogene Katalyse*, Verlag Chemie, 1972.
Taylor, H.S.: *Adv. Catal.*, **1**, 1 (1948).
Volkenstein, F. F.: *The Electronic Theory of Catalysis on Semi-Conductors*, English translation, Pergamon, New York, 1963.

[2]The fact that a heterogeneous system is inherently easier to separate than a homogeneous system was once illustrated to a lay audience by the analogy: It is easier to remove the olive from a martini than the vermouth.

CHAPTER

TWO

ADSORPTION

Two types of adsorption phenomena have been recognized in principle for many years: physical adsorption and chemical adsorption, or chemisorption. *Physical adsorption* is caused by secondary (van der Waals) attractive forces such as dipole-dipole interaction and induced dipoles and is similar in character to condensation of vapor molecules onto a liquid of the same composition. *Chemisorption* involves chemical bonding, is similar in character to a chemical reaction, and involves transfer of electrons between adsorbent and adsorbate. Borderline cases can also clearly exist since a highly unequal sharing of electrons may not be distinguishable from a high degree of distortion of an electron cloud. Physical adsorption is of particular interest here because it provides a method of measuring the surface area of a catalyst and determining its average pore size and pore-size distribution. Further, some reactions of interest are operated at pressures and temperatures only moderately above the boiling point of the mixture of reactants and products. Although the reaction might appear to occur in the vapor phase, pore condensation caused by physical adsorption may be a significant phenomenon.

Chemisorption is of concern since almost all reactions catalyzed by a solid are believed to involve, as an intermediate step in the overall reaction, the chemisorption of one or more of the reactants. The identification and knowledge of behavior of chemisorbed species are involved in obtaining an understanding of actual catalytic mechanisms. Chemisorption can also be used as a technique of determining the surface area of one particular catalyst component, e.g., a metal, in contrast to the total area, which is determined by physical adsorption.

The evidence that chemisorption is involved in almost all solid-catalyzed reactions stems from several kinds of observations. If a solid is found to affect the reaction of a fluid, this influence must have proceeded from molecules in the fluid coming into close proximity with the surface. Presumably adsorption of some type must have occurred for a finite time. Many catalytic reactions take place at temperatures far higher than those at which any significant physical adsorption could occur, suggesting that the adsorption must be chemical in nature. Further, there is a general correlation between catalytic activity and ability to chemisorb one or more of the reactants. Finally the forces involved in physical adsorption are much smaller than those involved in chemical bonding: it is hard to visualize that physical adsorption could cause distortion of the force fields around a molecule of sufficient magnitude to have an appreciable effect on its reactivity.

Physical adsorption may cause an increased rate of reaction, where the action is that of bringing molecules close together into a quasiliquid layer on the surface, rather than that of forming a chemisorbed intermediate. Some of the few reactions in which the catalytic effect may be of this nature are:

$$CO + Cl_2 \longrightarrow COCl_2 \text{ (phosgene)} \tag{2.1}$$

$$COCl_2 + H_2O \longrightarrow 2HCl + CO_2 \tag{2.2}$$

$$2NO + O_2 \longrightarrow N_2O_4 \rightleftharpoons 2NO_2 \tag{2.3}$$

For reactions (2.1) and (2.2), carbon is a common catalyst. Reaction (2.3) is typically carried out at 20 to 60°C on a silica gel catalyst.

2.1 CHARACTERIZATION OF TYPE OF SORPTION

Chemisorption is defined as involving electronic interaction between adsorbent (the solid) and adsorbate (the fluid). The problem is how to determine experimentally the extent to which this interaction occurs in any specific situation. Some methods, principally of interest for specialized laboratory studies, may be used for directly studying electronic interaction in contrast to secondary fields:

1. Surface electric potential (work function).
2. Surface electrical conductivity.
3. Collective paramagnetism (Selwood, 1975). (This method requires that the adsorbent be paramagnetic and is therefore limited essentially to adsorption on nickel, cobalt, or iron, but industrial-type catalysts may be examined.)

The principal difficulties in using these techniques are the detection of the small number of adsorbed molecules relative to the size of the sample, and apparatus limitations. Aside from these methods, evidence of chemisorption is indirect and rests on a number of kinds of observations. No single one of these will by itself indicate clearly in all cases whether a particular adsorption being studied is physical adsorption or chemisorption, but several taken together will usually be indicative.

2.1.1 Heat Effect

The magnitude of the heat effect is the most important criterion for differentiation. In physical adsorption the average heat of adsorption per mole for formation of a monolayer of adsorbed vapor usually somewhat exceeds that of liquefaction, but seldom by more than a factor of about 2. Perhaps the greatest heat effects observed with physical adsorption are with molecular sieves (zeolites) or certain forms of carbon where passageways are little larger than the molecular size of the adsorbate and the adsorbate is surrounded by the solid on all sides.

For relatively small molecules (for example, CO, N_2, CH_4) the heat of physical adsorption is typically of the order of 10 kJ/mol. (The heat of adsorption expressed on a molar basis would be expected to increase approximately proportional to molecular weight for a homologous series, as of the paraffins.) The heats of chemisorption are frequently comparable to those of chemical reactions (80 to 200 kJ/mol) and may be as high as 600 kJ/mol. Very occasionally, however, as with hydrogen under some conditions, a chemisorption may show a heat effect comparable to that of physical adsorption. The heat of adsorption may vary considerably with surface coverage in both types of adsorption (see Sec. 2.3).

Physical adsorption is always exothermic; chemisorption is usually exothermic, but it is possible for it to be endothermic, like a chemical reaction. For a spontaneous process to occur, the free energy must decrease, and from the relationship $\Delta G = \Delta H - T\,\Delta S$, it follows that $(\Delta H - T\,\Delta S) < 0$ and $\Delta H < T\,\Delta S$. (In other words, ΔH is a larger negative number than $T\,\Delta S$). If adsorption occurs without reaction on a substance whose properties are not altered by the process, a more ordered system is formed that corresponds to a decrease in the number of degrees of freedom. Therefore ΔS will be negative and ΔH also must be negative; that is, the process must be exothermic. However, deBoer has shown that if a molecule dissociates on adsorption and complete two-dimensional mobility of the adsorbate occurs, the number of degrees of freedom can increase. Hence ΔS can be positive, in which event ΔH also can be positive. In this unusual circumstance, if a diatomic molecule dissociates into two adsorbed atoms upon chemisorption, the dissociation energy of the molecule must be greater than the energy of formation of the bonds with the adsorbate.

Endothermic adsorption has been observed for several cases, e.g., when hydrogen is adsorbed onto iron contaminated with sulfide, and has been suspected in a number of others (deBoer, 1956, 1957). Even when the entropy of the adsorbed species decreases because of a decrease in the number of degrees of freedom, which is the usual case, this may be more than offset by an increase in entropy of the adsorbent itself, which might expand. Cases of endothermic adsorption are nevertheless rare.

2.1.2 Rate of Adsorption

Physical adsorption, like condensation, requires no activation energy and therefore can occur nearly as fast as molecules strike a surface. However, on a finely porous adsorbent such as a zeolite or some carbons a slow uptake of a vapor may be observed in which the rate is actually limited by the rate of diffusion of vapor into fine crevices or pores rather than by that of a sorption process as such. Many types of chemisorption exhibit an activation energy and therefore proceed at an appreciable rate only above certain minimum temperatures. Some surfaces are so active, however, that chemisorption occurs rapidly even at very low temperatures; e.g., hydrogen on tungsten metal at −183°C shows little or no activation energy. Rate measurements as such are thus of limited value in distinguishing between the two types of adsorption.

2.1.3 Effect of Temperature on Amount Adsorbed

The amount of gas physically adsorbed always decreases monotonically as temperature is increased. The amount is usually correlated with the relative pressure, P/P_0, where P is the partial pressure of the vapor in the system and P_0 is the vapor pressure that would exist above pure liquid at the same temperature. When P/P_0 is about 0.01 or less, the amount of physical adsorption is negligible except with solids possessing fine pores. At values of P/P_0 in the region of 0.1, the amount adsorbed corresponds to a monolayer, and as P/P_0 is increased, multilayer adsorption occurs until essentially a bulk liquid is reached at $P/P_0 = 1.0$ (but see Sec. 2.2).

With chemisorption a long period of time may be required for equilibrium to be established, especially at lower temperatures. The effect of temperature on the amount of material chemisorbed at equilibrium varies in a complex way with different systems. Little can be said of general value, although unlike physical adsorption, the amounts of vapor chemisorbed can be substantial at temperatures greatly above the boiling point, or indeed above the critical point. Frequently the amount chemisorbed is fairly constant over a certain temperature range at which the surface is saturated (Fig. 2.1, 78 kPa). At lower temperatures the amount observed to be chemisorbed is frequently less than this because the rate of adsorption is so low that saturation is not reached. It may be difficult to determine the true equilibrium amount chemisorbed over a sub-

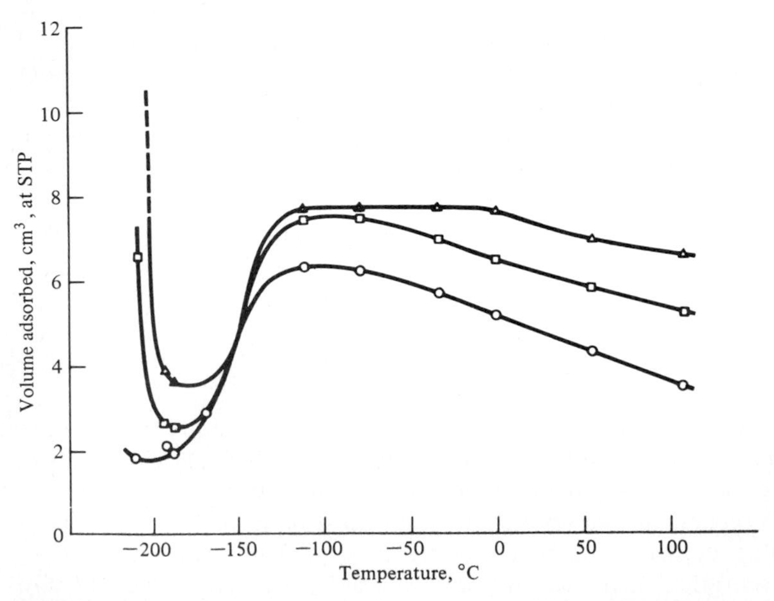

Figure 2.1 Adsorption isobars of hydrogen on nickel at 3.3 (○), 26 (□), and 78 (△) kPa. *(Benton and White, 1930.)*

stantial temperature range. More than one kind of chemisorption can occur simultaneously, one being rapid and a second slow (see Probs. 1, 2, and 3 in App. B).

In some cases it may be difficult to make a distinction between strong chemisorption and reaction with the solid. In general if the structural arrangement of the atoms in the bulk of the solid is not affected, the process is termed chemisorption; if the atoms in the solid are displaced or rearranged, it is termed reaction. If the adsorbate is confined to a layer not exceeding one molecule thick, this is generally termed chemisorption. In the chemisorption of oxygen onto a metal it may be difficult to limit the degree of sorption to a monolayer of oxygen, and the oxygen may readily penetrate the surface, as in the rusting of iron, sometimes termed *corrosive chemisorption.*

2.1.4 Extent of Adsorption

Physical adsorption becomes multilayered at P/P_0 values above approximately 0.1 to 0.3. Chemisorption is limited to a maximum of one layer of molecules on the surface, but the maximum is frequently much less, perhaps a small fraction of a monolayer. Both physical adsorption and chemisorption can occur together, but any adsorbed layers beyond the first must presumably be physically adsorbed. It is possible for a gas to be physically adsorbed initially and then more slowly form a chemisorbed species with the surface and/or for physical adsorption to occur over a chemisorbed layer. Hydrogen is readily *ab*sorbed into the interior of some metals such as nickel and palladium, and with the latter it forms a hydride. In these cases the amount of gas sorbed can be greatly in excess of that corresponding to *ad*sorption.

2.1.5 Reversibility

Physically adsorption is completely reversible, and equilibrium is established very rapidly unless diffusion through a finely porous structure occurs. Cycling of adsorption and desorption, as by alternately raising and lowering the pressure or temperature, can be performed repeatedly without changing the nature of the adsorbate.

Chemisorption may or may not be reversible. A chemical change in the adsorbate upon desorption is good evidence indeed that chemisorption in fact occurred. Thus oxygen chemisorbed on charcoal may be desorbed as carbon monoxide or carbon dioxide upon heating, and hydrogen adsorbed on an oxide may yield water upon heating. Ethylene adsorbed on nickel may yield other hydrocarbons on desorption. Hydrogen-deuterium exchange is also a useful diagnostic test. If HD is found in desorbed gas after H_2 and D_2 have been adsorbed onto the substrate, chemisorption must have occurred. Some chemisorbed substances are held very tenaciously. Chemisorbed oxygen on many metals, for example, can be removed only by extremely high temperature, by ion bombardment, or by reaction.

2.1.6 Specificity

Physical adsorption is relatively nonspecific. It will occur with all vapors or gases and on all surfaces, provided P/P_0 is sufficiently large. (However, this does *not* mean that the amount adsorbed at a given value of P/P_0 is independent of the nature of the adsorbate or adsorbent.) Chemisorption is highly specific: it will occur only if the adsorbate is capable of forming a chemical bond with the adsorbent. The extent of chemisorption may vary greatly with the nature of the surface and its previous treatment.

If chemisorption indeed occurs, an isobar or isotherm obtained on most catalysts will usually be complex, reflecting the heterogeneous nature of the surface. More than one type of chemisorption may be observed, and some rates may be so slow that it is questionable whether true equilibrium was reached. The sorption may be irreversible. As temperature is increased at constant pressure, more than one maxima may appear, as illustrated in Figs. 2.1 and 2.2. Data such as these may be obtained by conventional volumetric or gravimetric methods, in which the quantity of vapor transferred between the gas phase and the solid is followed as pressure or temperature is changed. Alternately, a nonadsorbing gas such as helium may be passed continuously through a sample, the gas being dosed with pulses of the adsorbate. The amount of material not adsorbed is conveniently determined by gas chromatography.

Figure 2.1 (Benton and White, 1930) shows isobars for hydrogen on nickel powder at 3.3, 26, and 78 kPa pressure. The low-temperature process is ascribed to physical adsorption or to a nonactivated (i.e., a very fast) chemisorption that is weak and reversible since the adsorbed volume decreases with increased temperature. (Since the temperature is above the critical temperature, the use of the value of relative pressure, P/P_0, to suggest degree of physical adsorption cannot be applied here.) The higher-temperature process is a type of chemisorption that must be activated at least in part (i.e., the rate increases with increased temperature). Presumably at temperatures below about $-100°C$ true equilibrium was not obtained with respect to this type of sorption. The decrease in amount adsorbed with increased temperature is consistent with that

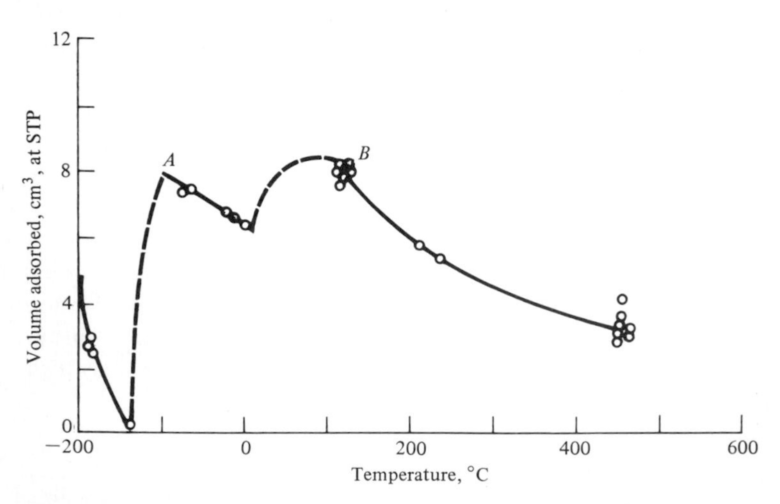

Figure 2.2 Adsorption isobar of hydrogen at atmospheric pressure on an iron catalyst used in the synthesis of ammonia, showing maxima corresponding to two types of chemisorptions. (*Emmett and Harkness, 1935.*)

expected for a reversible exothermic process, and the plateau in the 79-kPa study suggests saturation of the surface under those conditions.

Figure 2.2 (Emmett and Harkness, 1935) presents an adsorption isobar for hydrogen, at atmospheric pressure, on an iron catalyst used for synthesis of ammonia. This shows three maxima. The low-temperature maximum is again physical adsorption or a weak chemisorption, and the maxima at A and B reflect two types of activated chemisorption. The rate of type-A sorption was inappreciable below about $-100^\circ C$, and the rate of type-B sorption did not become appreciable until about $100^\circ C$. Strictly, Fig. 2.2 is not a true isobar in that, at least at temperatures below $100^\circ C$, true equilibrium was not obtained.

A study of hydrogen adsorption on zinc oxide is taken up in Probs. 1 and 2 (App. B). In the region of 111 to $154^\circ C$ the volume of hydrogen adsorbed with time was measured as temperature was alternately increased and decreased. The results show that two types of chemisorption were occurring simultaneously and independently of one another, suggesting that each type of sorption occurred on a different type of catalyst site. One chemisorption was extremely fast, was reversible, and decreased in amount with increased temperature; the other was slow, was irreversible, and reached a steady-state value only after many hours.

2.2 PHYSICAL ADSORPTION ISOTHERMS

An adsorption isotherm is the relationship at constant temperature between the partial pressure of the adsorbate and the amount adsorbed at equilibrium. This varies from zero at $P/P_0 = 0$ to infinity as P/P_0 reaches 1 provided that the contact angle of the condensed vapor is zero, i.e., the surface is completely wetted. [If it is greater than zero, condensed vapor can form drops, and in both theory and experiment the condensed layer is of finite thickness at $P/P_0 = 1$ (Gregg and Sing, 1967, p. 153). However, in practice a slight increase in vapor pressure at $P/P_0 \approx 1$ (or slight decrease in temperature) should be sufficient to cause complete condensation to occur.] If the isotherm asymptotically approaches the vertical line corresponding to $P/P_0 = 1$, this implies that the angle of contact is zero.

The shape of the isotherm may vary substantially depending upon the nature of the adsorbent and the adsorbate, as illustrated by Figs. 2.3, 2.4, and 2.5. *n*-Pentane adsorption on three solids and on liquid water is shown in Fig. 2.3 (Kiselev and Eltekov, 1957). With porous substances a hysteresis loop, associated with capillary condensation, is frequently observed as shown in Fig. 2.4 for argon, nitrogen, or *n*-butane on porous glass (Emmett and Cines, 1947). Figure 2.5 shows hysteresis loops for adsorption of nitrogen on silica gel, activated carbon, and clay cracking catalyst (Ries and Johnson in Barrett et al., 1951). The lower portion of the loop is traced out on *ad*sorption, the upper portion upon *de*sorption. The process is completely reproducible if the two ends of the loop are reached. For a substance such as porous glass in which all pores are fairly small and a narrow pore-size distribution exists, the curve may reach a virtual plateau at a value of P/P_0 significantly less than 1.0. Here all the pores have become filled with condensed vapor, but the amount of vapor adsorbed on

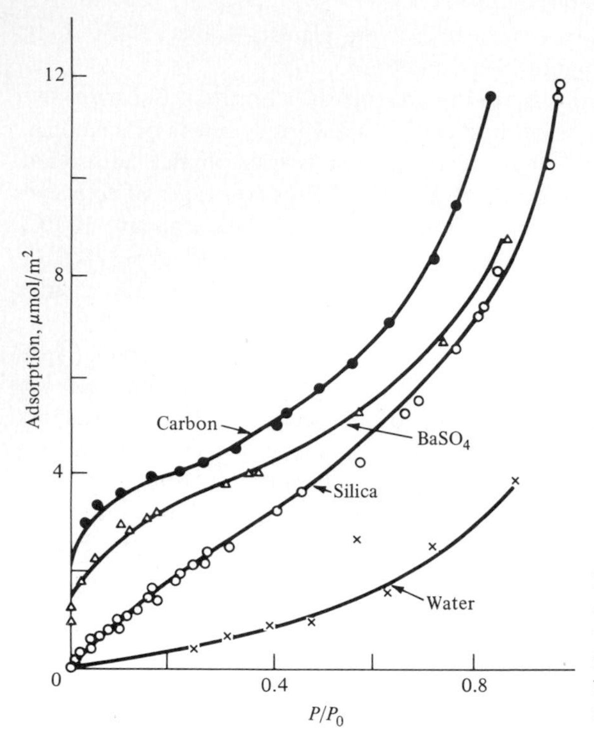

Figure 2.3 The effect of the nature of the adsorbent (marked on isoterm) on the shape of the adsorption isotherm of *n*-pentane. The adsorbent water was in the liquid form. (*Kiselev and Eltekov, 1957.*)

the exterior surface with increasing values of P/P_0 is small relative to that condensed in the pores.

Capillary condensation can occur in fine pores at values of $P/P_0 < 1$ since the value of P/P_0 at which this occurs is a function of the radius of curvature according to the Kelvin equation (Sec. 5.3.1). The hysteresis is caused by geometrical effects in that the surface curvature in contact with the vapor at a specified value of P/P_0 as vapor pressure is increased is different from that as the vapor pressure is decreased. As an example, in the "ink-bottle" hypothesis pores are visualized as being shaped like an old-fashioned round ink bottle with a narrow neck. The value of P/P_0 at which condensation occurs upon adsorption is determined by the (larger) effective radius of curvature of the body of the bottle. Evaporation from a filled bottle, which occurs upon desorption, is determined by the (smaller) effective radius of curvature of the neck. Other geometries can also cause hysteresis, and a substantial body of literature exists relating the shape and nature of hysteresis loops to pore geometry.

This topic is addressed again in Chap. 5 in connection with methods of determining surface area and pore-size distributions.

2.3 HEAT OF ADSORPTION

The heat of adsorption is a significant property for characterization of the type of sorption and of the degree of heterogeneity of a surface. If truly reversible isotherms

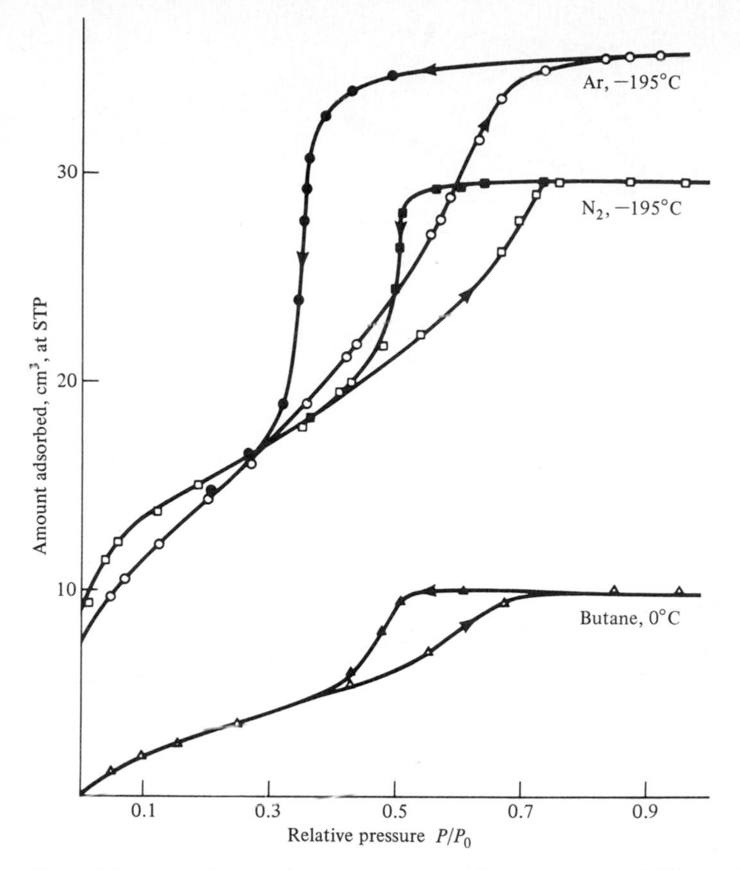

Figure 2.4 Adsorption isotherms for argon and nitrogen at −195°C and for *n*-butane at 0°C on porous glass. Open symbols, adsorption; solid symbols, desorption. (*Emmett and Cines, 1947.*) (*Reprinted with permission from the Journal of Physical Chemistry. Copyright 1947 by The Williams & Wilkens Company, Baltimore.*)

are obtainable, the differential heat of adsorption may be calculated as a function of volume of gas adsorbed, v, by cross plotting, utilizing the Clausius-Clapeyron equation:

$$\left(\frac{\partial \ln P}{\partial T}\right)_v = \frac{q}{RT^2} \tag{2.4}$$

These values of q are called *isosteric heats* of adsorption. Alternately, a calorimetric method gives an integral value of the heat of adsorption, which is the average value over the degree of surface coverage studied. Differential heats of adsorption can also be determined calorimetrically by admitting small quantities of vapor at a time or by differentiating integral data.

In general the differential heat of adsorption will decrease with increased surface coverage, although a large variety of results may be observed. Figures 2.6 and 2.7 illustrate the kinds of information that can be obtained (Joyner and Emmett, 1948). They

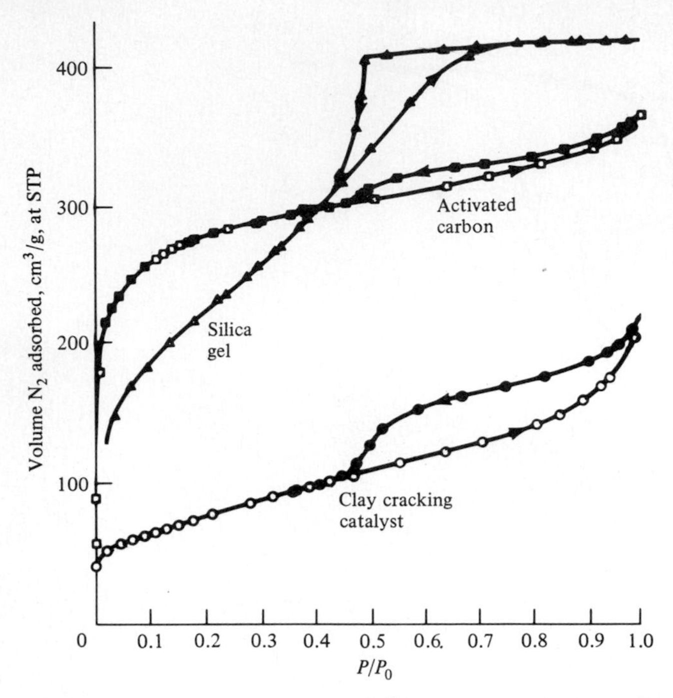

Figure 2.5 Adsorption of N_2 at $-195°C$ on porous substances. (*Ries and Johnson in Barrett et al., 1951.*) (*Reprinted with permission from the Journal of the American Chemical Society. Copyright by the American Chemical Society.*)

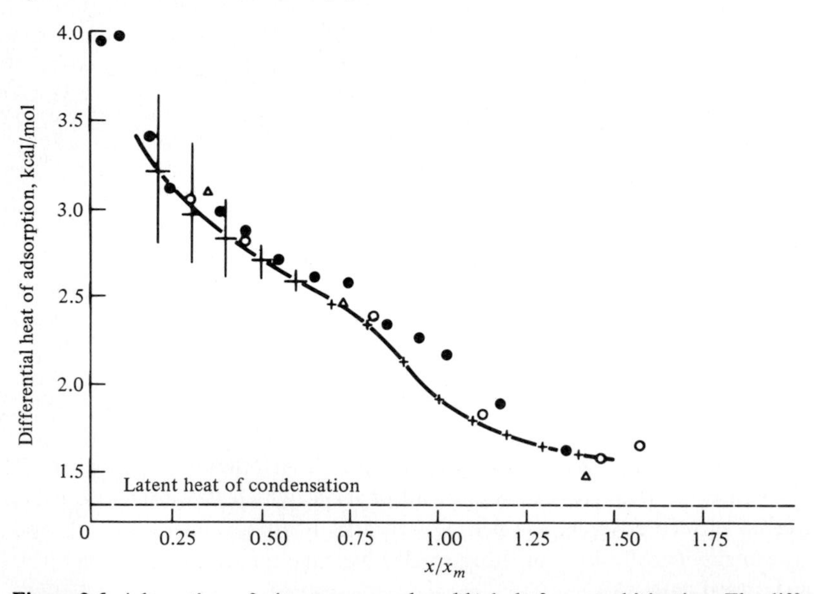

Figure 2.6 Adsorption of nitrogen on carbon black, before graphitization. The differential heat of adsorption q is plotted against x/x_m. q was determined calorimetrically at $-195°C$ (○, ●, △), and isosterically (+, -194.6 to $-183.1°C$). Vertical lines indicate the maximum variation observed. (*Joyner and Emmett, 1948.*) (*Reprinted with permission from the Journal of the American Chemical Society. Copyright by the American Chemical Society.*)

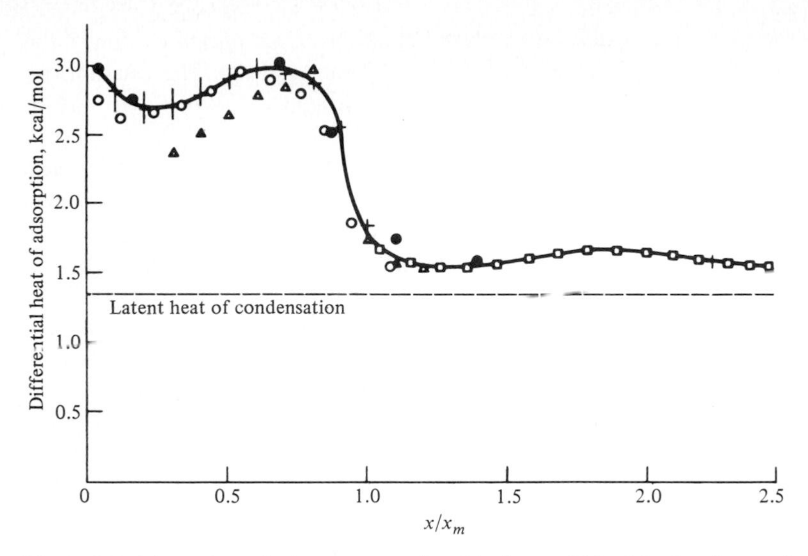

Figure 2.7 Adsorption of nitrogen on carbon black, graphitized by heating at about 3200°C. The differential heat of adsorption q is plotted against x/x_m. q was determined calorimetrically at −195°C (○, ●) and also isosterically (+, −194.8 to −182.8°C; △, −204.7 to −182.8°C; □, −204.7 to −194.8°C). Vertical lines indicate maximum variation observed. (*Joyner and Emmett, 1948.*) (*Reprinted with permission from the Journal of the American Chemical Society. Copyright by the American Chemical Society.*)

show differential heats of adsorption of nitrogen on a "medium processing" carbon black made by the channel process, before and after graphitization, as a function of x/x_m. x is the quantity of nitrogen adsorbed, and x_m the quantity corresponding to a monolayer. It is evident that graphitization has made the surface much more uniform energetically. In Fig. 2.7, the slight minimum at $x/x_m \approx 0.2$ is attributed to residual surface heterogeneity, the slight maximum to lateral interaction of adsorbed molecules. The initial heat of adsorption on carbon is higher before graphitization than afterwards, attributable to a few highly active centers that were removed by graphitization. Before treatment, surface heterogeneity was so great as to overshadow any effect of lateral interaction.

2.4 MODELS FOR ADSORPTION ISOTHERMS

The derivations that follow can be applied to either physical adsorption or chemisorption, provided that equilibrium is truly reached and that the sorption process is reversible in the sense that no change in adsorbate occurs upon cycling between sorption and desorption. Theoretical derivations of isotherms may be based on kinetics, statistics, or thermodynamics.

2.4.1 Langmuir Isotherm

The Langmuir adsorption isotherm is of the greatest general utility in application to catalysis because of its simplicity and because it serves as a point of departure for formulating many kinetic expressions for catalyzed reactions. A simplified version of the kinetic approach, which was originally used by Langmuir, follows. The assumptions are:

1. The adsorbed species are held onto definite points of attachment on the surface. (This implies that the maximum adsorption possible corresponds to a monolayer.) Each site can accommodate only one adsorbed species.
2. The differential energy of adsorption is independent of surface coverage. (This implies that the surface is completely uniform so that there is the same probability of adsorption on all sites. A further implication is that adsorbed molecules are localized.) Attractive or repulsive forces between adjacent adsorbed molecules are taken to be negligible, so the energy of an adsorbed species or the probability of adsorption onto an empty site are independent of whether or not an adjacent site is occupied.

Mathematical derivation Consider a single pure vapor A at a pressure P_A that adsorbs without dissociation onto a surface. Let the occupied fraction of sites on which adsorption is possible be θ_A. The rate of adsorption dn_A/dt is proportional to the rate of molecular collisions with unoccupied sites:

$$\left(\frac{dn_A}{dt}\right)_{ads} = k(1 - \theta_A) P_A \tag{2.5}$$

The rate of desorption is proportional to the number of molecules adsorbed.

$$\left(\frac{dn_A}{dt}\right)_{des} = k'\theta_A \tag{2.6}$$

At equilibrium, the rate of adsorption equals the rate of desorption, so that

$$k(1 - \theta_A) P_A = k'\theta_A \tag{2.7}$$

$$\theta_A = \frac{kP_A}{k' + kP_A} = \frac{K_A P_A}{1 + K_A P_A} \tag{2.8}$$

where the adsorption equilibrium constant $K_A = k/k'$. K_A can be expressed in the form $K = Ae^{+\lambda/RT}$, where λ $(= -\Delta H)$ is the heat of adsorption. A large value of K implies strong bonding. The larger the value of K, the greater is the fractional surface coverage at a fixed temperature and at a fixed value of P_A, or the higher is the temperature required for a specified fractional surface coverage at fixed P_A. At low values of P_A, the fraction of the sites covered is directly proportional to P_A, but at high values of P_A, the fraction approaches unity asymptotically and becomes independent of pressure.

A similar approach can be applied to derive isotherms for two or more gases competing for adsorption on the same kind of sites, for adsorption on different kinds of sites, for dissociation of gases upon adsorption, etc. Some of these cases are developed in Chap. 3 in conjunction with the formulation of correlations for kinetic data.

The Langmuir isotherm can also be derived by a thermodynamic or statistical approach, which avoids the necessity of assuming that forward and reverse rates of adsorption and desorption follow a particular postulated mechanism. A conclusion of the statistical derivation is that the assumption of no interaction between adsorbed species implies that they are immobile. This is in the sense that there is no translational motion of adsorbed species in the plane of the surface. The thermodynamic derivation assumes that adsorption is nonlocalized and arrives at the same mathematical form of the equation, but the constant K has a different theoretical interpretation. The Langmuir approach has also been applied to a heterogeneous surface. This involves assuming some distribution of energies of sites and summing up Langmuir isotherms over the total.

Few chemisorption isotherms correspond to the Langmuir equation over the whole range of surface coverage. A major objection to using the Langmuir model is that the heat of adsorption in fact generally decreases with increased surface coverage, contrary to the assumption above. This is caused by one or more of the following:

1. Repulsive forces between adjacent adsorbed molecules
2. The heterogeneous character of all but the most meticulously prepared surfaces
3. The fact that more than one type of bonding may occur between the adsorbed molecules and the surface

The first two effects will cause a range of adsorption energies, and this is, in fact, commonly encountered. The initial adsorption occurs most readily on the most energetic sites with accompanying high differential heats of adsorption, but this decreases as the less active sites become covered. In the application of the Langmuir isotherm to correlation of kinetic data, however, neglect of the variable heat of adsorption may not be serious. Molecules adsorbed on the most active sites may be held so firmly that they do not participate in the reaction, while the energy of interaction on weak sites may be too low to cause reaction to occur. Only those sites of an intermediate activity would thus actually participate in a catalytic reaction.

2.4.2 Freundlich Isotherm

A classic isotherm generally associated with the name of Freundlich, and originally empirical, is:

$$\theta_A = cP_A{}^{1/n} \tag{2.9}$$

Here $n > 1$ and the parameters n and c usually both decrease with increasing temperature. Often this equation satisfactorily represents data over a wide range of values of θ and for systems that do not follow the Langmuir isotherm. Even for a system which does follow the Langmuir isotherm, over a moderate range of coverage intermediate between the extremes of $\theta = 0$ and $\theta = 1$ the Langmuir isotherm is nearly equivalent to θ being proportional to a fractional power of P. The Freundlich isotherm, by suitable adjustment of the two constants, thus can be made to fit the data almost as well.

The Freundlich isotherm can also be derived theoretically by a statistical approach in which the Langmuir adsorption isotherm is applied to a distribution of energies among the sites such that the heat of adsorption decreases logarithmically with coverage. In its empirical form the Freundlich isotherm gives no limiting value of θ, which is contrary to chemisorptions, but the statistical derivation sets a maximum value of θ that is related to the heat of adsorption. A thermodynamic derivation leads to the theoretical interpretation of n as a constant representing the mutual interaction of adsorbed species. A value of n greater than unity, the usual case, is interpreted to mean that adsorbed molecules repulse one another.

2.4.3 Temkin (Slygin-Frumkin) Isotherm

The decrease in differential heat of adsorption with coverage is more likely to be linear than logarithmic, and this assumption is made in deriving the Temkin isotherm, namely,

$$q = q_0(1 - \alpha\theta) \tag{2.10}$$

Applying the Langmuir adsorption isotherm to this distribution of energies, we can show that in the middle range of coverage between $\theta = 0$ and $\theta = 1$,

$$\theta = \frac{RT}{q_0\alpha} \ln A_0 P \tag{2.11}$$

where q_0 = differential heat of adsorption at zero surface coverage, $A_0 = a_0 e^{-q_0/RT}$ (A_0 is independent of surface coverage), and a_0 and α are constants. The same mathematical expression is obtained if the drop in q is caused either by repulsive forces on a uniform surface or from surface heterogeneity. This equation, like that of Freundlich, provides two adjustable constants, α and a_0. If q_0 is unknown, data may be correlated using A_0 and the product $(q_0\alpha)$.

More details on the mathematical derivation of the various isotherms are given by Hayward and Trapnell (1964) and by Adamson (1976).

2.5 CHEMISORPTION

The belief that chemisorption of one or more of the reactants is involved, as an intermediate step, in essentially all solid-catalyzed reactions leads to the hope that an understanding of chemisorption phenomena on catalysts would illuminate and clarify the mechanisms of catalytic action. A vast literature exists on chemisorption and the relationships between chemisorption and catalysis, but most generalizations as have emerged must be hedged with such qualifications that little can be said in summary that is helpful without being misleading. Nevertheless, several aspects of chemisorption are of particular interest in catalysis.

1. The *rates* of chemical adsorption of reactants or desorption of products, studied individually, may indicate the slow and therefore rate-limiting step in the catalytic reaction. They may also help characterize surface heterogeneities.
2. The *heat of chemisorption* is a measure of the strength of the bonds formed between adsorbent and adsorbate. The variation of heat of adsorption with surface coverage is a measure of surface heterogeneity.

3. The *nature* of the actual chemisorbed species as revealed, for example, by infrared absorption gives direct evidence of possible chemical intermediates in the reaction.

Chemisorption phenomena may also be deceptive. Many chemisorption studies have been performed on catalysts under different conditions of pressure and temperature than those used for reaction, so it is far from clear that the phenomena observed also occur as an intermediate step in the actual reaction. Frequently, two or three kinds of chemisorption have been observed between one adsorbent and one adsorbate, but it is doubtful that more than one is actually involved in the chemical reaction; other chemisorbed molecules may just "sit" on the catalyst. Chemisorption on pure metals is simpler than chemisorption on oxides, and numerous studies have been reported using pure metal films, wires, etc. However the behavior of the same metal when supported on a carrier, as in a commercial catalyst, may be significantly different. Although conclusions from chemisorption studies are usually debatable, such studies do provide clues as to the cause of catalyst behavior.

It is sometimes useful, especially in developing kinetic mechanisms, to distinguish between *nondissociative chemisorption* in which the molecule adsorbs without fragmentation (also termed *associative chemisorption*) and *dissociative chemisorption* in which two or more fragments are formed, all of which remain momentarily adsorbed on the surface. Hydrogen generally dissociates upon adsorption on metals, a process that can be symbolized as:

$$\mathrm{H{-}H}(g) + 2* \longrightarrow 2\mathrm{H}*$$

where the asterisk represents a surface site.

Propylene adsorbs on certain metal oxides by dissociation, splitting off a H atom to form a π-allyl complex with the surface:

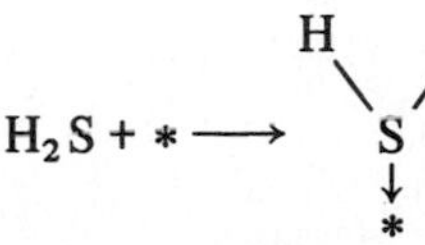

$$\begin{array}{c} \mathrm{H} \\ | \\ \mathrm{H_2C{\cdots}C{\cdots}CH_2} \\ \downarrow \\ * \end{array}$$

Hydrogen sulfide adsorbs on a metal site without dissociation, which may be symbolized as

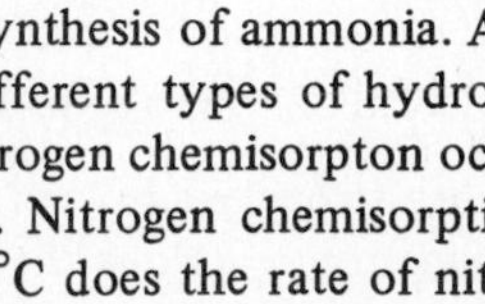

$$\mathrm{H_2S} + * \longrightarrow \begin{array}{c} \mathrm{H}\quad\mathrm{H} \\ \diagdown\;\diagup \\ \mathrm{S} \\ \downarrow \\ * \end{array}$$

The classic example of the study of chemisorption rates to indicate the slow step in a catalyzed reaction is the chemisorption of hydrogen or of nitrogen on the promoted iron catalyst used for synthesis of ammonia. A brief discussion illustrates some typical complexities. Three different types of hydrogen chemisorption on iron were identified by Emmett, but hydrogen chemisorpton occurs rapidly at temperatures well below those used in synthesis. Nitrogen chemisorption, on the other hand, is much slower, and only at about 450°C does the rate of nitrogen adsorption during the first few seconds of exposure to the catalyst become approximately equal to the rate of synthesis of ammonia from a 3:1 H_2/N_2 mixture at the same temperature and on the same catalyst. It would appear that the rate of nitrogen adsorption is rate-limiting.

The form of the observed rate equation, however, does not agree completely with this mechanism, and experiments by Tamaru showing that the presence of hydrogen accelerates the chemisorption of nitrogen do not fit into this simple explanation. The problem is of long standing and has undergone scrutiny by many investigators with no consensus as yet. One possibility is that the surface is covered with adsorbed NH in equilibrium with hydrogen and ammonia; another explanation lies in the possibility of a nonuniform surface; a third possibility is that there may not be one rate-limiting step under the conditions of the experiment. This situation is typical of heterogeneous catalysis: hardly any catalyzed reaction is yet understood in such detail that all the experimental facts concerning it can be brought together into a completely consistent framework.

The fact that chemisorption and chemidesorption rates may be quite slow even at elevated temperatures of reaction means that a long time lag is sometimes required for the adsorbed phase to come to equilibrium with the gas phase. This can amount to hours or days. In a continuous-flow chemical reactor, an appreciable time may be required for a steady-state exit-gas composition to be reached when a reactant is present in the feed stream in very low concentrations. The absolute quantity of reactant entering the reactor per unit time may then be small relative to the surface area of catalyst present. Nonsteady-state methods of studying catalytic reactions in the laboratory, as by pulse-type reactors, may lead to confusing and misleading results for this and other reasons.

A perhaps extreme case occurs in practice in the use of catalysts on automobiles for combustion of carbon monoxide and hydrocarbons in the exhaust to form carbon dioxide and water. Small amounts of sulfur in the fuel are converted to sulfur oxides, some of which adsorb on the alumina-based catalyst that is commonly used. Unlike the operation of most commercial catalytic reactors, here the temperature and flow rate of gases through the catalyst bed vary erratically as the automobile is driven. The result is that sulfur oxides may be stored on and emitted from the catalyst by adsorption and desorption in a different erratic pattern.

The relative adsorptivities of different species on a catalyst can have a profound effect on activity, selectivity, poisoning, and the form of the rate expression, as developed in Chap. 3. In a mixture of two reactants, a strongly adsorbed reactant may react to the almost complete exclusion of the second if the latter is weakly adsorbed. If a product is more strongly adsorbed than a reactant, the rate of reaction may drop very markedly as degree of conversion is increased.

A few generalizations help to provide orientation, although many exceptions will arise. Within groups of hydrocarbons, the strength of adsorption typically is in the order:

$$\text{Acetylenes} > \text{diolefins} > \text{olefins} > \text{paraffins}$$

Polar substances are generally more strongly adsorbed than nonpolar substances, with the consequence that in hydrogenation reactions the product is usually less strongly adsorbed than the reactant. The opposite is frequently the case in partial oxidation reactions. Other factors being equal, the degree of adsorptivity increases with molecular weight. Adsorption can also be affected by various specific interactions. Thus aromatic compounds may be held relatively strongly onto some metals by π bonding. On most metals, Bond (1974) finds that in general the strength of adsorption for some simple gases and vapors falls in the sequence:

$$O_2 > C_2H_2 > C_2H_4 > CO > H_2 > CO_2 > N_2$$

Oxygen is so strongly adsorbed that it frequently leads to reaction in which oxygen is no longer confined to a chemisorbed surface layer. In that event the catalyst structure becomes altered or even perhaps destroyed. Similarly, metal carbides may be formed from a metal, as occurs with iron or nickel.

REFERENCES

Adamson, A. W.: *Physical Chemistry of Surfaces*, 3d ed., Wiley, New York, 1976.

Barrett, E. P., L. G. Joyner, and P. P. Halenda: *J. Am. Chem. Soc.*, **73**, 373 (1951).

Benton, A. F., and T. A. White: *J. Am. Chem. Soc.*, **52**, 2325 (1930).

Bond, G. C.: *Heterogeneous Catalysis: Principles and Applications*, Oxford, 1974, p. 21.

deBoer, J. H.: *Adv. Catal.*, **8**, 17 (1956).

——: *Adv. Catal.*, **9**, 472 (1957).

Emmett, P. H., and M. Cines: *J. Phys. Chem.*, **51**, 1248 (1947).

—— and R. W. Harkness: *J. Am. Chem. Soc.*, **57**, 1631 (1935).

Gregg, S. J., and K. S. W. Sing: *Adsorption, Surface Area and Porosity*, Academic, New York, 1967.

Hayward, D. O., and B. M. W. Trapnell: *Chemisorption*, 2d ed., Butterworth, London, 1964.

Joyner, L. G., and P. H. Emmett: *J. Am. Chem. Soc.*, **70**, 2353 (1948).

Kiselev, A. V., and Y. A. Eltekov: *International Congress Surface Activity, II*, Butterworth, London, 1957, p. 228. (See also Gregg and Sing, 1967, p. 79.)

Selwood, P. W.: *Chemisorption and Magnetization*, Academic, New York, 1975. (An earlier edition was published as *Adsorption and Collective Paramagnetism*, Academic, New York, 1962.)

Taylor, H. S., and C. O. Strother: *J. Am. Chem. Soc.*, **56**, 586 (1934).

CHAPTER

THREE

RATES AND KINETIC MODELS OF CATALYTIC REACTIONS

3.1 INTRODUCTION

Correlations of rate data may be sought for any of several purposes. The process engineer may wish to develop a model for a specific reaction so as to be able to predict the effect of reactor operating changes on performance. The fundamental investigator may wish to determine how the rate of a particular reaction varies as catalyst composition is systematically varied, with the aim of relating the results in a fundamental fashion to specific physical or chemical properties of the catalyst. A study of the detailed kinetics of one particular reaction has been a traditional approach to obtaining some understanding, even though indirect, of its mechanism. Regardless of the objective, the investigator desires a mathematical model to represent the data.

Consider a reaction occurring between a fluid and a porous solid catalyst. In order for reaction to occur, the reactants in the fluid must first be transported to the outer surface of the solid, and then they must diffuse through the pores of the solid to catalytically active sites. At least one of the reactant species must usually be chemisorbed onto the surface of the solid. Subsequently, reaction occurs among chemisorbed species or between a chemisorbed species and another species that is either physically adsorbed or that collides with the chemisorbed species directly from the fluid phase. After reaction, products are desorbed and diffuse out through the pores of the catalyst to the bulk fluid. Because the rates of these various steps respond in a different way to experimental variables such as pressure, temperature, bulk-fluid velocity, and chemical and physical structure of the catalyst, it is convenient to classify them as follows:

1. Mass transfer of reactants and products by counterdiffusion between the bulk fluid and the outer surface of the catalyst particle

2. Mass transfer of reactants and products by counterdiffusion through the porous structure of the catalyst
3. Adsorption of reactants onto the catalyst surface and desorption of products
4. Chemical reaction involving one or more chemisorbed species

One or more of these steps may be rate-limiting in the sense that it consumes the major portion of the chemical potential available for carrying out the process.

Quantitative methods of determining the extent to which mass-transfer effects are significantly rate-limiting (steps 1 and 2) are described elsewhere (Satterfield, 1970), as well as the effect of temperature gradients that frequently accompany concentration gradients. In what follows it will be assumed that the rate of reaction is proportional to surface area (or number of catalytically active sites) and that the entire catalyst surface inside a pellet is exposed to reactant of uniform composition and temperature; i.e., effects of transfer of heat or mass to or within a porous catalyst are insignificant.

The true mechanism in all its details is not known for even the simplest catalytic reaction. The closer a model reflects actuality, of course, the more reliable it is, but an attempt to allow for the complex nature of a heterogeneous reaction may easily lead to a complicated formulation containing many parameters which must be empirically adjusted. In this event the model loses theoretical justification. If an industrial reaction proceeds by a complex and little-known mechanism, the process engineer may find it adequate to use an essentially empirical correlation. Conveniently, this is the Arrhenius expression with power functions of reactant concentrations, the exponents being arbitrary adjustable constants. In any event, the basic guiding principle should be based on a maxim enunciated by an English philosopher, William of Occam, in the fourteenth century: "Entities ought not to be multiplied except out of necessity." Since it is a mental paring device for pruning away "entities" in the sense of unnecessary hypotheses and complexities in explaining observations or experiments, it is known as *Occam's razor*. In the present context it suggests that mathematical formulations should be no more complicated than those necessary to explain the facts and to be consistent with well-established theory.

3.2 EMPIRICAL CORRELATIONS

For studies of homogeneous gas-phase reactions, the rate of an elementary bimolecular reaction (one occurring at the instant of collision of two molecules, free radicals, or other species) between two species A and B is given by

$$\text{Rate, } \frac{\text{molecules reacted}}{\text{(time)(volume)}} = kC_A C_B$$

$$= Ae^{-E/RT} C_A C_B \tag{3.1}$$

Equation (3.1) is known as the *Arrhenius expression* when A, the *preexponential factor*, is taken to be independent of temperature. From collision theory, A varies as the square root of the temperature. In transition-state theory the effect of temperature on A varies somewhat with the structure of the reactant molecules and the nature of the intermediate complex formed. Since in any event the effect of temperature on A is small relative to its effect on the exponential term, one may with little error take A to be independent of temperature.

By analogy a simple expression for the rate r of a heterogeneously catalyzed reaction between A and B is

$$\text{Rate, } \frac{\text{molecules reacted}}{(\text{time})(\text{area})} = k_0 e^{-E/RT} \cdot f(C_A C_B) \tag{3.2}$$

where k_0 is taken to be independent of temperature and surface area of the catalyst. The function of the concentrations which usually is easiest to use in correlating rate data, consists of simple power functions: $C_A{}^a \cdot C_B{}^b$, where a and b are empirically adjusted constants. Hence

$$-r = k_0 e^{-E/RT} C_A{}^a C_B{}^b \tag{3.3}$$

More generally this may be expressed as

$$-r = k \prod_i C_i^{a_i} \tag{3.4}$$

where a_i is termed the order of the reaction with respect to C_i.

Equation (3.3) is an example of a *power-rate law*. For this expression to be useful, k_0 and E should indeed be functions only of the catalyst and the reacting system, and not of temperature or concentration. Likewise the function of concentration should be independent of temperature and of composition; for example, a reaction first order with respect to A should follow that relationship over the range of concentrations of A that are of interest. For the most precise studies the rate should be expressed per number of active sites, termed the *turnover number* (Sec. 1.3.7). Sometimes this can be determined quantitatively, as with some supported metal catalysts, but usually the number of active sites is unknown. The rate may then be expressed per unit total area, r_a; per weight or volume of catalyst, r_w or r_v; or per volume of packed reactor. For most industrial catalysts the rate per unit weight of catalyst is most customary. The IUPAC recommendation is that r_a be termed the *areal rate of reaction*, but this usage is not currently widespread.

The term k_0 in Eq. (3.3) will usually have no theoretical significance, and the exponents may be integral or fractional, positive, zero, or negative. However many catalytic reactions follow a simple relationship of this type over a sufficiently wide range of conditions as to make the correlation useful. The development of a power-law expression for the water-gas shift reaction under conditions where the approach to equilibrium must be taken into account is described in detail by Bohlbro (1966). Theoretically derived models may also reduce to power-law forms in which a and b are integers or half-integers.

Power-law kinetics can often be used to develop useful correlations, such as when a given reaction is studied on a series of catalysts, or when a series of related reactions is studied on one catalyst. It is sometimes found that as one proceeds from catalyst to catalyst, or reaction to reaction, either the preexponential factor k_0 remains constant and the activation energy changes or the activation energy remains constant and the preexponential factor changes. Although such correlations may be essentially empirical, they may provide a basis for modest extrapolation of rate data, or for estimation of the rate of reaction of a substance that is a member of a homologous series for which information is available on other compounds in the group.

A more confusing series of reacting systems are also frequently found in which the concentration function in Eq. (3.3) remains constant but E and k_0 both change. When each change has the same algebraic sign (both positive or both negative), this is termed *compensation* since the reaction rate is affected less than it would be if either E or k_0 alone varied (Sec. 3.7). Finally, the reaction rate may not fit a relationship of the form of Eq. (3.2) or Eq. (3.3). The apparent order of the reaction and the apparent activation energy may change with temperature, which requires developing a different kinetic model, such as the Langmuir-Hinshelwood formulation discussed in Sec. 3.3.1.

3.3 FORMAL KINETIC MODELS

It is frequently observed that if it is attempted to express the rate as a simple power function, for example, of the form of Eq. (3.3), a and b may not be integers and their values, as well as the value of E, may change with temperature. In part this is because it has been assumed that the driving force for reaction is a function of the concentration of reacting species in the fluid phase. A more logical driving force is the concentration of adsorbed species on the catalyst. However in most cases neither the exact nature of these species nor their concentrations are known. In spite of this ignorance it is clear that a model will be somewhat closer to reality if it is possible on some rational basis to formulate rates in terms of concentration of species believed to exist on the surface. Relating these surface concentrations to those existing in the bulk phase allows the rate to be formulated in terms of readily measurable concentrations. These relationships are developed from knowledge of adsorption phenomena.

Relatively less guidance is available to indicate the actual form of the adsorbed species. In some cases it is clear from calorimetric or other studies that dissociation occurs on adsorption, as usually occurs with hydrogen on metals. Studies by infrared adsorption of surface species also give clues to the form of the adsorbed species. In the absence of any positive evidence to the contrary it is sometimes assumed that the adsorbed species has the same molecular structure as that in the fluid phase. However, some degree of charge transfer may occur; that is, an adsorbed species may be an ion rather than a neutral molecule.

For tractability in mathematical analysis and in theoretical understanding it is customary to assume that one step in the reaction is *rate-limiting* or *rate-controlling*. This may be the rate of adsorption of one reactant, the rate of a surface reaction between adsorbed species, or the rate of desorption of a product. All the other steps are assumed to be in equilibrium with one another. The concept of a rate-controlling step can be confusing. Since all the processes occur in series under steady-state conditions, they must all actually have the same rate, but the rate-limiting step is the one that consumes essentially all of the driving force (chemical potential) available. In an electrical analogy, a current passing through several resistances in a series is the same in each resistor, but if the conductivity of one is much less than that of the others, it is the rate-limiting resistor.

The evidence identifying the rate-controlling step is frequently tentative and stems from various kinds of studies. The rates of adsorption of individual reactants onto the catalyst surface (and/or desorption of products therefrom) may be studied in the absence of reaction. Adsorption and desorption rates as determined from studies with isotropic tracers may be compared with the reaction rate under reaction conditions. The rate-limiting step may also be indicated by formulating the kinetic expression for the rate of reaction for each different, plausible rate-limiting step. Each formulation is then compared with experimental data, and that mathematical form which best fits the experimental facts suggests a possible mechanism for the reaction. Frequently the same mathematical form may be derived from more than one different postulated mechanism, in which case the parameters may have considerably different theoretical interpretations. Hence the fitting of data to a particular mathematical expression seldom, of itself, proves much concerning the true mechanism. The extent to which these approaches lead to reliable and useful conclusions varies greatly from case to case and calls for astute judgment on the part of the investigator.

In many cases it appears that the rate of reaction of one or more chemisorbed species is the rate-limiting step, rather than rate of adsorption or desorption as such. The kinetic formulations based on this assumption usually bear the term *Langmuir-Hinshelwood*. The term *Langmuir-Rideal*, *Rideal*, or *Rideal-Eley* is applied if reaction is assumed to be between a chemisorbed species and a molecule reacting with it directly from the fluid phase or from a physically adsorbed layer.

3.3.1 Langmuir-Hinshelwood Model

The assumptions underlying the Langmuir adsorption isotherm are retained. Further, adsorption equilibrium is assumed to be established at all times; for example, the rate of reaction is taken to be much less than the potential rate of adsorption or desorption. The concentrations of adsorbed species are therefore determined by adsorption equilibria as given by the Langmuir isotherm. If two or more species are present, they compete with each other for adsorption on a fixed number of active sites.

Reaction is assumed to occur between adsorbed species on the catalyst. If a single reactant is decomposed, the process may be assumed to be either unimolecular or bimolecular, depending upon the number of product molecules formed per reactant molecule and whether or not the products are adsorbed. A simple decomposition in which products are not adsorbed is usually taken to be unimolecular (Case 3.1 below). If two adsorbed product molecules are formed for each reactant molecule decomposed, it is postulated that an empty site must be adjacent to the adsorbed reactant molecule to accommodate the additional molecule formed. The reaction is then "bimolecular" in the sense that it is proportional to the product of the concentration of adsorbed reactants and of empty sites.

If reaction takes place between adsorbed A and adsorbed B and these species are immobile, they must be adsorbed on neighboring sites in order for reaction to occur. The mechanism may be visualized as follows:

$$\mathrm{A + B + 2*} \underset{\text{adsorption}}{\rightleftharpoons} \begin{matrix} \mathrm{A} & \mathrm{B} \\ | & | \\ -*- & *- \end{matrix}$$

$$\Updownarrow$$

$$\text{Products} + 2* \underset{\text{desorption}}{\rightleftharpoons} \begin{matrix} \mathrm{A} - & \mathrm{B} \\ | & | \\ -*- & *- \end{matrix} \qquad \text{(activated complex)}$$

The probability of reaction here is taken to be proportional to the product $\theta_A \theta_B$.

By the above and analogous procedures, rate expressions can be derived for any type of postulated mechanism. The form and complexity of the expression depend on the assumptions made concerning this mechanism. A few cases are presented below:

Case 3.1 Decomposition, products not adsorbed.

$$\mathrm{A} \longrightarrow \text{products}$$

The reaction rate is taken to be proportional to the quantity of adsorbed A molecules. Then,

$$-r, \frac{\text{(moles)}}{\text{(time)(area)}} = k\theta_A \tag{3.5}$$

The value of θ_A is given by the Langmuir adsorption isotherm:

$$\theta_A = \frac{KP_A}{1 + KP_A} \tag{3.6}$$

Combining these two equations,

$$-r = \frac{kKP_A}{1 + KP_A} \tag{3.7}$$

If the system follows this model, the reaction rate should be first order at sufficiently low values of P_A. As P_A increases, the order of reaction should gradually drop and become zero order. Similarly the reaction rate should be first order if A is weakly adsorbed–for example, K is small–and zero order if A is strongly adsorbed. This type of behavior is indeed found for a number of decompositions.

Case 3.2 Decomposition, products adsorbed.

$$\mathrm{A} \longrightarrow \mathrm{B + C}$$

Assume:

1. A, B, C all may be appreciably adsorbed.
2. The reaction rate is proportional to the quantity of adsorbed A molecules.
3. No dissociation of A molecules occurs on adsorption.
4. Reverse reaction is negligible.

Again using the Langmuir adsorption isotherm, the fraction of surface covered by A, B, and C can be derived as follows:

$$k_A[1 - \Sigma\theta]P_A = k'_A\theta_A$$

where $\Sigma\theta$ is the fraction of available sites covered with A, B, and C.

$$\theta_A = K_AP_A[1 - (\theta_A + \theta_B + \theta_C)] = K_AP_A(1 - \Sigma\theta) \tag{3.8}$$

$$\theta_B = K_BP_B(1 - \Sigma\theta) \tag{3.9}$$

$$\theta_C = K_CP_C(1 - \Sigma\theta) \tag{3.10}$$

Adding Eqs. (3.8), (3.9), and (3.10),

$$\Sigma\theta = (1 - \Sigma\theta)[K_AP_A + K_BP_B + K_CP_C] \tag{3.11}$$

Subtracting both sides of Eq. (3.11) from unity and rearranging,

$$(1 - \Sigma\theta) = \frac{1}{1 + K_AP_A + K_BP_B + K_CP_C} \tag{3.12}$$

Since two molecules are formed for each one that reacts, and it is postulated that both product molecules are adsorbed, it would seem plausible that it is necessary for an empty site to be present adjacent to the reacting molecule to accommodate one of the product molecules.

In that event

$$-r = k\theta_A(1 - \Sigma\theta) \tag{3.13}$$

Combining Eqs. (3.8), (3.12), and (3.13) gives

$$-r = \frac{kK_AP_A}{(1 + K_AP_A + K_BP_B + K_CP_C)^2} \tag{3.14}$$

If an inert material X is present that is significantly adsorbed, then a term K_XP_X must be added in the denominator, and Eq. (3.14) would become

$$-r = \frac{kK_AP_A}{(1 + K_AP_A + K_BP_B + K_CP_C + K_XP_X)^2} \tag{3.15}$$

Case 3.3 Bimolecular reaction.

$$A + B \longrightarrow C$$

The same assumptions as in Case 3.2 are made except that the reaction rate now is assumed to be proportional to the product of the concentration of adsorbed A and adsorbed B. The rate expression then becomes

$$-r = k\theta_A\theta_B \tag{3.16}$$

Combining Eqs. (3.8), (3.9), (3.12), and (3.16),

$$-r = \frac{kK_AK_BP_AP_B}{(1 + K_AP_A + K_BP_B + K_CP_C)^2} \tag{3.17}$$

Case 3.4 Adsorption-desorption with dissociation.

$$A_2 \rightleftharpoons 2A \longrightarrow \text{products}$$

Assume that A dissociates upon adsorption and associates on desorption. In order for dissociation to occur, a gas molecule must plausibly impinge on the surface at a location where two sites are adjacent to one another. Up to fairly high fractional coverages, the number of pairs of adjacent sites is proportional to the square of the number of single sites. Then the rate of adsorption is given by

$$\left(\frac{dn}{dt}\right)_{\text{ads}} = kP_A(1 - \theta_A)^2 \tag{3.18}$$

Assuming desorption involves interaction of two neighboring adsorbed atoms,

$$\left(\frac{dn}{dt}\right)_{\text{des}} = k'\theta_A{}^2 \tag{3.19}$$

At equilibrium, $kP_A(1 - \theta_A)^2 = k'\theta_A{}^2$ and

$$\theta_A = \frac{(K_AP_A)^{1/2}}{1 + (K_AP_A)^{1/2}} \tag{3.20}$$

This simple equation applies to mobile adsorbed atoms at all degrees of surface coverage or to immobile adsorbed atoms at small values of θ_A. At high coverage, some unused single sites exist in the "fully covered" region, and they are not available for chemisorbing a molecule as atoms if adsorbed atoms are immobile. The exact equation for any case depends upon whether or not individual atoms are mobile and the extent of coverage.

The rate of reaction might plausibly be either first order or second order with respect to dissociated A, depending on circumstances.

If $-r = k\theta_A$, upon substituting in Eq. (3.20),

$$-r = \frac{k(K_AP_A)^{1/2}}{1 + (K_AP_A)^{1/2}} \tag{3.21}$$

Alternately,

$$-r = k\theta_A{}^2 = \frac{kK_AP_A}{[1 + (K_AP_A)^{1/2}]^2} \tag{3.22}$$

If two atoms of dissociated A react simultaneously with B, and product adsorption is negligible, then

$$-r = k\theta_A{}^2 \cdot \theta_B = \frac{kK_AP_AK_BP_B}{(1 + \sqrt{K_AP_A} + K_BP_B)^3} \tag{3.23}$$

The most common example of dissociative adsorption is encountered with hydrogen on most metals. However it is frequently observed that hydrogenation reactions are approximately first order in hydrogen rather than half order. Equation (3.23) shows that even if the hydrogen dissociates, a first-order process with respect to hydrogen (for example, A) will be observed if it is not strongly adsorbed relative to B. A first-order expression in the numerator with respect to hydrogen is also obtained if the addition of one of the hydrogen atoms to B is an equilibrium reaction and the addition of the other is the rate-limiting process.

Case 3.5 Adsorption of two gases on separate sites.

$$\mathrm{A} + \mathrm{B} \longrightarrow \text{products}$$

In this case, A and B molecules are assumed to adsorb independently on different sites. Applying the usual assumptions,

$$\theta_\mathrm{A} = \frac{K_\mathrm{A} P_\mathrm{A}}{1 + K_\mathrm{A} P_\mathrm{A}} \tag{3.24}$$

$$\theta_\mathrm{B} = \frac{K_\mathrm{B} P_\mathrm{B}}{1 + K_\mathrm{B} P_\mathrm{B}} \tag{3.25}$$

If the reaction rate is proportional to the product of adsorbed A and B molecules and sites are randomly distributed,

$$-r = \frac{k K_\mathrm{A} P_\mathrm{A} K_\mathrm{B} P_\mathrm{B}}{(1 + K_\mathrm{A} P_\mathrm{A})(1 + K_\mathrm{B} P_\mathrm{B})} \tag{3.26}$$

This type of behavior appears to be less common than competition for the same type of site. The behavior of some systems, however, suggests a mixture of independent adsorption and competitive adsorption. For example a detailed kinetic study of the hydrodesulfurization of dibenzothiophene (DBT) led to the conclusion that DBT and reaction products competed for one type of site while hydrogen adsorbed independently on a second type of site (Espino et al., 1978). Poisoning experiments also suggest that in some cases two kinds of sites exist on which competitive adsorption occurs, one kind being active but easily poisoned, the second less active but more resistant to poisoning. Such a hypothesis was advanced to explain some of the effects of pyridine poisoning in the hydrodesulfurization of thiophene (Satterfield et al., 1975).

3.3.2 Apparent Activation Energies

If the kinetic expression is of a complex form such as Eqs. (3.14) or (3.17), the overall apparent activation energy as determined from the effect of temperature on reaction rate at constant reactant composition will change with temperature. An Arrhenius-type plot of ln rate versus $1/T$ will not be a straight line. For some simpler kinetic expressions, the apparent activation energy will be independent of temperature, as in Case 3.1 for low surface coverage. Then the kinetic expression reduces to

$$-r = k_s K_\mathrm{A} P_\mathrm{A} \tag{3.27}$$

Here the experimentally observed reaction rate constant k_{exp} equals $k_s K_\mathrm{A}$, where k_s is the reaction rate constant for the surface reaction, assumed to follow the Arrhenius expression. K_A is the adsorption equilibrium constant, which decreases exponentially with increased temperature by the factor $e^{\lambda/RT}$. Here λ is the heat of chemisorption, taken to be independent of temperature. (An adsorption process is almost always exothermic, in which case λ is a positive number.) The effect of temperature upon the experimentally observed reaction rate constant for a rate expression of the type in (3.27) is given by

$$k_{\text{exp}} = Ae^{-(E_s - \lambda)/RT} \tag{3.28}$$

The apparent activation energy E_a as calculated from a plot of $\ln k_{\text{exp}}$ versus $1/T$ equals $E_s - \lambda$. Since E_s and λ will normally both have positive values, E_a will be less than the so-called true activation energy for the surface process, E_s.

In corresponding fashion consider the reaction $A \rightarrow B + C$ for the case in which C is strongly adsorbed and A and B are not. The rate expression given by Eq. (3.14) then reduces to

$$-r = \frac{kK_A P_A}{(K_C P_C)^2} \tag{3.29}$$

and the apparent activation energy will be given by

$$-E_a = -E_s + \lambda_A - 2\lambda_C \tag{3.30}$$

where λ_A is the heat of chemisorption of A and λ_C is the heat of adsorption of C, both normally positive.

The apparent activation energies for some bimolecular surface reactions may similarly be formulated for specific forms of Case 3.3. Consider some examples in which product adsorption is unimportant.

$$-r = \frac{kK_A K_B P_A P_B}{(1 + K_A P_A + K_B P_B)^2} \tag{3.31}$$

Case 3.3a A and B are weakly adsorbed.

$$-r = kK_A K_B P_A P_B \tag{3.32}$$

$$-E_a = -E_s + \lambda_A + \lambda_B \tag{3.33}$$

Case 3.3b A weakly adsorbed, B strongly adsorbed.

$$-r = \frac{kK_A K_B P_A P_B}{(K_B P_B)^2} = \frac{kK_A P_A}{K_B P_B} \tag{3.34}$$

$$-E_a = -E_s + \lambda_A - \lambda_B \tag{3.35}$$

Case 3.3c A and B weakly adsorbed: a poison X strongly adsorbed.

$$-r = \frac{kK_A K_B P_A P_B}{(K_X P_X)^2} \tag{3.36}$$

$$-E_a = -E_s + \lambda_A + \lambda_B - 2\lambda_X \tag{3.37}$$

This expression shows that adsorption of a poison increases the apparent activation energy of the reaction.

Return now to the more complex kinetic expressions when they do not reduce to power-law equations. The fact that the values of E_s and of the various λ's will usually be different from one another means that the relative importance of the various terms will shift with a change in temperature. They will also, of course, shift with gas composition. If data for such a case are forced into a power function, this will shift as pressure, temperature, and composition are changed.

3.3.3 Maximum in Rate with Increased Temperature

It is not unusual for the rate of a catalyzed reaction to reach a maximum with increased temperature. This is observed, for example, with hydrogenation of ethylene on various catalysts and with the decomposition of hydrogen peroxide vapor on platinum. (See Fig. 11.8.) Such a temperature maximum is not consistent with a power-law rate expression. However, most of the Langmuir-Hinshelwood expressions can be made to accommodate this type of behavior by invoking plausible effects of temperature on the individual adsorption and kinetic constants. For example, consider the irreversible reaction:

$$A \longrightarrow B$$

Assume B is not adsorbed (Case 3.1) and low surface coverage of A. If $\lambda > E_s$, then k_{exp} will *decrease* with increasing temperature [Eq. (3.28)].

If both A and B are adsorbed, the rate expression becomes

$$-r = \frac{kK_A P_A}{1 + K_A P_A + K_B P_B} \tag{3.38}$$

and

$$\frac{d \ln r}{d(-1/RT)} = E_s - \lambda_A + \frac{K_A P_A \lambda_A + K_B P_B \lambda_B}{(1 + K_A P_A + K_B P_B)} \tag{3.39}$$

A maximum in the rate with increased temperature can then occur if $\lambda_A > E_s$ and if the third term on the right-hand side of Eq. (3.39) is predominant at low temperatures and insignificant at high temperatures. Since K_A and K_B will both decrease with increased temperature, the latter can readily happen. A physical interpretation of the overall effect is that with increased temperature the increasing reactivity of the adsorbed intermediate complex is more than offset by a decrease in its concentration.

The explanation for such temperature maxima may also lie in the inapplicability of the assumption in the Langmuir-Hinshelwood model that the number and activity of catalytic sites is constant with temperature. Indeed a range of energies of sites is probably almost always present, so the number of sites of requisite energy for high reactivity may well change with temperature.

3.3.4 Rideal Model

From studies of the catalytic activity of platinum at 500 to 1050 K, Langmuir in 1921 concluded that in the oxidation of carbon monoxide with oxygen the reaction occurred by a carbon monoxide molecule striking an adsorbed oxygen atom, even though much of the surface was covered by adsorbed carbon monoxide molecules at the lower range of these temperatures. Likewise he concluded that the reaction of hydrogen and oxygen occurred by a similar mechanism, between a striking hydrogen molecule and an adsorbed oxygen atom.

The idea was revived by Rideal in 1939 who proposed that a simple molecular mechanism for heterogeneous catalytic reactions was reaction between a chemisorbed radical or atom and a molecule, the latter impacting directly from the gas phase or held in a deep van der Waals layer. In either case a new chemisorbed species was formed on the surface. Specifically, Eley and Rideal (1941) concluded that the conversion of *p*-hydrogen to *o*-hydrogen on tungsten and the exchange of hydrogen and deuterium occurred between a hydrogen molecule in a loosely bound layer and a strongly bound chemisorbed hydrogen atom. However it now appears that the para-to-ortho conversion occurs instead by adsorption of a hydrogen molecule with dissociation, a mechanism proposed by Bonhoeffer and Farkas (1931).

Reactions of this general type are referred to variously as Langmuir-Rideal, Rideal, or Rideal-Eley mechanisms. Only a few reactions have been clearly shown to proceed in this manner, although the evidence is strong in the case of the oxidation of ethylene to ethylene oxide on silver. The hydrogenation of ethylene has been intensively studied (for a detailed critical review, see Horiuti and Miyahara, 1967), and the results on some catalysts such as nickel under some conditions are consistent with a Rideal mechanism. On other catalysts such as copper a Langmuir-Hinshelwood mechanism seems more probable.

To formulate the Rideal mechanism for a simple reaction between A and B, retain all the other assumptions in the Langmuir-Hinshelwood model and consider reaction occurring between adsorbed molecules of B and gas-phase molecules of A. The rate is

$$-r = k\theta_B \cdot P_A \tag{3.40}$$

Performing the usual substitutions, one obtains

$$-r = \frac{kP_A K_B P_B}{1 + K_A P_A + K_B P_B + K_C P_C} \tag{3.41}$$

If the mechanism were assumed to be reaction between adsorbed A and gas-phase B, an equation of the same form is obtained, but with K_A replacing K_B.

$$-r = \frac{kP_B K_A P_A}{1 + K_A P_A + K_B P_B + K_C P_C} \tag{3.42}$$

The mechanism expressed by Eq. (3.40) could occur either with or without significant adsorption of A onto the catalyst surface (Langmuir, 1921), so further information is needed to determine whether or not the term $K_A P_A$ should be included in the denominator of Eq. (3.41).

3.3.5 Adsorption Control

In principle it may be possible to distinguish between surface-rate control and adsorption control by the form of the kinetic expression. The Elovich (Roginskii-Zeldovich) equation is frequently used to express the rate of chemisorption. This is a two-parameter expression which provides for a fall in adsorption energy with increased surface coverage.

$$\frac{dn}{dt} = ae^{-\alpha\theta/RT} \tag{3.43}$$

Here a is a constant and α relates activation energy to surface coverage by the expression

$$E_a = (E_a)_0 + \alpha\theta \tag{3.44}$$

The classic example of a kinetic expression in which the rate of adsorption is the rate-determining step is the Temkin-Pyzhev equation for synthesis of ammonia from the elements (Sec. 10.5). The use of the Elovich equation for deriving kinetic models is reviewed by Aharoni and Tompkins (1970).

3.3.6 Two-Step Kinetic Models

Many real reactions probably involve the formation and disappearance of several intermediates on the surface of the catalyst, and it is evident that a mechanistically rigorous formulation can become so complex as to lose most of its utility. Boudart has addressed this problem of formulating the simplest rate equation that can significantly represent a multistep reaction. He suggests that many cases can be usefully treated by making two simplifying assumptions: (1) that one step is the rate-determining step and (2) that one surface intermediate is the dominant one, i.e., that all other species are present in relatively insignificant amounts (Boudart, 1972). Under steady state all steps must occur at the same rate, but it is not necessary to assume details of the overall mechanism. For a given reaction the formulation depends on the assumptions made as to the rate-determining step and the most abundant surface intermediate, and, as with other approaches, the same formal expression may be obtained from more than one set of assumptions.

Example 3.1 The approach may be illustrated by the kinetic expression found to represent initial rate data on the dehydrogenation of methylcyclohexane (M) to toluene (T) on a platinum/alumina catalyst, reported by Sinfelt, et al. (1960). The rate law followed the form of Eq. (3.7), which was derived here previously by a Langmuir-Hinshelwood model assuming adsorption equilibrium and the rate-determining step to be the rate of reaction of adsorbed reactant (Case 3.1). However the rate was only slightly decreased in the presence of aromatics, which would be adsorbed preferentially to M. This and other observations suggest that instead M was adsorbed irreversibly and that the concentration of adsorbed M, θ_M was low. This also implies that hydrogen dissociates from M as it adsorbs. If it is assumed that any other surface intermediates formed by dehydrogenation before the appearance of toluene are not present in significant amounts, and that toluene desorbs irreversibly, then

$$-r = k_1 P_M (1 - \theta_T) = k_2 \theta_T \tag{3.45}$$

$$\theta_T = \frac{k_1 P_M}{k_1 P_M + k_2} \tag{3.46}$$

$$-r = \frac{k_1 P_M}{1 + (k_1/k_2) P_M} \tag{3.47}$$

This is formally the same as Eq. (3.7), but now k_1 represents the rate of adsorption of M and k_2 the rate of desorption of toluene, rather than the ratio k_1/k_2 representing an adsorption equilibrium constant.

Which of two mechanisms is the more likely in a general case may be indicated by the magnitude of the parameters and the effect of temperature upon them, that is, whether the values are more reasonable viewed as adsorption-desorption processes in contrast to an equilibrium constant. The use of isotopic tracers may also reveal the adsorption-desorption characteristics of the system and the relative rates of different steps. In this particular case, if the Langmuir-Hinshelwood mechanism indeed occurred, toluene would be expected to be more strongly adsorbed than methylcyclohexane, in which event Eq. (3.7) would not in fact express the results. Carberry (1976, p. 414) treats this example in considerable detail, illustrating how the data can be examined in terms of various models and methods of testing the reasonableness of the values of the parameters in terms of what they represent physically.

3.4 SOME USES AND LIMITATIONS OF KINETIC MODELS

Formulations of kinetic models for a variety of cases are given in books by Laidler (1965), Smith (1970), Carberry (1976), and Rase (1977) and in various other texts on kinetics. Hougen and Watson in a pioneering book (1947) developed expressions for a wide variety of postulated mechanisms, and rate expressions thus derived are frequently referred to by their names, especially in the chemical engineering literature. Thomas and Thomas (1967, p. 458) give a detailed listing of formulations for reactions of the types $A \rightleftharpoons P$ or $A + B \rightleftharpoons P$. The rate-controlling steps considered are adsorption, desorption, or surface reaction, and models are included in which A does or does not dissociate upon adsorption. They thus develop six possible formulations for the first type of reaction and seven for the second. Nagy in the book by Szabo' and Kallo' (1976, vol. 2, pp. 480–505) lists expressions for the types $A \rightleftharpoons B$ and $A \rightarrow B$, $A \rightleftharpoons B_1 + B_2$, $A \rightarrow B_1 + B_2$, and $A_1 + A_2 \rightarrow B$ plus some suggested procedures for selecting the most appropriate formulation in light of experimental data. The various mechanisms and rate-controlling steps considered lead to six different formulations for the first type of reaction and ten for each of the others.

A useful generalized way of formulating various cases is presented by Yang and Hougen (1950). (See also Rase, 1977, p. 183.) The rate equation is expressed in the general form:

$$-r = \frac{(\text{kinetic term})(\text{potential term})}{(\text{adsorption term})^n} \tag{3.48}$$

They present tables giving expressions for each of the three terms and for n for various adsorption or desorption processes as the rate-limiting steps, with or without dissociation, and for surface reaction as the rate-controlling step. Table 3.1 gives the expressions for the Langmuir-Hinshelwood model where all reactants are taken to be adsorbed, adsorption-desorption is taken to be in equilibrium, and all species are assumed to compete for the same sites. K is the equilibrium constant for a reversible reaction, and driving force is expressed in terms of partial pressure. If deviations from ideality are significant, fugacities or activities should be used instead. The adsorption term is, in the most general case,

$$[1 + K_A P_A + K_B P_B + K_R P_R + K_S P_S + K_X P_X]^n$$

If equilibrium adsorption of A occurs with dissociation of A, the term $K_A P_A$ in the denominator is replaced by $\sqrt{K_A P_A}$.

Some of the kinetic equations thus derived from different postulated mechanisms have the same mathematical form, but the constants have a different elementary meaning depending upon the assumptions made. Given some experimental results, various methods may be used to select the equation best fitting the data. These depend in part upon the manner in which reaction rate data were obtained, e.g., whether they are initial, differential, or integral rate data, the last being much more difficult to handle. Selection among possible equations typically starts by determining which species (reactants and/or products) have a significant effect on the rate when their concentration is varied. The mathematical formulations thus not eliminated may then be linearized if possible, and the experimental data for rate as a function of composition at a constant temperature are tested by a linearized plot.

There are further tests of whether a model that correlates experimental data is consistent with the underlying assumptions made in its derivation. In general, adsorption equilibrium constants in the proposed model, as determined from measurements at different temperatures, should decrease with increased temperature, and they are expected to be positive in sign. Dissociative adsorption can be endothermic (Chap. 2),

Table 3.1 Terms in the generalized formulation of Langmuir-Hinshelwood kinetic models*

	Reaction			
	$A \rightleftharpoons R$	$A \rightleftharpoons R + S$	$A + B \rightleftharpoons R$	$A + B \rightleftharpoons R + S$
Kinetic term, with or without dissociation of A	$k_s K_A$	$k_s K_A$	$k_s K_A K_B$	$k_s K_A K_B$
Potential term	$P_A - P_R/K$	$P_A - P_R P_S/K$	$P_A P_B - P_R/K$	$P_A P_B - P_R P_S/K$
Value of n				
A undissociated	1	2	2	2
A dissociated	2	2	3	3

*The general form is $-r$ = (kinetic term) (potential term)/(adsorption term)^n.

but this rarely occurs. Weller (1975) has noted that in mixed systems, adsorption of one species may enhance the adsorption of a second, and thus if the data are fitted to the Langmuir form, at least one of the K's will have a negative value. However, Carberry (1976, p. 391) further shows that positive values of all K's can be restored by invoking a step in which one gas is assumed to be chemisorbed onto a previously adsorbed gas to form a complex between both gases and one site. Rules have also been suggested for elimination of models on the basis of the calculated entropy of adsorption. Adsorption of a molecule without dissociation results in a loss of entropy since the number of degrees of freedom becomes less, but to go much beyond this point seems to push theory beyond its usefulness, except for some simple and very well characterized systems.

Kittrell (1970), Hofmann (1972), and Froment (1975), in detailed reviews describe various methods of discriminating among kinetic models and of estimating the values of the parameters for a given model by linear and nonlinear methods, and they provide suggestions for statistical design of experiments to determine the best model. The book by Box et al. (1978) emphasizes methods for design of experiments as well as analysis of results and construction of models. Weller (1975) discusses some of the pros and cons of using power-law relationships in contrast to Langmuir-Hinshelwood formulations. A number of experimenters have used pulse techniques under reaction conditions to estimate the values of the adsorption parameters in kinetic expressions. However, the assumptions that have frequently been made for mathematical tractability—such as reversibility of adsorption and linear processes—plus the fact that adsorption parameters are determined at low coverage severely limit the usefulness of this approach. Moreover, the behavior of a system may be considerably different under nonsteady-state than under steady-state conditions.

There is much room for judgment concerning the degree of precision with which it is appropriate to analyze or to correlate experimental kinetic data by these various procedures. The fact that various specific models can be proposed, each leading to a specific mathematical (but not necessarily unique) formulation, tempts the experimenter to correlate data by each of a variety of mathematical expressions and then to conclude that the true kinetic mechanism is the one which leads to the mathematics which best fits the data. The utilization of this approach has sometimes led to conclusions that are more enthusiastic than realistic. Some of the more complex mechanisms, including reversible reactions, can easily lead to equations containing so many adjustable constants that the flexibility at one's disposal may allow one to obtain good fit of data to a mathematical expression while proving little concerning the actual mechanism.

Rate expressions for various specific catalytic reactions of industrial interest are discussed in separate sections later, but these must be treated with caution. Seldom is enough information available to develop a mathematical model applicable over the entire range of composition, temperature, and pressure of interest, and extrapolation beyond the range of conditions studied experimentally is hazardous. Even for the same set of experimental conditions, the form of the equation may well vary substantially with the nature of the catalyst, since this affects adsorptivity and reactivity of the various species present. In some cases the rate of reaction under industrial conditions is so fast that temperature and concentration gradients introduce great uncertainties into the validity of the data. For a well-characterized catalyst used for a fairly simple reaction, e.g., a supported metal for a simple hydrogenation, the *form* of the published kinetic model is frequently useful as the starting point for development of a mathematical model. However the numerical values may vary considerably with the detailed formulation of the catalyst and such factors as degree of loading of an active ingredient on the support.

In a few cases the bulk composition of the catalyst changes significantly with reaction conditions, as is the case with the usual industrial catalyst used for oxidation of SO_2 to SO_3 based on vanadia. Under most reaction conditions this is actually a melt on a porous support. The catalytic activity may vary significantly with bulk composition, highly unusual kinetic behavior may be observed (see, for example, Figs. 8.7 and 8.8), and a long period may be required for a steady-state rate to be achieved.

Although many of the constants in the Langmuir-Hinshelwood expression are so-called adsorption equilibrium constants, it is not possible in most cases to obtain predictions of kinetic behavior using adsorption equilibrium constants obtained from a nonreacting system. One reason is that probably only a fraction of the adsorbed molecules actually participate in reaction. On intuitive grounds, molecules adsorbed on highly active sites would be expected to be so strongly held that they could not react further; instead they would act as poisons. Sites that are too unenergetic would be unable to cause any shifting of bonds. Consequently only those sites of intermediate activity would presumably be effective.

The fact that the heat of adsorption generally decreases with surface coverage has suggested to some investigators that a more fundamental correlation of kinetic data might proceed by using the Freundlich or Temkin isotherm to relate vapor concentration to surface concentration. Nevertheless, the probability that only those sites within a fairly narrow energy range are effective in reaction suggests retention of the simpler Langmuir expressions unless compelling arguments suggest otherwise.

In spite of the various difficulties, the relationships suggested by the simpler Langmuir-Hinshelwood models are helpful in ordering data, and they may give a physical interpretation to a correlation developed from experimental measurements that can suggest how reaction conditions might be changed or the composition modified so as to improve the reactivity or selectivity of a catalyst. Inhibition of a reaction by species X is identified with a high adsorptivity of the catalyst for X. A zero-order or negative-order reaction suggests strong adsorption of one or more reactants. If the system follows the Langmuir model, such "self-inhibition" may be overcome by raising the temperature, and the increase in temperature may have a more marked effect in increasing rate than for reactions of higher order.

In principle, for a reaction of the type $A \rightarrow B + C$, the apparent activation energy is $E_a = E_s + \lambda_a$ when A is strongly adsorbed relative to other species. This drops to $E_s - \lambda$ at higher temperatures where A is not strongly adsorbed, assuming the denominator is squared in the relevant expression, as in Eq. (3.15). If the Rideal mechanism were postulated for the same reaction, the kinetic expression is given by Eq. (3.41) and the apparent activation energy E_a equals E_s at higher temperatures. Similarly if A decomposes unimolecularly to nonadsorbed products (Case 3.1), the apparent activation energy E_a equals E_s at lower temperatures where **A** is strongly adsorbed, dropping to $E_s - \lambda$ as temperature is raised. It is seen that the apparent activation energy can frequently be expected to drop with increased temperature.

In the development of a new process, a target consisting of some set of combinations of high value of space-time yield (STY), selectivity, and percent conversion is usually dictated by economic considerations. The highest space-time yield (quantity of product formed per unit time per unit volume of reactor) is usually associated with the lowest percent conversion (and shortest contact time) since the rate of reaction usually drops as reactant concentration drops. It is sometimes found that increasing contact time does little to increase conversion and instead just decreases the space-time yield even when conditions are far from equilibrium. The above framework of theory suggests that this may be caused by strong adsorption of one or more products, or perhaps of undesired by products. Theory also suggests ways in which this may possibly be overcome:

1. Increased temperature may be effective, although this may cause a decrease in selectivity or in useful life of the catalyst.
2. Reformulation of the catalyst to reduce adsorptivity of the species suspected as the culprit may help.
3. It may be possible to add a species to the feed stream that is strongly adsorbed on the catalyst and displaces the product. This may, however, decrease catalyst activity, and continuously adding an extraneous material may be costly or may contaminate the product.

Langmuir-Hinshelwood formulations also show that a ranking of catalysts in order of increasing reactivity might change with experimental conditions, even at one fixed temperature, e.g., as between a low percentage conversion and a high percentage conversion. For example, over platinum the rate of oxidation of carbon monoxide is first order with respect to oxygen but is an inverse function of carbon monoxide concentration, whereas on base metals under some sets of circumstances the rate is proportional to carbon monoxide concentration but independent of oxygen concentration. Final test conditions must therefore be very similar to those in an application. Although in a simple homogeneous system the maximum rate of reaction occurs with a stoichiometric ratio of reactants, Langmuir-Hinshelwood formulations show that this does not necessarily occur on a heterogeneous catalyst.

Although these possible complexities need to be borne in mind, in some cases, such as rapid screening of a series of catalysts varying substantially in activity, an approximate ranking may be obtained by using as a correlation function the temperature T_R at which the conversion, for fixed feed rate, reaches some convenient value such as 50 percent. The most reactive system then has the lowest value of T_R. This is illustrated in Fig. 3.1, which compares the relative activity of a number of metals for the decomposition of formic acid vapor reported by Fahrenfort et al. (1960). This also shows how the Langmuir-Hinshelwood model can be used to obtain insight into the behavior of a relatively simple reaction. The activity of a metal is expressed here in terms of the temperature T_R at which a specified rate of reaction is reached; hence the

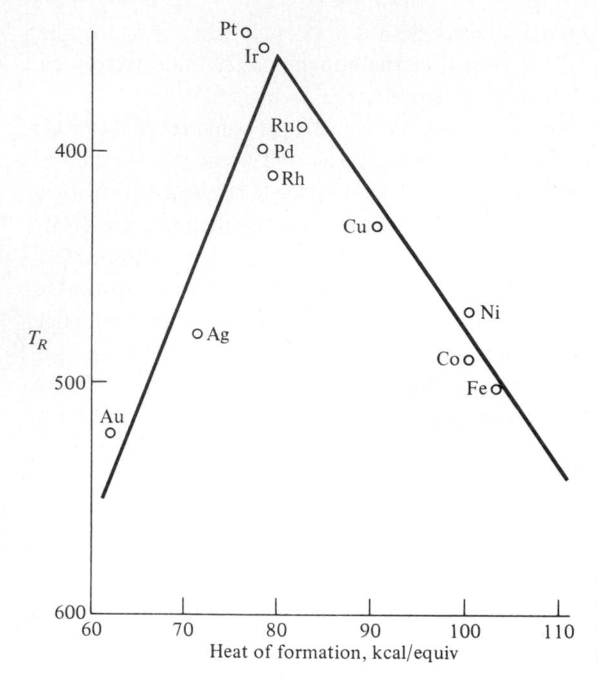

Figure 3.1 Activity of various metals for the decomposition of formic acid as a function of the heat of formation of their formates. (*Fahrenfort et al., 1960.*)

metal with the lowest value of T_R is the most active. This is plotted against the heat of formation of the metal formate, taken as a measure of the heat of adsorption of formic acid on the specific metal. The maximum activity is exhibited for the metal for which formic acid presumably has an intermediate strength of adsorption, which is in consonance with the general idea that the fastest rate is achieved when the bonds between the adsorbed intermediate complex and the catalyst are neither too strong nor too weak.

The reaction on nickel was zero order and that on gold approximately first order, which again is consistent with theory since formic acid is presumably strongly adsorbed on nickel and weakly adsorbed on gold. This also leads to a broad generalization that in comparing a number of catalysts for a specified reaction the most active catalyst is that which is about half covered with the most stable adsorption complex under reaction conditions. Strictly, catalysts should be compared on the basis of equal turnover numbers, but often insufficient information is available.

The manner in which data are presented in Fig. 3.1 is frequently called a *volcano curve*, and it has been applied with varying degrees of success to a number of reactions. The concept was developed in great detail by Balandin in his multiplet theory of catalysis (1958, 1969), but the idea has sometimes been pushed far beyond its basic limitations. A volcano curve is sometimes interpreted in terms of the semiempirical Polanyi relationship, which states that for a series of simple related reactions the change in activation energy from one to a second reaction may be proportional to the change in heat of reaction; that is,

$$E_2 - E_1 = -\alpha(q_2 - q_1) \tag{3.49}$$

where α is a fraction between 0 and 1 (Boudart, 1968, p. 167). If the heat of reaction q (positive for an exothermic step) increases by an amount Δq, the activation energy decreases by the fraction α of Δq.

As applied to catalysis, Polanyi relationship may be used to compare the activation energy and heat of adsorption with those for breakup of the adsorbed complex for, say, a specified reaction on a series of catalysts. Weak adsorption is associated with a high activation energy for adsorption. Then the rate of adsorption is the rate-limiting process, and the surface concentration of adsorbed complex is low. Strong adsorption is associated with a low activation energy for adsorption, the rate of breakup of the adsorbed complex is the rate-limiting process, and surface concentration of the complex is high. The same approach is also applied to the Mars–van Krevelen mechanism of oxidation reactions (Sec. 8.1).

3.5 MIXTURES OF REACTANTS

It is often found that on a specified catalyst the rate of reaction of a mixture of reactants is far different than would be predicted by some type of additive rule from measurements on the same catalyst with the reactants taken one at a time. Moreover, the order of increasing reactivity of individual species may be greatly different in mixtures than when studied separately. This behavior can be readily understood in terms of competitive adsorption of different reactants for the same catalyst sites.

Example 3.2 Beecher et al. (1968) studied the hydrocracking of n-decane, Decalin (decahydronaphthalene), and mixtures of the two at 255°C and 3 MPa using a great excess of hydrogen on a catalyst consisting of palladium supported on a zeolite. For a specified feed rate (moles of total hydrocarbon per unit time) the conversion of Decalin alone was 21 percent and that of n-decane alone was 48 percent. With a 50:50 mixture of the two the reaction rate of Decalin, expressed in moles per unit of time, was essentially unchanged but the amount of n-decane converted was essentially zero. This behavior suggests that the Decalin was substantially the more strongly adsorbed, such that little of the n-decane could have access to the catalyst surface. The fact that the rate of reaction of Decalin was unchanged when its partial pressure was cut in half means that the rate was zero order with respect to Decalin, which is consistent with the hypothesis that it was strongly adsorbed.

Example 3.3 The hydrodenitrogenation of heterocyclic nitrogen compounds on $CoMo/Al_2O_3$ or $NiMo/Al_2O_3$ to convert them to NH_3 is of concern in the processing of various liquid fuels such as those derived from low-grade crude petroleum, coal, or shale oil. In studies with individual compounds quinoline is less reactive than indole, but in real shale-oil mixtures indole-type compounds (includes pyrroles) are less reactive than quinoline-type compounds (includes pyridines) [Koros et al. (1967)]. This again can be attributed to competitive adsorption effects in mixtures in which it can be plausibly postulated that the more basic quinoline-type compounds would probably be preferentially adsorbed and converted.

It can be readily understood that mixtures can show highly unusual kinetic behavior, far different from the usual textbook cases, again stemming from the relative degree of adsorptivity of individual reactants and products and the relative rate of reaction of adsorbed species. Thus, the rate of a reaction may increase with conversion, sometimes termed *autocatalysis*, although this term may be misleading. In the case of homogeneous free-radical reactions, autocatalysis can be readily interpreted in terms of a branched chain reaction. Here the mechanisms involved are quite different, but the mathematical formulations for the two quite different types of mechanisms may be similar in form. Boudart (1968) develops the mathematical approach in considerable detail, treating "active centers" as a class, be they free radicals, surface complexes, or other active intermediates.

Example 3.4 An acceleration of rate with extent of reaction is illustrated in the study by Wauquier and Jungers (1957) of the liquid-phase hydrogenation of a mixture of *p*-xylene (species 1) and tetralin (species 2) over a Raney nickel catalyst in a batch process. Assuming that the two species compete for one kind of site and that all active sites are occupied (since the catalyst is in contact with a liquid), the rate of reaction in terms of overall rate of uptake of hydrogen can be expressed as

$$r = \frac{k_1 K_1 (A_1) + k_2 K_2 (A_2)}{K_1 (A_1) + K_2 (A_2)} \tag{3.50}$$

If $k_1 > k_2$, it can be shown that the rate of reaction will increase as total reactant concentration drops wherever $k_2 K_2 > k_1 K_1$. The mechanistic interpretation is that *p*-xylene, once adsorbed, reacts more rapidly than does adsorbed tetralin, but the relative adsorptivity of tetralin is so great that most of the active sites are occupied by it rather than *p*-xylene. The increase in rate with extent of reaction must, of course, eventually reach a maximum, and a more detailed rate expression to cover the entire range of conversion should allow for this by including terms for adsorption of products. However, reactants are in general more strongly adsorbed than products in hydrogenations, so it was not surprising to find this accelerating effect over a wide range of conversion.

Similar effects can occur in gas-phase, continuous packed-bed reactors. As an example, the small amount of acetylene present in ethylene produced by thermal cracking of various hydrocarbon feedstocks may be removed by selective hydrogenation. As the gas passes through the catalyst bed, first acetylene is removed with high selectivity. Then the rate of hydrogenation increases substantially and the reaction becomes hydrogenation of ethylene, a reaction which is undesirable (Sec. 6.7.2).

In hydrogenation reactions in general the product is often less strongly adsorbed than the reactant. Olefins and aromatics are more strongly adsorbed than paraffins, and oxygenated species are usually more strongly adsorbed than hydrocarbons. The chemical nature of the solid is, of course, also important, but useful generalities are lacking.

3.6 POISONING AND INDUCTION PERIODS

Many of the phenomena of poisoning and induction periods in heterogeneous catalysis can be readily understood within the above framework of competitive adsorption for active sites. If a species in the reactant mixture is strongly adsorbed and is nonreactive, it is a poison and removes active sites from reaction. It is sometimes observed in, for example, a well-stirred liquid-phase batch reactor in which, with successive additions of a catalyst, the rate of reaction increases more than proportionately to the amount of catalyst added. In such a case the first additions of catalyst have become largely inactivated by adsorption of impurities. Sometimes a catalyst which is spent and unsuitable for further use as such retains its adsorptive powers for impurities and may be used deliberately for this purpose. The same principle is also applied in continuous fixed-bed reactors in which a *guard catalyst* is placed in the entrance section of the reactor or in a separate vessel upstream of the reactor. The guard catalyst may be a nearly spent catalyst, possibly a lower-cost version of the main catalyst, or perhaps a bed of adsorbent.

In a liquid-phase batch reactor an induction period may be encountered before the reaction assumes a significant rate. Excluding those cases in which free radicals are involved, it is seen that the phenomenon can be caused by a poison that is strongly adsorbed but is somewhat reactive to form less strongly adsorbed products. The kinetic behavior is like that observed by Wauquier and Jungers but in an even more extreme form.

The rate of deactivation of a catalyst may vary greatly with circumstances because of the variety of phenomena which may be responsible (Sec. 1.3.10). Two limiting cases may be recognized. In one the deactivation is caused by adsorption of a poison and the reduction in rate (as measured, for example, in a differential reactor) is a function of the quantity of poison adsorbed onto the catalyst but is independent of the time of exposure. At the other extreme are cases in which the deactivation increases with time on stream, but is relatively independent of feedstock composition, as might be caused, for example, by sintering. This rate of deactivation can sometimes be expressed adequately by a relatively simple two-parameter empirical equation of the form

$$-\frac{dk}{dt} = Ak^b \tag{3.51}$$

in which t is the time on stream and A and b are positive constants, which, however, may vary with temperature and feed composition. Catalyst deactivation in general and the development of suitable rate expressions are reviewed by Butt (1972).

3.7 COMPENSATION

Consider one reaction studied with a series of catalysts or with a catalyst activated by a variety of means; or consider a series of reactions studied on one catalyst. If each of the data in the set is fitted to the Arrhenius expression, it is sometimes found that E and A both increase or both decrease. Consequently k changes less than it would if only E or only A changed. A change in one is "compensated" for in whole or in part by a change in the other. The phenomenon can merely represent a false correlation (Sec. 3.7.1), or it may be a real effect. In some cases the relationship between A and E is of the form $\ln A = \alpha + \beta E$, in which case the rate constant is the same for all reactions at one particular temperature, sometimes termed the *isokinetic temperature* T_θ, or the *theta temperature*. Likewise the compensation effect is sometimes termed the *theta effect.* Above the isokinetic temperature the faster reaction has the higher activation energy; below the isokinetic temperature the faster reaction has the lower activation energy. This is illustrated in Fig. 3.2 (Cremer, 1955), which shows the decomposition of formic acid on magnesite that had been previously heated to the temperature specified in the range of 370 to 800°C.

If the isokinetic temperature occurs in the middle of the range of temperatures covered by the measurements, the compensation effect may be simply false correlation caused by scatter of data. If the theta temperature is well below or above the range of temperatures covered by the measurements and if the variation in activation energy is large, then the correlation is significant, although the interpretation is uncertain.

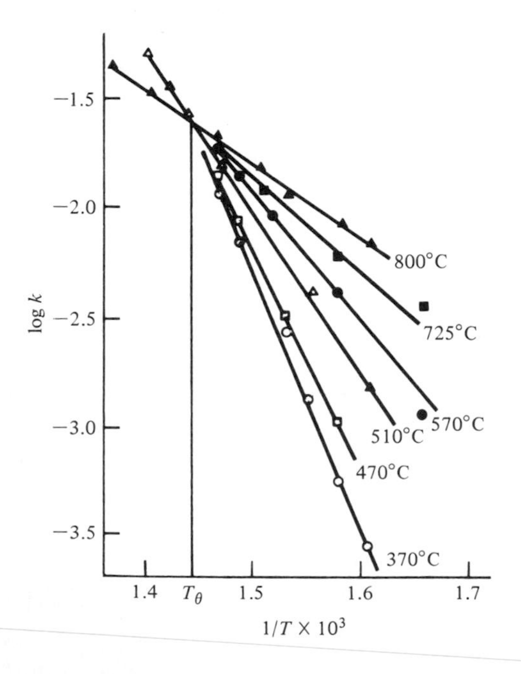

Figure 3.2 Arrhenius plots for decomposition of formic acid vapor over magnesite ($MgCO_3$-MgO). (*Cremer and Kullich in Cremer, 1955.*) (*Reprinted with permission from Advances in Catalysis. Copyright by Academic Press.*)

From the absolute theory of reaction rates,

$$k = \frac{\mathrm{k}T}{h} e^{\Delta S^{\ddagger}/R} \cdot e^{-\Delta H^{\ddagger}/RT} \tag{3.52}$$

where $\Delta S^{\ddagger}$ and $\Delta H^{\ddagger}$ are the differences between the values of the activated complex and those of the reactants. Compensation occurs if an increase in $\Delta H^{\ddagger}$ is associated with an increase in $\Delta S^{\ddagger}$. In homogeneous systems this plausibly may happen since in order for reaction to occur it is necessary both that a sufficiently low energy barrier exists and that a favorable configuration is made possible, accompanied by a change in entropy. $\Delta S^{\ddagger}$ is usually a negative number, and often one finds an approximate linear relationship between $\Delta H^{\ddagger}$ and $\Delta S^{\ddagger}$.

True compensation is frequently observed for homogeneous liquid-phase reactions in a series of solvents or for a homologous series in which substituents are introduced into a reactant. The general explanation is in terms of solvent-solute interactions. Stronger bonding between solute molecules and solvent will lower the enthalpy (and increase the enthalpy change to form the intermediate complex $\Delta H^{\ddagger}$) and, by restricting freedom of vibration and rotation, will lower the entropy (and increase the entropy change $\Delta S^{\ddagger}$).

In many cases of heterogeneous catalysis a linear relationship may likewise exist between enthalpy change and entropy change upon chemisorption over a series of catalysts, but other explanations may also be invoked. If a reaction were studied on a series of catalysts identical except for, say, calcination temperature, it is reasonable to accept that changing the calcination temperature might alter both the number of sites (to which the preexponential factor is proportional) and the activation energy. It is also possible that molecules may surface-diffuse to the reaction site of a surrounding area, and as the temperature is increased, the rate of surface diffusion may increase. But the rate of desorption also increases, resulting in an effective decrease in the preexponential factor with increasing temperature. Compensation in heterogeneous catalysis was earlier reviewed by Cremer (1955) and more recently by Galwey (1977).

3.7.1 False Compensation

If reaction rates are measured over only a fairly small range of temperature and the activation energy is determined from an Arrhenius plot or its equivalent, the value thus determined may be subject to considerable error. Substitution into the Arrhenius expression will then yield a value of A that is likewise in error. A positive error in E results in a positive error in A, and similarly for a negative error. This results in *false compensation.* Consider reaction at two temperatures, T_1 and T_2, with corresponding measured reaction rate constants k_1 and k_2. Then

$$\ln \frac{k_2}{k_1} = \frac{-E}{R} \left(\frac{T_1 - T_2}{T_1 T_2} \right) \tag{3.53}$$

If T_1 and T_2 are close together in value so that $T_1 \approx T_2 \approx T$, then

$$\frac{dk_2}{k_2} - \frac{dk_1}{k_1} = \frac{\Delta T}{RT^2} dE \tag{3.54}$$

where $\Delta T = T_2 - T_1$. The maximum error in E occurs when $dk_2 = -dk_1$. Hence

$$\partial E = \frac{2RT^2}{\Delta T} \frac{\partial k}{k} \tag{3.55}$$

The error in E is seen to be very sensitive to ΔT. For example, if k is subject to an error of ± 20 percent, at $T = 673$ K and for $\Delta T = 30°C$, the error in E may be as large as 12 kcal.

A considerable number of correlations in the literature report the effect upon activation energy of such variables as changing the crystal habit of a catalyst, doping a semiconductor to change its electronic properties, etc. In some cases these effects are real, but in many instances insufficient information is published to determine whether the relationship was truly significant.

REFERENCES

Aharoni, C., and F. C. Tompkins: *Adv. Catal.*, **21**, 1 (1970).

Balandin, A. A., in A. A. Balandin et al. (eds.): *Catalysis and Chemical Kinetics*, English translation, Academic, New York, 1964, p. 33.

——: *Adv. Catal.*, **10**, 96 (1958); **19**, 1 (1969).

Beecher, R., A. Voorhies, Jr., and P. Eberly, Jr.: *Ind. Eng. Chem., Prod. Res. Dev.*, **7**, 203 (1968).

Bohlbro, H.: *An Investigation on the Kinetics of the Conversion of Carbon Monoxide with Water Vapour over Iron Oxide Based Catalysts*, Gjellerup, Copenhagen, 1966.

Bonhoeffer, K. F., and A. Farkas: *Z. Phys. Chem., Abt. B*, **12**, 231 (1931).

Boudart, M.: *Kinetics of Chemical Processes*, Prentice-Hall, Englewood Cliffs, N.J., 1968.

——: *AIChE J.*, **18**, 465 (1972).

Box, G. E. P., W. G. Hunter, and J. S. Hunter: *Statistics for Experimenters: An Introduction to Design, Data Analysis and Model Building*, Wiley, New York, 1978.

Butt, J. B.: *Adv. Chem. Ser.*, **109**, 259 (1972).

Carberry, J. J.: *Chemical and Catalytic Reaction Engineering*, McGraw-Hill, New York, 1976.

Cremer, E.: *Adv. Catal.*, **7**, 75 (1955).

Eley, D. D., and E. K. Rideal: *Proc. R. Soc. London, Ser. A*, **178**, 429, 452 (1941).

Espino, R. L., J. E. Sobel, G. H. Singhal, and G. A. Huff, Jr.: *Prepr., Pet. Div. Am. Chem. Soc.*, **23**, 46 (1978).

Fahrenfort, J., L. L. van Reyen, and W. H. M. Sachtler in J. H. deBoer (ed.): *The Mechanism of Heterogeneous Catalysis*, Elsevier, Amsterdam, 1960, p. 23.

Froment, G. F.: *AIChE J.*, **21**, 104 (1975).

Galwey, A. K.: *Adv. Catal.*, **26**, 247 (1977).

Hofmann, H.: *Adv. Chem. Ser.*, **109**, 519 (1972).

Horiuti, J., and K. Miyahara: *Hydrogenation of Ethylene on Metallic Catalysts*, U.S. Department of Commerce, National Bureau of Standards, Report NSRDS-NBS 13, 1967. (62 pp.)

Hougen, O. A., and K. M. Watson: *Chemical Process Principles*, part III: *Kinetics and Catalysis*, Wiley, New York, 1947.

Kittrell, J. R.: *Adv. Chem. Eng.*, **8**, 97 (1970).

Koros, R. M., S. Bank, J. E. Hofmann, and M. I. Kay: *Prepr., Pet. Div. Am. Chem. Soc.*, **12(4)**, B-165 (1967).

Laidler, K. J.: *Chemical Kinetics*, 2d ed., McGraw-Hill, New York, 1965.

Langmuir, I.: *Trans. Faraday Soc.*, **17**, 621 (1921). [See also *Trans. Faraday Soc.*, **17**, 607 (1921).]

Rase, H. F.: *Chemical Reactor Design for Process Plants*, vol. 1: *Principles and Techniques*, Wiley, New York, 1977.

Rideal, E. K.: *Proc. Cambridge Philos. Soc.*, **35**, 130 (1939).

Satterfield, C. N.: *Mass Transfer in Heterogeneous Catalysis*, M.I.T., Cambridge, Mass., 1970. (Reprint edition available from M.I.T. Department of Chemical Engineering.)

——, M. Modell, and J. F. Mayer: *AIChE J.*, **21**, 1100 (1975).

Sinfelt, J. H., H. Hurwitz, R. A. Shulman: *J. Phys. Chem.*, **64**, 1559 (1960)

Smith, J. M.: *Chemical Engineering Kinetics*, 2d ed., McGraw-Hill, New York, 1970.

Szabo', Z. G., and D. Kallo' (eds.): *Contact Catalysis*, Elsevier, Amsterdam, 1976. (2 vols.)

Thomas, J. M., and W. J. Thomas: *Introduction to the Principles Of Heterogeneous Catalysis*, Academic, New York, 1967.

Yang, K. H., and O. A. Hougen: *Chem. Eng. Prog.*, **46**, 146 (1950).

Wauquier, I. P., and I. C. Jungers: *Bull. Soc. Chim. Fr.*, **10**, 1280 (1957).

Weller, S. W.: *Adv. Chem. Ser.*, **148**, 26 (1975).

CHAPTER

FOUR

CATALYST PREPARATION AND MANUFACTURE

The discussion in this chapter is directed to three objectives:

1. To describe general methods of preparation for readers who desire to make their own catalysts
2. To indicate some of the methods used by and economic constraints on catalyst manufacturers as they scale up a catalyst manufacturing process
3. To provide some understanding of the properties and characteristics of industrial catalysts that may be purchased on the market

An enormous variety of substances exhibit catalytic activity of some type. Many of these, as studied in the laboratory in fundamental investigations, are prepared so as to have simple, uniform, or known structure rather than high area, high activity, or good mechanical strength. Consequently they may often represent elements, materials, or methods of preparation of little immediate industrial interest, and will not be discussed here. Industrial catalysts themselves comprise a wide variety of materials and are manufactured by a variety of methods. As in the manufacture of any other substance, several alternative procedures are usually available, and the process chosen represents a balance between the cost of preparation and the degree to which the ideal chemical and physical properties are achieved.

The preparation of catalysts is frequently described as an art, and a catalyst recipe may specify detailed and arcane procedures that appear to be necessary in order to achieve reproducibility and the desired properties. Even though the relationship between formulation procedures and ultimate catalyst behavior may in many cases be obscure, an understanding of some of the effects produced with typical catalyst ingredients by manipulations such as precipitation, washing, drying, heating, and so on as described below, helps to clarify the reasons for suggested procedures and to indicate possible improved methods for preparing a specific catalyst. In all these manipulations the usual laws of chemistry apply, but are made more complicated by the complex nature of the substances of interest.

For an industrial catalyst the chemical composition is the most overriding consideration, but other factors, primarily of a physical nature, are usually also of major importance.

Surface area. High surface area is usually desirable for high activity per unit volume or unit weight, so most catalysts are made to be porous, with internal surface areas ranging from about 10 m^2/g to as high as 1000 m^2/g. However the porous structure in the catalyst and pore-size distribution may cause diffusional resistances that affect the ease of access of reactants to catalyst sites and removal of products, thereby affecting the rate and selectivity of the reaction (Chap. 11). For very fast reactions a large-pore-size (and therefore low-area) catalyst may be desired.

Stability. This includes stability to heat, to poisons, to fluctuations in process conditions, and to such common components of reacting mixtures as water vapor. If a catalyst can be regenerated or reconstituted instead of being discarded, important savings may be realized, so stability to regeneration conditions may be important.

Mechanical properties. Attrition resistance, hardness, and compressive strength are of particular concern.

The desired final catalyst particle size is determined by the process in which it is to be used. For fluidized-bed reactors or slurry reactors (catalyst suspended in a liquid), particles usually range from about 20 to 300 μm in size. In fluidized beds, the lower limit is set by the difficulty of preventing excessive carryover of finely divided solid through cyclone separators in the reactors; the upper limit is set by the poorer fluidization characteristics of larger particles and possible diffusion limitations. In slurry reactors, powders that are too coarse may be difficult to suspend and may be less effective per unit mass; powders that are too fine are difficult to remove by filtration. Powdered catalysts for use in slurry reactors are typically similar in size to those for fluidized-bed reactors, but materials suitable for use in slurry reactors may be too soft and friable for use in fluidized-bed reactors.

For typical high-area catalysts the outside surface of a particle is a negligible portion of the total area, and in the absence of diffusional resistances, changing particle size as such has no significant effect on reactivity per unit volume. For a spherical particle the outside surface per unit weight is given by $S/W = 6/d \cdot \rho$, where d is the diameter and ρ is the density. For a particle as fine as 20 μm and of unit density the outside area is only 0.3 m^2/g. The enormous area present in a typical high-area catalyst can be visualized by recognizing that the quantity of such a catalyst that can easily be held in one's hand has a total surface area greater than that of a football field. Whether a catalyst has high or low internal area is completely unrelated to its appearance to the eye.

For use in fixed beds, catalysts generally range from about 1.5 to 10 mm in diameter and have about the same length. With the larger sizes in this range, diffusion resistances may reduce the rate of reaction in the center of the particles and hence decrease the activity of the catalyst per unit mass and/or affect selectivity adversely. Hence the pellets frequently may be formed as rings. Sizes smaller than about 1 to 2 mm may cause excessive pressure drop through the bed. Furthermore the cost of forming tablets or extrudates per unit weight or volume of product increases for smaller sizes so an economic limitation also applies. Catalysts in the form of spheres are sometimes claimed to allow a lower pressure drop in a fixed bed than the "equivalent" size of tablets or extrudates. It is not clear whether this may be an inherent geometrical characteristic or whether it is because spheres are generally less susceptible to formation of fines by spalling and erosion of edges and corners. Some unusual shapes are reported to provide better performance in a trickle-bed reactor (Chap. 11), but this may be caused more by improved wetting or better liquid contacting rather than by a reduction in diffusion resistance.

In principal the overall task in catalyst manufacture is to identify the particular chemical and physical properties of the greatest importance in any specific application and then to develop means of achieving or approaching these properties by preparative methods that can be utilized economically on a large scale. In some cases, e.g., catalysts for simple reactions that do not produce byproducts and that have been used industrially for a long time, the desired properties and means of achieving them may be reasonably well known. In contrast, with a reaction such as a partial oxidation of an organic compound to produce one desired intermediate from many possible products on a multicomponent oxide catalyst, the structure of the catalyst found to be most effective may be obscure, and the method of preparation may have been developed largely by trial and error.

4.1 GENERAL METHODS OF MANUFACTURE

Most catalysts are either a finely divided metal supported on a carrier such as alumina or silica, or a metal oxide either on a carrier or unsupported. Metal sulfide catalysts are usually prepared first as the oxide and then treated with hydrogen sulfide or another sulfur compound to convert it to the sulfide. Either of two types of processes, generally termed the *precipitation method* and the *impregnation method*, is commonly used for making catalysts. The first involves in its initial stages the mixing of two or more solutions or suspensions of material, causing precipitation; this is followed by filtration, washing, drying, forming, and heating. Simple wet mixing without precipitation is occasionally used, but it seldom provides the degree of intimate contact between species that is usually desired, unless very high temperatures can subsequently be applied to provide the desired homogeneity and compound formation by thermal diffusion and solid-state reaction.

If a carrier is to be incorporated in the final catalyst, the original precipitation is usually carried out in the presence of a suspension of the finely divided support or a compound or a suspension that will eventually be converted to the support may be initially present in solution. Thus a soluble aluminum salt may be converted to aluminum hydroxide during precipitation, and ultimately to alumina; or a supported nickel catalyst could be prepared from nickel nitrate and a suspension of alumina by precipitation with ammonium hydroxide. In general, starting with aluminum in a soluble form is probably less desirable than starting with a calcined alumina, since using a soluble aluminum salt as a reactant increases the probability of an undesirable reaction of catalyst and carrier. Silica is less reactive, and this problem may not be encountered with, e.g., a silica sol suspension such as Ludox. Binders, cements, die lubricants, thixotropic agents, etc., may also be added at this or a later stage. The final size and shape of the catalyst particles are determined by the forming process which may also affect pore size and pore-size distribution. Large pores can be introduced into a catalyst by incorporating into the mixture wood flour, finely divided carbon such as carbon black, α-cellulose, or other fine organic powders that can be burned out later. With a gelatinous precipitate mechanical manipulation may have a significant effect.

After it is dried and formed, the catalyst is *activated*; that is, it is converted into its active form through physical and chemical changes. This typically involves heating to cause calcination or decomposition, followed by reduction if a metallic catalyst is desired. In many cases a supported-metal catalyst is pyrophoric, and reduction is carried out in the plant reactor rather than by the catalyst manufacturer to avoid hazards upon shipping and reactor loading.

Some advantages of the precipitation method are that it generally provides more uniform mixing on a molecular scale of the various catalyst ingredients, the distribution of active species through the final catalyst pellet is uniform, and the ultimate sizes and shapes are not limited to the forms in which desired carriers are available. Also, more control may be available over pore size and pore-size distribution. One problem is that if two or more metal compounds are present, they may precipitate at different rates or in sequence rather than simultaneously, thus affecting the final structure of the solid.

Impregnation is the easiest method of making a catalyst. A carrier, usually porous, is contacted with a solution, usually aqueous, of one or more suitable metallic compounds. The carrier is then dried, and the catalyst is activated as in the case of precipitated catalysts. The size and shape of the catalyst particles are that of the carrier. The impregnation technique requires less equipment since the filtering and forming steps are eliminated and washing may not be needed. It is the preferred process in preparing supported noble metal catalysts for which it is usually economically desirable to spread out the metal in as finely divided a form as possible. The noble metal is usually present in the order of 1 wt % or less of the total. This makes maximum use of a very expensive ingredient; in contrast, in a precipitated catalyst some of the active ingredient will usually be enclosed by other material present and thus unavailable for reaction.

With catalysts consisting of supported base metals, such as copper and nickel, it is frequently desirable to incorporate a high percentage of the metal, up to 20 to 40 wt %, onto the support. It may be very difficult to obtain such high loadings by impregnation or even by multiple impregnations, so such catalysts are usually prepared by a precipitation process. Although seemingly the same chemical structure may frequently be prepared by either process, the final catalysts produced by the two routes may have substantially different physical and chemical properties.

4.2 PRECIPITATION METHOD

4.2.1 Precipitation

In the usual procedure an aqueous metal salt solution is contacted with an aqueous alkali, ammonium hydroxide, or ammonium carbonate to cause the precipitation of an insoluble metal hydroxide or carbonate. These can be readily converted to oxides by heating. The starting compounds are generally chosen because of their availability and high water solubility, and in some cases to avoid introducing elements that may be deleterious in the final catalyst. Thus halogens are common poisons, and sodium may cause sintering in the ultimate catalyst. If the final catalyst is to be a supported metal, sulfate may be undesirable since it can be reduced to a sulfide, which is a common poison for metal catalysts. The metal nitrate salt is often preferred because it usually is highly water-soluble, generally available, and cheap, but the nitrogen oxides evolved on heating must be controlled. An organic compound such as a formate or oxalate may be used, although these are more expensive and organic fragments from their decomposition on heating may absorb on the catalyst to cause partial inactivation. Also, the average ultimate metal particle size may be considerably different if it is formed by decomposition of a compound rather than by reduction of an oxide. Sulfates and chlorides are generally water-soluble, but the anions must usually be removed by washing and disposal of waste water may be a problem. The preferred alkali is usually ammonium hydroxide since it leaves no cation residue.

If a relatively crystalline precipitate is formed, the size of the crystals may affect the ultimate particle size of a supported metal catalyst. Thus fine crystals may be desired to produce high surface area of a supported metal catalyst, but crystals that are too fine may be difficult to filter. The size of such crystals may be controlled by a variety of techniques. In a multicomponent catalyst, crystals may be smaller if the metals are truly coprecipitated rather than precipitated in sequence. Crystal size may also be affected by temperature and by stirring since this affects nucleation and the degree of supersaturation. *Ripening*, in which a precipitate is allowed to stand for a period, can allow for recrystallization in which small and/or amorphous particles dissolve and crystalline particles grow. This may convert a gelatinous precipitate to a more crystalline and filterable solid.

Silicic acid and a number of metal hydroxides, e.g., those of aluminum, iron, and titanium, form gelatinous colloids. This can make them extremely difficult to filter or to purify by washing. Such gels may be coagulated by electrolytes, but the process of washing to remove electrolyte impurities may cause them to redisperse into colloidal solution, termed *peptization.* Hence a silicic acid gel may be washed with hydrochloric acid or an aluminum hydroxide gel with ammonium nitrate, to maintain an ionic environment and hence the coagulated form. Either hydrochloric acid or ammonium nitrate can be subsequently removed by heating. Gels readily occlude ionic impurities, which may be extremely difficult to remove by washing. The possibility of reaction between carrier and reagents should also be considered in this step. Thus acidic solutions of reagents may react with basic carriers, and vice versa (Sec. 4.5).

4.2.2 Forming Operations

The nature of the forming operations to be used is determined by a balance among several factors, including rheological properties of the mixture, the necessity to achieve satisfactory strength, an open-pore structure and high activity in the ultimate catalyst, and economics. Relatively hard materials, which typically have high melting points, cannot be made into pellets without suitable additives. Operations causing an increase in crushing strength usually also decrease pore volume, and hence may cause diffusion limitations. Typically, commercial catalysts have a void fraction of about 0.5 cm^3 of voids per cubic centimeter of pellet.

Granules Granules are produced simply by grinding and screening, but granulated catalysts are less frequently used since fines represent an economic loss and may be difficult to reprocess. Granules also usually cause a higher pressure drop in a packed bed than pellets of the same size, and they may dust and fragment readily. Grinding may also cause thermal effects that can destroy some materials, such as some zeolites.

Spheres Spherical shapes may be produced by spray-drying a slurry or solution, as in the manufacture of catalytic-cracking catalysts in a microspheroidal form, used in fluidized beds.

Larger spheres may be formed in a continuously rotating granulator (Fig. 4.1). Fine powder and a spray of liquid are brought together in a horizontal rotating cylinder or in a tilted rotating pan. Granules are formed, and they roll over one another and over powder in a snowballing effect. With a tilted pan, spheres may be continuously ejected by centrifugal force once they have reached a critical size. This process is typically used for large-volume operations such as preparing catalyst supports.

If the material has suitable rheological properties, it can be extruded and cut into short cylinders (see below). These can be rolled or tumbled into a semispherical form. Smaller spheres can also be produced if precipitation or coagulation can be caused to occur while a liquid is falling or rising through a second immiscible liquid. The process is used to form spherical bead catalysts for use in a moving-bed reactor. Coagulation of a gel may be produced by dropping it into a bath of hot oil or by altering pH.

Tabletting Powder is compressed in a die that shapes it into pellets or rings. Plasticizing agents and die lubricants such as stearic acid are usually added to the mixture, and the process can only be used with those powder mixtures that are free flowing and that cohere upon pressing. This is often a more expensive process than extrusion. The tabletting area in a representative manufacturing plant is shown in Fig. 4.2. Three rotary tablet presses are visible behind the drums of catalyst in the foreground. Each of these machines has two hoppers into which powder is fed, and pellets are ejected through the chute below. Two tablets are made in each die during one rotation. The powder is compressed by punches from both the top and bottom of the die. To make rings the bottom punch is designed with a core.

Figure 4.1 Sphere-forming operation. (*Courtesy of United Catalysts, Inc., Louisville, Ky.*)

Extrusion A thick paste is extruded through a die pierced with multiple holes, and the spaghetti-shaped extrudate is cut off on the opposite side to form short cylinders. Suitable rheological properties are usually obtained by incorporating methylcellulose, stearates, small amounts of clay, colloidal silica or alumina, etc. The water content may be very important in that the ultimate mechanical strength typically increases as water content is reduced, but below some critical concentration extrusion can become very difficult.

Either the extrusion or tabletting process may produce a skin effect such that the pores at the surface are smaller than those in the interior thus causing diffusion limitations. The gases evolved upon drying and calcining usually prevent a skin effect from being significant, but it may occur with experimental preparations that have not been calcined or if excessive temperatures cause sintering.

Figure 4.2 Tabletting area. (*Courtesy of United Catalysts, Inc. Louisville, Ky.*)

4.2.3 Calcination

This may have several purposes. One is to eliminate extraneous material such as binders and die lubricants, as well as volatile and unstable anions and cations that have been previously introduced, but are not desired in the final catalyst. A substantially elevated temperature is usually needed to increase the strength of the final pellet or extrudate by causing incipient sintering. Excessive sintering will reduce the catalyst activity by reducing surface area, and it may also cause diffusion limitations by reduction of pore size. Thus, a little sintering is usually desired, but not excessive sintering. If a metallic catalyst is the ultimate goal, conversion to the oxide form is frequently sought prior to reduction. If a mixed oxide catalyst is the goal, a substantially elevated firing temperature may be required to cause mixing by diffusion of individual species to form a desired compound or crystal phase. In any event the catalyst should be heated under controlled conditions to a temperature at least as high as will be encountered in the plant reactor to remove bound water, carbon dioxide, etc. If these decompositions occur to a significant extent in the plant, they may cause structural weaknesses in pellets, leading to breakup, dusting, and so on, that may cause excessive pressure drop and premature reactor shutdown. (See Fig. 4.3.)

Possible reactions with the carrier during precipitation, washing, drying, and heating must also be borne in mind. The oxide or other metal compound that is the active catalyst may form a stoichiometric compound with the carrier or may dissolve in the

Figure 4.3 Hot, supported nickel, steam reforming catalysts leaving a calciner. (*Courtesy of United Catalysts, Inc., Louisville, Ky.*)

carrier to form a solid solution. At the usual calcination temperatures, up to 500 to 600°C, silica does not react appreciably with most metals. The commonly used carriers γ- and η-alumina can react with a divalent metal oxide to form a metal aluminate, $MeAl_2O_4$, which of itself may be relatively unreactive. The extent of its formation depends upon the degree of diffusion of the metal ions into the alumina, so the structure in many cases exists only on the surface of the alumina. The process can occur with such elements as Mn, Co, Ni, and Cu, and since the ion must be in the +2 oxidation state, the extent of formation also depends upon the firing atmosphere. The ratio of metal oxide to alumina is also important since with a low concentration of metal much of it may be in the form of an aluminate but at high concentrations most of the oxide form may be retained (Jacono and Schiavello, in Delmon et al., 1976, p. 473). Similarly, molybdenum oxide and tungsten oxide can react with alumina to form

$Al_2(MoO_4)_3$ and $Al_2(WO_4)_3$, respectively. V_2O_5 can react to form $AlVO_4$, and Re_2O_7 to form $Al_2(ReO_4)_3$ (Pott and Stork, in Delmon et al. 1976, p. 537).

The reduction of nickel aluminate incorporated with alumina to form metallic nickel may require temperatures of the order of 500°C, considerably higher than that needed for reduction of nickel oxide, and this may cause excessive sintering. However, the formation of a solid solution or intermediate compound with the carrier or other components present is not necessarily undesirable. Indeed it may be deliberately sought after, at least to some degree, if the substance can be reduced to the metal at a temperature not so high that excessive sintering or other deleterious effects are encountered. Such metal crystallites may be small, well dispersed, and well stabilized by a textural promoter effect (Sec. 4.6.1).

An example is shown in a purification process for H_2 as produced by the water-gas shift reaction on synthesis gas. This may contain up to about 0.5 percent CO and CO_2. The carbon oxides are removed by hydrogenation to CH_4 over a Ni catalyst at about 315-365°C and pressures up to 3 MPa (Sec. 10.6). In one manufacturing process the Ni catalyst is supported on alumina with which a small amount of a magnesium compound is included. During the calcination of $NiCO_3$, the catalyst precursor,

Figure 4.4 High-temperature water-gas shift catalyst, iron oxide promoted with chromia; 5 × 5 mm pellets. (*Courtesy of United Catalysts, Inc., Louisville, Ky.*)

the presence of MgO in the support retards the growth of NiO crystals, and a solid solution of NiO-MgO having crystallites considerably smaller than those of NiO is formed. The NiO-MgO solid solution is more difficult to reduce than NiO, but it results in a more active catalyst with considerable sintering resistance at reaction temperature. An optimum is found here in which the addition of small amounts of MgO (2 to 3 percent) increases the required reduction temperature by only about 15°C, so there is little additional crystal growth. The NiO-MgO may not be completely reducible except at extremely high temperatures, so excessive formation is undesirable. One consequence is that reduction to the active form may continue for some time after the reactor is on stream, so catalyst activity may climb for a substantial period (Bridger and Woodward in Delmon, 1976, p. 311).

In general, heavy metal ions are more readily incorporated into the support when a catalyst is prepared by coprecipitation, e.g., of aluminum and nickel hydroxides, than if nickel is adsorbed by impregnation onto an alumina carrier. Although silica, in general, is more inert than alumina, silicates of nickel and other metals may be formed to some extent. Coprecipitated catalysts, with either alumina or silica, may be more difficult to reduce than the equivalent impregnated catalysts.

Supported nickel catalysts have been studied in considerable detail. (See for example the review by Morikawa et al. 1969, and the discussion by Anderson, 1975.) Cobalt and iron catalysts behave similarly to nickel although the ease of reduction decreases in the order: Ni, Co, Fe. Figures 4.4, 4.5, and 4.6 are photographs of representative commercial catalysts prepared by pelletization. The sides of the pellets may be shiny, as in Fig. 4.4, caused by the deformation characteristics of the powder.

Figure 4.5 Supported nickel oxide catalyst for removal of carbon oxides by methanation. *(Courtesy of United Catalysts, Inc., Louisville, Ky.)*

Figure 4.6 Supported nickel oxide catalyst for steam reforming of hydrocarbons, 16 × 9.5 × 6.4 mm. (*Courtesy of United Catalysts, Louisville, Ky.*)

4.2.4 Reduction to the Metal

Most commonly a metal is formed by reduction of the oxide at an elevated temperature by contact with flowing hydrogen or hydrogen diluted with nitrogen, the latter for safety reasons. A considerable excess of hydrogen may be required to sweep away the product water. If present in too high a concentration, water vapor may accelerate the sintering of an oxide, and it can also retard the rate of the reduction reaction by forming a hydroxylated surface.

Thermodynamic calculations (Anderson, 1975, p. 166) show that at temperatures in the region of 570 to 770 K the reduction of the metal oxide to the metal is highly favored for all group VIII elements, plus copper, silver, gold, and rhenium. For the elements chromium, vanadium, tantalum, titanium, and manganese, the oxide form is highly favored. Tungsten and molybdenum oxides are reducible to the metallic form over a range of conditions, but the equilibrium is much less favorable than for the group VIII elements. They may also react readily with high-area alumina and silica to form compounds more difficult to reduce than the oxides. Thermodynamically, metal chlorides are slightly easier to reduce to the metal than the oxides, but the hydrogen chloride formed is highly corrosive in the presence of even small concentrations of water vapor.

Some metal compounds can be reduced by chemical reagents, such as formaldehyde or hydrazine. Alternately, the metal may be formed by decomposition of an organic compound, as by decomposition of nickel formate to yield deposited nickel. In this case, however, organic fragments may become absorbed onto the metallic catalyst, possibly giving rise to considerably different properties than that of a catalyst reduced in hydrogen or with hydrazine, N_2H_4. Moreover the ultimate metal particle size is determined by the sintering characteristics of the metal produced, whereas in oxide reduction particle size may be determined in part, in a rather complicated way, by the formation of metal nuclei and their growth from the oxide particles. For minimum particle size and hence maximum surface area with hydrogen reduction, it is generally desirable to use initially as high a reduction temperature as is feasibly compatible with avoidance of significant sintering, and as low a partial pressure of water vapor as possible. This maximizes the number of metal nuclei formed (Anderson, 1975, p. 171). In this initial stage hydrogen reacts with O^{2-} or OH^- ions, a process which is relatively slow. After metal nuclei are formed, the dominating mechanism in most cases changes to dissociative chemisorption of hydrogen onto the metal, from which atomic hydrogen migrates to the metal–metal oxide interface. This mechanism is much more rapid. In the reduction of nickel oxide the rate of reduction can be accelerated by adding small amounts of a compound of a metal such as platinum or copper. This is more readily reduced, thereby providing metal nuclei that shorten the induction period.

Sometimes reduction is carried out in situ in the plant reactor, where the large quantities of hydrogen required may be more readily available than on the catalyst manufacturers' premises. This also avoids hazards associated with handling a pyrophoric material. This is the usual procedure with the iron catalyst used for ammonia synthesis and for some nickel catalysts, as those used for methanation. To reduce the time required for reduction in the plant reactor, the catalyst may be reduced to the metal by the catalyst manufacturer and then stabilized for shipment by converting a thin surface layer of the metal to the oxide by controlled oxidation. This thin oxide layer may be rapidly removed in the plant reactor. A nickel catalyst for hydrogenation of edible oils is often coated with a hydrogenated product that is solid at room temperature but readily dissolves in the reaction mixture.

The maximum temperature for reduction obtainable in the plant reactor is usually not much more than the normal reaction temperature, if that, and reduction conditions may be more difficult to control in the plant. These may be important reasons to have the reduction carried out by the catalyst manufacturer. Spent catalyst may be more pyrophoric than fresh catalyst because of absorbed organic material which may spontaneously smoulder or inflame upon contact with air. Such material may be doused with water or an oil with low volatility immediately upon removal.

A supported metal oxide formed by calcining a metal salt may be considerably more difficult to reduce than a nonsupported oxide, and the nature of the support may have a marked effect. This is illustrated in Figs. 4.7, 4.8 (Sieg et al. in Eischens, 1975), and 4.9 for nickel (Webb in Eischens, 1975). Reaction may occur between the oxide and the support, e.g., a nickel hydrosilicate may be formed between a nickel compound and silica. With a high degree of dispersion much of the metal oxide may be in the form of a surface oxide on the support, which is difficult to reduce.

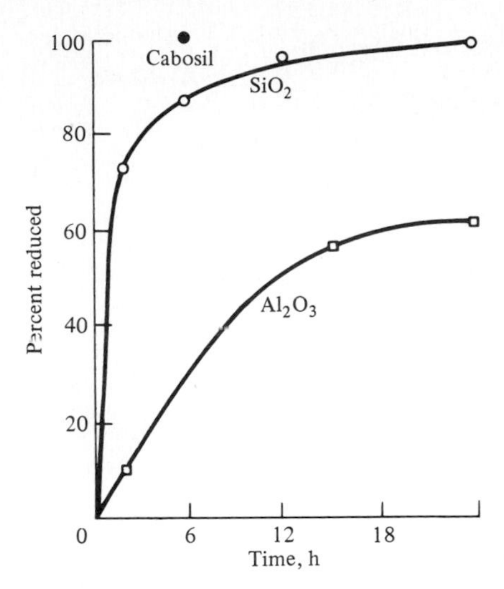

Figure 4.7 Effect of support on reducibility of nickel. Calcination temperature = 538°C. (*Sieg et al. in Eischens, 1975.*)

Fig. 4.7 shows that after calcination at 538°C, nickel oxide formed on Cabosil (a nonporous silica prepared by a flame process) is more readily reduced than nickel oxide on a silica gel. This in turn is more readily reduced than nickel oxide on alumina. The nickel content was 7 wt %, expressed as the metal, and reduction was carried out at 371°C. Figure 4.8 shows that a higher calcination temperature decreases the reducibility of nickel in all cases, and for alumina it appears that the reduction would not exceed 10 percent under practical conditions. For the data in Fig. 4.9, a nickel salt on silica-alumina was first calcined at 330°C for 3 h and then reduced at 400°C for in-

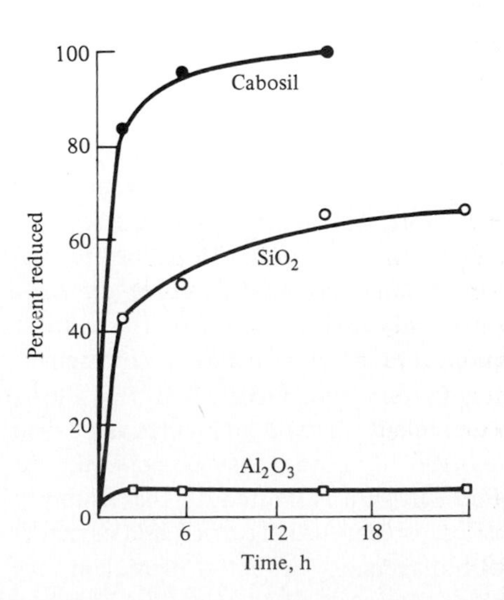

Figure 4.8 Effect of support on reducibility of nickel. Calcination temperature = 732°C. Compare with Fig. 4.7. (*Sieg et al. in Eischens, 1975.*)

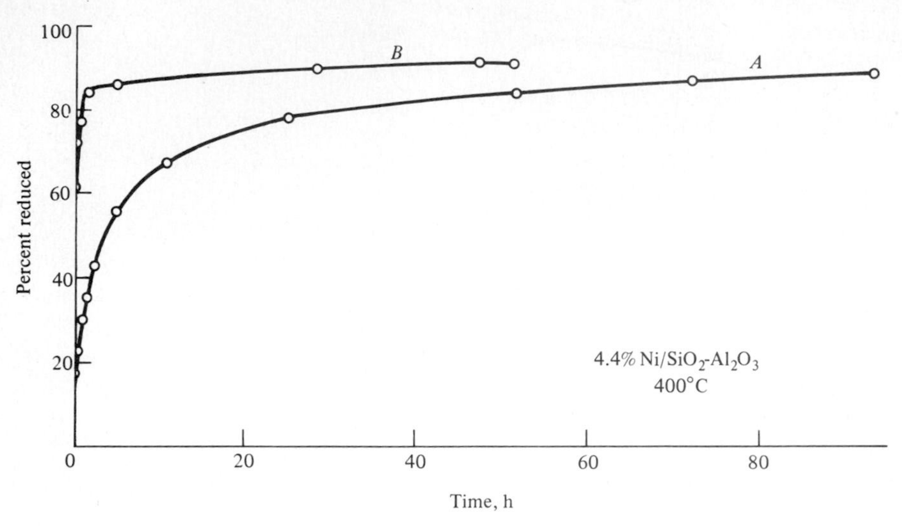

Figure 4.9 Reduction of nickel supported on silica-alumina. See text. (*Webb, in Eischens, 1975.*)

creasing lengths of time (curve *A*). The same sample was then reoxidized at 400°C and then re-reduced at 400°C (curve *B*). Curve *B* is similar to the more rapid reduction that is observed with an unsupported catalyst and shows that there was little interaction between the support and the nickel oxide upon reoxidation. The ease of reduction of selected supported metals of interest is in the order Pt > Ni > Co > Fe.

The above results emphasize the uncertainties that can be introduced into the study of a catalytic reaction if the supported metal element is only partly in the metallic form and partly in other combined forms that may or may not have significant catalytic activity of their own.

4.3 IMPREGNATION

Two methods of contacting may be distinguished. The support is sometimes dipped into an excess quantity of solution, but the composition of a batch of solution will change as additional supports are impregnated. The release of debris into the treating solution can form a mud, which makes it difficult to completely utilize the treating solution. More precise control is achieved by a technique termed *dry impregnation*, or *impregnation to incipient wetness*, which is commonly used industrially. The support is contacted, as by spraying, with a solution of appropriate concentration, corresponding in quantity to the total known pore volume (Sec. 5.2) or slightly less. This allows accurate control of the amount of the active ingredient that will be incorporated into the catalyst. Good reproducibility from one particle to another may be achieved but the maximum loading obtainable in a single impregnation is limited by the solubility of the reagent. In any event, the resulting catalyst is then usually dried and calcined. In a few cases, in order to obtain more uniform dispersion, the active ingredient may be fixed inside the catalyst by immersing the impregnated catalyst in a reagent to cause precipitation to occur.

Oxide supports such as alumina and silica are readily wet by aqueous solutions, as are most activated carbons, which have a layer of chemisorbed oxygen on them. Capillary forces then ensure that liquid will be sucked up into the entire porous structure. Because of capillary pressure, even pores closed at one end will be nearly filled, and the solution of gas in the liquid assists the process. If the support is not readily wetted, e.g., a carbon that is highly graphitized or without chemisorbed oxygen, an organic solvent may be used or the support may be impregnated under vacuum. These procedures are somewhat more costly to use in the plant than incipient impregnation.

The time required for liquid penetration into a pore may be calculated by equating the capillary force to the viscous drag. For a wetting liquid where the contact angle is zero, the time required for liquid to penetrate a distance x into a capillary is given by: $t = 4\eta x^2/\gamma d$, where η is the liquid viscosity and γ is the surface tension. Anderson (1975, p. 172) notes that for pore diameters d of 2 to 50 nm, the time required to penetrate a distance of 2 mm is typically in the range of 115 to 5 s.

4.3.1 Distribution Through Pellet

Most metal reagents are absorbed to varying degrees on most supports, but the characteristics of the process are complicated since various types of adsorption are possible. Metal ions may be cation-exchanged with a surface containing hydroxyl groups or containing alkali or alkaline earth metal ions, or they may be held by coordination. The surface structure of the carrier may also be altered by the impregnation procedure, thus changing its adsorption characteristics. The ultimate degree of dispersion of metal through the catalyst pellet is also determined by the interplay of a large number of factors whose relative importance varies with circumstances. These include the method of impregnation, the strength of adsorption, the extent to which the metal compound is present as occluded solute (that in the bulk liquid in the pores) in contrast to adsorbed species on pore walls, and the chemical reactions that occur upon heating and drying. The situation may be further complicated by attack on the support. Silica gel is attacked at high pH, and high-area aluminas are attacked at a pH that is either too high or too low. Indeed during the impregnation of an alumina support with an acidic liquid, some solution of alumina may first occur, followed by precipitation as the pH increases. It may be desirable to control this effect by using a buffer.

As an example of the interplay of some of these effects, chloroplatinic acid, H_2PtCl_6, a commonly used platinum reagent, is strongly adsorbed on alumina or activated carbon but not on silica gel. Its application to alumina by the incipient-wetness method leads to the deposition of a thin shell of platinum on the outer portions of the particle (this may be desirable in order to avoid diffusion limitations in a fast reaction). To obtain a more uniform dispersion, the adsorptivity of $PtCl_6^{2-}$ ions may be reduced by competitive adsorption by adding nitric or hydrochloric acid to the solution, resulting in a more uniform deposit. Alternatively, platinum could be applied to alumina as $Pt(NH_3)_4Cl_2$, in which case platinum is in the form of a cation. It is then less readily adsorbed on alumina, but more strongly adsorbed on silica gel. If a halogen-free catalyst preparation is desired, a compound such as platinum diaminodinitrite, $Pt(NH_3)_2(NO_2)_2$, may be used. It is also possible to embed the catalytically active material as a layer slightly inside the catalyst particle by adding an organic acid such as citric acid to the impregnating solution (Hoekstra, 1968). Such a structure may be desired for prolongation of catalyst life in an application where poisons are deposited on the outside surface of a porous catalyst support. An example is supported platinum catalysts for oxidation of pollutants in automobile-engine exhaust.

In general, use of the dipping method with a great excess of solution should lead to an essentially uniform deposit of adsorbed material if sufficient time is allowed for diffusion of reagent species to the interior and if side reactions are unimportant. If adsorption is initially nonuniform and not too strong, redistribution will continue even after the pellet is removed from solution, leading to a more uniform distribution. A dipping process is not commonly used commercially since the concentration of the solution will usually change with use.

The drying process can also affect the distribution of an active ingredient. The crystallite size of a resulting supported metal catalyst may also be altered if a considerable portion of the soluble metal is occluded rather than adsorbed. Again the effects are complex, and little can be said of general guidance. Initially, evaporation occurs at the outer surface of the particle, but liquid evaporated from small pores will be replaced by liquid drawn from large pores by capillarity. The places where crystallization begins and the ultimate distribution of metal depends upon such factors as the initial degree of saturation of the liquid, the rate of nucleation, the rate of heating, the degree of connection of liquid paths between pores at the time of crystallization, and the possibility of surface migration. More detailed analyses of impregnation and drying effects are given by Moss (1976) and Anderson (1975). Commercially available impregnated catalysts will usually be found to have a higher concentration of metal at the outside than at the center, even when a more or less uniformly deposited catalyst is desired. This may be in part because of the common use of the incipient-wetness method of impregnation.

4.4 SPECIAL PREPARATIVE METHODS

A number of special techniques have limited application, but they illustrate the wide variety of ways in which catalysts may be formed and utilized.

4.4.1 Massive-Metal Catalysts

Metal catalysts are sometimes used as massive metal, as in the form of wire screens or granules. In ammonia oxidation and in the Andrussow process for synthesis of hydrogen cyanide the catalyst is typically a number of layers of 80-mesh screen, made of a 90% platinum–10% rhodium alloy consisting of wires 0.075 mm in diameter. In one of the processes for partial oxidation of methanol to formaldehyde, metallic silver is used in the form of screens or granules. For economy it is tempting to electrolytically plate an expensive catalyst such as platinum onto a less expensive base, but in high-temperature processes such deposits may readily flake off or structural rearrangements upon reaction may expose and destroy the base material. Finely divided metal powders may be compacted into desired shapes and sizes by powder metallurgy, as in the preparation of porous electrodes for fuel cells. However, sintering may occur in high-temperature operations or at lower temperatures in the presence of hydrogen or other reactants, and much higher metal dispersion can usually be obtained by use of a carrier.

4.4.2 Thermal Fusion

The usual promoted iron catalyst for ammonia synthesis is made by fusion of naturally occurring magnetite, Fe_3O_4, with small amounts of potassium carbonate, alumina, and other ingredients. The melt is cast, and after being allowed to cool, it is crushed and sieved into the desired particle sizes. The optimum size depends on the reactor configuration used (Sec. 10.5). Figure 4.10 shows three sizes of commercial interest. Since finely divided iron is pyrophoric, the catalyst is reduced in situ in the reactor, or, if prereduced, is stabilized for shipment by forming a surface oxide layer that is then reduced in the reactor. (See Sec. 4.2.4.)

4.4.3 Leaching Processes

The best known example of a metal catalyst prepared by leaching is Raney nickel, which is highly active for hydrogenation reactions. It is named after Murray Raney, who patented the method of preparation in 1925. Its earlier uses were reviewed by Lieber and Morritz (1953). The catalyst is prepared from a nickel-aluminum alloy by leaching out much of the aluminum with caustic solution to leave behind a porous nickel catalyst. Typically a 50:50 nickel-aluminum alloy is reacted with a 20% solution of sodium hydroxide. To achieve maximum activity and structural stability some aluminum must be left behind, and some hydrated alumina is also formed and retained. This may act as a textural promoter (Sec. 4.6.1). The surface area of the catalyst may range up to 80 to 100 m^2/g. After deterioration from use, the catalyst may be reactivated a few times by further leaching. Raney nickel is pyrophoric, so it must be handled carefully; it is usually stored under water or under an organic solvent.

A leaching method of making catalysts can also be applied to other metals alloyed with aluminum to make, e.g., Raney cobalt, Raney iron, etc., and the method may

Figure 4.10 Topsøe ammonia synthesis catalyst KMI. The three sizes are about 1.5 to 3 mm, 6 to 10 mm, and 12 to 21 mm. (*Courtesy of Haldor Topsøe A/S.*)

also be used to produce metal catalysts in unusual forms. Thus, in a process in which it is desired to remove heat directly from a catalyst, a metal aluminum alloy may be flame-sprayed directly onto a metal heat-exchange surface and the active catalyst formed by leaching.

Considerable hydrogen is evolved when Raney nickel is heated, greater than that corresponding to adsorption and solution. In the earlier literature the high activity of Raney nickel was sometimes attributed to the existence of an unusual form of hydrogen. However, it now appears that, upon heating, the hydrogen is formed instead by reaction between bound water in the alumina and the aluminum metal.

Clay catalysts, formerly used for catalytic cracking processes, were prepared by leaching certain clays with a mineral acid such as sulfuric. An acidic structure is required for cracking activity, and this process replaces alkali and alkaline earth cations with hydrogen ions. It also increases the surface area of the catalyst and removes undesired impurities such as iron. These catalysts have now been superseded by synthetic silica-alumina, usually incorporating several percent of molecular-sieve zeolites (crystalline silica-aluminas).

The preparation of supported and unsupported metal catalysts in general from a scientific point of view is discussed by Anderson (1975), who also gives a number of specific procedures. A briefer discussion is given by Moss (1976). Ciapetta and Plank (1954) give specific preparation methods for a variety of catalysts and give extensive references to methods of preparation in the earlier literature. Innes (1955) presents a classification of heterogeneous catalytic vapor-phase reactions, with extensive references, and discusses catalysts and methods of preparation for each type of reaction. Gil'debrand (1966) gives a detailed review of methods of preparation, catalytic behavior, and characterization of various catalysts present in low concentration (up to 1 to 2 wt %) on carriers, with particular emphasis on platinum.

4.5 CATALYST SUPPORTS

The early concept of a support or a carrier was that of an inert substance that provided a means of spreading out an expensive catalyst ingredient such as platinum for its most effective use, or a means of improving the mechanical strength of an inherently weak catalyst. However, the carrier may actually contribute catalytic activity, depending upon the reaction and reaction conditions, and it may react to some extent with other catalyst ingredients during the manufacturing process. The carrier may be used as pellets or powders to be impregnated, a powdered carrier may be incorporated into a precipitated mixture, or the carrier may itself be precipitated from solution in the manufacturing process. Some substances such as colloidal alumina or colloidal silica may play a double role, acting as a binding agent in catalyst manufacture and as a carrier in the ultimate product. Alumina in the η and γ forms is intrinsically weakly acidic, but such a substance may be a truly inert carrier for many reactions. In other cases it can be used by itself as a catalyst. An example is a reaction easily catalyzed by an acid catalyst, such as dehydration of an alcohol. High-area carriers are sometimes loosely referred to as "active" carriers in contrast to low-area "inert" carriers.

The selection of a carrier is based on its having certain desirable characteristics. Principally they are:

1. Inertness.
2. Desirable mechanical properties, including attrition resistance, hardness, and compressive strength.
3. Stability under reaction and regeneration conditions.
4. Surface area. High surface area is usually, but not always, desirable.
5. Porosity, including average pore size and pore-size distribution. High area implies fine pores, but relatively small pores (for example, <5 nm) may become plugged in catalyst preparation, especially if high loadings are sought.
6. Low cost.

Of a wide variety of possible materials, only three combine the above characteristics in an optimum way, and therefore they account for most uses. These are alumina, silica, and activated carbon. Of the three, alumina is the most widely used industrially. Magnesia generally has poor strength, and zinc oxide tends to be reduced. Chromia tends to cause dehydration, and its acidity can cause undesirable reactions to occur. Zirconia, although more expensive, is stable at high temperatures, and titania has some limited uses.

A necessary requirement for any carrier is resistance to sintering under reaction conditions. The temperature at which lattices begin to be appreciably mobile is sometimes termed the *Tammann temperature*, and that at which surface atoms become significantly mobile, the *Hüttig temperature*. For simple compounds without phase changes and of low vapor pressure the Tammann temperature is very approximately 0.5 T_m and the Hüttig temperature about 0.3 T_m, where T_m is the melting point in absolute units. Consequently, suitable carriers must usually have fairly high melting points as a minimum. Appreciable mobility appears at about $T_m/3$ for metals, so group 1B metals (Cu, Ag, Au) with melting points in the neighborhood of 1300 K must almost always be supported or have textural promoters incorporated with them in order for high area to be maintained. The transition metals Fe, Co, and Ni, with melting points of about 1800 K, will become mobile at temperatures above roughly 250 to 300°C. The platinum-group metals melt at high temperatures, but are usually supported for economy.

The above are but approximate guides to behavior. Metal crystallites may grow by other mechanisms, and they may be stabilized by keeping them from contact with one another by use of textural promoters.

4.5.1 Alumina

A large variety of aluminas exist, in most cases distinguishable from one another by X-ray diffraction. However the nomenclature used by different researchers for the various aluminum hydroxides and aluminas may vary somewhat, even in recent literature. That used in the following paragraphs follows the Alcoa Research Laboratory designations, which are probably the most widely accepted. A comparison of the various nomenclatures is given by Wefers and Bell (1972).

Aluminas are generally prepared by dehydration of various aluminum hydroxides, but even if the hydroxide is a gel, it is readily converted to a crystalline form on aging and/or heating. The particular crystalline form obtained depends in a rather complicated way upon the time-temperature-environmental history to which the hydroxide is subjected, and this may be difficult to control, especially on a large scale. The aluminas contain water of constitution, which is slowly removed by heating, but it may amount to several tenths of a percent even at 1000°C. Aluminas may also contain various amounts of impurities, such as sodium and iron, as a result of the manufacturing process. For some catalyst uses these impurities are detrimental, and the catalyst manufacturers or their suppliers may go to considerable pains to make their own active alumina by a special process starting with high-purity aluminum metal or aluminum compounds.

The most important aluminas for use as a carrier are γ-Al_2O_3 or η-Al_2O_3, which have high area and are relatively stable over the temperature range of interest for most catalytic reactions. They are very similar in structure and indeed sometimes are not easily distinguishable. Both have a crystallographic form in which the oxygen atoms are arranged similarly to that in spinel ($MgAl_2O_4$), but the η form is more distorted than the γ form. The η form is inherently more acidic than the γ form, which makes it more active for many acid-catalyzed reactions such as olefin isomerization (Chap. 7). All the oxygen ions in a spinel structure are equivalent, forming a close-packed cubic arrangement. The oxygen ions are much larger than the cations, and the latter fit into two kinds of gaps, octahedral (surrounded by six atoms) and tetrahedral (surrounded by four atoms), that exist between the oxygen ions in the structure (see Fig. 4.11, Ryshkewitch, 1960, p. 259). A large number of mixed oxides exist in the spinel structure, which is expressed in the general form $M^{II}M_2{}^{III}O_4$. Some single oxides also form this structure, for example, Mn_3O_4, Fe_3O_4, and Co_3O_4.

Crystallographically, the atomic ratio of total metal atoms to oxygen atoms is 3:4 for a spinel but only 2:3 for alumina. Hence for aluminas such as η and γ a portion of the gaps are vacant and there are varying degrees of disorder. This may be the

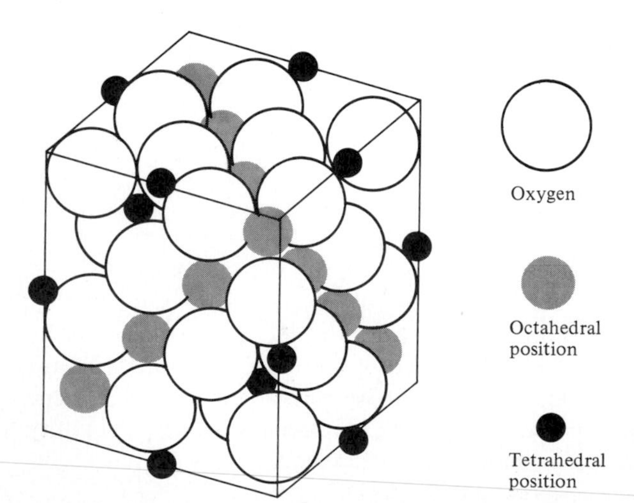

Figure 4.11 Lattice structure of spinel, $MgAl_2O_4$. (*Ryshkewitch, 1960, p. 259.*) (*Reprinted with permission. Copyright by Academic Press.*)

principal reason for the greater solubility and compound formation between heavy metal cations and γ-alumina than between heavy metal cations and silica, which occur upon heating. In particular, aluminates and spinel-type structures may be formed between γ-alumina and a supported catalyst. These may have little catalytic activity. Aluminates in general are formed more readily than silicates. Aluminas react more readily with other species present during a phase change, a phenomenon that seems to be observed for solids in general and is sometimes termed the *Hedvall effect* (1956).

A common manufacturing process for aluminas starts with sodium aluminate produced in conjunction with the Bayer process for purification of bauxites prior to their reduction to aluminum metal. Bauxite is dissolved in sodium hydroxide to form sodium aluminate. After separation of undissolved impurities, the solution is diluted with water to cause hydrolysis and precipitation of α-alumina trihydrate (gibbsite).

$$2NaAlO_2 + 4H_2O \longrightarrow Al_2O_3 \cdot 3H_2O + 2NaOH$$

Gibbsite can also be prepared by other procedures, but it always contains at least 0.2 to 0.3% Na_2O, even after washing with hydrochloric acid (Lippens and Steggerda in Linsen, 1970, p. 171).

Transformation sequences Figure 4.12 (Wefers and Bell, 1972) outlines the principal decomposition sequences that occur upon heating the aluminum hydroxides. Under suitable conditions gibbsite is converted first to the α-monohydrate (boehmite) and,

Conditions favoring transformations

Conditions	Path *a*	Path *b*
Pressure, atm	> 1	1
Atmosphere	Moist air	Dry air
Heating rate, °C/min	> 1	< 1
Particle size, μm	> 100	< 10

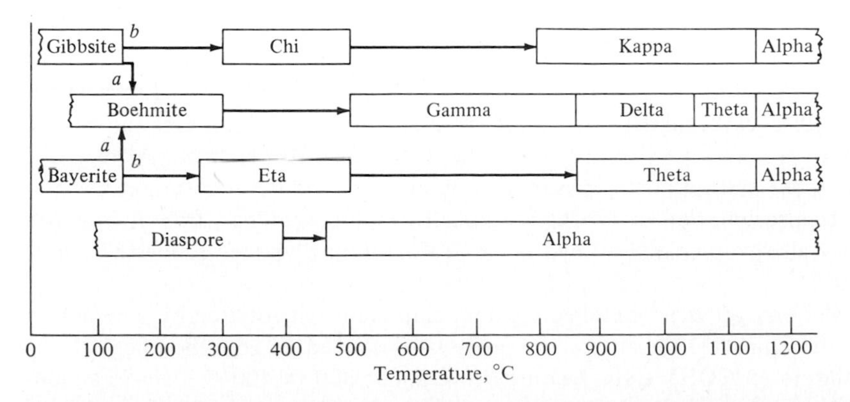

Figure 4.12 Decomposition sequence of aluminum hydroxides. (Enclosed area indicates range of occurrence. Open area indicates transition range.) (*Wefers and Bell, 1972, p. 43.*)

at about 400 to 500°C, to γ-alumina which contains about 0.5% water of constitution. There is considerable controversy over the reaction paths leading to the formation of the various alumina forms. Mixtures of different types are frequently formed since the structures obtained are determined by kinetic considerations rather than thermodynamics. Thus alumina commercially prepared from gibbsite from the Bayer process is probably a mixture of boehmite and the χ form when heated at lower temperatures and χ and γ forms when heated at higher temperatures. It is not surprising that various aluminas supplied by different manufacturers may be found to vary considerably in their behavior as catalyst supports, in ways which may be difficult to relate to identifiable differences.

The path to the γ form from gibbsite via boehmite is favored by using a rapid heating rate, moist air, the presence of small amounts of alkali, and larger particle sizes. With larger sizes the average rate of loss of water by diffusion from the system is reduced, and this promotes hydrothermal reactions. γ-Alumina has a high surface area (typically about 250 to 350 m^2/g). Further heating to about 850°C either in use or in regeneration converts this to δ-alumina. Above about 1100°C this is converted to the θ form, which then rapidly goes to α-alumina with collapse of the structure and the formation of a dense, low-area material. The θ form can be stabilized by addition of rare earth oxides or other additives. The nature of various additives and impurities present as well as the environment can have a marked effect on high temperature stability (Gauguin et al., 1975). Thermodynamically, hexagonal α-alumina is the stable form at all temperatures, and structures such as γ, η, and so on, are frequently referred to as *transition aluminas.* These are not hydrated structures, and they are more or less crystalline.

Another set of routes to various transition aluminas starts with β-alumina trihydrate (bayerite). A rather dense bayerite may be prepared by carbon dioxide precipitation from sodium aluminate. Bayerite can also be formed by reacting water with amalgamated aluminum metal. In this case it is a finely divided light powder with considerable surface area before calcination, and it has greater stability on heating than the dense form. This procedure is used by some catalyst or support manufacturers to prepare a highly pure form of alumina. Heating it under conditions similar to those utilized for converting α-alumina trihydrate to boehmite likewise converts bayerite to boehmite and then to γ-alumina. The β-alumina trihydrate may be converted instead to η-alumina by an opposite set of reaction conditions: slow heating, dry air, the absence of alkali, and the use of small particles to maximize the rate of elimination of water. A powdered form of the β-trihydrate (for example, Alcoa C-37) is a useful binder and support and is frequently incorporated in catalyst mixtures prepared by the precipitation or mixing method. Ultrapure aluminas, free of iron and sodium, may also be prepared by hydrolysis of aluminum alkoxides, followed by suitable heating.

If an alumina gel is precipitated from an aluminum salt solution by ammonium hydroxide, it forms a high-surface-area material described as an α-monohydrate or pseudoboehmite (AlOOH). Upon heating above about 300 to 400°C it forms an alumina usually described as pseudo-γ-alumina. The colloidal material itself consists of solid fibers about 100 nm × 5 to 6 nm, and it has a surface area of about 275 m^2/g, which is retained at high temperatures. Even after heating to 750°C (in which it is still in the γ form), the area is still about 240 m^2/g (Thomas, 1970). This material is both a good binder and a good catalyst support, and therefore it is a useful ingredient in catalyst preparation.

For most reactions the maximum temperature reached in the reactor or upon regeneration does not exceed about 500 to 600°C. The γ- and η-type aluminas have good stability under these conditions. For a few uses, such as in pollution control, where oxidation reactions at very high temperatures are required, catalyst supports that are reasonably stable up to temperatures as high as 1000°C may be needed. Gauguin et al. (1975) describe some of the transformations of aluminas under these high-temperature conditions.

The aluminas formed at temperatures of about 300°C or more are not hydrated but contain small amounts of water of constitution, which is gradually evolved on heating to higher temperatures. The conversion of OH groups to water on heating the aluminum hydroxides or other aluminas leaves behind a structure with exposed aluminum atoms that behaves like a Lewis acid. In addition, a Brönsted-type acidity exists, stemming from the OH groups. The high-area aluminas of interest as supports typically exhibit an intrinsic weak acidity, which may be enhanced by traces of common impurities such as chloride, iron oxide, or sulfates, as may be found in aluminas produced by the Bayer process. Stronger acidity may be produced by deliberately incorporating halogens such as chlorides or fluorides in the structure to produce a more acidic material for reactions of various hydrocarbons such as in catalytic reforming. The halide may be introduced inadvertently if a metal halogen compound such as $PdCl_2$ or H_2PtCl_6 is used in catalyst preparation. Some commercial aluminas are basic in the sense that a suspension in water exhibits a pH above 7, caused by the presence of sodium. This may be washed out to a considerable extent, whereupon the leached alumina may behave differently as an adsorbent because of a different ultimate pH reached upon impregnation.

The thermal stability of γ-alumina may be enhanced by two different procedures. Small amounts of silica (for example, 1 to 4 wt %) may be cogelled with it in its manufacture, but this may increase its acid strength (Chap. 7). It has been suggested that a viscous film of silicate is formed that isolates alumina crystallites. The thermal stability of alumina may also be enhanced by incorporating small amounts of divalent ions, such as calcium, magnesium, or barium into the alumina. These occupy tetrahedral voids in the spinel and retard the diffusion of Al^{3+} cations. Other divalent ions such as Cu^{2+} and Ni^{2+} have a similar stabilizing effect, but these may contribute an undesirable catalytic activity.

A finely divided alumina having a surface area of about 100 m^2/g is also made by a flame process and may have applications in catalyst formulations. Well crystallized diaspore occurs in nature and can be converted to α-alumina at much lower temperatures than those at which transition aluminas can be converted. Apparently diaspore is difficult to prepare synthetically, so this is as yet not a practicable route to α-alumina.

An alumina may be used by itself as a catalyst for dehydration of an alcohol, and alumina is a commercial catalyst in the Claus process for conversion of hydrogen sulfide to elemental sulfur (Sec. 9.8). The properties of aluminas for catalytic reactions of hydrocarbons and alcohols are reviewed by John and Scurrell (1977), with particular attention to the literature of 1970 to 1976.

4.5.2 Silica

Silica gel This is most commonly prepared by mixing an acid with a solution of "water glass," which consists of orthosilicates (Na_4SiO_4), metasilicates (Na_2SiO_3), and related compounds. As the pH is lowered, a polymerization and condensation process takes place, which can be visualized as starting with silicic acid [$Si(OH)_4$]. This polymerizes with condensation of silane groups (SiOH) to form an ill-defined polymer in which the primary bonds are the siloxane type (Si—O—Si). This will precipitate as a gel or as a colloid, the properties of which depend upon mixing procedures, the presence of nuclei, electrolytes, temperature, aging, etc.

By proper control a hydrogel, consisting of small micelles that are roughly spherical, is obtained. During drying the micelles do not coalesce appreciably, particularly if the liquid is removed at a temperature and pressure above the critical. Under these conditions, no interface forms that could otherwise collapse the structure by the forces of surface tension. The tiny size of the micelles and use of procedures to prevent coalescence can lead to a product of high surface area, and commercial material is available with surface area as high as about 700 m^2/g. The average pore diameter is correspondingly very low, typically in the range of 2.5 to 5 nm. (These pores are, however, considerably larger than those in zeolites and are substantially greater than the sizes of most reactant molecules of interest. See Tables 7.2 and 7.3.) Normally only micropores are present, in contrast with many aluminas that have macropores as well as high area. Consequently, diffusion problems may be more severe with many silica gels. By varying the manufacturing and aging procedures, one can manufacture silica gels with considerably higher pore diameter and correspondingly lower surface area. Silica gel is generally more difficult to form than alumina. It is not as mechanically rugged, but is generally more inert. Commercial products typically contain small amounts of impurities such as sodium, calcium, iron, and alumina in the concentration range of hundreds of parts per million.

The final dry product should be referred to strictly as a xerogel or porous silica, but the term *silica gel* is in common usage. At ambient temperature the surface consists of a layer of silanol groups (SiOH) plus physically adsorbed water. Most of the water is removed upon drying in air at 150 to 200°C. Silanol groups are left on the surface, and these are progressively lost with increased temperature. Some siloxane groups

```
     O
    / \
—Si—Si—
  |   |
```

may also be present on the surface.

Colloidal silica A variety of colloidal silicas are available as articles of commerce (for example, Ludox), containing up to 40 wt % SiO_2 in the form of spherical, nonporous particles. The colloid is stabilized by ammonium or sodium ions. It may be gelled by increased temperature or by altering pH, and since it is also a good binder, it is useful in formulating catalysts.

Kieselguhr Kieselguhr (diatomaceous earth) is a naturally occurring, finely divided silica consisting of the skeletal remains of diatoms. Depending upon the deposit, it typically contains small amounts of alumina and iron as part of the skeletal structure. The surface area is usually in the range of 20 to 40 m^2/g, and a rather broad range of pore sizes exists, mostly of the order of 100 nm or more.

Flamed silica A finely divided, nonporous, and highly pure silica powder (Cabosil, Aerosil) is manufactured by flame hydrolysis of $SiCl_4$. It may have use in catalyst formulations. Particle sizes are about 40 to 50 nm.

Comparison of alumina and silica carriers The combination of useful properties of aluminas make them generally the first choice for carriers. However, active aluminas can dissolve or become soft and mushy under acidic conditions–conditions under which silica is stable. The relative nonreactivity of silica upon calcination with other catalyst ingredients may also be a significant factor. If adsorption of products or reactants on alumina is deleterious, the nonadsorptive character of silica may be an improvement.

4.5.3 Activated Carbon

If a carbonaceous material such as coal, lignite, wood, or petroleum pitch is heated in the absence of air, much of the substance devolatilizes, leaving behind a porous structure of carbon that usually also contains some hydrogen. This may then be activated by controlled oxidation with steam or carbon dioxide to further open up the pores and increase total surface area. The activated carbon may contain up to about 10 wt % oxygen, which may cover a large fraction of the surface as chemisorbed oxygen in the form of ketones, hydroxyls, or carboxylic acids. These can cause its adsorptive properties to be considerably different than a carbon heated in inert gas or under reducing conditions. The surface area can range up to 1200 m^2/g.

Carbon supports can obviously be used for catalytic reactions only under conditions where the support itself is nonreactive. The catalyst usually cannot be regenerated except perhaps by washing, but the active ingredient can be recovered by burning the support. Carbon supports are used primarily for noble metals and for reactions in which the strong adsorption of carbon for organic molecules may be an asset. Activated carbon may contain considerable "ash" (from the mineral matter in the starting material) and various metals and sulfur compounds. These can have an important and generally undesirable catalytic effect. For a catalytic support a form of carbon essentially free of metal salts, low in sulfur, and with high surface area is preferred. Such a material may be several times more costly than a conventional activated carbon as used for removal of impurities by adsorption. An activated carbon from coal or lignite may be unacceptably soft and have an excessive impurity content. A petroleum-based coke is typically stronger than that from a coal, but it may contain some sulfur and small amounts of vanadium and nickel. These may make it unacceptable for use as a catalyst support for some applications. If these impurities are unacceptable, a wood charcoal may be used instead. Carbon from coconut shells is relatively hard and attrition-resistant and is therefore especially useful.

4.5.4 Other Supports

Historically, many early catalysts were supported on naturally occurring materials such as asbestos and pumice, but now these are seldom used. These and various other supports are described by Innes (1954). Several chapters in the book edited by Linsen (1970) describe active magnesia and hydrous zirconia and provide a more detailed discussion of alumina, silica, and carbon. Anderson (1975) discusses briefly the chemistry of a variety of support materials and gives extensive references to the literature. Silica-alumina and zeolites are discussed in Chap. 7.

An unusual support form known as a *monolith* or *honeycomb* is used in some automobile catalytic converters (Sec. 8.12), where a very low pressure drop is required to minimize power loss from the engine. This is a single block of material containing within it an array of parallel, uniform, straight, nonconnecting channels. In one process this is manufactured by extrusion of a thick inorganic dough through a die, followed by drying and firing. The ingredients are fused into cordierite, a magnesia alumina silicate. A variety of cell shapes and sizes can be manufactured, and monolith blocks can likewise be fabricated in a variety of cross-sectional shapes, as shown in Figs. 4.13, 4.14, and 4.15. The cell density may vary from 200 to 400 cells per square inch. Somewhat similar is a ceramic α-alumina honeycomb available with several different types of internal passageways (Torvex), which can be used at temperatures up to 1500°C.

4.6 PROMOTERS

The term *promoter* is used in a rather general sense to refer to a substance which, when added in relatively small amounts in the preparation of a catalyst, imparts

Figure 4.13 Representative monolith shapes. (*Courtesy of Corning Glass Works.*)

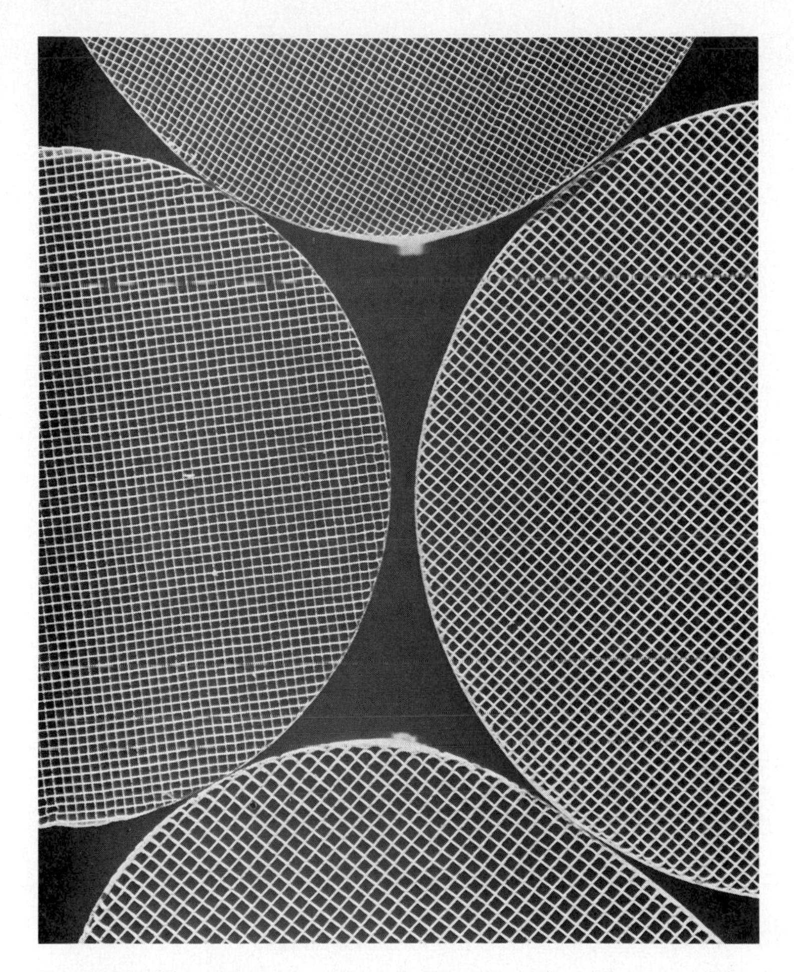

Figure 4.14 Monolith cross sections of different cell sizes. (*Courtesy of Corning Glass Works.*)

better activity, selectivity, or stability. The increased performance is greater than that attributable to the promoter acting independently. The term is used to cover a wide variety of basic phenomena, but most promoters can be classified as textural promoters or structural promoters. A *textural promoter* ("stabilizer") acts by a physical effect, a *structural promoter* by a chemical effect.

4.6.1 Textural Promoters

A textural promoter is an inert substance which inhibits the sintering of microcrystals of the active catalyst by being present in the form of very fine particles. These separate the true catalyst particles from contact with one another so they do not coalesce, thus preventing or minimizing loss of active catalyst area during service. An example is the incorporation of a small amount of alumina in the conventional iron catalyst

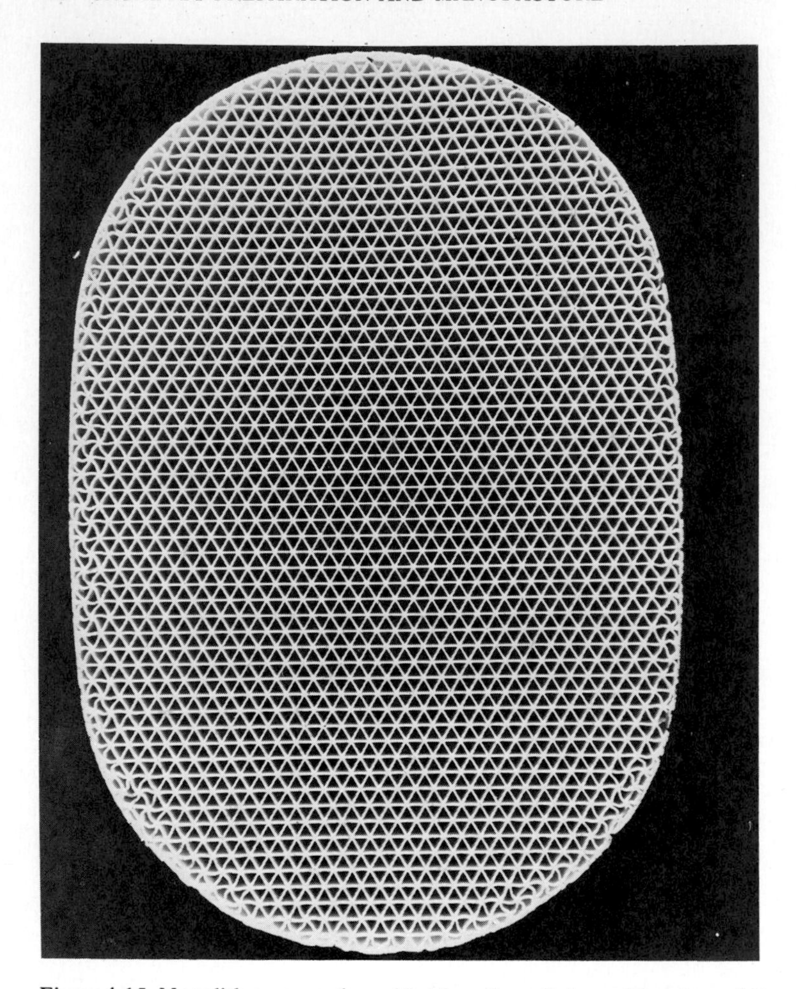

Figure 4.15 Monolith cross section with triangular cell shape. (*Courtesy of Corning Glass Works.*)

for the synthesis of ammonia, or the action of zinc chromite in the zinc oxide/chromia catalyst used for methanol synthesis. To be effective the textural promoter must generally be of considerably smaller particle size than that of the active species, it must be well dispersed, and it must not react with or form a solid solution with the active catalyst. As a minimum, it must have a relatively high melting point. Some useful substances with their melting points in degrees Celsius are as follows: Al_2O_3, 2027; SiO_2, 1700; ZrO_2, 2687; Cr_2O_3, 2435; CeO_2, ~2600; MgO, 2802; TiO_2, 1855. BeO also has a high melting point but is highly poisonous. Under reducing conditions carbon may also be useful since it is stable to extremely high temperatures.

That a substance acts as a textural promoter can sometimes be identified in the case of a metallic catalyst by comparing the specific metal surface area (e.g., by chemisorption) of a catalyst in the presence and absence of the promoter. Another method of identification is by the effective activation energy. This is unchanged by a textural promoter but may be markedly affected by a structural promoter.

In the case of mixed oxide catalysts a particular compound, crystal form, or defect structure may be desired for maximum selectivity, or even a certain mixed phase may be required. This may be stabilized by the presence of a particular material. Frequently it is uncertain here whether the promoter provides a structural or a chemical effect because of the complex nature of many of these catalysts.

4.6.2 Structural Promoters

In contrast to a textural promoter, a structural promoter causes a chemical effect–it changes the chemical composition of the catalyst. In many cases the effect of the structural promoter is clear but its mechanism of action is not. In other cases it is not clear whether the effect of the promoter is primarily a physical or a chemical effect. Some possible mechanisms are:

1. There may be a dual-function action. The promoter may catalyze the formation of an intermediate. Thus the addition of a chloride to a platinum/alumina reforming catalyst may enhance its activity and be termed "promotion," although the true mechanism is more clearly seen as an example of dual functionality.
2. The promoter may produce lattice defects or interstitial substitution.
3. The promoter may change the electronic structure of a catalyst, e.g., the ease of addition or removal of electrons from a metal, and hence the strength of chemisorption. Potassium oxide promoter in the iron catalyst for synthesis of ammonia may act in this fashion.

A substance may also be erroneously thought to act as a structural promoter when in fact it poisons some undesirable side reactions and/or accelerates desired side reactions. Thus potassium salts may be added to a chromia-alumina catalyst used in dehydrocyclization (for example, to convert *n*-hexane to benzene). Its function here is probably to poison acid sites that otherwise might accelerate undesirable cracking reactions, and it may also minimize the rate of coke formation.

In many cracking reactions as, for example, dehydrogenation of ethylbenzene to styrene, steam is added to the feed to minimize carbon deposition on the catalyst. Potassium oxide is usually added to the catalyst as a promoter (e.g., in the form of the carbonate) to accelerate the rate of reaction of steam with incipient carbonaceous deposits.

Some ways to identify a structural promoter are:

1. The effective activation energy of a reaction may be altered.
2. Adsorption isotherms may be altered.

Neither should be affected by a textural promoter.

Some examples of promoter effects are discussed later in conjunction with specific catalytic reactions. See for example, the sections on methanol synthesis (Sec. 10.4) and ammonia synthesis (Sec. 10.5).

REFERENCES

Anderson, J. R.: *Structure of Metallic Catalysts*, Academic, New York, 1975.
Augustine, R. L.: *Catalytic Hydrogenation: Techniques and Applications in Organic Synthesis*, Dekker, New York, 1965.
Ciapetta, F. G., and C. J. Plank in P. H. Emmett (ed.): *Catalysis*, vol. 1, Reinhold, New York, 1954, p. 315.
Delmon, B., P. A. Jacobs, and G. Poncelet (eds.): *Preparation of Catalysts*, Elsevier, Amsterdam, 1976.
Eischens, R. P., in E. Drauglis and R. I. Jaffee (eds.): *The Physical Basis for Heterogeneous Catalysis*, Plenum, New York, 1975, p. 485.
Emmett, P. H., in E. Drauglis and R. I. Jaffee (eds.): *The Physical Basis for Heterogeneous Catalysis*, Plenum, New York, 1975, p. 3.
Freifelder, M.: *Practical Catalytic Hydrogenation: Techniques and Applications*, Wiley, New York, 1971.
Gauguin, R., M. Graulier, and D. Papee: *Adv. Chem. Ser.*, **143**, 147 (1975).
Gil'debrand, E. I.: *Int. Chem. Eng.*, **6**, 449 (1966).
Hedvall, J. A.: *Adv. Catal.*, **8**, 1 (1956).
Higginson, G. W.: *Chem. Eng.*, Sept. 30, 1974, p. 98.
Hoekstra, J.: U.S. Patent 3,388,077 (to U.O.P.).
Innes, W. B., in P. H. Emmett (ed.): *Catalysis*, vol. I, Reinhold, New York, 1954, p. 245.
, in P. H. Emmett (ed.): *Catalysis*, vol. II, Reinhold, New York, 1955, p. 1.
John, C. S., and M. S. Scurrell: *Catalysis*, The Chemical Society, London, 1977, vol. I, chap. 4, p. 136.
Lieber, E., and F. L. Morritz: *Adv. Catal.*, **5**, 417 (1953).
Linsen, B. G. (ed.): *Physical and Chemical Aspects of Adsorbents and Catalysts*, Academic, New York, 1970.
Morikawa, K., T. Shirasaki, and M. Okada: *Adv. Catal.*, **20**, 98 (1969).
Moss, R. L., in R. B. Anderson and P. T. Dawson (eds.): *Experimental Methods in Catalytic Research*, vol. II, Academic, New York, 1976.
Natta, G., in P. H. Emmett (ed.): *Catalysis*, vol. III, Reinhold, New York, 1955, p. 349.
Rylander, P. N.: *Catalytic Hydrogenation over Platinum Metals*, Academic, New York, 1967.
Ryshkewitch, E.: *Oxide Ceramics*, Academic, New York, 1960.
Thomas, C. L.: *Catalytic Processes and Proven Catalysts*, Academic, New York, 1970.
Wefers, K., and G. M. Bell: *Oxides and Hydroxides of Aluminum*, Alcoa Research Labs., E. St. Louis, Ill., 1972. (A revision of "Alumina Properties," Tech. Paper No. 10.)

CHAPTER

FIVE

PHYSICAL CHARACTERIZATION AND EXAMINATION

Most practical catalysts are highly complex materials, and a basic problem is how to correlate catalyst behavior with physical and chemical structure. Only a few methods of characterization are standardized or nearly so. These include determination of total surface area by the Brunauer-Emmett-Teller (BET) method, void fraction, pore-size distribution, and in some cases specific metal area by selective chemisorption. Crystallite size may be determined by X-ray line broadening or by direct observation in electron microscopy. Many analytical procedures as developed for other branches of chemistry are, of course, also applicable here. Beyond these are an enormous variety of powerful instrumental techniques for examining and characterizing surface and adsorbed species. Many require expensive and elaborate apparatus and a high degree of sophistication on the part of the experimenter for interpretation of results. These techniques have been developing rapidly during the past decade or two, and in many cases their potential for application to problems in catalysis is only beginning to be explored.

In the following the more standardized procedures will be considered first, followed by a brief introduction to a few of the methods of characterizing surfaces that seem to be particularly applicable to practical catalysts, with examples of the kinds of information they may reveal. The objective here is to indicate some of the capabilities and limitations of these methods without entering into the sophisticated background which is required to work with these instruments and interpret the results. Most instrumental methods of characterizing surfaces operate at very high vacuum, and the structure and behavior of surfaces under such conditions may be far different than in a reacting environment.

Almost any instrument is capable of revealing some information of possible use concerning some practical catalyst. However a number of fairly well developed methods primarily used at present for fundamental research purposes will not be considered here. These include low-energy electron diffraction (LEED), infrared spectroscopy, collective paramagnetism, and many of the evolving surface spectroscopy methods.

5.1 MEASUREMENT OF SURFACE AREA

In comparing different catalysts or the effect of various treatments on catalytic activity, it is necessary to know the extent to which a change in activity is caused by a change in the area of a catalyst, in contrast to a change in intrinsic reactivity. Methods of measuring surface areas are of concern in many fields of science and technology and have received wide and detailed study. Direct observation by an optical or electron microscope is the most straightforward, although tedious, method of determining particle size and particle-size distribution. If the solid is impervious and the shape is well established, the total surface area can then be estimated closely. But most catalysts should be porous in order to maximize the catalytic area per unit volume of reactor. It is this total surface area, both interior and exterior, that is of concern.

The principal method of measuring total surface area of porous structures is by adsorption of a particular molecular species from a gas or liquid onto the surface. If the conditions under which a complete adsorbed layer, averaging one molecule thick, can be established and the area covered per molecule is known, then the quantity of adsorbed material gives directly the total surface area of the sample. Sorption from the liquid phase, as of fatty acids and dyes, is of limited use in catalysis because the size of these molecules is much greater than that of many reactant molecules. Therefore they may not have access to portions of a fine microporous structure which nevertheless may contribute to the total area of significance during reaction. The most useful measurements are by adsorption of a gas or vapor of sufficiently small molecular dimensions that interstices down to a few tenths of a nanometer are penetrated.

5.1.1 Physical Adsorption Isotherms

To measure total surface area nonspecific physical adsoption is required, but even with physical adsorption the isotherm varies somewhat with the nature of the adsorbent (the solid). Most physical adsorption isotherms may be grouped into five types, which are frequently referred to as the Brunauer, Deming, Deming, and Teller (BDDT) classification (Fig. 5.1). In all cases the amount of vapor adsorbed gradually increases as its partial pressure is increased, becoming at some point equivalent to a monolayer, but then increasing to a multilayer, which eventually merges into a condensed phase.

Type I is frequently called the Langmuir type. The asymptotic value was originally ascribed to a monolayer, as derived from the Langmuir equation. However, this isotherm is seldom encountered on nonporous materials. It is fairly common with certain activated carbons, silica gels, and zeolites that contain only very fine pores, and it is now generally believed that in these cases the asymptotic value represents the complete filling of micropores at a relative pressure substantially less than unity, rather

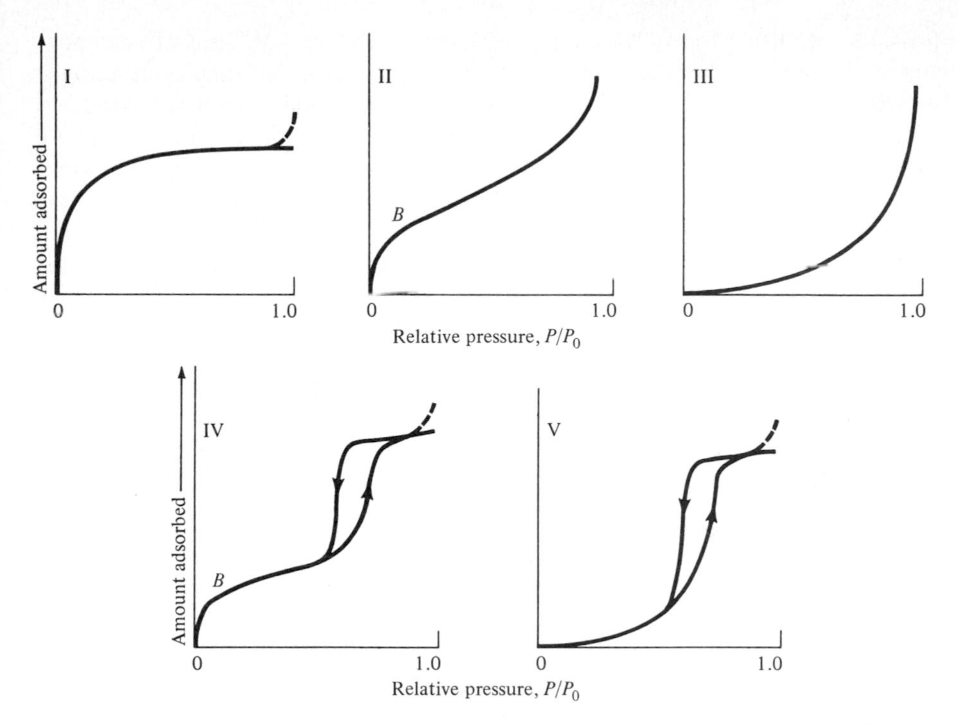

Figure 5.1 The five types of adsorption isotherms as classified by Brunauer et al. (frequently termed the BDDT classification).

than monolayer adsorption (see Figs. 2.4 and 2.5). This type of isotherm would also be expected for reversible chemisorption.

Type II, sometimes termed the *sigmoid* or *S-shaped isotherm*, is commonly encountered on nonporous structures. Point *B* occurs at a "knee" and is the stage at which monolayer coverage is complete.

The *Type III* isotherm is convex over the entire range and does not exhibit a point *B*. It is relatively rare and is typical of a system where the forces of adsorption are relatively weak, as when the adsorbate is not wetted by the surface, e.g., water vapor on graphite.

Type IV is encountered with porous materials. At low values of P/P_0 the isotherm is similar to type II, but then adsorption increases markedly at higher values of P/P_0 where pore (capillary) condensation takes place. A hysteresis effect associated with this pore condensation is frequently, but not always, observed (Figs. 2.4 and 2.5). Isotherms of this type are often encountered with industrial catalysts, and the capillary condensation curve may be used to determine a pore-size distribution (Sec. 5.3.1).

Type V is similar to type III, but with pore condensation taking place at higher values of P/P_0. It is also relatively rare.

A few other kinds of adsorption isotherms do not fit into the above classification, but there is yet no generally accepted characterization of these. Parfitt and Sing (1976) assign *type VI* to an isotherm in which adsorption occurs in steps. Each step represents the adsorption of an additional layer, the differential heat of adsorption being fairly constant during the buildup of one layer, but then dropping abruptly as the next layer begins to be formed. This behavior is encountered only with relatively uniform surfaces such as graphite. Adamson (1976, p. 567) assigns *types VI* and *VII* to two cases in which the bulk-liquid adsorbate has a finite contact angle and hence the amount of vapor adsorbed does not approach infinity asymptotically as P/P_0 approaches 1.

Any isotherm having only a gradual curvature at low values of P/P_0 represents a case in which adsorbent-adsorbate interaction is weak. These isotherms (such as types III and V) are difficult to use for determination of surface area because second and succeeding layers build up before the first is complete.

5.2.1 Brunauer-Emmett-Teller (BET) Method

The most common method of measuring surface area, and one used routinely in most catalyst studies, is that developed by Brunauer, Emmett, and Teller (1938). Early descriptions and evaluations are given by Emmett (1948, 1954). In essence, the Langmuir adsorption isotherm is extended to multilayer adsorption. As in the Langmuir approach, for the first layer the rate of evaporation is considered to be equal to the rate of condensation, and the heat of adsorption is taken to be independent of coverage. For layers beyond the first, the rate of adsorption is taken to be proportional to the fraction of the lower layer still vacant. The rate of desorption is taken to be proportional to the amount present in that layer. (These assumptions are made largely for mathematical convenience.) The heat of adsorption for all layers except the first layer is assumed to be equal to the heat of liquefaction of the adsorbed gas. Summation over an infinite number of adsorbed layers gives the final expression as follows:

$$\frac{P}{V(P_0 - P)} = \frac{1}{V_m C} + \frac{(C - 1)P}{V_m C P_0} \tag{5.1}$$

where V = volume of gas adsorbed at pressure P
V_m = volume of gas adsorbed in monolayer, same units as V
P_0 = saturation pressure of adsorbate gas at the experimental temperature
C = a constant related exponentially to the heats of adsorption and liquefaction of the gas

$$C = e^{(q_1 - q_L)/RT} \tag{5.2}$$

where q_1 = heat of adsorption on the first layer
q_L = heat of liquefaction of adsorbed gas on all other layers

The larger the value of C the more the isotherm approaches the type II form and the more accurately the surface area can be determined.

If Eq. (5.1) is obeyed, a graph of $P/V(P_0 - P)$ versus P/P_0 should give a straight line whose slope and intercept can be used to evaluate V_m and C. Many adsorption data show very good agreement with the BET equation (Fig. 5.2) over values of the

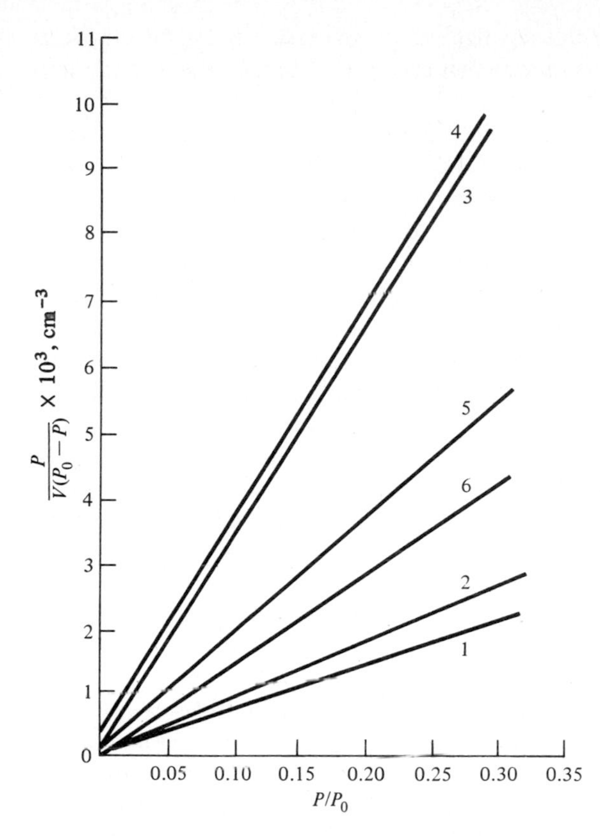

Figure 5.2 Linear plots of the Brunauer, Emmett, and Teller equation for nitrogen adsorption. Isotherms of various adsorbents. (*Brunauer et al., 1938*). Adsorption of nitrogen at 90 K on (1) Unpromoted Fe catalyst #937, per 489.0 g; (2) Al_2O_3 promoted Fe catalyst #424, per 49.8 g; (3) Al_2O_3–K_2O promoted Fe catalyst #957, per 54.5 g; (4) Fused Cu catalyst, per 550 g; (5) Chromium oxide gel, per 1.09 g; (6) Silica gel, per 0.606 g.

relative pressure P/P_0 between approximately 0.05 and 0.3, and this range is usually used for surface-area measurements. At higher P/P_0 values, complexities associated with the realities of multilayer adsorption and/or pore condensation cause increasing deviation. At values of P/P_0 much below about 0.05 the amount adsorbed in many cases is so low that the data become less accurate.

From Eq. (5.1), $V_m = 1/(S + I)$, where S is the slope and is equal to $(C - 1)/V_m C$ and I is the intercept and is equal to $1/V_m C$. This proceeds from the fact that

$$S + I = \frac{1}{V_m C} [(C - 1) + 1] = \frac{1}{V_m} \tag{5.3}$$

The surface of the catalyst may then be calculated from V_m if the average area occupied by an adsorbed molecule is known.

Any condensible inert vapor can be used in the BET method, but for the most reliable measurements, the molecules should be small and approximately spherical. The vapor also should be easy to handle at the required temperatures; for example, P/P_0 values of 0.05 to 0.3 should be conveniently attainable. Krypton, argon, and nitrogen are suitable choices in view of their commercial availability. Liquid nitrogen is a readily available coolant, but argon and krypton are expensive relative to nitrogen and must be highly purified. Consequently, nitrogen is usually used since it is relatively cheap and readily available in high purity. It yields well-defined type II curves on most surfaces, and the cross-sectional area per adsorbed molecule has been well established.

The partial pressures of nitrogen gas will be in the range of 10 to 100 kPa in order to obtain values of P/P_0 in the range of about 0.05 to 0.30. When the total surface area of the sample is less than a few square meters, the amounts of gas adsorbed become small relative to the total amount in the apparatus and the accuracy of the measurements becomes poor. By using a vapor of higher boiling point as an adsorbate, measurements at the temperature of liquid nitrogen can be made at much lower pressures to achieve the desired range of P/P_0 values; so the amount of gas adsorbed on the solid is now a much larger fraction of that present and can be more precisely measured. The gas most often used for such low-pressure measurements is krypton, which has a vapor pressure at this temperature of about 0.4 kPa. However, the experimental difficulties with low-pressure equipment are somewhat greater because the system must be completely leakproof.

With any adsorbate it is desirable to desorb water and other gases from the vessels and from the sample, typically by heating under vacuum, before making the measurements. If this is not done, slow desorption during determination of the isotherm can give misleading results. Measurements can be made either gravimetrically or volumetrically, and a variety of types of apparatus have been designed and used. In a representative procedure the sample is first degassed at 180 to 190°C for 10 to 15 min under vacuum. It is then cooled to liquid-nitrogen temperature, and a known quantity of nitrogen gas is then admitted and allowed to equilibrate. From the equilibrium pressure and *PVT* relationships the amount of nitrogen adsorbed is calculated. The procedure is repeated, yielding a series of values of the volume adsorbed corresponding to a set of increasing values of the equilibrium pressure.

The projected cross-sectional area of an adsorbed gas molecule, A_m, can be estimated from the liquid density, but there is no assurance that the packing in a monolayer will be the same as that in the bulk. Thus, for accuracy the method has to be standardized by measurements with adsorbents whose area can be determined directly, e.g., nonporous, finely divided, uniform crystals or spheres. By this means the value for nitrogen has been established as 0.162 nm^2, which, perhaps coincidentally, equals that calculated from the bulk density.

If the constant C is sufficiently large, for example, greater than about 50 (dimensionless), as it usually is with nitrogen adsorption, the isotherm should have a well-defined point B and the intercept is usually small relative to the slope. Hence a straight line can be drawn connecting the origin and one point obtained at a P/P_0 value of about 0.2 to 0.3 to obtain the slope for the BET equation. This is a simple, quick method requiring only one datum point, which is very useful for a surface of known properties. The calculated area will frequently be in error by only a few percent.

The use of one fixed value for the effective cross-sectional area of an adsorbed gas molecule assumes that this is unaffected by the nature of the solid. For most adsorbates on most solids the effective value of A_m varies somewhat from solid to solid because the lattice parameter of the solid causes some localization of the adsorption, i.e., preferential adsorption on certain sites. Hence, a very strongly adsorbed vapor, which corresponds to a high C value, is not desired, but a weakly adsorbed vapor is likewise not desired because of the loss in accuracy in determining the amount of adsorption corresponding to a monolayer. Nitrogen is unusual in that it produces a well-defined knee on most solids, but its adsorption is not excessively localized. Water vapor is not recommended for general use because it has a variety of specific interactions with oxide structures commonly encountered in catalysis and it tends to form an organized tetrahedral structure rather than a random orientation.

The theory underlying the BET model has been criticized for the fact that it postulates that adsorption can occur in the nth layer before the $(n - 1)$th layer is filled. This implies that molecules can be piled up on top of one another into a system of irregular vertical columns, whereas surface energy considerations indicate that there is probably little adsorption onto the nth layer until the $(n - 1)$th layer is largely filled. For this and other reasons the constant C should be treated as an empirical parameter rather than a quantity that can be calculated independently. Modifications to the BET model to bring it closer to reality, however, do not change the calculated surface area appreciably from that obtained from the simple theory, probably in part because multilayer adsorption is not great over the P/P_0 range usually used for BET area measurements.

A variety of other equations can also be used (see, e.g. Gregg and Sing, 1967, and Parfitt and Sing, 1976), but they seem to offer no significant advantage over the BET method. If much of the area is in the form of pores with diameters less than say 1 to 1.5 nm or so, as occurs with some activated carbons and zeolites, pore condensation may occur at relatively low values of P/P_0, and reported BET areas may be misleadingly high. The highest surface areas reliably obtained for porous substances are about 1000 to 1200 m^2/g, and values reported much higher than this range should be examined critically. The proceedings of an international symposium on surface-area determination critically compared and contrasted various methods (Everett and Ottewill, 1970).

5.1.3 Specific Area by Selective Chemisorption

With a supported metal catalyst it is frequently desirable to be able to determine the exposed metal area in distinction to the total surface area. This may be achieved by measuring the uptake of a gas that is chemisorbed on the metal but negligibly so on the support, under conditions that allow the coverage corresponding to a monolayer to be determined. Most useful for this purpose is hydrogen, but carbon monoxide and oxygen have also been used. Although simple in concept, the method can be complex in application. The most suitable experimental conditions vary considerably with the nature of the metal and the nature of the support. A number of factors rather specific to each system can introduce uncertainties.

As a prerequisite, it is necessary to know the chemisorption stoichiometry, that is, the number of surface atoms covered for each molecule of gas adsorbed, and the surface area occupied per metal atom. For hydrogen the stoichiometric number is almost always 2 since the hydrogen molecule usually dissociates upon adsorption and each hydrogen atom is adsorbed on one metal atom. Carbon monoxide can adsorb in either a linear form in which it covers one metal atom (stoichiometric number of 1) or in a bridged form covering two metal atoms (stoichiometric number of 2). The number of surface atoms per unit area of metal varies slightly with the crystallographic plane, but for all metals it is about 10^{19} atoms per square meter. Specific representative values are 1.5 to 1.6×10^{19} atoms per square meter for Fe, Co, and Ni; 1.25 to 1.33×10^{19} atoms per square meter for Pt, Pd, Ir, and Rh; and 1.15×10^{19} atoms per square meter for Ag (Anderson, 1975, p. 296).

Hydrogen has been studied the most extensively, especially on platinum and nickel. It is often the first choice for other metals, except for palladium, in which it dissolves. A slow chemisorption is sometimes observed that has been ascribed to the presence of some support material or contaminant on the surface of metal particles or possibly to the intrinsic properties of very small particles (e.g., those containing perhaps 50 atoms or less; see Chap. 6). On a carbon support or with carbon contamination adsorbed hydrogen can surface-diffuse from metal crystallites onto the support (termed *spillover*) to give erroneously high results. This does not seem to occur with oxide supports such as alumina or silica when they are clean.

The optimum temperature and pressure will vary with the system and must be established experimentally. Studies must also be made with the support by itself to establish its possible contribution. High-area supports contribute excessively to hydrogen adsorption at 77 to 90 K. Preferred conditions for a dispersed platinum are about 273 to 300 K and about 0.01 to 0.3 kPa, and for dispersed nickel, about 273 to 300 K and 20 kPa (Anderson, 1975, p. 317–322).

Hydrogen chemisorption cannot be used for palladium because it is adsorbed into the bulk. Oxygen adsorption at very low pressures is a possible method, but carbon monoxide chemisorption may be more satisfactory (Farrauto, 1974). For nickel, hydrogen has been most commonly used, and carbon monoxide has been used to some extent. For the iron surface area in catalysts used for ammonia synthesis, carbon monoxide adsorption has long been used. Hydrogen does not chemisorb on silver, and oxygen has been studied as a possibility.

A particular limitation to oxygen adsorption in general has been uncertainties concerning the stoichiometric number and the ease with which bulk oxidation can occur with many systems. In some cases a monolayer of oxygen can be chemisorbed under carefully controlled conditions. This can then be reduced with hydrogen, and the quantity of water formed is determined. This is termed *titration*.

Further details concerning this and other methods are given by Anderson (pp. 295–323) and in the review by Farrauto. The latter also discusses systems which have been considered for various metal oxides. Chemisorption also provides a means of determining an average particle size of a supported metal, if the quantity present is known, by assuming a particle shape (Sec. 6.2). This method can be applied over a wide range of particle sizes, including those beyond the range of the X-ray line-broadening method, but chemisorption studies require a high-vacuum, high-purity system and careful experimentation and interpretation of results.

5.2 PORE VOLUME

A direct and simple method of determining the total volume of the pores is by measuring the increase in weight when the pores are filled with a liquid of known density. The liquid should preferably be of low molecular weight so that fine pores are filled and water or various hydrocarbons may be used satisfactorily. A simple procedure is to boil a sample of dry catalyst pellets of known weight in distilled water for 2 to 5 min to cause the water to penetrate the pores. The entire sample should then be cooled to minimize subsequent vaporization; the pellets should then be transferred to a damp cloth, rolled to remove excess water, and reweighed. This will determine the total volume of pores between, approximately, 1 and 150 nm in diameter. The method is limited in accuracy by the fact that it is difficult to dry the external surface of the particles without removing liquid from the large pores, and some liquid tends to be held around the points of contact between the particles.

More accurate results are obtained by the *mercury-helium method*. This is based on the fact that mercury does not wet most surfaces, and therefore does not penetrate pores at atmospheric pressure, and that adsorption of helium gas is generally negligible at room temperature. A container of known volume V (in cubic centimeters) is filled with a known weight of pellets or powder W (in grams). After evacuation, helium is admitted, and from the gas laws the sum of the volume of the space between the pellets V' and the void volume inside the pellets V_g is calculated. The true density of the solid is

$$\rho_T = \frac{W}{V - (V' + V_g)} \tag{5.4}$$

The helium is then pumped out and the container filled with mercury at atmospheric pressure. Its volume is that of the space between the pellets V'.

The porosity or void fraction θ is the volume of voids in cubic centimeters per cubic centimeter of pellets and is given by

$$1 - \theta = \frac{V - (V' + V_g)}{V - V'} \tag{5.5}$$

The density of the pellets is given by

$$\rho_p = \frac{W}{V - V'} \quad \text{g/cm}^3 \tag{5.6}$$

These measurements are often made in a mercury porosimeter (Sec. 5.3.2) in conjunction with a measurement of pore-size distribution.

The total pore volume can also be determined from an adsorption measurement at a value of P/P_0 sufficiently high that all pores of interest are filled by condensed vapor (Sec. 5.3). For high-surface-area catalysts the amount of vapor adsorbed on the exterior of the particles is negligible compared to that condensed in the pores, so the equivalent liquid volume of the amount of vapor adsorbed is the same as the pore volume. The density is assumed to be the same as that of the liquefied vapor at its boiling point.

Operation at pressures slightly below saturation avoids condensation of liquid around the points of contact. A simple method developed by Benesi et al. (1955) achieves this by adding a small amount of nonvolatile solute to the saturating liquid to lower the vapor pressure of the latter to the desired degree, usually to 95 percent of saturation. In their method, carbon tetrachloride is used as the volatile solvent and cetane (n-$C_{16}H_{34}$) as the nonvolatile solute. Samples are dried and weighed and placed in a desiccator containing the solution in the bottom. The gain in weight combined with the known density of carbon tetrachloride gives the pore volume directly. Taking the carbon tetrachloride surface tension as 0.026 N/m at 25°C, molar volume as 97.1 cm^3, and assuming zero contact angle, at P/P_0 of 0.95 all pores below 80 nm in diameter should be filled. By using various ratios of carbon tetrachloride to cetane the partial pressure of carbon tetrachloride can be varied, and the method can be adapted to determine the total volume of pores having diameters below a specified desired value.

5.3 PORE-SIZE DISTRIBUTION

This is of interest primarily for prediction of the effective diffusivity in a porous catalyst in conjunction with calculations of the ease of access of reactant molecules to the interior of a catalyst pellet by diffusion. Two different methods may be used: physical adsorption of a gas, which is applicable to pores less than about 60 nm in diameter, and mercury porosimetry, applicable to pores larger than about 3.5 nm. The true pore structure is of almost infinite complexity, and a considerable literature exists on interpretation, in terms of pore shapes, of hysteresis loops from physical adsorption data and, to a lesser extent, from mercury porosimeter results.

The pore-size distribution reported depends upon the model assumed for interpretation. This is usually taken as an array of cylindrical capillaries of different radii, randomly oriented. If the pores are fairly close in size, a useful concept is that of the average pore radius defined as $\bar{r} = 2V_g/S_g$, where V_g is the pore volume per gram and S_g the surface area per gram; for example, $\bar{r}$ is the radius of a cylinder having the same volume/surface ratio as the real pore. If the pores vary substantially in size, the diffusion characteristics in the structure cannot be adequately represented by an average radius, and it is necessary to determine the pore-size distribution.

Pores larger than about 50 nm in diameter are generally termed *macropores*; those less than about 2 nm, *micropores*; and pores of intermediate size, *mesopores*.

5.3.1 Nitrogen Adsorption

Measurements of the amount of gas adsorbed or desorbed as a function of reduced pressure provide the most commonly used procedure for determining the pore-size distribution of fine pores. The basic principle is that the pressure at which vapor will condense (or evaporate) is determined by the curvature of the meniscus of the condensed liquid in the pores. This is given by the Kelvin equation for the variation of vapor pressure with surface curvature in a capillary tube closed at one end.[1]

$$\ln \frac{P}{P_0} = \frac{-2\sigma V_m \cos\theta}{r_K RT} \tag{5.7}$$

where P = vapor pressure of liquid over the curved surface
P_0 = vapor pressure of liquid over a plane surface

[1] For surface curvature in two dimensions (instead of three), Eq. (5.10) applies instead of the Kelvin equation.

σ = surface tension of the liquid adsorbate
V_m = molal volume of the liquid adsorbate
θ = contact angle
r_K = radius of curvature, or *Kelvin radius* (positive for a concave surface)
r_c = physical radius of cylindrical pore
R = gas constant
T = absolute temperature

Consider a porous solid in contact with a vapor at some relative pressure P/P_0. A vapor which wets the surface, such as nitrogen, is chosen so that $\cos\theta = 1$. An adsorbed layer of thickness t will be present on the walls of all unfilled capillaries. It is assumed that the radius of the meniscus in the unfilled pores is not the true physical radius r_c but rather that this has been diminished by the thickness of the adsorbed layer and therefore $r_K = r_c - t$. The critical radius r_c will be related to the reduced pressure by the expression

$$r_c = \frac{-2\sigma V_m}{RT \ln (P/P_0)} + t \tag{5.8}$$

Nitrogen adsorption has been used almost universally, and the value of t as a function of P/P_0 has been estimated by a number of investigators who have worked with nonporous substances of known area. Values of t are essentially independent of the chemical nature of the adsorbent for most systems at coverages greater than a monolayer. Table 5.1 gives the relationships published by three groups of investigators. DeBoer et al. (1966) state that a monolayer is 0.354 nm thick.

To determine pore-size distribution, the following procedure is used. Consider a slight increase in pressure and let $V_r\,\Delta r$ be the volume of pores having radii between

Table 5.1 Thickness of adsorbed layer of nitrogen on nonporous substrate as a function of P/P_0

	Thickness t,* nm		
P/P_0	Cranston and Inkley (1957)	deBoer et al. (1966)	Gregg and Sing (1967)
0	0		
0.05	0.339		
0.10	0.412	0.368	
0.20	0.485	0.436	
0.30	0.567	0.501	0.56
0.40	0.635		0.62
0.50	0.70		0.68
0.60	0.75	0.736	0.75
0.70	0.86		0.85
0.80	1.00		0.98
0.90	1.22		1.27
0.95	1.40		1.63

*t is the volume of N_2 adsorbed divided by the BET area.

r_c and $r_c + \Delta r_c$ that become filled with condensate. Simultaneously, in larger pores the thickness of the adsorbed layer increases by Δt. The total volume of nitrogen found to be adsorbed, $v_r \, \Delta r_c$ (calculated as liquid), during this increase in pressure is the sum of the two processes. This may be expressed as

$$v_r \, \Delta r_c = \frac{(r_c - t)^2}{r_c^2} V_r \, \Delta r_c + \Delta t \int_{r_c + \Delta r_c}^{\infty} \frac{(r_c - t)}{r_c} \frac{(2V_r \, dr_c)}{r_c} \tag{5.9}$$

The first term on the right-hand side is the volume of liquid nitrogen which has filled pores whose critical values of P/P_0 have been exceeded. The second term is the increase in volume of the adsorbed layer. The integral term is simply the surface area of all pores not filled by capillary condensation. For each increment of pressure, the average value of t may be obtained from Table 5.1 and that of r_c and of Δr_c may be calculated from Eq. (5.8). Note that $V_r \, \Delta r$ is a volume term, not V_r. Methods of using Eq. (5.9) may be somewhat involved, but a procedure is discussed in detail by Gregg and Sing. Roberts (1967) also gives a method which is relatively easy to use and can be readily programmed for a computer.

The method is described above for an adsorption curve obtained with increasing pressure, but clearly the same results should be achieved with a descending curve if the two curves coincide. The integral in the second term of Eq. (5.9) gives the cumulative area represented by all pores down to the radius r_c, and the value integrated over all pores is a measure of the surface area S_{cum}, which can be compared to the BET value S_{BET}. Since the BET method is based on the amount of vapor adsorbed onto a surface at P/P_0 values in the range of 0.05 to 0.30 whereas the above method involves condensation into pores at P/P_0 values primarily higher than 0.30, the two methods are essentially independent; however, S_{cum} is less reliable.

A reproducible hysteresis is usually observed, so two different values of S_{cum} will be calculated depending upon whether the adsorption branch or the desorption branch is used. For an incremental volume of nitrogen adsorbed, assigning it to the desorption branch identifies it with a lower value of P/P_0 than if it is assigned to the adsorption branch. Hence, from the Kelvin equation, the calculated area will, in general, be greater if the desorption branch is used instead of the adsorption branch. The desorption curve may be preferred, for various reasons. Some degree of supersaturation may be needed before a pore will fill with liquid, so that thermodynamic equilibrium, assumed in the Kelvin equation, would not be obtained in the adsorption branch. Also, the contact angle can be affected by surface properties, e.g., contamination, so this might change after being in contact with adsorbed vapors.

Even if the above considerations do not apply, S_{BET} will not necessarily equal S_{cum} from either the adsorption or desorption curve since in either case S_{cum} will depend upon the true pore geometry. As an example, consider two idealized pores, a cylinder of radius r and length L and an "ink bottle" (sphere) of neck of radius r and body of radius r', the two pores having equal volumes. Both pores will empty at the same value of P/P_0 corresponding to the radius r. For equal volumes, $\pi r^2 L = (4/3)\pi r'^3$ or $L = 4r'^3/3r^2$. Let the above volume be assigned to a cylinder, the usual assumption. If the true pore shape were a sphere, the calculated area would be greater than the true value if $2\pi rL > 4\pi r'^2$, or by substitution, $(2/3)r' > r$. If the inequality were reversed, the calculated area would be less than the true value. It is thus seen that S_{cum} can be either greater than or less than the true area. For most purposes the BET area is used, but comparison with the two values of S_{cum} may help reveal details of pore structure, useful for other purposes.

The size of the largest pores that can be measured by this method is limited by the marked change of meniscus radius with pressure as the reduced pressure nears unity. Therefore the method is usually applied to pores of about 60 nm in diameter and less, so surface area in pores larger than 60 nm would be neglected. Data reported at the upper end of this range become increasingly less accurate, depending upon the instrument and experimental procedures, and may be misleading. This procedure gives satisfactory results for high-area materials such as some aluminas and silica gels in which typically almost all of the area is in pores of the order of 30 nm or less; but to measure pore-size distribution for pores of the order of 60 nm and higher, it is necessary to use mercury porosimetry.

The smallest pore sizes that can be determined by the nitrogen adsorption method are about 1.5 to 2 nm in diameter. Although measurements corresponding to lower pore sizes can be reported, it becomes increasingly uncertain as to the extent to which the Kelvin equation applies. The derivation assumes that the properties of the condensed phase are the same as those of a bulk liquid, yet the concepts of surface tension or of a hemispherical surface must break down as pore size becomes of the order of magnitude of the size of adsorbate molecules. If such fine pores do in fact exist, capillary condensation can occur in the P/P_0 range used for BET measurements, thus causing S_{BET} to be too high. Also, a straight line may not be obtained on a BET-type plot. The data in Table 5.2 (Inkley, 1958) indicate the conditions under which this might occur.

Hysteresis effects These frequently occur in physical adsorption on porous substances. The desorption curve always lies to the left and above the adsorption curve: i.e., at a given value of P/P_0 more vapor is condensed in the pores if a specified pressure has been approached from a higher pressure than from a lower pressure. In some cases supersaturation may be required before condensation can begin, or the contact angle may change after being in contact with adsorbed vapors, but the principal explanation lies in the complex pore geometry of typical porous materials. As one simple example,

Table 5.2 Critical pore diameter as a function of reduced pressure*

Relative pressure, P/P_0	Thickness of adsorbed layer, nm	Kelvin radius, r_K, nm	Critical pore diameter = $2(t + r)$, nm
0.265	0.535	0.715	2.5
0.168	0.465	0.535	2.0
0.130	0.440	0.460	1.8
0.090	0.400	0.400	1.6
0.058	0.355	0.345	1.4

*Inkley (1958).

if pores consisted of cylindrically shaped capillaries open at both ends, adsorption would occur on a surface curved in only two dimensions and pores would thus become filled in the adsorption branch when P/P_0 reached the value given by

$$\ln \frac{P}{P_0} = \frac{-2\sigma V_m \cos\theta}{RT(2r_K)} \tag{5.10}$$

On the desorption branch, however, the meniscus at the pore opening is curved in three dimensions instead of two, and therefore it empties at a value of P/P_0 given by the Kelvin equation.

Another geometry which can lead to hysteresis effects is that of an "ink bottle" in which it is visualized that many pores will have necks smaller than the interior. (This is named after the old-fashioned round container with a narrow throat.) Condensation will not begin with increased pressure until it exceeds that corresponding to the effective radius of the interior. But after the pore is filled, pressure must be decreased to a value corresponding to the minimum radius of the neck before the pore will empty. Various pore structures are possible, and the kinds of hysteresis loops that might be obtained from each are discussed by de Boer and others.

In a series of interconnected pores of varying cross section, in general most of the pore volume empties at a pressure corresponding to the radius of the largest circle that can be inscribed in the throat. In predicting rates of diffusion within porous catalysts the restrictions contribute a resistance which is not offset by the enlargements, so for this purpose it is throat rather than cavity dimensions which are of importance and the desorption curve may be a better measure of pore size. The mercury penetration method, discussed below, also relates quantity of mercury forced into pores to throat dimension, so results from that method might be expected to agree more closely with desorption than with adsorption curves. If an interior, interconnected porous structure has several throats in parallel, both the desorption isotherm and the mercury porosimeter method relate the interior volume to the dimension of the largest throat, which again in a qualitative sense parallels the kind of relationship governing access of diffusing molecules.

5.3.2 Mercury Penetration

This method is based on the behavior of nonwetting liquids in capillaries, again usually assuming that pores can be represented as cylinders. If the contact angle between liquid and solid, θ, is greater than 90°, the interfacial tension opposes the entrance of liquid into the pore. This can be overcome by external pressure. For a cylindrical pore the force opposing entrance to the pore acts along the circumference and equals $-2\pi r\sigma \cos\theta$. The external pressure, which opposes this force, acts over the entire pore cross-sectional area and equals $\pi r^2 P$. At equilibrium the two forces are equal and

$$r = \frac{-2\sigma \cos\theta}{P} \tag{5.11}$$

The application of this principle for determination of pore-size distributions was developed by Ritter and Drake (1945) and Drake (1949). They found that the contact angle between mercury and a wide variety of materials such as charcoals and metal oxides varied between 135 and 142° and suggested that an average value of 140° could be used in general. Taking the surface tension of mercury as 0.48 N/m and an average value of the wetting angle as 140°, Eq. (5.11) reduces to

$$r = \frac{7500}{P} \tag{5.12}$$

where r is the pore radius in nanometers and P is the pressure in atmospheres. Different investigators have used slightly different values for surface tension and wetting angle, resulting in values of rP varying from 7500 to 6000. The method is sometimes referred to as the *Barrett-Joyner-Halenda (BJH) method* (Barrett et al., 1951).

The pore radius into which mercury is forced is inversely proportional to the pressure, so the smallest pore sizes that can be detected by this method depend on the pressure to which mercury can be subjected in a particular apparatus. Pore radii down to about 1.5 nm can be determined with available commercial apparatus. Only a few comparisons of the mercury porosimetry and nitrogen desorption methods have been published for the region in which they overlap, but these show generally good agreement.

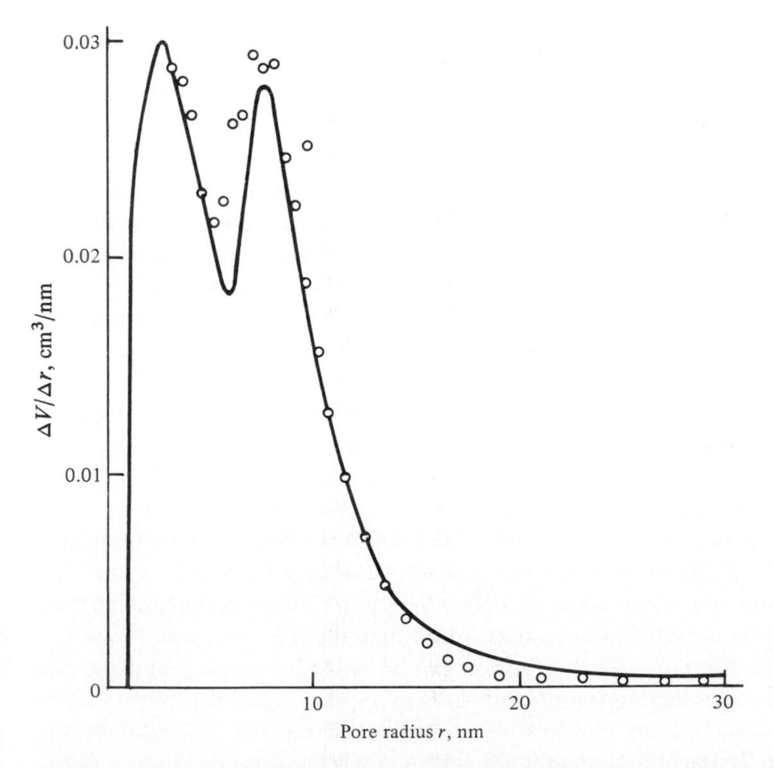

Figure 5.3 Pore-size distribution in a bone-char sample as calculated from a nitrogen adsorption isotherm. Experimental points were obtained on the same bone char by the mercury porosimeter method. (*Emmett, 1962.*)

Studies by Joyner et al. (1951) showed that for a variety of charcoals, over the pore-radius range up to 30 nm the pore-size distribution curve as determined by nitrogen desorption agreed very closely with that found by the mercury porosimeter method. Indeed, a double peak in one sample having maxima at about 2.5 and 7.5 nm was traced out by both methods (Fig. 5.3). Brown and Lard (1974) also showed good agreement between the two methods for aluminas, silicas, and related substances if the pore volume was less than about 1.2 cm^3/g. However, with a high-volume silica, the pressure of the mercury caused some of the large pore walls to collapse, forming smaller pores. This led to a spuriously low average pore radius but little change in the surface area. Mercury porosimetry studies can also show hysteresis, but relatively little has been published on interpretation of these effects.

In applying this knowledge of pore-size distribution to the prediction of diffusion rates in catalysts, the limit in accuracy is due in most cases not to the accuracy with which the pore-size distribution is known but rather to the uncertainty concerning the proper formulation of diffusion rates in such a complex structure.

5.3.3 Examples of Pore-Size Distributions

The pore-size distribution of a carrier or a catalyst is usually presented in the form of a plot of the increment of pore volume per increment in pore size, versus pore size. Fig. 5.4 gives an example for a particular alumina. The upper curve shows the adsorption isotherm, and the lower curve the calculated pore-size distribution. The upper curve at values of $P/P_0 < \sim 0.4$ is essentially the t curve. This particular alumina had a narrow pore-size distribution, most of the pores being about 2 nm in radius.

Many supports and catalysts have a more or less well-defined bimodal pore-size distribution in which the larger pores are the residual spaces between particles existing or formed in the catalyst preparation, and the fine pores are developed within the particles by the calcination and reduction procedures. The size of the large pores is decreased by increased tabletting pressure, but the fine pores are essentially unaffected. An example of such a bimodal pore-size distribution is shown in Fig. 5.5 for a Fe_3O_4-Cr_2O_3 commercial catalyst used for the water-gas shift reaction (Bohlbro, 1966), obtained by splicing together nitrogen adsorption and mercury porosimeter measurements. The micropore sizes cluster at about 3 nm radius and the macropores at about 100 nm. Figure 5.6 shows a similar plot for another iron-chromia catalyst before and after reduction, which shows a slight increase in micropore size upon reduction.

In a catalyst having a bimodal pore-size distribution the fine pores are frequently in the range of 1 to 10 nm in radius and account for most of the surface area. The macropores typically are in the range of 10^2 to 10^3 nm. The latter can provide rapid mass transfer into the interstices of the mass from which the fine pores lead to the ultimate reaction sites. The pore-size distribution can be varied by a variety of processing techniques. A wide size distribution of pore sizes in a gelled cracking catalyst may be produced if hard microporous powders are incorporated into the gel before drying. Extruded catalysts frequently contain a network of fine pores or cracks from drying and calcining operations. In a pelleted or extruded catalyst an organic material can be

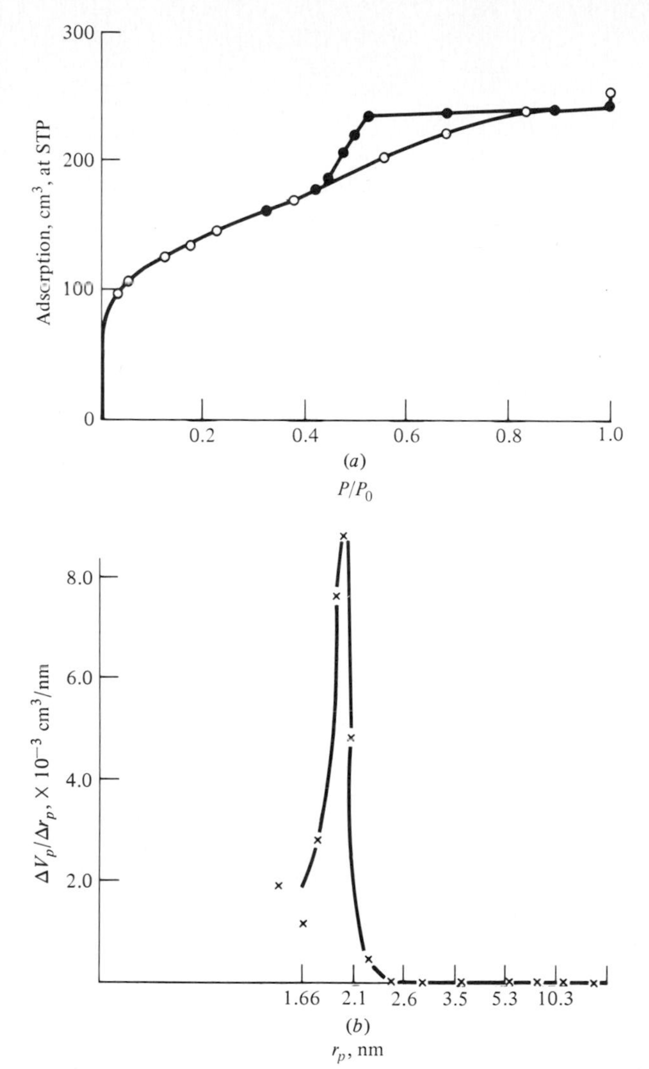

Figure 5.4 Pore-size distribution from the adsorption isotherm of nitrogen at −195° on alumina. (*a*) Adsorption isotherm. ○, adsorption; •, desorption. (*b*) Pore-size distribution curve. (*Harris, in Gregg and Sing, 1967, p. 169.*)

incorporated into the mix and then burned out after the final particles are shaped and dried.

5.4 MECHANICAL PROPERTIES

The resistance to crushing (e.g., in packed beds), to attrition (e.g., in a fluidized-bed reactor), and to breakup during shipping and reactor loading are important catalyst characteristics, but there are as yet no accepted standard tests analogous to the common use of the BET method for determining total surface area. This is in part because

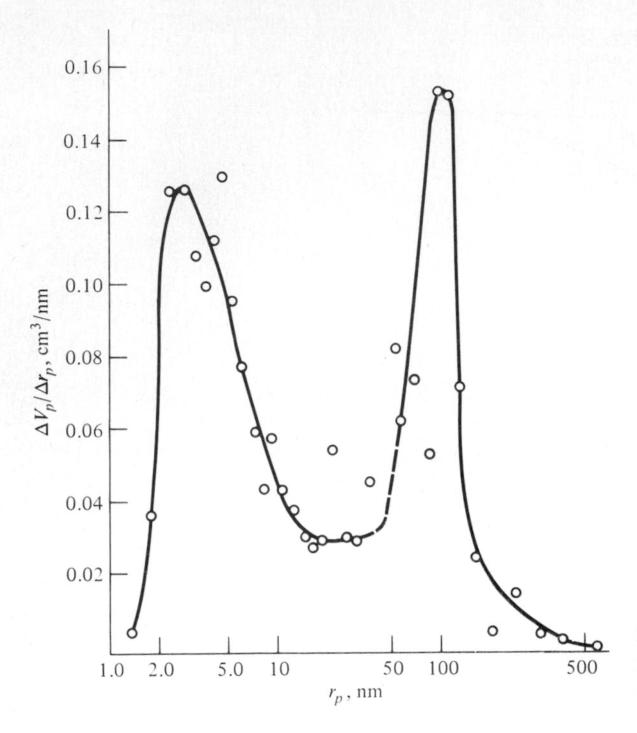

Figure 5.5 Pore-size distribution for a commercial Fe_3O_4-Cr_2O_3 catalyst. (*Bohlbro, 1966, p. 14.*)

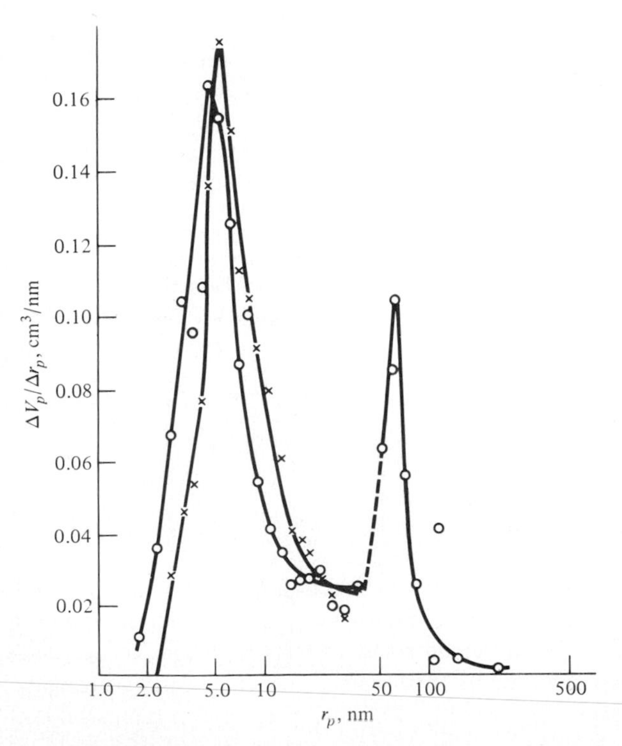

Figure 5.6 Pore-size distribution for a commercial iron-chromia catalyst, before (O) and after (X) reduction. (*Bohlbro, 1966, p, 16.*)

much reduction in strength may come about during service, as by chemical attack or thermal shock that is highly specific to the process. Typically, catalyst manufacturer and user agree upon a specified procedure, which for a specific application has been found from experience to lead to acceptable material. The American Society for Testing Materials (ASTM) has an active Committee on Catalysts (D32) that is concerned with development of standards for commercial purposes.

5.4.1 Crush Tests

Crush tests may be made in a variety of ways on individual pellets or on beds of catalyst (Beaver, 1974; Dart 1974; Adams, et al. 1974). A generally useful procedure is to slowly force a piston down onto a bed of catalyst pellets of several hundred cubic centimeters volume and measure displacement as a function of applied pressure. The flow rate of a gas through the bed is sometimes measured simultaneously to indicate the degree of increased flow resistance caused by catalyst breakup, since formation of a small amount of fines may be more deleterious to operation than a larger amount of partially fractured pellets. The increase in pressure drop with a fixed rate of gas flow or, more easily, the decrease in flow rate of a gas supplied at fixed pressure may be observed. The crush test may be applied to both fresh catalyst and catalyst that has been subjected to an accelerated deactivation procedure designed to simulate long-time use in the plant. The catalyst on the bottom of a fixed bed in operation must withstand not only the force from the catalyst above it but also the force applied to it by the pressure drop through the bed in use, and the second force may be much greater than the first. However the weight of the catalyst bed is not applied completely to the catalyst at the bottom but is distributed to a greater or lesser degree to the reactor walls by an arch effect. Pellets are usually made with a height/diameter ratio of about unity since long, thin pellets are generally weaker.

Pellets may also be crushed individually for testing. The radial strength is the important factor since it is usually much less than the axial strength so the force is usually applied to opposite sides of the curved surface.

Abrasion and attrition resistance are of importance for fixed-bed catalysts to minimize breakage that may occur on handling. A representative test is to place a quantity of pellets in a rotating horizontal cylinder, 15 to 30 cm in diameter, equipped with a single flight (baffle) which raises pellets and drops them, or a quantity of pellets may be dropped one at a time from a specified height, for example, 3 m. For a given catalyst composition maximum crush and attrition resistance typically is achieved by preparing a relatively dense pellet of low porosity, but this will reduce the effective diffusivity, so an optimum must be sought between this characteristic and mechanical properties.

For fluidized-bed reactors, attrition resistance of the fine catalyst is of paramount importance, and this is usually determined in an air-jet attrition test.

5.4.2 Particle-Size Distribution

The average particle size and particle-size distribution can markedly affect fluidization characteristics in a fluid-bed reactor and affect settling and filtering characteristics in a

Table 5.3 Tyler standard sieve series (TSSS)

Sieve size, Tyler designation	Sieve opening, mm
4	4.76
8	2.38
12	1.41
20	0.841
35	0.420
60	0.250
80	0.177
100	0.149
150	0.105
200	0.074
270	0.053
325	0.044

slurry reactor. Size fractions of a powder are commonly separated for measurement by shaking the powder through a stack of sieves of standard construction, the opening size decreasing from top to bottom. In the United States this is the Tyler Standard Sieve Series in which the sieve is identified by the nominal number of meshes per linear inch. The finer the sieve, the finer the wire diameter used. Table 5.3 gives the sieve opening for a number of sieves commonly used. A 60- to 80-mesh powder, for example, consists of particles that pass through a screen with openings of 0.250 mm but are retained on a screen of openings 0.177 mm. Since the finer particles in a container will settle towards the bottom upon handling, suitable procedures for obtaining a representative sample must be used.

Measurements of the particle sizes of interest in catalysis can also be made by microscopy (which is tedious), by sedimentation, or by electronic counting devices.

5.5 SELECTED INSTRUMENTAL METHODS

An enormous variety of instruments are capable of revealing information of value to some aspect of catalysis. Many instrumental methods are still evolving, and their capabilities are being explored. The following is intended to give a brief introduction to a few of the better-developed techniques which have been of particular value in working with technical catalysts. Methods of surface characterization are of particular importance and are treated in detail in the book edited by Kane and Larrabee (1974). Experimental methods for characterizing surface structure are also reviewed by McRae and Hagstrum (1976).

5.5.1 Microscopy

Light microscopy helps to characterize materials and to define problems by revealing such features as size, shape, surface markings, occurrence of occlusions or other discontinuities, and color. It is particularly helpful to identify the characteristics of single particles and to determine particle-size distribution.

Transmission electron microscopy The limit of resolution for a microscope is proportional to the wavelength of the illumination. For light microscopy this is about 200 nm; an electron microscope is usually used when resolution for sizes smaller than this is required. Electrons are emitted from an electron gun, which is a heated pointed cathode (filament), accelerated through two electrodes, the second of which is an anode. The electrons then pass through a condenser system, the specimen, and a magnetic lens system. Size or microstructure can be determined for materials in the range of about 1 nm to 30 μm. In general, any solid material can be studied, but if the section is thicker than about 20 nm (the penetrating power of 100-kV electrons), and one wishes to see more than an outline, then the specimen must be sectioned or replicated. A very thin section, for example, 30 to 40 nm thick, can be prepared by mounting the sample in an epoxy resin, cutting it slowly with a diamond knife, and floating the specimen off in a liquid. A replica may be made readily by putting a layer of polymer on a solid and depositing carbon or other material from a vapor onto the replica. A shadowing technique in which vapors are deposited from an angle provides an image looking much like a map replica.

The depth of field is typically about 150 nm, so usually the entire specimen will be in focus. Some precautions to note are:

1. The sample must be stable under vacuum.
2. The electron bombardment can cause volatile material to disappear.
3. One must watch out for various artifacts. Impurities such as dust, etc., may be incorporated into the sample during its preparation, and the process of slicing a thin layer may cause distortion of the structure.

Transmission electron microscopy is useful for indicating the size of supported metal crystallites and changes in their size, shape, and position with catalyst use. Replicas are useful for obtaining information on pore structure. Figure 5.7 shows an electron micrograph of a supported nickel catalyst, the nickel crystals appearing as small dark spots and the light material being the support. The nickel crystallites are in the size range of 10 to 50 nm. Figure 5.8 shows a thin section of an iron catalyst used for ammonia synthesis. The pores are relatively parallel to one another and are of the order of tens of nanometers. Figure 5.9 is a thin section at higher magnification. The light areas are pore mouths, and the pore cross sections appear to be quite irregular and, as in Fig. 5.8, vary considerably in size.

Scanning electron microscopy (SEM) In the scanning microscope the electron spot focused on the sample is moved over a small area by means of a set of deflecting coils. This area is displayed highly magnified on a cathode ray tube (CRT) by causing the currents passing through the scanning coils to pass through the corresponding deflecting coils of the cathode ray tube while the electrons emitted from the sample are collected, amplified, and used to modulate the brightness of the CRT. As of 1980, the

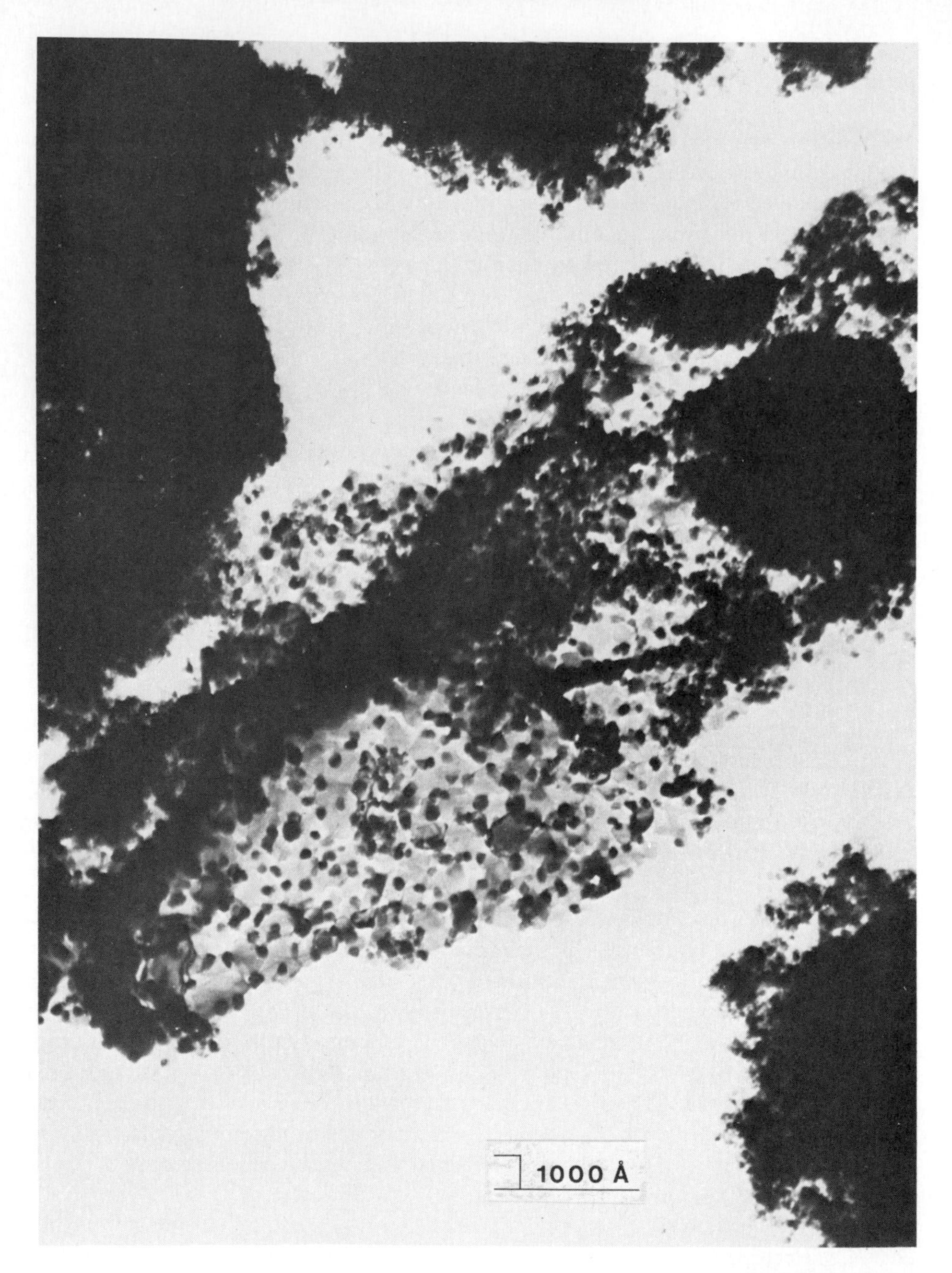

Figure 5.7 Electron micrograph of Topsøe RKNR catalyst for steam reforming of naptha, about 25% nickel on magnesia. (*Rostrup-Nielsen, 1975, p. 41.*)

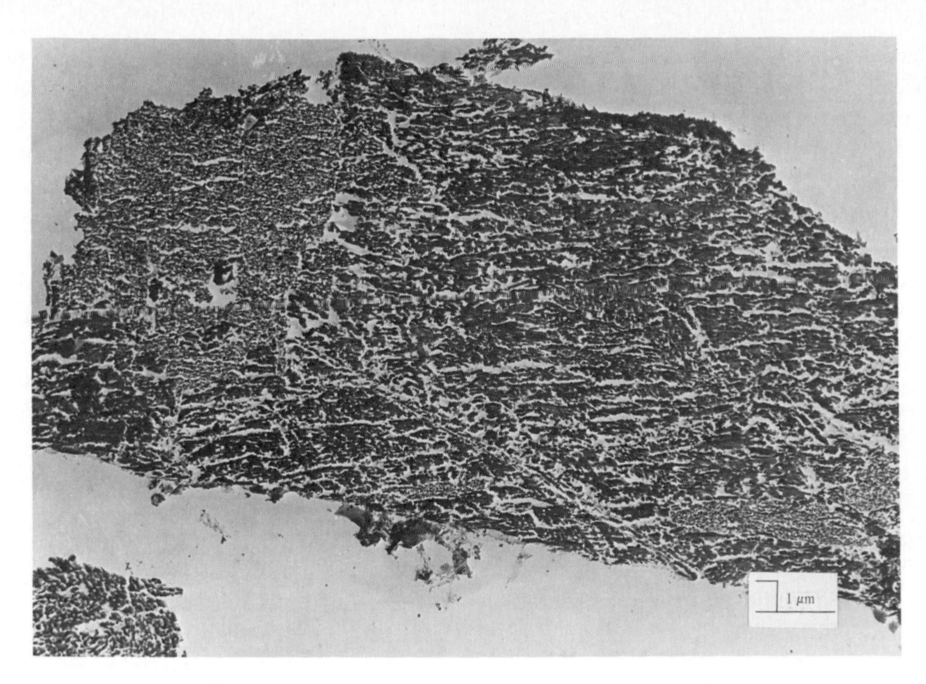

Figure 5.8 Electron micrograph of thin section of Topsøe ammonia-synthesis catalyst KMIR. (*Nielsen, 1968.*)

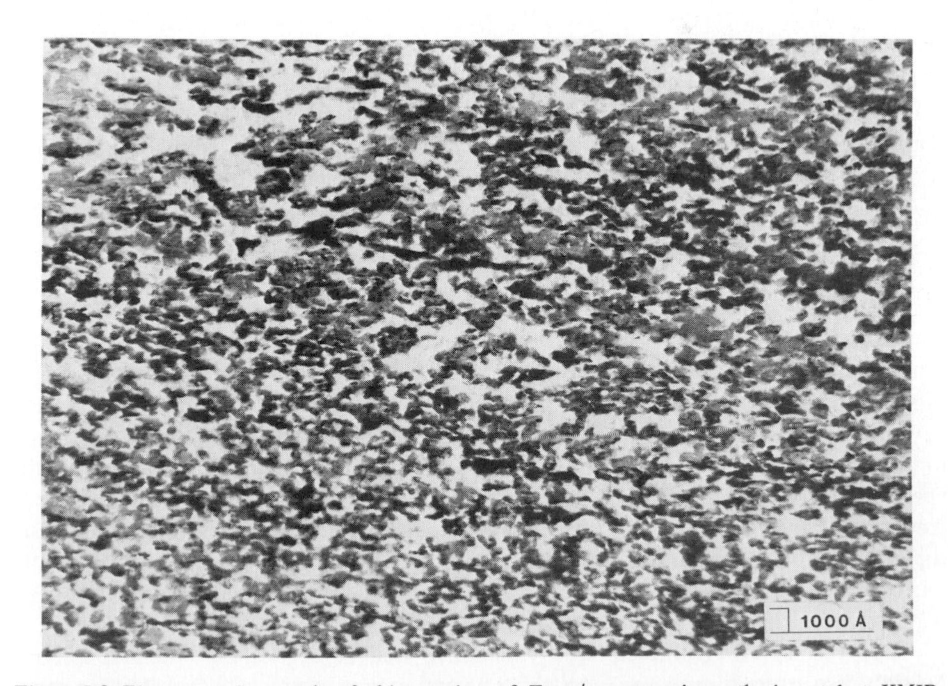

Figure 5.9 Electron micrograph of thin section of Topsøe ammonia-synthesis catalyst KMIR. (*Nielsen, 1968.*)

resolution of representative commercial instruments allowed markings of the order of 50 nm or greater to be distinguished. Below this size, the degree of definition is determined more by the characteristics of the sample rather than by the instrument; for example, a charge cannot be allowed to build up on the sample–it distorts the image. Some samples are conductive, but if a sample is not, a conductive layer must be applied.

Scanning microscopy has been used primarily for examination of the topology of catalyst surfaces, for example, characterization of platinum gauzes before and after use in a reactor and sintering of larger supported metal crystallites. By special procedures for enclosing the specimen it is also possible to observe dynamic events in a controlled atmosphere environment at total pressures up to a substantial fraction of an atmosphere. By such means the mobility of small particles and the growth of carbon filaments from metal surfaces in a reducing environment (Sec. 6.5) have been observed.

The incident electron beam causes the emission of secondary electrons, Auger electrons, and characteristic X rays that can in principle be analyzed to provide additional information about the same area of the sample on which a SEM picture is obtained. There has been much interest in recent years in instruments that combine several analytical methods. The gain from obtaining multiple information on a single specimen or from a small area on a heterogeneous sample is purchased at some reduction in performance from a single purpose instrument. One instrument may combine an electron gun and an X-ray source, and be capable of use for X-ray photoelectron spectroscopy (XPS or ESCA–electron spectroscopy for chemical analysis), for Auger electron spectroscopy (AES), and for scanning electron microscopy. AES has been used primarily for analysis of elements and is very sensitive to the top two to three layers of the surface. The exciting electrons can be focused down to 5 μm or less for spatial resolution of the AES spectrum. The X-ray excitation involved in ESCA (see below) cannot be focused very readily, and the analysis area is typically several millimeters across in order to gain sensitivity. However it provides chemical bonding information, about which little is revealed from AES.

5.5.2 Electron Spectroscopy for Chemical Analysis (ESCA)

This technique is also known as *X-ray photoelectron spectroscopy (XPS).* A sample is bombarded with monochromatic X rays, and photoelectrons are ejected from various atomic shells. (In a sense ESCA is the reverse of electron microprobe analysis.) The intensity of electrons emitted from the first few atomic monolayers is high, but that emitted from subsequent layers falls exponentially as a limit of about 10 nm is approached. The electrons are resolved for their kinetic energy, which gives their binding energy. This is characteristic of the atomic number of the element to which the electron is bound. Moreover, these binding energies are sensitive to the overall charge on the atom, and hence chemical shifts can also be observed. ESCA can be combined with ion sputtering to gradually remove surface layers and thus provide a description of composition as a function of depth.

ESCA is a powerful technique which can analyze for all the elements except possibly hydrogen. It gives the most accurate and precise atomic electron binding energies yet obtained, and thus it yields structural data in terms of the valence of the oxidation state as well as atomic charge densities. Also the analysis is chemically nondestructive. However it may not be as accurate for quantitative analysis as is frequently desired, especially for insulating solids.

5.5.3 Auger Electron Spectroscopy (AES)

Auger electrons or fluorescent X rays are emitted when an ionized atom returns to the neutral state after electron bombardment, and these may be analyzed. Data are typically obtained in the form of intensity of electrons emitted as a function of binding energy. The location of peaks may be used to indicate, for example, whether sulfur is present in the form of sulfite or sulfate. With a Co-Mo/Al_2O_3 catalyst, AES has been used to compare the concentration of cobalt on the surface to that in the bulk and the same for molybdenum, thereby revealing that in a particular catalyst preparation the ratio of Co/Mo varied between surface and bulk, although in the usual commercial catalyst it does not. In a Co/Al_2O_3 catalyst it revealed that the extent to which cobalt was present as an oxide in contrast to an aluminate was a function of preparation conditions. With a supported molybdenum catalyst, AES revealed the relative concentration of the +4, +5, and +6 valence forms as a function of the conditions to which the catalyst had been subjected.

5.5.4 Electron Microprobe

A beam of energetic electrons is focused onto the surface of a specimen. The inner-shell (K, L, M) ionizations of the atoms produce characteristic X rays that are representative of the nature of the elements, and the intensity of the X ray is proportional to concentration. Thus, an X-ray image can be produced that gives the nature and the distribution of a particular element to a high degree of resolution. The electron beam penetrates 200 nm or more, so the composition is averaged through most of a specimen as typically prepared. The limitation of degree of resolution is not the focusing of the electron beam but rather the scatter of emitted radiation and/or electrons. This is about equal to 1 μm or the thickness of sample, whichever is less; consequently the resolution can be as small as a few tens of nanometers. The output may be presented in the form of a raster scan (map) or an analysis profile (line scan).

As of 1980 the electron microprobe can be used for all but the very first elements. It is particularly useful for obtaining a profile of the distribution of heavy metals through a catalyst particle by working with a thin slice of the material, or for detecting the buildup of poisons and their distribution through individual pellets. It may also be used as an analytical method on a ground and representative sample of a catalyst, which may be as small as 1 μm^3. Electron microprobe analysis has been applied extensively to characterization of the promoted iron catalysts used in ammonia synthesis, to show how certain promoters dissolve in magnetite during fusion and migrate to crystallite boundaries during reduction to iron metal. Figure 8.12 (Sec. 8.10) shows its use to identify segregation of rhodium at the surface of a platinum-rhodium alloy gauze, used for the catalytic oxidation of ammonia.

An example of the power of the method is shown by the series of X-ray images

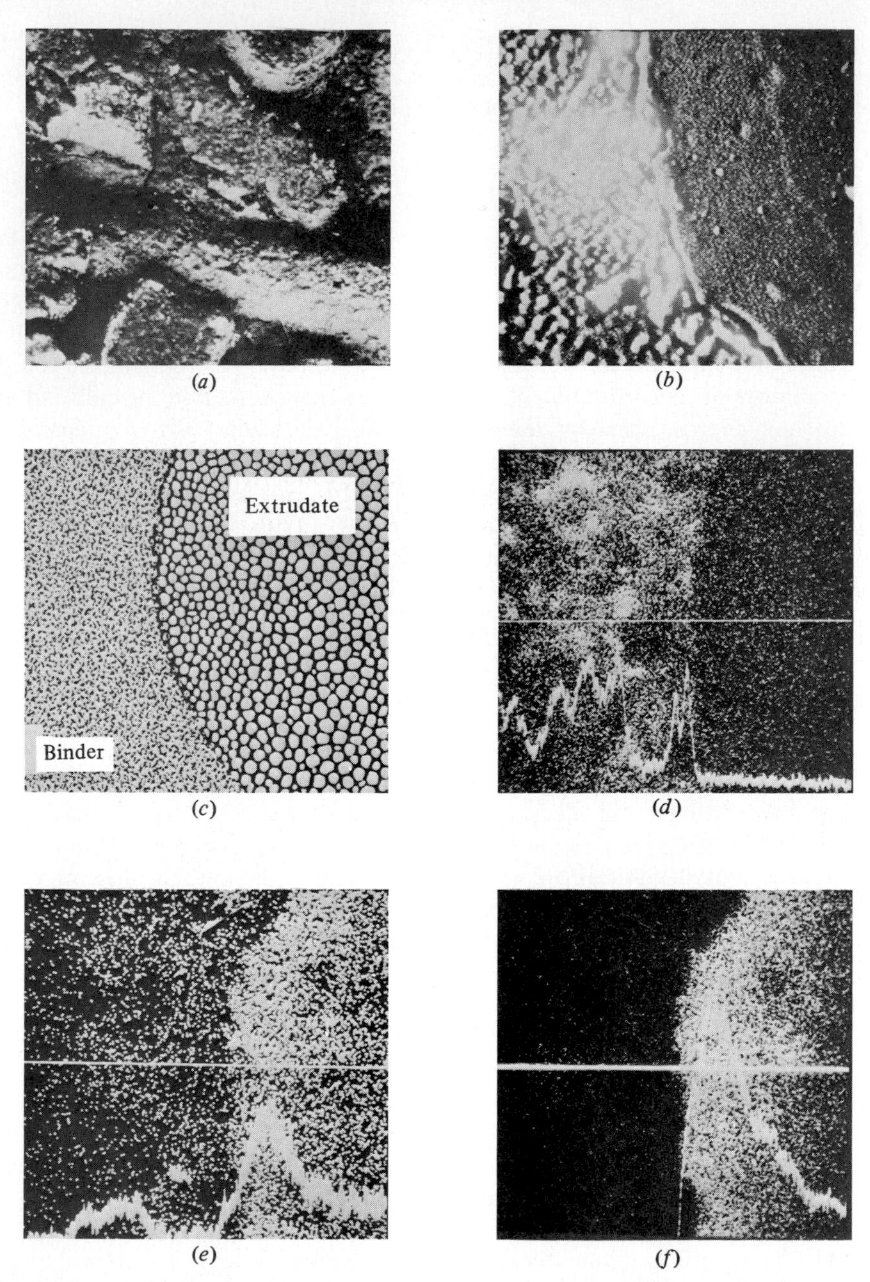

Figure 5.10 X-ray images by electron microprobe of a plugged catalyst section of a pilot plant reactor (see text). All images are of the same specimen. (*a*) Optical micrograph. (*b*) Back-scattered electron image. (*c*) Drawing of (*b*). (*d*) Fe X-ray image. (*e*) Ni X-ray image. (*f*) V X-ray image. (*g*) S X-ray image. (*h*) Na X-ray image. (*i*) Cl X-ray image. (*j*) Co X-ray image. (*k*) Mo X-ray image. (*l*) Al X-ray image. (*Work conducted by and photographs courtesy of G. W. Bailey, Exxon Research and Development Laboratories, Baton Rouge, La.*)

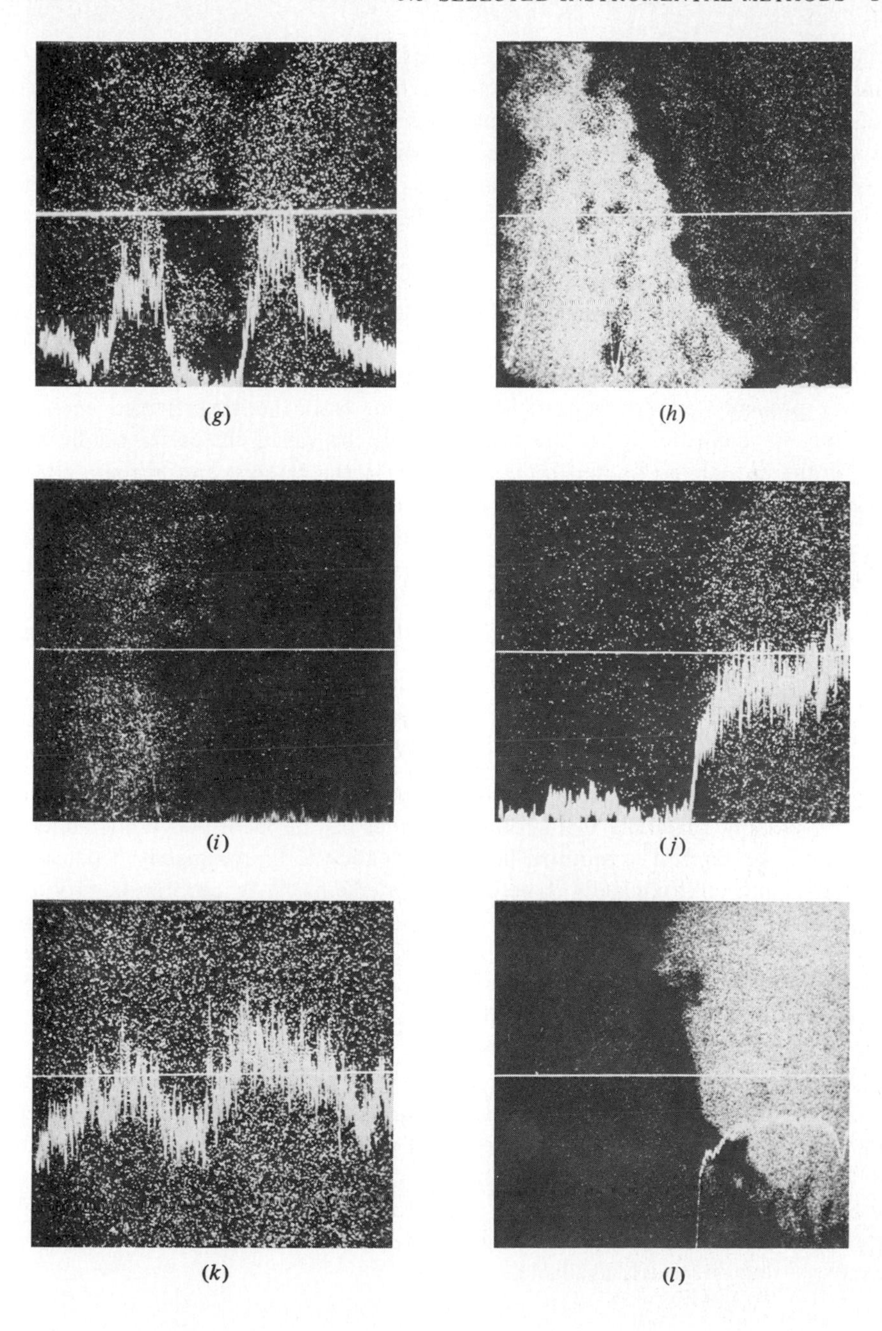

(*g*) (*h*)

(*i*) (*j*)

(*k*) (*l*)

for different elements shown in Fig. 5.10. These were taken and used to determine the causes of plugging of a pilot plant reactor packed with Co-Mo/Al_2O_3 extrudates and utilized for hydrodesulfurizing a residual crude-oil fraction. A section of the plugged bed was ground flat, coated with a conductive layer, and examined directly. Each X-ray map has superimposed on it an analysis profile showing the variation of concentration of the specified element along the straight line in the middle of the picture.

The plug was found to consist of a mass of catalyst extrudates embedded in a matrix of fine material acting as a binder. In Fig. 5.10, *a* is an optical micrograph, *b* a back-scattered electron image showing the topography of the sample, and Fig. 5.10*c* a drawing of *b* showing the boundary between binder and catalyst extrudate in *b* and subsequent X-ray images. The binder is seen to have a high concentration of iron (*d*), probably present as particulates, judging from the irregularities. Nickel and vanadium are deposited from the residual feed, the vanadium preferentially at the edge of the extrudate (*f*), and the nickel both in the catalyst and in the binder (*e*). Sulfur is deposited irregularly (*g*). The binder contains considerable NaCl (*h*, *i*). Cobalt does not appear in the binder (*j*); therefore the plugging was not caused by ground catalyst. Molybdenum is almost uniform inside and outside the catalyst (*k*), and therefore must have migrated out of the catalyst. The aluminum image (*l*) shows a sharp boundary at the catalyst surface, confirming the conclusion from *j* that the plugging was not caused by ground catalyst.

5.5.5 X-Ray Diffraction Crystallography

X-ray diffraction may be used to obtain information about the structure and composition of crystalline materials. Common compounds can be identified using tabulations of reference patterns. The minimal limit of detection is approximately 5 percent for compounds and approximately 1 percent for elements. With calibration procedures it is possible to obtain quantitative information and thus determine the approximate amount of a particular phase in a sample. Phases for which reference patterns are not available can sometimes be identified by other X-ray procedures. A change in the cell dimensions sometimes indicates the incorporation of an isomorphous material, for example iron into a cobalt nickel molybdate.

The mean crystallite size of a material can also be determined from the broadening of an X-ray diffraction peak. The line broadening is inversely proportional to crystallite size, and can be used to give the microcrystalline size in the range of about 3 to 10 nm.

Small-angle scattering (about 0.5° from the incident beam) may be used to give the particle-size distribution in the range of 5 to 100 nm, but as of 1980 the apparatus is expensive and not widely available.

5.5.6 Calorimetry

Calorimetry may be employed to observe various types of chemical transformations that are accompanied by significant energy changes. For example, if a catalyst or catalyst support undergoes several irreversible phase changes at different temperatures, such as alumina does, calorimetry may indicate the maximum temperature to which the material has been subjected. With a poisoned catalyst, the disappearance of a deposit at a low temperature may indicate that the poison is organic rather than inorganic. A *differential scanning calorimeter* may be used with small samples.

5.5.7 Gravimetric Methods

The change in weight of a catalyst with changing experimental conditions can be used for a variety of studies. Modern microbalance instruments can be used with catalyst samples of from a few milligrams up to gram quantities and can detect changes in weight of the order of 0.05 mg or less. Measurements can be made at temperatures up to 1000°C, which can be programmed, at constant or variable pressures, and under static or flow conditions. Massoth (1972) described representative apparatuses and examples of uses.

The method has been applied for adsorption-desorption studies and has been particularly useful for studying the rate of coking, dehydration, sorption of poisons, catalyst regeneration, etc., as a function of reaction conditions. It is also useful for catalyst characterization, e.g., of oxidation catalysts in which weight gain or loss may reveal the state of oxidation and hence stability as a function of environment. In many instances gravimetric methods are combined with other studies to develop a more comprehensive understanding of the processes under observation.

REFERENCES

Adams, C. R., A. F. Sartor, and J. G. Welch: *AIChE Symp. Ser.*, **70**(143), 49 (1974).
Adamson, A. W.: *Physical Chemistry of Surfaces*, 3d ed., Wiley, New York, 1976.
Anderson, J. R.: *Structure of Metallic Catalysts*, Academic, New York, 1975.
Barrett, E. P., L. G. Joyner, and P. C. Halenda: *J. Am. Chem. Soc.*, **73**, 373 (1951).
Beaver, E. R.: *AIChE Symp. Ser.*, **70** (143), 1 (1974).
Benesi, H. A., R. V. Bonnar, and C. F. Lee: *Anal. Chem.*, **27**, 1963 (1955).
Bohlbro, H.: *An Investigation on the Kinetics of the Conversion of Carbon Monoxide with Water Vapour over Iron Oxide Based Catalysts*, Gjellerup, Copenhagen, 1966.
Brown, S. M., and E. W. Lard: *Powder Technol.*, **9**, 187 (1974).
Brunauer, S., L. S. Deming, W. S. Deming, and E. Teller: *J. Am. Chem. Soc.*, **62**, 1723 (1940).
——, P. H. Emmett, and E. Teller: *J. Am. Chem. Soc.*, **60**, 309 (1938).
Cranston, R. W., and F. A. Inkley, *Adv. Catal.*, **9**, 143 (1957).
Dart, J. C., *AIChE Symp. Ser.*, **70** (143), 5 (1974).
de Boer, J. H., B. C. Lippens, B. G. Linsen, J. C. P. Broekhoff, A. van den Heuval, and Th. J. Osinga: *J. Colloid Interface Sci.*, **21**, 405 (1966).
Drake, L. E.: *Ind. Eng. Chem.*, **41**, 780 (1949).
Emmett, P. H.: *Adv. Catal.*, **1**, 65 (1948); *Catalysis*, vol. 1, Reinhold, New York, 1954, p. 31.
——: *36th Annual Priestley Lecture*, Penn. State Univ., University Park, 1962.
Everett, D. H., and R. H. Ottewill (eds.): "Surface Area Determination," *Proceedings of an International Symposium on Surface Area Determination, International Union of Pure and Applied Chemistry*, Butterworth, London, 1970.
Farrauto, R. J.: *AIChE Symp. Ser.*, **70** (143), 9 (1974).
Gregg, S. J., and K. S. W. Sing: *Adsorption, Surface Area, and Porosity*, Academic, New York, 1967.

Inkley, F. A., in D. H. Everett and F. S. Stone (eds.): *Structure and Properties of Porous Materials*, Academic, New York, 1958, p. 124.

Joyner, L. G., E. P. Barrett, and R. Skold: *J. Am. Chem. Soc.*, **73**, 3158 (1951).

Kane, P. F., and G. B. Larrabee (eds.): *Characterization of Solid Surfaces*, Plenum, New York, 1974.

Massoth, F. E.: *Chemtech*, May 1972, p. 285.

McRae, E. G., and H. D. Hagstrum in N. B. Hannay (ed.): *Treatise on Solid State Chemistry*, vol. 6A: *Surface Structure: Experimental Methods*, Plenum, New York, 1976, p. 57.

Nielsen, A.: *An Investigation on Promoted Iron Catalysts for the Synthesis of Ammonia*, 3d ed., Gjellerup, Copenhagen, 1968.

Parfitt, G. D., and K. S. W. Sing: *Characterization of Powder Surfaces*, Academic, New York, 1976.

Ritter, H. L., and L. E. Drake: *Ind. Eng. Chem., Anal. Ed.*, **17**, 782, 787 (1945).

Roberts, B. F.: *J. Colloid Interface Sci.*, **23**, 266 (1967).

CHAPTER

SIX

SUPPORTED METAL CATALYSTS

Metal catalysts are of particular interest for reactions involving hydrogen, such as hydrogenation, hydrogenolysis, and catalytic reforming. The latter two processes are discussed in Chap. 9. The use of iron for ammonia synthesis and for the Fischer-Tropsch reaction, and nickel for steam reforming and methanation is discussed in Chap. 10. Some representative industrial hydrogenation reactions are considered in Sec. 6.7. In many of these applications the metal is highly dispersed on a support, in aggregates so small that many or most of the atoms present are on the surface. A central question here is the extent to which the rate of a reaction per atom of exposed metal (turnover number) and, more importantly, the relative rates of parallel or sequential reactions are affected by the number and arrangement of sites on the catalyst. These in turn may be affected by particle size; by the use of alloys; by kinks, steps, and other crystal imperfections; and by the blockage of some of the sites by deliberately added poisons or by accumulation of carbonaceous deposits. There is also a possible role of the support in influencing the properties of the metal.

Metals also may be useful for oxidation reactions. Examples are the use of supported silver for oxidation of ethylene to ethylene oxide; platinum-rhodium wire gauze for the partial oxidation of ammonia to nitric oxide or for conversion of a mixture of methane, ammonia, and air to hydrogen cyanide, and bulk silver for the partial oxidation of methanol to formaldehyde. In oxidation reactions with metals, which characteristically involve relatively high temperatures, questions arise of crystal growth, migration, and the loss of active material by volatilization associated with chemisorbed oxygen and formation of metal oxides. These are discussed in Chap. 8

as well as Sec. 6.4. The discussion which follows relates primarily to uses of metals under reducing conditions.

6.1 METAL ACTIVITY

Maximum catalytic activity is associated with rapid but not too strong chemisorption of reactants. The heat of adsorption of gases such as oxygen, nitrogen, hydrogen and ammonia, ethylene, and acetylene on metals decreases in a continuous manner as one proceeds across groups VB, VIB, VIIB, and VIII_1, VIII_2, and VIII_3 of the metals in either the first, second or third transition series. Group IB metals do not readily adsorb some gases, e.g., hydrogen, since the dissociation of hydrogen is the rate-limiting process and this has a high activation energy. Hydrogen atoms can be chemisorbed under conditions where molecular hydrogen is not chemisorbed.

Hydrogenation reactions are carried out with a wide variety of reactants. The most active metallic catalysts are those in group VIII. This is readily rationalized in terms of adsorption effects; the strength of adsorption on, for example, group VB and VIB metals is too strong, that on IB metals is too weak or nonexistent. Within group VIII relative activity differences vary with the nature of the reaction and require a more detailed examination of the mechanism.

Hydrogenolysis of a hydrocarbon is the cleavage of a C—C bond accompanied by hydrogenation to form two molecules from one. Hydrogenolysis of paraffins is of great importance in petroleum processing (there termed as "hydrocracking") in which it is sometimes a desired reaction, as in commercial hydrocracking, and sometimes not desired, as in catalytic reforming. The mechanism probably involves as the first step adsorption of the paraffin on a group of sites with dissociation of a hydrogen atom. A commonly studied model of the reaction is that of ethane and hydrogen to form methane. The relative activity of metals changes more with position in the periodic table for ethane hydrogenolysis than it does for hydrogenation. Group VIII metals are again the most active. In the first transition series maximum activity is shown for the third subgroup element, nickel, but in the second and third transition series it is shown for the first subgroup element, ruthenium or osmium.

The distribution of primary products from hydrogenolysis varies substantially with the nature of the metal. For paraffins on nonnoble group VIII metals (Fe, Co, Ni) cracking occurs at terminal C—C bonds, producing successive demethylation of the carbon chain with accompanying formation of methane. With the platinum-group metals, the initial cracking pattern is relatively nonselective on platinum and iridium, but terminal C—C bonds are attacked almost exclusively on palladium and rhodium. The successive demethylation scheme does not occur on the platinum-group metals.

Paraffin isomerization reactions as carried out commercially require the presence of an acidic component. The reaction can also occur on platinum by itself, but it is not observed on other platinum-group catalysts in which hydrocracking is observed instead. Under industrial conditions utilizing a bifunctional catalyst, however, the purely platinum-catalyzed process does not contribute significantly to the total.

The above generalizations are developed in more detail by Sinfelt (1975) and by Dowden (1978).

6.2 METAL DISPERSION (PERCENTAGE EXPOSED)

The extent of dispersion is defined as the ratio of the number of surface metal atoms in a catalyst to the total number present. A value of unity means that all metal atoms are exposed to reactants. The IUPAC recommendation is that the term *percentage exposed* be used instead of *dispersion*, but the former term is not widely used at present. For crystals of platinum in the shape of regular octahedra, the fraction exposed is 0.78 for an edge length of 1.4 nm, 0.49 for an edge length of 2.8 nm, and 0.30 for an edge length of 5.0 nm. The fraction exposed for a commercial platinum or platinum alloy reforming catalyst can readily exceed 0.5. In contrast, even a very finely divided nonsupported metal catalyst has a low fraction exposed. For a metal particle 1 μm in size, the fraction exposed is about 0.001.

The average fraction exposed is measured most directly by determining the number of surface atoms present by selective chemisorption (Sec. 5.1.3), combined with a knowledge of the total amount of metal present. X-ray line broadening for determination of crystal size becomes too diffuse to be of much value below dimensions of about 2 to 3 nm. Small-angle scattering of X rays is useful, but the apparatus is expensive and not widely available. Transmission electron microscopy may give direct measurements of particle sizes and is a useful method for comparison with results obtained by chemisorption. Some assumption concerning particle shape is necessary. A spherical or cubical shape is usually taken, and this is generally consistent with other observations; but where there is strong metal-support interaction, the metal may be spread out in the form of a layer or "raft" rather than in a compact three-dimensional form.

Burton (1974) has shown that the structure of very small microcrystals is different than that of macrocrystals, so that different types of surface faces are present. The lattice parameter is slightly smaller (about 3 percent less for a 1-nm particle), and the microcrystals melt at low temperatures compared to those of bulk solids–typically at about one-half of the normal melting point for a cluster of 55 atoms (about 1 nm). The liquidlike behavior of very small particles may also be observed directly in an electron microscope. Even if the same crystal habit were preserved, it is evident that with particles roughly 2.5 nm or smaller a substantial and increasingly larger fraction of the total atoms present will be on edges and corners. Hence they will have lower coordination numbers than those in a crystal face. (For a regular octahedron of platinum atoms the fraction of surface atoms that are on edges or corners is 0.64 for an edge length of 1.4 nm and 0.32 for an edge length of 2.8 nm. Thus it is reasonable to expect that some reactions would change their character as particle size is reduced in this range.)

The term *structure-sensitive* (or *demanding*) is applied to a reaction whose specific activity (same as turnover number) varies with the percentage exposed or, more generally, with the structure of the active sites. These may be altered in other ways, as by alloying, introducing crystal imperfections, or poisoning (Boudart, 1969). This is in contrast to *structure-insensitive* (or *facile* reactions). For some reactions the active site may be a single atom, but for others it may involve several surface atoms whose arrangement relative to one another is critical. Hence methods of catalyst preparation may be much more critically important for structure-sensitive than for structure-insensitive reactions. A considerable number of studies now indicate that simple hydrogenation reactions on various metals in general are structure-insensitive but that reactions in which C—C bonds are broken, such as hydrogenolysis and skeletal isomerization, are structure-sensitive.

Calculations applying molecular orbital theory to small clusters of metal atoms also provide a theoretical perspective on the above. One may consider how the electronic structure is predicted to vary as one goes from an isolated atom to a small cluster or group of atoms. Studies show, for example, that a cluster of silver atoms approaches metallike properties at a grouping of 55 atoms. However, the values of such properties as the work function differ from those of the bulk metal and the cohesive energy density is only one-third of that expected for bulk material (Baetzold, 1976). Clusters of palladium differ from bulk palladium much as silver clusters differ from bulk silver. Measurements of the Curie temperature as a function of crystal size for nickel/silica show that this property varies with crystallite size over the range of about 1.0 to 10 nm. For crystallites larger than about 10 nm the magnetic properties are essentially the same as for bulk nickel (Sinfelt, 1972). (The *Curie temperature* is a transition temperature above which a ferromagnetic metal becomes paramagnetic.)

6.3 ALLOY CATALYSTS

Studies of alloy catalysts were of considerable interest in the 1950s in conjunction with the then burgeoning electronic theories of catalysis, but then fell into disfavor. More recently they have been revived, stimulated by the industrial importance of bimetallic catalytic reforming catalysts and a more sophisticated fundamental understanding of the structure of alloys and the factors affecting the distribution of alloy components between the surface and the bulk. Of particular interest have been studies of a binary alloy of an active metal and of an essentially inert metal such as nickel-copper alloys and, more generally, of a mixture of a group VIII metal and a group IB metal.

6.3.1 Surface Composition

The nature of the surface composition determines the catalytic properties of an alloy, and the surface composition may be much different than that of the bulk. In nickel-copper alloys copper is highly segregated at the surface, as shown by hydrogen chemisorption studies. Hydrogen is strongly adsorbed on nickel but not on copper. Copper is the dominant element on the surface even with copper-nickel alloys containing as little as 5 at. % copper (Sinfelt, 1975).

The surface composition of alloys can also be observed by Auger electron spectroscopy or photoelectron spectroscopy; but the observed electrons escape from layers below the surface, so the results represent a weighted average of a layer a number of atoms deep (Sec. 5.5). Work function measurements are a good representation of the surface, as is chemisorption, but the alloy components must have substantially different work functions or chemisorption characteristics for the measurements to be useful.

The degree of surface enrichment can also be a function of particle size for some alloys. Sinfelt has shown that some metallic alloy compositions (e.g., ruthenium and copper) can be obtained in the form of very small particles even when there is little miscibility of the two metals in the form of bulk alloys. The high fraction exposed in such compositions can be demonstrated by the absence of an X-ray diffraction pattern and by the reaction behavior. Thus the ratio of the rate of dehydrogenation to that of hydrogenolysis is different than would be expected if the two metals were present as separate bulk components (see Sec. 6.3.3). Sinfelt uses the term *bimetallic cluster* to refer to a metallic entity containing atoms of two or more metals highly dispersed on a support, the composition of which may or may not exist as an alloy in the bulk form.

The surface composition of an alloy is determined by a number of factors. These have been recently reviewed by Sachtler (1976), Sachtler and van Santen (1977), and Ponec (1975), who also discuss the effect of surface composition on selectivity. The latter is also reviewed by Clarke (1975). It is first necessary to ensure that the alloy is in its equilibrium composition, which for thin metal films requires heating for about an hour at a temperature of about 0.3 T_m (the Hüttig temperature) or higher. (T_m is the melting point in absolute units.) If the alloy exhibits a miscibility gap, then over a particular composition range it appears that one phase—that having the lower surface energy—will envelop the other (the "cherry model" of Sachtler).

The behavior of a catalytic reaction is determined primarily by the composition of the outer layer and not by that of the interior (the "pit"). This kind of structure has been adduced by chemisorption studies and hydrogen isotope exchange for the system platinum-gold, which has a wide miscibility gap. As the quantity of the outer phase is lowered, the structure changes to one in which patches of the outer phase exist but do not surround the interior, and ultimately a single phase develops. Under the two-phase conditions the surface and the interior of the outer phase may assume different compositions.

In one-phase alloys or in the cherry model described above, the surface tends to be enriched by the component with the lower surface energy (lower heat of sublimation), but the situation is complicated if gases in the environment can react with or be chemisorbed on the alloy. The component of the alloy with the greater affinity for the gas (e.g., higher heat of adsorption) tends to segregate at the surface. With group VIII–group IB alloys in the absence of a gas phase, the group IB metal has the lower heat of sublimation. It thus moves preferentially to the surface, as shown for example, by nickel-copper, nickel-gold, and palladium-silver. In the presence of oxygen the surface of a nickel-gold system is enriched in nickel rather than gold, and in the presence of carbon monoxide, the surface of palladium-silver becomes enriched with palladium instead of silver.

6.3.2 Reactions on Alloys

Some of the early work with alloy catalysts was based on the concept that an alloy might exhibit some type of averaged property of its components. More specifically, in the rigid-band electronic model of binary alloys it was assumed that a common band would be formed from the two constituents. At an alloy composition at which holes in the d band became filled, there might occur a sudden shift in catalytic properties since the adsorption characteristics of the alloy might change markedly at this point. This viewpoint is now clearly incorrect, at least for those systems most widely studied, such as nickel-copper, palladium-silver, and palladium-gold, and most probably for group VIII–group IB systems in general. This is shown by alloy spectra and a variety of reaction studies.

In these alloys the atoms do not lose their individuality, and it is more fruitful to consider the structure as a group of individual atoms, at least as a first approximation. With nickel-copper, for example, the activity is ascribed to the arrangement of nickel atoms only, and the copper is regarded as an inert diluent. This essentially geometrical concept may be modified to some extent, to consider that the nearest neighbors to a nickel atom may influence its adsorptive properties and hence catalytic behavior; i.e., there may be an electronic effect. Sachtler terms the first the *ensemble* effect and the second the *ligand* effect. In many cases it is difficult to separate the contributions of the two. The effects observed with alloy catalysts can be rationalized primarily in terms of ensemble (geometrical) effects, but it seems necessary to involve ligand effects to explain some of the changes observed in reactivity.

6.3.3 Site-Number Requirements (Geometrical Effects)

A underlying concept in catalysis for the last several decades is that some reactions require adsorption on a group of sites or a "multiple site" in order to occur. Such a group of sites has been termed an *ensemble* by Kobozev and Dowden and a *multiplet* by Balandin, and the concept has been visualized and developed in a variety of ways by different investigators. Of great significance is that this allows a rational approach to the tailoring of catalysts to achieve high selectivity when more than one reaction pathway may occur; it also explains a variety of poisoning effects on selectivity. If geometrical considerations are the predominent effect, then dilution of an active metal with an inactive metal should decrease the rate of those reactions requiring the greatest number of nearby sites relative to the rate of a reaction requiring the least number of sites. That this indeed occurs has been shown notably in a series of studies by Sinfelt and coworkers (Sinfelt, 1974), and this approach has been applied in the design of industrial reforming catalysts.

As an example of this effect, Fig. 6.1 shows the striking contrast between the effect of nickel-copper alloy composition on the dehydrogenation of cyclohexane and on the hydrogenolysis of ethane, at 316°C. The catalyst is finely divided metal granules with surface areas of 1 to 2 m^2/g, prepared by coprecipitation of the metals as carbonates followed by calcination and reduction. As the copper concentration of the alloy is increased, the specific activity for cyclohexane dehydrogenation remains unchanged over a wide composition range whereas that for ethane hydrogenolysis drops precipitously. Other workers have likewise reported that addition of small amounts of copper to nickel greatly reduces the activity for paraffin hydrogenolysis. It is concluded that two or more adjacent nickel atoms must be required for this reaction whereas probably one suffices for dehydrogenation of cyclohexane to benzene. The reasons for the decrease in rate of the latter reaction at the two extremes of the composition range are not certain, but it may be caused by a change in the rate-limiting

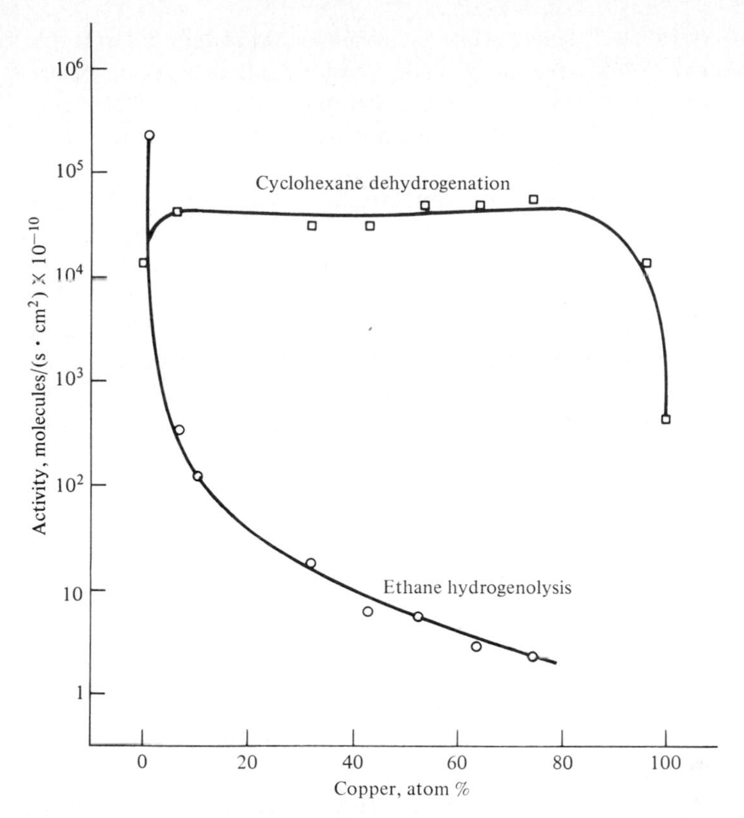

Figure 6.1 Activities of copper-nickel alloy catalysts for the hydrogenolysis of ethane to methane and the dehydrogenation of cyclohexane to benzene. Ethane hydrogenolysis activities were obtained at ethane and hydrogen pressures of 3 and 20 kPa, respectively. Cyclohexane dehydrogenation activities were obtained at cyclohexane and hydrogen pressures of 17 and 83 kPa, respectively. (*Sinfelt et al., 1972.*) (*Reprinted with permission from the Journal of Catalysis. Copyright by Academic Press.*)

step of the reaction. At the reaction temperature, the copper-nickel alloy system is completely miscible, but at lower temperatures a miscibility gap may occur.

Studies by Sinfelt with cyclohexane dehydrogenation on ruthenium-copper or osmium-copper at 316°C showed behavior similar to that on nickel-copper. Hydrogenolysis of cyclohexane decreased with increased copper content, but the effect was not as pronounced as that observed with hydrogenolysis of ethane. A complicating, but significant, factor in these types of studies is that self-poisoning can occur by the accumulation of carbonaceous residues on the catalyst. This process, which is sometimes described as a polymerizing of acetylenic residues or a total or deep hydrogenolysis of paraffins, plausibly involves a group of several sites. If so, it should be decreased by alloying. This appears to indeed be the case, at least for cyclohexane reactions on alloy systems. In some cases in which improved activity upon alloying has been attributed to the intrinsic properties of the alloy, the effect may have been caused instead primarily by a decrease in self-poisoning.

A number of reactions have been classified as to the number of adjacent sites that are required (Clarke, 1975). Hydrogenation of olefins, hydrogenation of cyclopropane, and many reactions involving C—H bond breaking or formation seem to proceed on a single site. Those involving rupture of the carbon skeleton require two or more sites. Clarke discusses the possible mechanisms on alloys of a variety of hydrocarbon transformations such as dehydrocyclization and aromatization of paraffins, and hydrocracking.

There are some indications that carbonaceous deposits form preferentially at certain stepped regions or rougher portions of metals such as nickel and platinum. Thus the tendency for a catalyst to cause carbonaceous deposits may be diminished by pretreatment with a poison that is preferentially adsorbed on these types of sites. For example, platinum catalysts may be pretreated with hydrogen sulfide for "activation," which may operate by decreasing its activity for hydrogenolysis and for formation of carbonaceous deposits. The formation of a small degree of carbonaceous residues may also be desirable because this may cause a partial poisoning of a catalyst surface, thereby causing the preferential blocking of reactions requiring multiple sites. The silver catalyst used industrially for conversion of ethylene to ethylene oxide is partially poisoned with chloride to enhance its selectivity (Sec. 8.3).

A number of cases occur in which an alloy of group VIII metals is more active than either metal by itself (Clarke, 1975). The effects are apparently not major and their interpretation is uncertain, but an electronic approach is perhaps most useful. Methods of preparation of alloy catalysts are outlined by Ponec (1975) and discussed by Sinfelt (1972). Sinfelt, and Burton and Garten (1977) also discuss methods of characterization. Some of the complications in working with alloy film catalysts are reviewed by Moss and Whalley (1972). In an earlier review Boudart (1969) analyzed information then available on structure-sensitive and structure-insensitive reactions. Structure sensitivity was more recently reviewed by Cinneide and Clarke (1973) and Boudart (1977).

There has been little application of these geometrical or ensemble concepts to reactions on catalysts other than metals, largely because of difficulty in characterization. Magnetic methods have been applied to transition metal oxide catalysts to determine the state of dispersion, and the dispersion of chromia in chromia/alumina catalysts in particular has received considerable study. For the dehydrocyclization of *n*-heptane to toluene it has been shown that the activity per chromium atom in the catalyst increased as the chromium concentration decreased and that this was associated with an increase in the fraction exposed.

6.4 SINTERING AND MOBILITY

To a first approximation the rate of reaction on a metal catalyst is proportional to the total surface area effectively exposed to reactants. A nonsupported high-surface-area structure of a metal, such as may be made by compacting a metal powder, is seldom used in catalysis. The mobility of metal atoms over metal surfaces is very high, so such a compact sinters much more readily than if the individual metal particles are separated from one another by inert fine particles of a high-melting-point material, a so-called textural promoter (Sec. 4.6). Individual metal particles may also be spread out from one another on an inert support. Porous metals may show enhanced sintering in the presence of hydrogen, even at low temperatures. In a study of finely divided nickel powder compacted into a thin disk, having a surface area of 5 m^2/g, Satterfield and Iino (1968) observed significant sintering at 62°C, as determined by a change in permeability. This occurred in the presence of hydrogen but not helium. By use of a support, the fraction exposed can approach unity, much higher than that obtainable with nonsupported forms such as metal blacks or sponges. This is of particular importance with expensive materials such as the platinum-group metals.

The sintering of supported metal crystallites is a complicated phenomenon. The behavior pattern is quite different in an inert or hydrogen atmosphere on the one hand in contrast to one in which the metal may form transitory and mobile molecular intermediates.

With the platinum-group metals, sintering in an oxidizing atmosphere is much different than in a reducing or inert atmosphere. Slightly volatile oxides are formed that may diffuse along the surface of the support or through the vapor from higher-energy sites to lower-energy sites. With base metals such as iron, cobalt, and nickel, particle growth can occur in the presence of tiny concentrations of chloride, because of the formation of volatile metal chlorides. At low temperatures in the presence of carbon monoxide, growth occurs via the formation of metal carbonyls. For these mechanisms to operate, the bulk metal must be more stable than the metal compound, either isolated in the vapor or as a surface compound.

The mechanisms whereby supported individual metal atoms and metal crystallites can grow are surveyed by Geus (1975). Mechanistic models are reviewed by Wanke and Flynn (1975) in conjunction with examination of a large number of sintering studies. Wynblatt and Gjostein (1974) analyze these models from a more theoretical point of view, with particular attention to the Pt/Al_2O_3 system.

The various stages of sintering are shown in Fig. 6.2. With 100 percent fraction exposed the metal is present in the form of separate atoms or in the form of a two-dimensional cluster one atom thick. (This is also termed a *raft* of atoms by others.) Generally the two-dimensional cluster is more stable than individual atoms, so individual atoms (or molecules) can surface-diffuse to form two-dimensional rafts. Larger two-dimensional clusters are more stable than smaller ones since edge atoms have a higher energy than those in the interior, so smaller two-dimensional clusters can grow to larger two-dimensional clusters by surface diffusion of atoms. This can be limited by the rate of detachment of atoms from a two-dimensional cluster or by rate of surface diffusion.

Two-dimensional clusters can rearrange into three-dimensional particles, which will be the more stable form if metal-metal bond energies exceed metal-support energies. A larger three-dimensional particle (crystallite) is more stable than a smaller one, and growth may occur by either of two mechanisms. Atoms may detach from particle A and move to particle B, a mechanism that is sometimes termed *Ostwald ripening* by analogy to the growth of particles suspended in a gas or liquid by movement of individual atoms or molecules. The second mechanism is the movement of

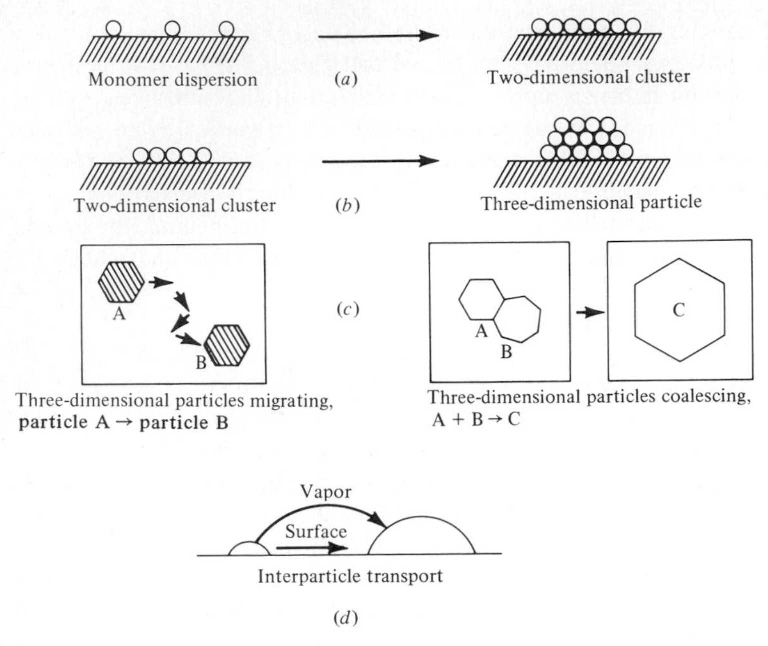

Figure 6.2 Schematic of the various stages in the formation and growth of particles from a monomer dispersion. (*Wynblatt and Gjostein, 1974.*)

individual crystallites as such along the surface to cause growth by coalescence, analogous to coagulation by Brownian motion. The rate-limiting process here can be that of particle migration as such or the process of coalescence, involving the growth of a neck between two particles and the gradual change in shape of the merged pairs into the lowest-energy configuration. For systems of interest in catalysis the migration step rather than coalescence is generally the rate-limiting process.

The rates of these different processes are determined by a number of factors. Basically these are the values of metal-metal bond energies (and that of surface, edge, and corner atoms in contrast to those in bulk metal) relative to bond strength between metal and support. Dissociation of atoms from a crystallite occurs more readily if the atom-surface bonding is appreciable, but if this is too great, surface diffusion will be slow. Some portion of the detached atoms must be mobile for surface diffusion-coalescence to occur. It appears that individual metal atoms are weakly bonded to ionic surfaces and are therefore highly mobile. The rate of detachment of metal atoms from a crystallite would seem to be very slow except at very high temperatures, unless some sort of species more strongly bonded with the surface is formed, or unless the metal is held in a reactive atmosphere.

The migration and collision of crystallites cannot operate after the particles grow to a limiting size where they become substantially immobile. For Pt/Al_2O_3, Wynblatt and Gjostein estimate that particle migration is rapid up to a particle radius of about 5 nm but that particles larger than this must grow by some form of interparticle transport. This may be either surface diffusion of atoms or molecules or vapor transport.

Surface diffusion should be rapid on smooth surfaces and with low interaction energy with the support, but supports of interest in catalysis are rough and often complex in composition. Hydrocarbon deposits increase adhesion, and the formation of surface compounds, e.g., a nickel aluminate, can affect surface properties. Atoms are more strongly adsorbed at various steps and dislocations on the substrate, so surface-diffusing metal atoms can nucleate and grow into crystallites at these locations, a process known as *decoration*. The degree of dispersion of platinum on graphite made heterogeneous by partial gasification was found to be greater than on the original graphite (Ehrburger et al., 1976).

In principle, whether Ostwald ripening or particle coalescence is the dominating mechanism may be determined by the change in particle-size distribution with time and the effect of number of particles per unit area. The complex nature of real catalysts, however, introduces other effects that can readily overshadow prediction from this theoretical treatment. One of these is the pore structure (Wynblatt and Ahn, 1975). Particles of metal dispersed into the finest pores so as to fill them, or to fill crevices within the pores, will be stabilized. This is analogous to the preferential filling of fine pores by condensation from a vapor, as predicted by the Kelvin equation (Sec. 5.3.1).

In general, stabilization should be increased with lower particle densities and with lower values of the surface diffusivity. A catalyst with a broad pore-size distribution is also predicted to sinter more rapidly than one with a narrow pore-size distribution (Flynn and Wanke, 1975). In an inert or reducing atmosphere particle growth is predicted to be related inversely to the strength of the cohesive forces in a metal crystallite. For a given substrate the stability of a number of metals of interest should then increase in the order of increased melting point (given in degrees Celsius):

$$\text{Ag (916)} < \text{Au (1063)} < \text{Pd (1552)} < \text{Pt (1770)} < \text{Rh (1966)} < \text{Ru (2250)} < \text{Ir (2410)} < \text{Os (3000)}$$

This generally agrees with experimental results. In studies on alumina Fiedorow et al. (1978) reported the sequence in hydrogen as

$$\text{Pt} < \text{Rh} < \text{Ru} < \text{Ir}$$

From examination of a large amount of experimental data, mostly on alumina, Wanke and Flynn (1975) reported the sequence

$$\text{Ni} < \text{Pd} < \text{Pt} < \text{Rh}$$

Alloying a metal with a second metal of higher melting point should also increase stability, but it may have other catalytic effects.

A useful method of correlation of sintering data is by a power-law rate function

$$\frac{-dD}{dt} = kD^n \tag{6.1}$$

where D is the dispersion (percent exposed) and t is time. Data at different temperatures can be used to calculate an apparent activation energy for k, but this depends on the value of n used for correlation. The scatter of data makes the activation energy values reported subject to a large error. Equation (6.1) can also be derived from some simple sintering models, but it should be regarded for most systems as an essentially empirical relationship. Wanke and Flynn cast a large number of experimental studies into this form. For a number of studies with Pt/Al_2O_3, they reported that higher initial dispersions generally gave higher values of n. However values of n varied from 2 to as high as 16, and the values can also change as sintering progresses. The order in reducing atmospheres was larger than in oxygen. A high value of n means that the crystallites reached a stable size relatively quickly.

In the presence of oxygen, the rate of sintering of metals may be markedly changed from that in hydrogen, by transport of metal oxide molecules through the vapor or along the surface. The vapor pressure of the platinum-group metals as such at most reaction temperatures of interest (up to 1000°C) is insufficient for transport of metal atoms to be a significant mechanism. It has been known for some time that an *increase* in dispersion of platinum on alumina can be caused to occur in oxygen atmospheres in the temperature range of about 400 to 600°C. This is highly desired for regeneration of deactivated catalysts such as reforming catalysts, and a number of procedures are listed in the patent literature. Systematic studies of the effect of an oxygen atmosphere, in contrast to a hydrogen atmosphere, on sintering of platinum, iridium, rhodium, and ruthenium on alumina show that supported iridium can also be redispersed in oxygen by heating for a short time, although rhodium was not redispersed (Fiedorow and Wanke, 1976; Fiedorow et al., 1978). In this particular study the optimum temperature for redispersion was about 400°C for iridium in contrast with about 550°C for platinum, which can be attributed to the higher volatility of iridium oxides.

The mechanism of redispersion is still uncertain. Suggestions that a crystallite of a metal splits are unsubstantiated, although conceivably with an alloy one element might oxidize and split the crystallite while a second element remains in the metallic form. It is most likely that a metal oxide molecule is first detached from the crystallite, migrates to a site on the surface, and becomes fixed by forming a surface complex with the support. Upon subsequent reduction, some reagglomeration occurs.

With an iridium/alumina catalyst onto which barium, calcium, or strontium oxides had been impregnated, sintering of iridium was inhibited compared with sintering that occurred using iridum/alumina alone. It appeared that iridium oxide species were detached from the original crystallites and trapped by the group IIA oxides to form stable, immobile, surface iridates (McVicker, et al., 1978). These were subsequently reduced to the metal.

With platinum or iridium on alumina alone at sufficiently high temperatures or with prolonged treatment the percentage exposed decreases. Crystal growth outweighs the redispersion phenomenon, and the growth rate increases with oxygen partial pressure. In the presence of hydrogen, sintering of all the metals follows the normal pattern, increasing with increased temperature and with time.

In oxygen the order of increasing stability to crystallite growth was found to be

$$\mathrm{Ru} < \mathrm{Ir} < \mathrm{Pt} < \mathrm{Rh}$$

which agrees with the order predicted by Wynblatt and Gjostein,

$$\mathrm{Os} < \mathrm{Ru} < \mathrm{Ir} < \mathrm{Pt} < \mathrm{Pd} = \mathrm{Rh}$$

The assumption here is that the principal governing factor is the enthalpy of formation of the oxide from the metal.

In the presence of a liquid phase, crystallite growth of a supported metal may occur much more readily than in a hydrogen environment. Bett et al. (1976) reported a decrease in surface area of a 5% platinum on graphitized carbon in 24 h at 100°C; the decrease was from 7 to 28 percent in the presence of such different liquids as bromobenzene, 96% phosphoric acid, water, and toluene. No decrease occurred under the same conditions with hydrogen in the gas phase. Many metals dissolve to a very slight extent in a variety of nonaqueous liquids.

6.5 CARBON FORMATION

Carbonaceous deposits (coke) can be formed on catalysts under a wide variety of conditions in a reducing environment. The factors involved in carbon formation on nonmetallic catalysts such as acid catalysts are substantially different than those involved with metals. On nonmetallic catalysts the deposit may contain considerable hydrogen, represented by an empirical formula CH_x in which x may vary between about 0.5 and 1. Carbon deposits on metals generally contain little or no hydrogen, depending in part on the reaction temperature. Many reactions can cause carbon deposits, but the process can be visualized in terms of the decomposition of CO ($2CO \rightarrow C + CO_2$), of CH_4 ($CH_4 \rightarrow C + 2H_2$), or of other gaseous reactions (for example, $2H_2 + CO_2 \rightarrow C + 2H_2O$).

The thermodynamic equilibrium conditions under which solid carbon will form are now well established. Storch et al. cover this in some detail in their book on Fischer-Tropsch synthesis (1951). Gruber (1975) gives a detailed treatment, as do Mills and Steffgen (1973). The analyses usually assume that carbon is present as graphite. A more active form of carbon, sometimes termed *Dent carbon* may also be formed, especially at temperatures below about 700°C or so. Calculations made assuming graphitic carbon is the solid phase are conservative, in the sense that the composition zone for which graphite deposition is predicted to occur is broader than that for a more active carbon.

Even outside the composition-temperature zone in which solid carbon is thermodynamically stable, it may be found if the carbon-forming reactions are inherently faster than the carbon-removal reactions. This is a function of the nature of the reactants and of the catalyst. Thus carbon seems to deposit on iron more readily than on nickel. Carbon formation and its prevention is discussed in more detail with respect to steam reforming on a nickel catalyst in Sec. 10.1.

Carbon is often formed as filaments that appear in some cases like hollow tubes, termed *filamentous carbon.* The head of the growing filament contains a small crystal or particle of metal (or in some cases possibly a metal carbide) of about the same diameter as the filament. This may be seen in electron photomicrographs, such as Fig. 6.3. The metal particle at the head may disintegrate as the filament grows, and the filaments themselves may disintegrate, leaving behind a finely divided carbon containing some metal. The carbon deposition mechanism causes attack on the metal and disperses it through a mass of finely divided carbon, which thus becomes very cata-

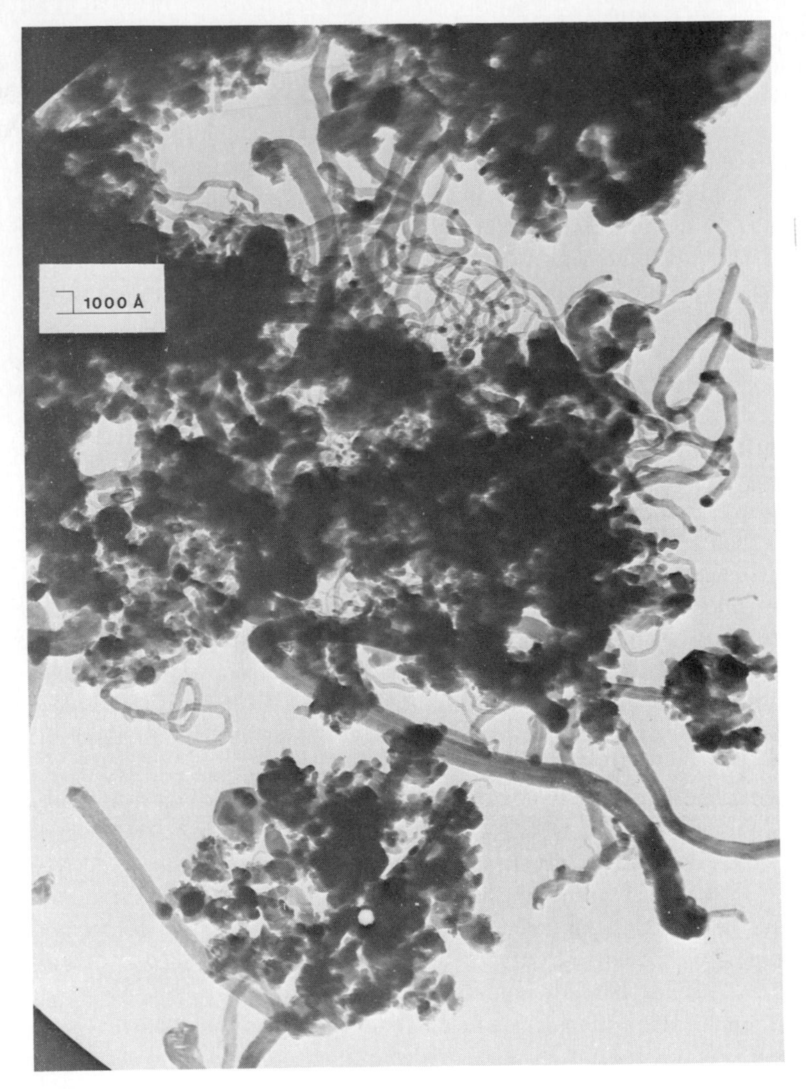

Figure 6.3 Carbon filaments formed by decomposition of methane at 500°C on a supported nickel reforming catalyst. (*Rostrup-Nielsen, 1975, p. 135.*)

lytically active because the metal is finely divided. With time the fibrous carbon structure may sinter and become more compact, whereupon it becomes less reactive; for example, it is less easily removed by contact with steam. The effect has been observed on a variety of metals including iron, cobalt, nickel, and platinum, both in the form of massive metal and of supported crystallites, and with a variety of gases including carbon monoxide, methane, benzene, and acetylene (Baker et al., 1972, 1973; Baker and Harris, 1978).

The mechanism of carbon-filament formation is unclear. Presumably carbon is deposited in some fashion on the metal particle, possibly diffusing through it to the rear, where it is nucleated on the growing filament. From studies with acetylene it has been suggested that a thermal gradient exists between a hot metal crystallite and a cooler carbon filament down to which carbon diffuses, but this is untenable for carbon deposition in general since filament formation occurs with both endothermic and exothermic reactions.

With iron the effect was apparently first observed by Fischer in the early work on catalysts for the Fischer-Tropsch synthesis. It is described in some detail by Shultz et al. (1961), who refer to it as a new type of catalyst termed *carbon-expanded iron*. They report studies with a carbon steel–tube wall reactor "activated" by carbon deposition in which a quantity of carbon-expanded iron formed. On a volume basis the reactor had an activity for the Fischer-Tropsch synthesis reaction at 316°C that was about one-fifth that of a reactor packed with ammonia-synthesis catalyst (which is primarily iron) and operated at 250°C. Per weight of iron present in the carbon deposit, the activity was about 200 times greater than that of the ammonia-synthesis catalyst, which had a surface area of about 10 m^2/g. The comparison of activities is at different temperatures, but even so the iron in the carbon deposit must have been very finely dispersed as well as highly active.

Ruston et al. (1969) made a detailed study of carbon filaments formed on electrolytic iron by contact with carbon monoxide at 550°C. They report that the carbon monoxide decomposition reaction came to a halt when the iron content in the deposited carbon fell below about 3 wt %. In the study of Shultz et al. it was reported that at the end of the run (about 2000 h) there was about 2 g of solid deposit, consisting of 0.07 g iron and 1.93 g carbon. Perhaps coincidentally this is very close to the 3 percent figure reported by Ruston. Hofer (1956) discusses the effect briefly in a chapter on crystalline phases in Fischer-Tropsch catalysts and also cites early work by Von Wangenheim on the carbon phases produced by reaction of carbon monoxide on iron at 350 to 500°C. Von Wangenheim reported that carbon could never be produced free of iron regardless of the form of the iron catalyst.

This phenomenon is a mechanism of attack on bulk metal and supported metal catalysts that can have serious consequences. Especially in a cyclic or erratic situation where carbon is alternately formed and reacted, a metal catalyst may deteriorate and the metal may be dispersed. Carbon can also be formed homogeneously in the gas phase and deposited physically onto a surface, in which case it would not be expected to contain metal. This subject is reviewed by Palmer and Cullis (1965).

6.6 POISONING OF METAL CATALYSTS

A catalyst may become deactivated for a variety of reasons (Sec. 1.3.10). In the case of metals, two of the important mechanisms are agglomeration (Sec. 4.6 and Sec. 6.4) and poisoning. The latter occurs whenever an impurity in the feed stream alters the surface composition of the metal. This may be by chemisorption; by reaction, in which the process is not confined to the surface; or by alloy formation. Obviously a large variety of possibilities are possible. Some guidance comes from a general knowledge of chemical reactivities coupled with the ability to anticipate impurities likely to be introduced into the process, even in trace concentrations. In an early review Maxted (1951) discusses poisoning of metallic catalysts in part from the point of view of "detoxication" of poisons by reacting them so as to alter their adsorption characteristics.

A poison which acts by chemisorption generally does so by being more strongly adsorbed than the reactant. The effect may be substantially reversible or irreversible, depending upon the strength of adsorption and other factors, although many cases will be intermediate. The rate of catalyst recovery after a poison has been eliminated from a feed stream may be conveniently formulated in terms of a rate of desorption.

Sulfur is a commonly encountered impurity and in the bivalent form is readily adsorbed onto metals. In the presence of hydrogen it may also be desorbed as hydrogen sulfide. It may be helpful to think of chemisorbed poisons in terms of a potentially reversible reaction in which one of the species is a surface compound rather than a bulk solid. The surface layer is more reactive than the bulk, especially for small particles and highly disorganized systems. Thus a chemisorbed species such as a surface nickel sulfide can appear on metallic nickel at a partial pressure ratio P_{H_2S}/P_{H_2} much less than that necessary for a bulk sulfide to form.

A comparison of the ease of formation of bulk compounds in a series, e.g., sulfides of iron, cobalt, and nickel, indicates the relative likelihood of a chemisorbed species (surface compound) being formed.

Other compounds that have an unshared pair of electrons, such as carbon monoxide and phosphorus trihydride, can also readily coordinate with metals to form a chemisorbed species. Carbon monoxide is of concern since the hydrogen supplied for hydrogenation is frequently made from carbonaceous sources and the carbon monoxide content must be decreased to a low level to avoid poisoning of metals. At low temperatures, a volatile metal carbonyl may form. A partial poisoning may sometimes be desired to achieve improved selectivity or for long-term activity.

Oxygen and water are common poisons for metals used as catalysts under reducing conditions (e.g., the iron catalyst used for ammonia synthesis) by forming chemisorbed oxygen. Under oxidizing conditions, however, chemisorbed oxygen on metals may be required for catalytic activity, as in the silver-catalyzed oxidation of ethylene to ethylene oxide. Whether or not poisoning is reversible if the poison is removed from the feed stream depends on the strength of the chemisorbed bond and the considerations above. Even if the metallic form can be restored, catalytic activity may be diminished by the rearrangement of atoms or by the growth of crystallites, which reduces surface area.

Chloride is a commonly encountered impurity, e.g., from salt contamination or from chlorine in water. It may poison a metal by forming a surface metal chloride, or it may enhance sintering via the formation of volatile metal chlorides. Another source of poisoning is from other metals, or compounds reducible to metals under reaction conditions. These may alloy with the surface of the catalytically active metal and reduce its effectiveness. Arsenic is present in trace amounts in various feedstocks and can cause poisoning by this process. Iron is ubiquitous as a material of construction and can be a serious poison to platinum-group catalysts. Phosphorus compounds are typically serious poisons and may be introduced from additives to lubricating oils used in pumps, blowers, fans, and other machinery.

In a fixed-bed reactor a poison, being generally strongly adsorbed, is usually deposited first on the upstream portions of the catalyst and on the outermost portion of the pellets. With time the poison distributes itself progressively towards the centers of pellets and downstream in the bed. In an industrial fixed-bed reactor this progressive poisoning can usually be followed by an axial temperature profile. In the case of an exothermic reaction, the point of highest reaction rate—as monitored by maximum temperature in a multitube cooled reactor or by highest rate of increase of temperature with distance in an adiabatic reactor—will slowly move downstream with time. It is usually desirable to maintain a fixed rate of production, so typically the slow decrease in inherent activity is compensated for by gradually raising the reactor temperature. Operation is finally discontinued because of approaching failure of the catalyst, equipment limitations, an increasingly unacceptable selectivity, or other reasons.

6.7 HYDROGENATION REACTIONS

A great variety of hydrogenation reactions are carried out industrially. These range from large-scale, continuous catalytic operations in petroleum refineries (Chap. 9), dealing with streams of complex composition, to small-scale batch operations in the pharmaceutical and fine chemical industry where a very precise hydrogenation step is often desired, starting with a relatively pure reactant. The most common catalyst is some form of nickel. Next is palladium, where its higher activity and/or selectivity more than compensates for its higher cost. Both of these metals readily absorb hydrogen into the interstices between metal atoms. Other metals such as iron, copper, and platinum are less absorptive, which may account for the high catalytic activity of palladium and nickel. Hydrogen diffuses readily through palladium and, to a lesser extent, nickel. A commercial process for separation of hydrogen from gaseous streams utilizes palladium membranes at an elevated temperature and pressure. With palladium there is crystallographic and other evidence that, in contrast to what might be termed a solid solution, two specific hydride phases can be formed, a β phase at higher hydrogen partial pressures and an α phase at lower pressures. The two coexist under some circumstances. These two phases may have different catalytic properties.

Sulfur compounds poison nickel; in a few cases nickel or other metallic catalysts may be partly poisoned, deliberately, with, for example, hydrogen sulfide. This improves selectivity at the expense of activity, or tempers an initial activity that might otherwise cause a rapid rate of deposition of coke (Sec. 6.5). Morikawa et al. (1969) discuss some of the relationships between methods of preparation of supported metal catalysts, especially nickel and palladium, and their catalytic activity. Gully and Ballard (1963) review the hydrogenation of aromatics, as found in catalytic-cracking charge stocks, with emphasis on equilibrium considerations and relative rates of reaction. Books on hydrogenation are summarized in App. A.4.

Except for the fuels-processing industry, only a relatively small number of industrial hydrogenation operations are carried out on a substantial scale. A discussion of three of these follows.

6.7.1 Edible Oils

Edible oils are triglycerides of fatty acids, primarily n-C_{18}, having various degrees of unsaturation. (Stearic acid is saturated; oleic acid has one C=C bond; linoleic, two C=C bonds; linolenic, three C=C bonds.) These oils are produced from a variety of natural sources throughout the world. Some of the major oils are extracted from soybeans, sunflower seeds, rapeseed, cottonseed, coconut, and palm. Oils containing highly unsaturated fatty acids are particularly susceptible to autoxidation, which affects their flavor. Consequently it is desirable to partially hydrogenate them to improve their stability. The melting-point range is simultaneously increased. For most products the desired softening characteristics correspond to oils which are only partially hydrogenated, so selective reaction is desired. During hydrogenation cis-trans isomerization of the remaining double bonds and double-bond migration occur. Both of these have an effect on the desired softening properties of the product.

The overall series of reactions is complex. This is in large part because of the varying nature of the raw materials, but also because different fatty acids in one triglyceride may behave independently. The reaction is carried out on carefully refined oils utilizing a finely divided and high-area (for example, 50 to 100 m^2/g) supported nickel or Raney nickel catalyst and high-purity hydrogen, at temperatures in the range of 150 to 200°C and pressures of 0.1 to 0.7 MPa. All oils, even after prepurification, contain small amounts of sulfur compounds and other substances which poison the catalyst, so it is necessary to replace the catalyst fairly frequently. Commercially the nickel is usually prereduced, stabilized in hydrogenated fat, and supplied for easy handling in the form of granules or flakes (sometimes termed *Rufert flakes*). Most processes are operated in batches in stirred autoclaves because of the variation in raw materials and desired product characteristics, but some continuous operations utilize several stirred tanks in series.

An unusual feature of the reaction is that it is preferably operated with a high degree of diffusional limitations with respect to hydrogen. This is achieved by an optimum combination of low pressure, moderate agitation, high temperature, and high catalyst loading. This results in a low concentration of dissolved hydrogen at the surface of the catalyst, which minimizes overhydrogenation. However it is desirable that the reaction *not* be pore diffusion–limited with respect to the triglycerides, which would reduce the rate of transfer of partially hydrogenated products out of the catalyst particles. Consequently very fine pores are not desired. A packed column reactor utilizing larger pellets, as is sometimes proposed, would probably cause poorer selectivity.

Copper catalysts have also been used to a limited extent, but they are much less active than nickel, they are more sensitive to poisoning, and traces of copper in the product catalyze autoxidation reactions.

Coenen (1976) reviews the subject in general; Coenen et al. (1965) and van der Plank et al. (1972) discuss the mechanism of the reaction with particular emphasis on the effect of diffusion on selectivity. Commercial processes are reviewed in a paper by Albright (1967).

6.7.2 Selective Hydrogenation of Acetylenes

The effluent from steam cracking of hydrocarbons to produce ethylene, other olefins, and other products contains small amounts of acetylenic compounds. In some cases acetylene and other species may be economically recovered; in others, acetylenes and dienes may be removed by selective hydrogenation, using a nickel or a palladium catalyst. The acetylenes and dienes are so much more strongly adsorbed than olefins that high selectivity with minimum hydrogenation of olefins is achievable down to very low acetylene concentrations. The choice of the catalyst depends on the nature of the feedstock and whether impurity removal is to be achieved before or after fractionation of products. Steam cracking is carried out at substantially atmospheric pressure, but the product is then compressed to about 1.5 MPa for fractionation, so hydrogenation is carried out at this pressure. Hydrogenation may be applied to a mixed product stream before fractionation or, more commonly, to separate streams after fractionation, for example, to C_2, C_3, and C_4 streams from a naphtha cracker. In the case of a C_3 stream it is usually desired to remove acetylenes and propadiene from propylene. With a C_4 stream it may be desirable to remove acetylenic compounds selectively in the presence of butadiene, which is then recovered.

If sulfur poisons are at a level below a few parts per million or if hydrogenation is to be carried out after fractionation, a supported palladium catalyst is usually used with the products from an ethane-propane feed to the steam cracker. An example is I.C.I. 38-1 consisting of 0.04 wt % Pd/Al_2O_3, which has a surface area of 18 m^2/g (*Oil Gas J.*: 1972). Typically the acetylene concentration can be reduced from an initial value in the region of 5000 ppm to below 5 ppm with hydrogenation of no more than 1 percent of the ethylene, operating at temperatures in the range of about 60 to 70°C. The palladium catalyst is more active than a nickel catalyst. Potentially the reaction can run away, especially with high partial pressures of hydrogen and/or acetylenes, in which a temperature increase could initiate a self-accelerating exothermic hydrogenation of olefins. With ethane-propane feed, the hydrogen concentration in the product from the cracker may reach 30 mol %. With naphtha, a value in the range of 10 to 15 percent is representative. If acetylene is to be removed before substantial fractionation, a nickel catalyst is usually used, and a small amount of hydrogen sulfide or a mercaptan is added to the feed to minimize hydrogenation of ethylene. This partial poisoning improves selectivity but lowers activity, and a reaction temperature in the neighborhood of 200°C is representative (see also Sec. 6.8). Livingston (1973) gives more details.

6.7.3 Cyclohexane

About 95 percent of the cyclohexane produced is used to make adipic acid or caprolactam, intermediates to nylon 66 and nylon 6, respectively. In the United States a small portion of the total (about 15 percent) is extracted from natural gas condensate, but most is made by catalytic hydrogenation of benzene. A nickel, platinum, or palladium catalyst is generally used, in a continuous process. The benzene may be in either the vapor or liquid phase.

The temperatures and pressure depend on the catalyst used. The range of 150 to 200°C and a pressure of about 3 MPa have been quoted for a nickel/alumina catalyst and about 450°C and 30 MPa for a sulfided nickel or palladium catalyst. The reaction is highly exothermic.

$$C_6H_6 + 3H_2 \longrightarrow C_6H_{12} \qquad -\Delta H = 206 \text{ kJ/mol} \tag{6.2}$$

In one version of this process (Institut Français du Pétrole) a high degree of conversion is obtained in the liquid phase in a slurry-bed reactor using an external pumparound system to keep the catalyst in suspension. A small fixed-bed finishing reactor is then used to obtain the final desired conversion, if this is not achieved in the main reactor.

A commonly used procedure to control the temperature in a highly exothermic reaction carried out in a fixed-bed reactor is to cool and recycle some of the product to the inlet and mix it with fresh feed. This reduces the temperature rise down the reactor. This design is applied in a liquid-phase process for cyclohexane in which several adiabatic beds are used in series. In a vapor-phase process using a noble metal catalyst (Arco) heat is removed during reaction by steam generation. Cyclohexane can isomerize to form methylcyclopentane, and in these processes catalyst and reaction conditions are chosen to inhibit this reaction. Conversion is nearly 100 percent.

Kinetic information on this reaction has been published by many investigators; see, for example, van Meerten and Coenen (1975). There are several reports that the reaction rate reaches a maximum at a temperature of about 180 to 200°C, above which the rate falls with increasing temperature (see also Sec. 3.3.3).

6.8 SULFIDE CATALYSTS

Sulfide catalysts are of particular interest in two different types of applications. In processing a feedstock containing sulfur compounds, a metallic or oxide catalyst will usually become converted to a sulfide either as a bulk compound or as a chemisorbed surface sulfide. The most important application is hydrodesulfurization (Sec. 9.8). Generally superior performance is obtained by converting the catalyst to the sulfide form under controlled conditions rather than allowing the catalyst to become sulfided by contact with sulfur compounds in the reaction mixture.

Metal sulfides also possess catalytic activity for hydrogenation and dehydrogenation reactions. They may be more resistant than metallic catalysts to the formation of coke deposits, so that under practical conditions a nickel sulfide catalyst may, for example, be more active than a metallic nickel catalyst for the hydrogenation of a hydrocarbon. They may also resist poisoning by sulfur compounds more readily than metallic catalysts.

The catalytic activity of a metallic catalyst may be tempered by the controlled addition of a small concentration of hydrogen sulfide or other sulfur compound to the feedstock. This attenuates the initial high activity of a metallic catalyst such as nickel or platinum, which might otherwise cause rapid deactivation by coke formation via hydrogenolysis or other reactions. The long-term activity of the catalyst is thus stabilized at a higher level than it would be otherwise. In the presence of hydrogen the degree of sulfiding is usually reversible. Aside from the effects of coke formation, sulfide catalysts (as a bulk or surface compound) are generally less active for hydrogenation than the most active metal catalysts. Hence they must usually be used at higher temperatures for practicable rates of reaction. However the sulfide form may permit improved selectivity. In the selective hydrogenation of acetylene in the presence of ethylene, the degree of selectivity on a nickel catalyst is improved by the presence of a tiny concentration of hydrogen sulfide in the reacting mixture, and this is deliberately added if necessary (see Sec. 6.7.2).

Sulfide catalysts are considered in detail in the book by Weisser and Landa (1973). Recent literature on the various kinds of reactions they catalyze is reviewed by Mitchell (1977). These catalysts are essentially bifunctional in nature, having hydrogenation-dehydrogenation capability as well as acidity.

REFERENCES

Albright, L. F.: *Chem. Eng.*, Oct. 9, 1967, p. 249.
Augustine, R. L.: *Catalytic Hydrogenation*, Dekker, New York, 1965.
Baetzold, R. C.: *Adv. Catal.*, **25**, 1 (1976).
Baker, R. T. K., M. A. Barber, P. S. Harris, F. S. Feates, and R. J. Waite: *J. Catal.*, **26**, 51 (1972).
——, and P. S. Harris in P. L. Walker, Jr., and P. A. Thrower (eds.): *Chemistry and Physics of Carbon*, vol. 14, Dekker, New York, 1978.
——, P. S. Harris, R. B. Thomas, and R. J. Waite: *J. Catal.*, **30**, 86 (1973).
Bett, J. A. S., K. Kinoshita, and P. Stonehart: *J. Catal.*, **41**, 124 (1976).
Boudart, M.: *Adv. Catal.*, **20**, 153 (1969).
——: *Proceedings of the Sixth International Congress on Catalysis*, The Chemical Society, London, 1977, p. 1.
Burton, J. J.: *Catal. Rev.*, **9**, 209 (1974).
——, and R. L. Garten: *Advanced Materials in Catalysis*, Academic, New York, 1977, p. 33.
Cinneide, A. D. O., and J. K. A. Clarke: *Catal. Rev.*, **7**, 213 (1973).
Clarke, J. K. A.: *Chem. Rev.*, **75**, 291 (1975).
Coenen, J. W. E.: *J. Am. Oil Chem. Soc.*, **53**, 382 (1976).
——, H. Boerma, B. G. Linsen, and B. De Vries: *Proceedings of the Third International Congress on Catalysis*, North-Holland Publishing Company, Amsterdam, 1965, p. 1387.
Dowden, D. A.: *Catalysis*, vol. 2, The Chemical Society, London, 1978, chap. 1, p. 1.
Ehrburger, P., O. P. Mahajan, and P. L. Walker, Jr.: *J. Catal.*, **43**, 61 (1976).
Fiedorow, R. M. J., B. S. Chahar, and S. E. Wanke: *J. Catal.*, **51**, 193 (1978).
——, and S. E. Wanke: *J. Catal.*, **43**, 34 (1976).
Flynn, P. C., and S. E. Wanke: *J. Catal.*, **37**, 432 (1975).
Freifelder, M.: *Practical Catalytic Hydrogenation: Techniques and Applications*, Wiley, New York, 1971.
Geus, J. W., in G. C. Kucynski (ed.): *Sintering and Catalysis*, Plenum, New York, 1975, p. 29.
Gruber, G.: *Adv. Chem. Ser.*, **146**, 31 (1975).
Gully, A. J., and W. P. Ballard in J. J. McKetta (ed.): *Advances in Petroleum Chemistry and Refining*, vol. 7, Wiley, New York, 1963, p. 240.
Hofer, L. J. E., in P. H. Emmett (ed.): *Catalysis*, vol. IV, Reinhold, New York, 1956, p. 373.
Livingston, J. Y.: *Chem. Eng. Prog.*, **69**(5), 65 (1973).
McVicker, G. B., R. L. Garten, and R. T. K. Baker: *J. Catal.*, **54**, 129 (1978).
Maxted, E. B.: *Adv. Catal.*, **III**, 129 (1951).
Mills, G. A., and F. W. Steffgen: *Catal. Rev.*, **8**, 159 (1973).

Mitchell, P. C. H.: *Catalysis*, vol. 1, The Chemical Society, London, 1977, chap. 6, p. 204.
Morikawa, K., T. Shirasaki, and M. Okada: *Adv. Catal.*, **20**, 98 (1969).
Moss, R. L., and L. Whalley: *Adv. Catal.*, **22**, 115 (1972).
Oil Gas J., March 27, 1972.
Palmer, H. B., and C. F. Cullis in P. L. Walker, Jr., (ed.): *Chemistry and Physics of Carbon*, vol. 1, Dekker, New York, 1965.
Ponec, V.: *Cat. Rev. Sci. Eng.*, **11**, 41 (1975).
Rostrup-Nielsen, J. R.: *Steam Reforming Catalysts*, Danish Technical Press, Copenhagen, 1975.
Ruston, W. R., M. Warzee, J. Hennaut, and J. Waty: *Carbon*, **7**, 47 (1969).
Rylander, P. N.: *Catalytic Hydrogenation Over Platinum Metals*, Academic, New York, 1967.
——: *Organic Syntheses with Noble Metal Catalysts*, Academic, New York, 1973.
Sachtler, W. M. H.: *Cat. Rev. Sci. Eng.*, **14**, 193 (1976).
——, and R. A. van Santen: *Adv. Catal.*, **26**, 69 (1977).
Satterfield, C. N., and H. Iino: *Ind. Eng. Chem., Fundam.*, **7**, 214 (1968).
Shultz, J. F., F. S. Karn, R. B. Anderson, and L. J. E. Hofer: *Fuel*, **40**, 181 (1961).
Sinfelt, J. H.: *Ann. Rev. Mater. Sci.*, **2**, 641 (1972).
——: *Cat. Rev. Sci. Eng.*, **9**, 147 (1974).
——: *Prog. Solid State Chem.*, **10**, 55 (1975). [See also J. H. Sinfelt, *Adv. Catal.*, **23**, 91 (1973); *Catal. Rev.*, **3**, 175 (1969).]
——, J. L. Carter, and D. J. C. Yates: *J. Catal.*, **24**, 283 (1972).
Somorjai, G. A.: *Adv. Catal.*, **26**, 1 (1977). See also *Catal. Rev.*, **7**, 87 (1972).
Storch, H. H., Jr., N. Golumbic, and R. B. Anderson: *The Fischer-Tropsch and Related Syntheses*, Wiley, New York, 1951.
van der Plank, P., B. G. Linsen, and H. J. van den Berg: *Proceedings of the Fifth European/Second International Symposium on Chem. React. Engineering*, Elsevier, Amsterdam, 1972, p. B 6.
van Meerten, R. Z. C., and J. W. E. Coenen: *J. Catal.*, **37**, 37 (1975).
Wanke, S. E., and P. C. Flynn: *Catal. Rev. Sci. Eng.*, **12**, 93 (1975).
Weisser, O., and S. Landa: *Sulphide Catalysts, Their Properties and Applications*, Pergamon, New York, 1973.
Wynblatt, P., and T. M. Ahn in G. C. Kucynski (ed.): *Sintering and Catalysis*, Plenum, New York, 1975, p. 83.
——, and N. A. Gjostein: *Prog. Solid State Chem.*, **9**, 21 (1974).

CHAPTER

SEVEN

ACID AND ZEOLITE CATALYSTS

The concept that solid surfaces may be acidic arose from the observation that hydrocarbon reactions such as cracking that are catalyzed by acid-treated clays or silica-alumina, give rise to a much different product distribution than those obtained by thermal reaction. These solid-catalyzed reactions exhibit features similar to reactions catalyzed by mineral acids. Furthermore, it was shown that cracking catalysts could be titrated with a base and that they could be inactivated by adsorption of basic nitrogenous compounds or by inorganic basic ions.

By analogy to solution chemistry, it is postulated that the primary requirement for catalytic activity is that the solid be acidic and be capable of forming carbonium ions by reaction with a hydrocarbon. Carbonium ions are intermediates in such reactions as cracking, polymerization, and isomerization. An acid site may be of the Brönsted type in which it donates a proton to an unsaturated hydrocarbon, or of the Lewis type in which it acts as an electron acceptor, removing a hydride ion from a hydrocarbon. Many methods of measuring acidity do not distinguish between the two types of sites and simply report the total acidity.

7.1 SOURCE OF ACIDITY

The structures that give rise to acidity and indeed to catalytic activity are subject to some controversy. In the case of silica-alumina or similar mixed-oxide catalysts, the source of acidity may be rationalized in terms of a theory developed largely by Pauling. If an aluminum ion, which is trivalent, is substituted isomorphously for a silicon ion, which is quadrivalent, in a silica lattice comprising silica tetrahedra, the net negative charge must be stabilized by a nearby positive ion such as a proton. This can be produced by the dissociation of water, forming a hydroxyl group on the aluminum

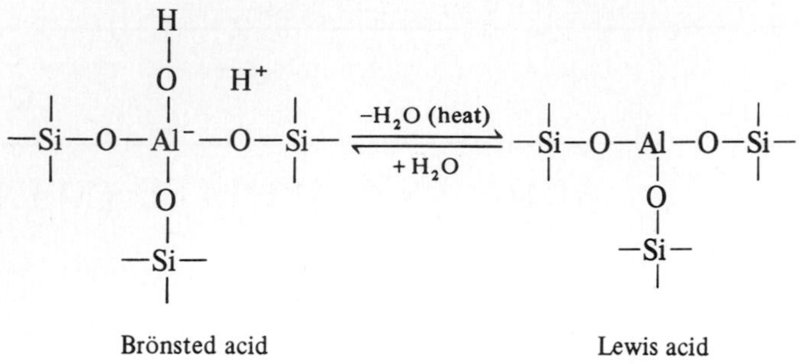

$$
-\mathrm{Si}-\mathrm{O}-\underset{\substack{|\\ \mathrm{O}\\ |\\ -\mathrm{Si}-\\ |}}{\overset{\substack{\mathrm{H}\\ |\\ \mathrm{O}\\ |}}{\mathrm{Al}^-}}\overset{\mathrm{H}^+}{}-\mathrm{O}-\mathrm{Si}- \underset{+\mathrm{H_2O}}{\overset{-\mathrm{H_2O\ (heat)}}{\rightleftharpoons}} -\mathrm{Si}-\mathrm{O}-\underset{\substack{|\\ \mathrm{O}\\ |\\ -\mathrm{Si}-\\ |}}{\mathrm{Al}}-\mathrm{O}-\mathrm{Si}-
$$

Brönsted acid Lewis acid

Figure 7.1 Postulated structures of silica-alumina causing Brönsted and Lewis activity.

atom. The resulting structure, in which the aluminum and the silicon are both tetrahedrally coordinated, is a Brönsted acid.

If this is heated, water of constitution is driven off and Brönsted acid sites are converted to Lewis acid sites as shown in Fig. 7.1. Some metal atoms are now three-coordinated and some four-coordinated. The reverse can also occur. The addition of water and heating can convert Lewis acid sites back to Brönsted acid sites. The aluminum atom is electrophilic and can react with hydrocarbons to form an adsorbed carbonium ion, as illustrated below for the two kinds of sites:

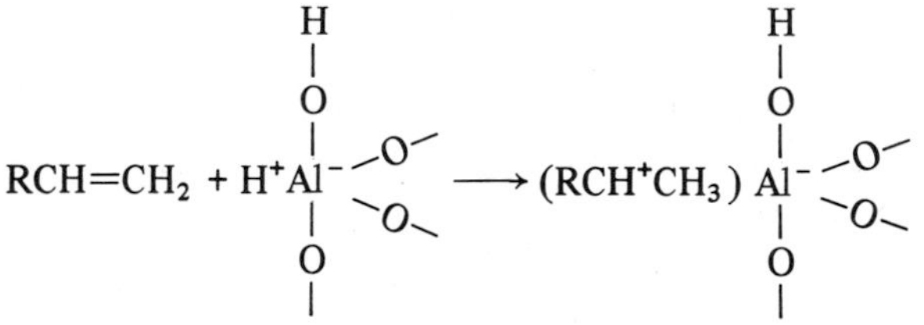

$$
\mathrm{RCH{=}CH_2} + \mathrm{H^+}\underset{\substack{|\\ \mathrm{O}\\ |}}{\overset{\substack{\mathrm{H}\\ |\\ \mathrm{O}\\ |}}{\mathrm{Al}^-}}\!\begin{smallmatrix}\diagup \mathrm{O}\diagup\\ \diagdown \mathrm{O}\diagdown\end{smallmatrix} \longrightarrow (\mathrm{RCH^+CH_3})\ \underset{\substack{|\\ \mathrm{O}\\ |}}{\overset{\substack{\mathrm{H}\\ |\\ \mathrm{O}\\ |}}{\mathrm{Al}^-}}\!\begin{smallmatrix}\diagup \mathrm{O}\diagup\\ \diagdown \mathrm{O}\diagdown\end{smallmatrix} \qquad (7.1)
$$

and

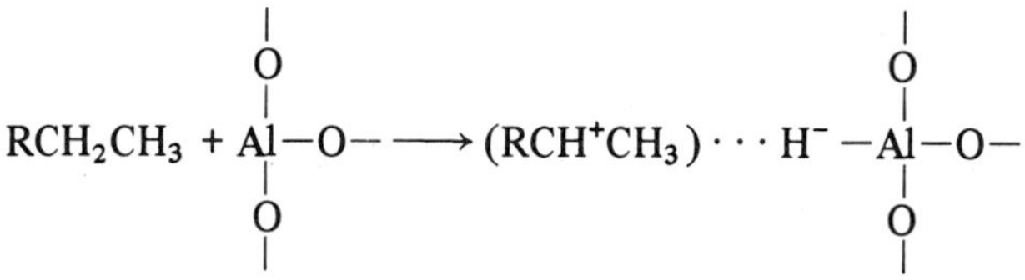

$$
\mathrm{RCH_2CH_3} + \underset{\substack{|\\ \mathrm{O}\\ |}}{\overset{\substack{|\\ \mathrm{O}\\ |}}{\mathrm{Al}}}-\mathrm{O}- \longrightarrow (\mathrm{RCH^+CH_3})\cdots\mathrm{H^-}-\underset{\substack{|\\ \mathrm{O}\\ |}}{\overset{\substack{|\\ \mathrm{O}\\ |}}{\mathrm{Al}}}-\mathrm{O}- \qquad (7.2)
$$

Similar arguments can be advanced to explain the acidity of various other mixed oxides containing metal atoms of different valence, such as $SiO_2 \cdot MgO$, $SiO_2 \cdot ZrO$, and $Al_2O_3 \cdot MgO$. Even when the cations have the same valence, as in $Al_2O_3 \cdot B_2O_3$, acidity may be observed, which can be rationalized in terms of differences in the electronegativities of the different metals. Brönsted acid sites are probably responsible for much of the activity of acid catalysts, but there is still considerable uncertainty over their role relative to that of Lewis acid sites in each of various acid-catalyzed reactions (Sec. 7.5). This is caused in large part by the difficulty of distinguishing between the two under reaction conditions. In any event it is evident that the removal of water or its reincorporation into the structure plays an important role in the acidity and catalytic activity of acid catalysts, as will be shown.

Silica-alumina, which is amorphous, typically has a maximum degree of acidity and activity at an Al/Si atomic ratio of less than unity (See Fig. 7.7 and Sec. 7.5). This can be rationalized by the concept that —Al—O—Si— type of bonds are desired but not the —Al—O—Al— type and that formation of the former is enhanced by an excess of silica gel. This further implies that the detailed procedures used in manufacture may have a significant effect on activity at a specified Al/Si ratio. Alumina is more expensive than silica, so an excess of silica is also preferred for economic reasons.

Pure silica of itself shows no acidic or basic properties, although commercial samples may exhibit a low amount of acidity because of the presence of impurities. A pure alumina may exhibit an acid concentration comparable to that of silica-alumina but of weaker acid strength, depending in large part on the nature of heat treatments (Chap. 4) and resulting changes in structure (Sec. 7.4). Commercial aluminas are also often contaminated by foreign ions from the solutions used in their preparation, and catalysts supported on alumina may have various anions or cations introduced during catalyst preparation. These may be difficult to remove by washing.

The acid strength of alumina may be deliberately increased by incorporation of halogen ions such as chloride and fluoride. The effect is to increase the acid strength, although not the total number of acid sites. Direct treatment with an aqueous mineral acid may cause partial solution or other undesirable alterations in the alumina structure. Hence the alumina may instead be contacted with an organohalogen compound in the vapor phase at an elevated temperature. This decomposes to provide the acid vapor in a dilute form, which is adsorbed onto the alumina. Alternately, alumina might be impregnated with, say, NH_4F, and then calcined to incorporate up to several percent of fluoride into the structure. The halogen replaces a hydroxyl group and, having a higher electron affinity than a hydroxyl group, causes the residual hydrogen on the surface to be more acidic.

$$
\begin{array}{ccc}
\begin{array}{c} \mathrm{H} \\ | \\ \mathrm{H^+} \quad \mathrm{O} \\ \diagdown \; | \\ -\mathrm{O}-\mathrm{Al}^{-}-\mathrm{O}- \\ | \\ \mathrm{O} \\ | \end{array}
& \longrightarrow &
\begin{array}{c} \mathrm{H^+} \searrow \\ \mathrm{Cl} \\ | \\ -\mathrm{O}-\mathrm{Al}^{-}-\mathrm{O}- \\ | \\ \mathrm{O} \\ | \end{array}
\end{array}
\tag{7.3}
$$

Some solids exhibit both acidic and basic properties, and therefore under some circumstances may act as bifunctional acid-base catalysts. Many aluminas contain small amounts of sodium and hence may have both acidic and basic sites. Although the role of the sodium sites is far from clear, high temperature stability is enhanced by low sodium content, and a low-sodium alumina seems to have some intrinsically desirable properties for a number of reactions.

7.2 ACID STRENGTH

The acid strength of a solid may be determined by its ability to change a neutral organic base, adsorbed on the solid, into its conjugate acid form. This may occur by transfer of a proton from a Brönsted acid site to the adsorbed base, or by transfer of an electron pair from the adsorbed molecule to a Lewis acid site, thus forming an acidic addition product. The acid strength can be expressed by the Hammett acidity function H_0 as

$$H_0 = pK_a + \log \frac{[B]}{[BH^+]} \tag{7.4}$$

or

$$H_0 = pK_a + \log \frac{[B]}{[AB]} \tag{7.5}$$

where K_a is the equilibrium constant of dissociation of the acid and $pK_a = -\log K_a$. [B] and [BH^+] are the concentrations of the neutral base and of its conjugated acid, and [AB] is the concentration of the addition product formed by adsorption of B onto a Lewis site.

Many compounds have a different color in the neutral base form than in the conjugate acid form and are termed *color indicators.* If such a compound upon adsorption assumes the color of its acid form, at least some of the surface sites have an H_0 value less than or equal to the pK_a value of the indicator. The lower the value of H_0 the more acidic the surface. Samples of a solid suspended, for example, in a ground form in a nonaqueous inert liquid may be tested with a battery of neutral basic indicators, each of which turns color at a different pK_a value. This gives an approximate value of the acid strength, which typically can be measured over the range of about +4 (very weak) to about −8 (very strong). Following are some representative basic indicators with their pK_a value and the wt % H_2SO_4 in aqueous solution having the acid strength corresponding to the pK_a value: *p*-dimethylaminoazobenzene (butter yellow), +3.3 (3×10^{-4} wt % H_2SO_4); dicinnamalacetone, −3.0 (48 wt % H_2SO_4); anthraquinone, −8.2 (90 wt % H_2SO_4) (Tanabe, 1970).

Some disadvantages of the method are that it is necessary that water be rigorously excluded, since water of itself may react with the surface and alter the acidic character of the solid; the time required for equilibrium may be long, amounting sometimes to days; and measurements are under conditions much different than those occurring during reaction. With dark or highly colored solids alternate procedures to determine the end point are available and may be more precise. These include spectrophotometric techniques, which are used to determine the change in spectrum of the base indicator as it becomes adsorbed in the acid form.

The amount of a gaseous base such as ammonia, pyridine, or quinoline adsorbed at elevated temperatures under a specified set of conditions is another measure of acid strength. This has the advantage of allowing the study of a catalyst under conditions more nearly similar to those of reaction. Ammonia has been studied extensively. Catalysts may be compared in terms of the amount of ammonia adsorbed as a function of temperature over a range such as 150 to 500°C. A minimum temperature of about 150°C is necessary to eliminate physical adsorption. By infrared spectra it is possible to distinguish between Brönsted and Lewis acid sites; for example, pyridine may be adsorbed as the pyridinium ion or as coordinately bonded pyridine, respectively. The acid strength of some solids as determined by the indicator method are given in Table 7.1 (Tanabe, 1970).

Table 7.1 Acid strength of some solids*

Solid acids	H_0
Original kaolinite	$-3.0 \sim -5.6$
Hydrogen kaolinite	$-5.6 \sim -8.2$
Original montmorillonite	$+1.5 \sim -3.0$
Hydrogen montmorillonite	$-5.6 \sim -8.2$
Silica-alumina	<-8.2
$Al_2O_3 \cdot B_2O_3$	<-8.2
Silica-magnesia	$+1.5 \sim -3.0$
1.0 mmol/g H_3BO_3/SiO_2	$+1.5 \sim -3.0$
1.0 mmol/g H_3PO_4/SiO_2	$-5.6 \sim -8.2$
1.0 mmol/g H_2SO_4/SiO_2	<-8.2
$NiSO_4 \cdot xH_2O$ heat-treated (350°C)	$+6.8 \sim -3.0$
$NiSO_4 \cdot xH_2O$ heat-treated (460°C)	$+6.8 \sim +1.5$
ZnS heat-treated (300°C)	$+6.8 \sim +4.0$
ZnS heat-treated (500°C)	$+6.8 \sim +3.3$
ZnO heat-treated (300°C)	$+6.8 \sim +3.3$
TiO_2 heat-treated (400°C)	$+6.8 \sim +1.5$

*Tanabe, 1970, p. 7.

7.3 ACID AMOUNT

A sample of the solid acid as a powder is suspended in an inert nonaqueous liquid, e.g., benzene, and is titrated with a base, utilizing an indicator. The titrating base must be a stronger base than the indicator, and *n*-butylamine, $pK \sim +10$, is often used for this purpose. As the base is added it adsorbs on acid sites, the strongest ones first, and ultimately it displaces indicator molecules from the solid. When the indicator has been substantially replaced, the color changes. At this equivalence point $[B]/[BH^+] \approx 1$. If the pK_a value of the indicator is, say +3.3, then the amount of base added is equivalent to the amount of acid sites having $H_0 \leqslant +3.3$. By amine titration with indicators having different pK_a values, the amount of acid sites having strengths exceeding the various corresponding values of pK_a $(=H_0)$ may be determined. As with the determination of acid strength, this method gives the sum of Brönsted and Lewis acid sites.

The method is not suitable for molecular-sieve zeolite catalysts in which the pore size is so small that the indicator molecules cannot penetrate into the interior. No systematic study appears to have been made of this effect, but indicator molecules can probably enter the pores of only the most open zeolites such as zeolite Y and possibly some acid-leached mordenites (see Tables 7.2 and 7.3, pp. 166, 167).

A number of methods have been reported for measuring Brönsted acids alone or Lewis acids alone, but some of these are in dispute. A series of so-called H_R indicators, consisting of various aromatic alcohols, are reported to be specific for protonic acids (Hirschler, 1963) and are used in the same way as H_0 indicators. Infrared spectroscopic studies of adsorbed species such as ammonia or pyridine in the vapor phase are perhaps the most conclusive, since the spectrum of, e.g., coordinately bonded pyridine is much different from that of the pyridinium ion formed by proton transfer from a Brönsted acid. However at reaction conditions the ratio of Lewis to Brönsted acid sites may be far different than at the conditions under which spectroscopic studies are made, especially if water vapor is present under one set of conditions but not the other. Methods for characterizing acidity of solid catalysts are described in more detail in reviews by Goldstein (1968), by Forni (1974), and by Benesi and Winquist (1978).

7.4 ACID PROPERTIES OF REPRESENTATIVE SOLIDS

Figure 7.2 (Benesi, 1957) shows the amount of acid as a function of acid strength for three fluid-bed cracking catalysts, as determined by amine titration. The vertical lines indicate titer uncertainties. The solids had been previously calcined at 550°C. Filtrol SR is a commercial cracking catalyst prepared from a clay. With silica-alumina and Filtrol almost all the sites are very acidic, with $H_0 \leqslant -8.2$. Silica-magnesia has the highest concentration of acid sites per gram, but none exceeding $H_0 = -3$ in strength.

Figure 7.3 (Benesi, 1957) shows the acid amount versus acid strength for three natural clays that had previously been dried at 120°C for 16 h. These consist of oxides of silicon and aluminum and were the source of the original cracking catalysts. They clearly have a lower acid concentration and acid strength than synthetic silica-alumina. Present-day cracking catalysts usually consist of several percent of a synthetic zeolite incorporated in a silica-alumina matrix. Figure 7.4 (Ukihashi et al. in Tanabe, 1970, p. 75) shows that Y zeolite, which is often used in catalyst preparations, has a high acid concentration and acid strength in the H^+ form or when exchanged with calcium or lanthanum.

The acidic properties of silica-alumina are substantially dependent on the method of preparation, temperature of dehydration, and treatment with steam as well as on the Al/Si ratio present. An extensive literature exists on relationships between acid properties, various treatments, and catalytic activity, much of it relatively empirical.

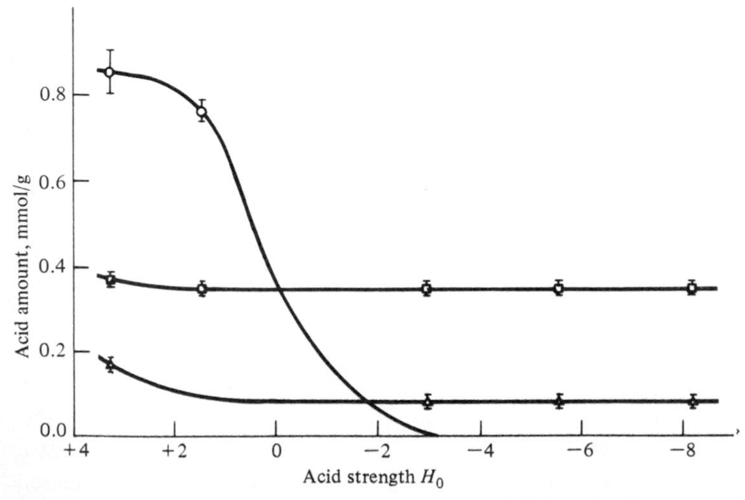

Figure 7.2 Acid amount versus acid strength for three solid acids: $SiO_2 \cdot MgO$ (○), $SiO_2 \cdot Al_2O_3$ (□), and Filtrol (△). *(Benesi, 1957.)*

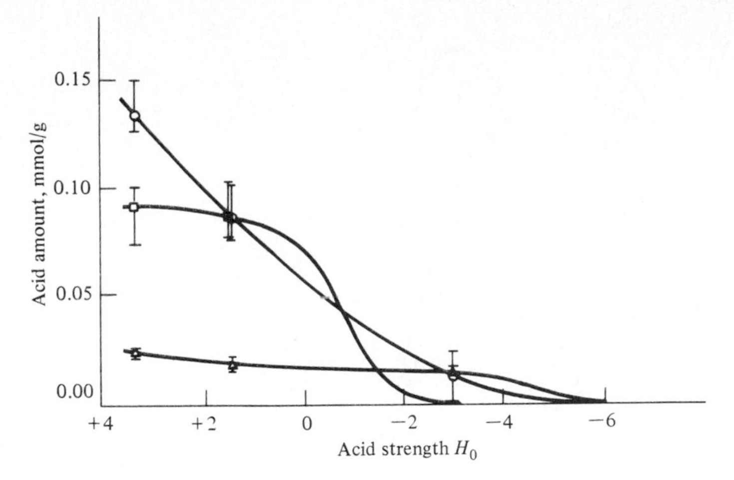

Figure 7.3 Acid amount versus acid strength for various natural clays: attapulgite (○), montmorillonite (□), and kaolinite (△). *(Benesi, 1957.)*

Figure 7.5 (Ito et al. in Tanabe, 1970, p. 46) shows the amount of acid exceeding various acid strengths for a pure alumina prepared from aluminum isopropoxide, as a function of calcination temperature. Two maxima in acidity are exhibited, at about 500°C and again at about 800°C. X-ray analysis indicated that at 450 to 500°C an η-alumina phase of low crystallinity was present, being converted to a highly crystalline η-alumina at 600°C. After calcination at 800°C a mixture of the η and θ phases existed, but at 1000°C the alumina was in the α form. These results show that the total amount of acidity present is comparable to that of silica-alumina (Fig. 7.2). The

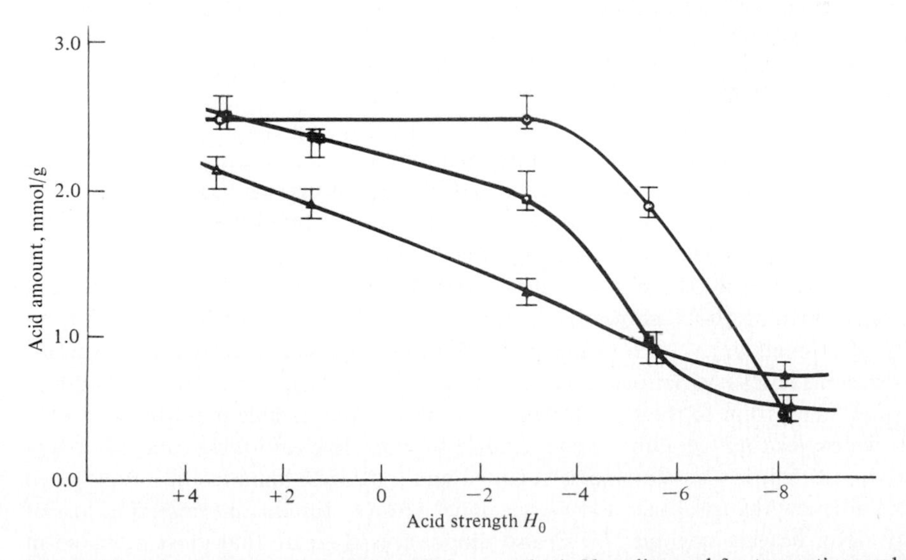

Figure 7.4 Acid amount versus acid strength for a synthetic Y zeolite, and for two cation-exchanged catalysts: the synthetic Y zeolite (H^+) (○), calcium cation-exchanged (Ca^{2+}) (□), and lanthanum cation-exchanged (La^{3+}) (△). *(Ukihashi et al. in Tanabe, 1970, p. 75.)*

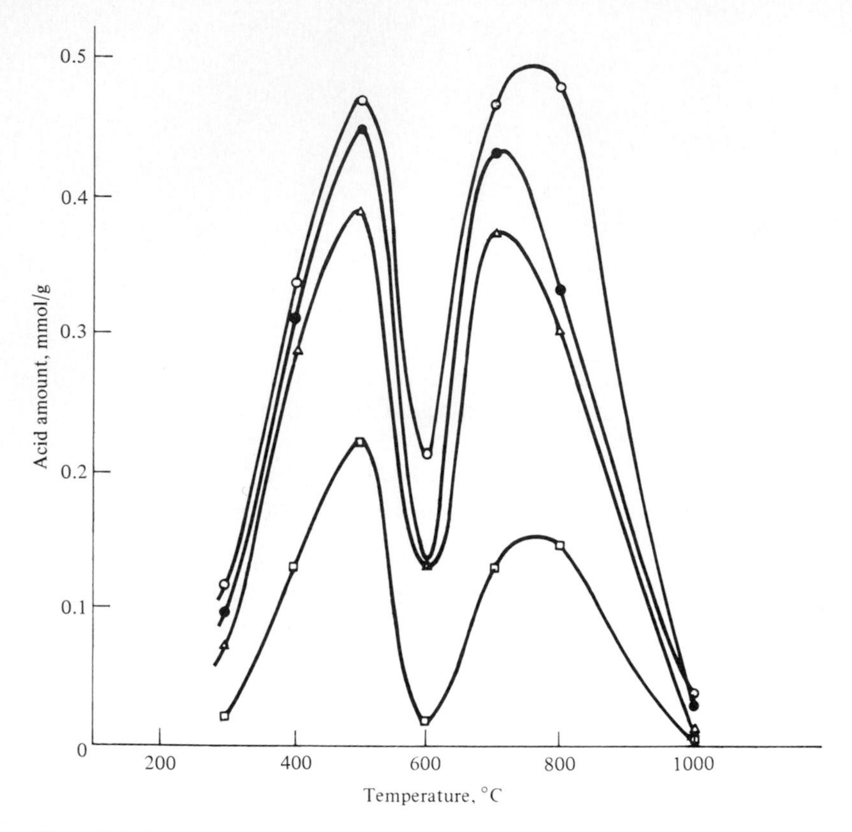

Figure 7.5 Amount of acid on Al_2O_3 at various acid strengths versus calcination temperature; acid strength $H_0 \leqq 3.3$ (○), $H_0 \leqq 1.5$ (●), $H_0 \leqq -3.0$ (△), and $H_0 \leqq -5.6$ (□). X-ray analysis: 450–500°C, η-Al_2O_3 (low crystallinity); 600°C, η-Al_2O_3 (high crystallinity); 800°C, η-Al_2O_3 + θ-Al_2O_3; 1000°C, α-Al_2O_3. (*Ito et al. in Tanabe, 1970, p. 46.*)

same conclusion is reached by various other methods of characterizing alumina acidity, such as chemisorption of gaseous ammonia and of various organic bases. However, the acid strength is lower, and the alumina does not seem to have any Brönsted-type acid sites, regardless of the calcination temperature.

In contrast to the pure alumina characterized in Fig. 7.5, a particular commercial alumina calcined at 500°C showed an acidity of 0.287 mmol/g for $H_0 \leqslant +1.5$ but no acidity of strength $H_0 \leqslant -5.6$ (Tanabe, 1970, p. 46). These and related observations showing considerable variations between the acidic properties of different aluminas, even after calcination to the same temperature, are of considerable importance in view of the widespread use of alumina as a catalyst support. In a reforming catalyst such as Pt/Al_2O_3, acidity in the alumina is required and a catalyst manufacturer may go to considerable lengths to obtain a low-crystalline η-Al_2O_3 for this purpose. The loss of acidity upon heating beyond 500°C also emphasizes the care that must be taken in catalyst regeneration procedures where it is desired to maintain intrinsic acidity of the alumina.

Ordinary metal sulfates or metal phosphates have no intrinsic acidity, but they acquire moderate acid strength and catalytic activity after heat treatment. Other processes, such as irradiation, that produce crystal imperfections, also may create catalytic activity. With nickel sulfate a maximum of acidity and catalytic acidity is obtained by calcination at about 375°C. The development of acidity is associated with the gradual removal of water of constitution, and maximum acidity occurs at about 0.5 mol H_2O/mol $NiSO_4$. This is an intrinsic property and not due to impurities. Solid metal sulfates are of considerable theoretical interest in that a wide range of acid amounts and acid strengths can be obtained. Catalysis by these substances is reviewed by Takeshita et al. (1974). Acid concentrations are comparable (on an area basis) to those of other solid acidic catalysts, but the strengths are moderate relative to such acids as silica-alumina.

Solid metal phosphates may be used to catalyze the polymerization of olefins to low-molecular-weight polymers, and they exhibit acidity comparable to that of the metal sulfates. Zinc oxide and titanium oxide exhibit only weak acidic properties, but if prepared from the chlorides they may exhibit fairly high acid strength, again depending on calcination conditions. Compounds such as Cr_2O_3, MoO_3, and ZnS also exhibit some acidity.

Mineral acids may be *mounted*, i.e., supported, on inert substances such as diatomaceous earth or silica gel. The order of acid strength is $H_2SO_4 > H_3PO_4 > H_3BO_3$. Mounted phosphoric acid is used to catalyze the polymerization of olefins to low-molecular-weight products, as in the conversion of a refinery stream containing C_3 and C_4 olefins to a gasoline fraction. The process has been used since the 1930s and typically operates at about 200 to 230°C and a pressure of about 3 to 7 MPa using an acid strength of 100 to 115 percent (H_3PO_4 plus some dissolved P_2O_5, which is a viscous liquid under these conditions). The properties of the catalyst are markedly affected by concentration, and water is added to the feed to maintain the acid concentration at the desired level. The process is reviewed by McMahon et al. (1963), Schaad (1955), and Villadsen and Livbjerg (1978). Figure 7.6 is a photograph of a representative industrial catalyst. It may also be used for olefin hydration.

7.5 CORRELATIONS BETWEEN ACIDITY AND CATALYTIC ACTIVITY

A substantial number of correlations have been established between the total amount of acid present, as measured by butylamine titration, and catalytic activity. Figure 7.7 (Johnson, 1955) shows, for example, a linear correlation between the amount of acid ($H_0 \leqslant +3.3$) in a series of silica-alumina catalysts and activity for polymerization of propylene, at a reaction temperature of 200°C. The maximum amount of acidity was obtained with a catalyst containing 10.3 wt % Al_2O_3; that containing 25.1 wt % Al_2O_3 was less active and showed less acidity, possibly because fewer bonds of the —Al—O—Si— type were formed (Sec. 7.1). Similarly, the activity for cumene decomposition was shown to increase with amount of acid, although here the correlation was not linear. With silica-alumina essentially all the acid sites are strong. There are also good correlations between the acidity of alumina (as measured by adsorption of ammonia or pyridine) and activity for polymerization of ethylene or propylene. Cor-

Figure 7.6 Phosphoric acid "mounted" on clay catalyst. (*Courtesy of United Catalysts, Inc., Louisville, Ky.*)

relations are less successful if the number or strength of acid sites are varied by greater changes in the nature of the solid, as in comparing silica-aluminas and silica-magnesias.

It seems plausible that a certain minimum acid strength should be needed to cause catalysis and that the threshold acid strength would vary with the reaction. This has indeed been shown by selective poisoning experiments. The polymerization of propylene at 100°C is proportional to the amount of acid sites on nickel sulfate having

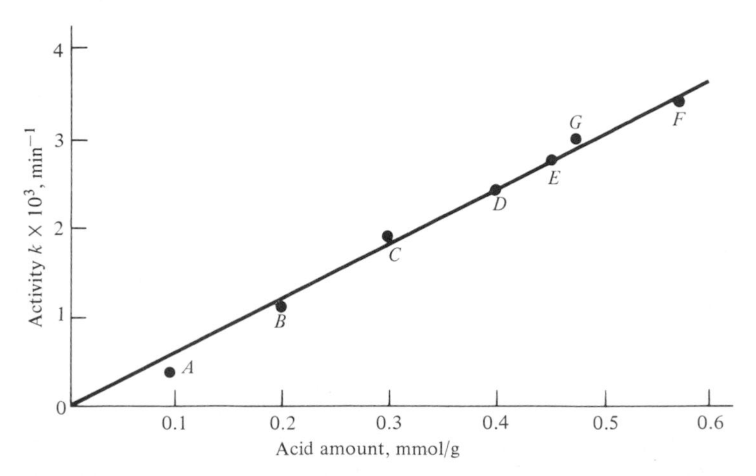

Figure 7.7 Propylene polymerization activity versus acid amount for a series of $SiO_2 \cdot Al_2O_3$ catalysts. (*Johnson, 1955.*)

Catalyst	A	B	C	D	E	F	G
Al_2O_3 (wt %)	0.12	0.32	1.04	2.05	3.56	10.3	25.1

$H_0 \leqslant +1.5$. From studies with other solid acids it appears that sites of this or greater acid strength are required for this reaction regardless of the type of solid. Silica-alumina can be treated with a weak base, such as potassium acetate, which neutralizes the stronger acid sites so that no cracking occurs. However butylamine titration indicates that some acidity is still present. Pines and Haag (1960) studied the dehydration of butanol, the skeletal isomerization of 3,3 dimethylbutene, and the skeletal isomerization of cyclohexene over an alumina catalyst poisoned to various degrees. Butanol dehydration occurred on weaker sites than did the olefin reactions, and they concluded that the concentration of sites active for this reaction was about 10^{14} sites per square centimeter versus about 10^{13} sites per square centimeter for cyclohexene isomerization. The products formed can also vary with the strength of the acid site. High acid strength may be undesirable in that it can cause undesired side reaction, or it may decrease the rate of a reaction by strong adsorption of products or reactants. Very strong acidity is undesirable since it causes an excessive rate of formation of carbonaceous deposits (coke).

The gradual poisoning of an acid catalyst, either deliberate or inadvertent, makes it inactive first for those reactions requiring the highest acid strengths and then progressively inactive for those requiring weaker acid strengths. Studies by Misano et al. (Tanabe, 1970, p. 121) on silica-alumina poisoned by increasing amounts of pyridine showed the following listing of reactions in order of increasing difficulty.

1. Dehydration of *tert*-butanol to butenes
2. Depolymerization of diisobutylene to butenes
3. Double-bond migration and trans-cis isomerization of *n*-butenes
4. Dealkylation of *tert*-butylbenzene (cracking)
5. Skeletal isomerization of isobutylene to *n*-butenes

One consequence is that as a solid catalyst ages by gradual accumulation of deposits and poisons, the most difficult reactions disappear first. Disproportionation reactions, as of toluene to benzene and xylene, require highly acidic sites. If dimethylnaphthalene is passed over a particular acidic zeolite catalyst, initially considerable disproportionation occurs. This gradually dies off and the reaction becomes predominately isomerization, after which the entire activity dies down.

Other generalizations may be developed, based in part on the relative stability of various carbonium ions as determined from studies with mineral or liquid-phase Lewis acids. In an homologous series, reactivity increases with molecular size. Thus in the paraffin series, keeping acid strength and reaction conditions constant, with *n*-butane one may observe only isomerization, with *n*-hexane other products in addition to isomers start to appear, and with *n*-heptane significant cracking occurs.

A number of correlations have been reported between catalytic activity and type of acid site. The activity of silica-alumina catalysts for polymerization of propylene or isobutylene and for the cracking of cumene is closely correlated with the amount of Brönsted acid sites, as shown by several investigators (Tanabe, 1970, pp. 125-133). Some isomerization reactions have been correlated with Lewis acid sites. In some early studies it was reported that maximum activity of silica-alumina catalysts occurred when the composition corresponded to an Al/Si atomic ratio of 1, but it now appears that for many reactions catalyzed by silica-alumina, maximum activity occurs at an alumina content in the region of about 10 to 30 wt % Al_2O_3, which corresponds to Al/Si atomic ratios considerably less than unity. A maximum in the concentration of Brönsted acid sites occurs in the general vicinity of about 20 wt % Al_2O_3, so this composition range of maximum activity may only apply to reactions catalyzed by Brönsted sites.

The various kinds of hydrocarbon reactions that have been studied make it apparent, however, that acidity as such, either of the Brönsted and/or Lewis type, is not the unique determinant of catalyst behavior. Basic sites on so-called acid catalysts may cause reaction to occur by a dual-function mechanism. Some relatively nonacidic zeolites exhibit high activity but reaction behavior more similar to that encountered with noncharged species (free radicals). For several reactions maximum activity is observed if the catalyst is calcined at an intermediate temperature, for example, 500 to 700°C, which suggests that perhaps some ideal balance of Lewis and Brönsted sites is optimum.

Tanabe (1970) gives a more detailed discussion of correlations between acid-base properties and catalyst activity and selectivity. Cracking catalysts are discussed in early reviews by Oblad et al. (1951) and by Ryland et al. (1960). Physical properties and sintering characteristics are reviewed by Ries (1952). Base-catalyzed reactions of hydrocarbons were reviewed by Pines and Schaap (1960), more recently by Pines (1972), and are treated in detail in a recent book by Pines and Stalick (1977). The emphasis here is on metals such as metallic sodium and potassium rather than on basic oxides and hydroxides.

7.6 MECHANISM OF CATALYTIC CRACKING

The reactions occurring in catalytic cracking are highly complex, and many of the primary products undergo secondary reactions. The reaction pattern can be interpreted in broad outline in terms of the carbonium-ion theory following the original suggestions of Whitmore to interpret certain hydrocarbon reactions in solution. The theory as applied to catalytic cracking was developed by many investigators, including Greensfelder et al. (1949), C. L. Thomas (1949), Hansford (1952), and others. It is reviewed by Oblad et al. (1951), by Emmett (1965, p. 89), and more recently by Gates et al. (1979).

A carbonium ion can be formed in a number of ways, but one of the simplest is via abstraction by an olefin of a proton from a Brönsted acid.

$$H_2C{=}CHR + H^+ \longrightarrow CH_3CH^+R \tag{7.6}$$

This can undergo a variety of reactions, including the reverse process to cause double-bond shift, as well as skeletal isomerization, cracking, and hydrogen exchange. The observed product distribution indicates that a large carbonium ion formed from a paraffin undergoes fission (cracking) at the β position, for example,

$$CH_3CH^+CH_2CH_2(CH_2)_nCH_3 \longrightarrow CH_3CH{=}CH_2 + CH_2{}^+(CH_2)_nCH_3 \tag{7.7}$$

The carbonium ion formed will then undergo rapid isomerization to a more stable form. By this mechanism no C_1 or C_2 products should be formed from a paraffin reactant, in general accord with the facts. The net effect in cracking is the conversion of a paraffin to give a smaller paraffin and an olefin. The olefin, in turn, may crack to give two smaller olefins.

Industrially the feedstock to a catalytic-cracking reactor may consist of a mixture of paraffins, naphthenes, aromatics with alkyl side chains, and more complex molecules. Recycled feed streams may contain considerable olefins and a higher aromatic content. For hydrocarbons of the same carbon number the order of decreasing reactivity is about as follows on silica-alumina and REXH catalysts (zeolite X exchanged with rare earths and in the hydrogen form) (Sec. 7.7.6):

1. Olefins.
2. Alkylbenzenes with C_3 or larger-size groups. (Side groups are removed to form benzene and corresponding olefins.)
3. Naphthenes (rupture of ring and rather complex products).
4. Polymethyl aromatics.
5. Paraffins.
6. Unsubstituted aromatics. (These compounds are highly stable and little or no cracking occurs.)

The rate of cracking increases with molecular weight. It is faster with tertiary-carbon structures and molecules in which an aromatic ring is attached to a side chain of sufficient length.

Some reactions that are thermodynamically possible do not occur to a significant extent. These include dehydrogenation of paraffins to olefins, dehydrogenation of naphthenes to aromatics, dehydrocyclization of paraffins to ring compounds, isomerization of saturated hydrocarbons (paraffins or naphthenes), decomposition of hydrocarbons to carbon and hydrogen or to methane.

Some reactions subsequent to the initial cracking may also be significant. Double-bond isomerization of olefins occurs so rapidly that the products are in chemical equilibrium with respect to this reaction. Hydrogen transfer (hydrogen exchange) may be quite rapid. This is exemplified by a reaction such as

$$C_6H_{14} + C_3H_7^+ \longrightarrow C_6H_{13}^+ + C_3H_8 \tag{7.8}$$

Hydrogen exchange is particularly important since it leads to a variety of coreaction effects. For example, in reaction (7.8) a carbonium ion of the original reactant is formed, and simultaneously a smaller carbon skeleton can be stabilized from further cracking. Hydrogen transfer is especially rapid to tertiary olefins, converting them to the corresponding isoparaffins. Thus the ratio of isoparaffins to *n*-paraffins in the product from industrial catalytic cracking may exceed that calculated for thermodynamic equilibrium at the reaction temperature. This is desirable because of the high octane number of isoparaffins.

Aromatics may also be formed by dimerization of olefins and cyclization of diolefins. Adsorbed aromatics and the products of polymerization and condensation reactions of olefins are probably the precursors of coke.

The slowest step in the overall series of reactions appears to be the initial formation of the carbonium ion. Reaction (7.6) is readily conceivable as a mechanism if an olefin is present. If a paraffin alone were being reacted, it may react as a weak Lewis base with a Lewis acid site.

$$\mathrm{RH} + \begin{array}{c} | \\ \mathrm{O} \\ | \\ \mathrm{Al{-}O{-}} \\ | \\ \mathrm{O} \\ | \end{array} \longrightarrow \mathrm{R}^+ + \mathrm{H{-}}\begin{array}{c} | \\ \mathrm{O} \\ | \\ \mathrm{Al^-{-}O{-}} \\ | \\ \mathrm{O} \\ | \end{array} \qquad (7.9)$$

Under industrial conditions it appears that the primary initiating mechanism is proton donation from a Brönsted acid site to an olefin, which, if not present in the feed, could readily be formed by thermal decomposition. Studies by Weisz (1973) showed that the accelerating effect of an olefin (butene) could be observed on the rate of cracking of pure butane on hydrogen mordenite at olefin concentrations as low as 0.001 percent. There is some evidence from isotope exchange studies that the actual proton transfer agent may be an essentially irreversibly adsorbed carbonium ion rather than a proton attached directly to the solid catalyst. The mechanisms of alkylation, isomerization, polymerization, and cyclization in terms of carbonium ion chemistry are reviewed by Emmett (1965, p. 104) and by Gates et al. (1979).

7.7 ZEOLITES

Zeolites are highly crystalline, hydrated aluminosilicates that upon dehydration develop in the ideal crystal a uniform pore structure having minimum channel diameters (apertures) of from about 0.3 to 1.0 nm. The size depends primarily upon the type of zeolite and secondarily upon the cations present and the nature of treatments such as calcination and leaching. Zeolites have been of intense interest as catalysts for some two decades because of the high activity and unusual selectivity they provide in a variety of acid-catalyzed types of reactions. In many cases, but not all, the unusual selectivity is associated with the extremely fine pore structure. This permits only certain molecules to penetrate into the interior of the catalyst particles, or only certain products to escape from the interior.

The structure of a zeolite consists of a three-dimensional framework of SiO_4 and AlO_4 tetrahedra, each of which contains a silicon or aluminum atom in the center. The oxygen atoms are shared between adjoining tetrahedra, which can be present in various ratios and arranged in a variety of ways. Zeolites may be represented by the empirical formula

$$M_{2/n} \cdot Al_2O_3 \cdot xSiO_2 \cdot yH_2O \qquad (7.10)$$

or by a structural formula

$$M_{x'/n}[(AlO_2)_{x'}(SiO_2)_{y'}] \cdot wH_2O \qquad (7.11)$$

where the bracketed term is the crystallographic unit cell. The metal cation (of valence *n*) is present to produce electrical neutrality since for each aluminum tetrahedron in the lattice there is an overall charge of −1. Access to the channels is limited by apertures consisting of a ring of oxygen atoms of connected tetrahedra. There may be 4, 5, 6, 8, 10, or 12 oxygen atoms in the ring. The largest apertures occur in the faujasite-type zeolites (types X and Y) and mordenite, which are of high current interest as catalysts, and also types L and Ω (Table 7.2). Zeolites are often prepared in the sodium form, and this can be replaced by various other cations or by a hydrogen ion. At least 34 species of zeolite minerals and over 100 types of synthetic zeolites are known. The number of papers published and patents issued on zeolites has grown exponentially, and as of 1980 well exceeds 10,000. A guide to the literature is given at the end of this section.

Classification and nomenclature is still undergoing change and refinement (Breck, 1974, p. 19). For some groups it is confused by inadequate characterization of previously named zeolites. Naturally occurring materials or their synthetic equivalents are usually described by a mineral name, e.g., mordenite. New synthetic types are usually designated by a letter or group of letters assigned by the original investigators, for example, A, X, Y, Ω, and ZSM. Types X and Y are structurally and topologically related to the mineral faujasite and are frequently referred to as *faujasite-type* zeolites. The zeolites are sometimes classified into seven groups having a common sub unit of structure of aluminum and silicon tetrahedra. The atomic ratio of Si/Al in the zeolites as originally prepared varies between 1 and 5, so bonds of the type Al—O—Al are not formed. Alumina can be selectively removed from some of the zeolites such as mordenite, which has an original Si/Al ratio of about 5, to produce a stable structure having a much higher Si/Al ratio. Identification and characterization of the zeolites containing potassium was recently clarified by Sherman (1977).

Zeolites are of practical interest for a variety of reasons. The fine pore structure permits adsorption separations to be carried out on the basis of molecular size and shape, so-called *molecular sieving*, as in the separation of *n*-paraffins from isoparaffins. The ability to alter zeolite properties by ion exchange permits the synthesis of adsorbents of unusual selectivity, even when all molecules have free access to the interior pores of the zeolite. The ion-exchange properties also allow a high degree of flexibility in synthesizing catalysts, for example, the ability to produce a highly dispersed metal. From the catalytic point of view zeolites are of especial interest in that they exhibit unusually high activity for various acid-catalyzed reactions such as cracking, the ability to combine a molecular-sieving property with catalysis, and unusual selectivity behavior.

7.7.1 Pore Structure

Table 7.2, abstracted from a larger list by Breck (1974), gives some well-characterized zeolites of interest in catalysis. Of these, especial industrial interest lies in type Y, mordenite, erionite and offretite (which are closely related and frequently intergrown), and zeolites synthesized in the presence of various quaternary alkylammonium

Table 7.2 Selected well-characterized zeolites of interest in catalysis

Group	Name	Void fraction (from H_2O content)	Channel geometry*	Kinetic diameter of dehydrated form[†]
6	Mordenite (Na form), large port	0.28	0.67 × 0.7, one-dimensional interconnecting with 0.29 × 0.57, one-dimensional	0.62
3	Type A (Na form)	0.47	0.42, three-dimensional	0.36–0.39
4	Type X Type Y (both in Na form)	0.50 0.48	0.74, three-dimensional	0.81 0.81
4	Chabazite (Ca form)	0.47	0.36 × 0.37, three-dimensional	0.43
2	Erionite	0.35	0.36 × 0.52, three-dimensional (but tortuous)	0.43
2	Offretite[‡§]	0.40	0.64, one-dimensional interconnecting with 0.35 × 0.52, one-dimensional	0.6
4	Type L, $K_9[(AlO_2)_9(SiO_2)_{27}] \cdot 22H_2O$	0.32	0.71, one-dimensional	0.81
2	Omega, $Na_{6.8}TMA_{1.6}[(AlO_2)_8(SiO_2)_{28}] \cdot 21H_2O$	0.38	0.75, one-dimensional	1.0

*Breck, 1974, p. 48. Based on structure of hydrated zeolite. Units in nanometers.

[†]Breck, 1974, pp. 133–180. Commercial NaX has a reported pore size of about 1.0 nm, commercial Na mordenite about 0.7, commercial H mordenite about 0.8–0.9 nm (Breck, p. 747).

[‡]TMA offretite is also known as Zeolite O.

[§]Zeolite T is mostly offretite intergrown with some erionite with free aperture sizes probably similar to that of erionite.

Table 7.3 Values of selected critical molecular diameters, nm

Compound	From structure †	‡	Kinetic diameter*
C_3H_8	–	–	0.43
n-C_4H_{10}	–	–	0.43
iso-C_4H_{10}	0.56	–	0.50
n-C_5H_{12} and all higher n-paraffins	0.49		
2,2-dimethyl propane (neopentane)	–	–	0.62
2,2,4-trimethyl pentane	0.67		
benzene	0.63	0.675	0.585
toluene	–	0.675	
cumene	0.67	0.675	
cyclohexane	0.65	0.69	0.60
m-xylene	–	0.74	0.71
p-xylene	–	0.675	
1,3,5-trimethylbenzene (mesitylene)	0.84	0.84	0.85
1,3,5-triethylbenzene	0.92	0.92	
1,3-diethylbenzene	–	0.74	
1-methylnaphthalene	–	0.79	
$(C_2F_5)_3N$	–	–	0.80
$(C_4H_9)_3N$	–	–	0.81
$(C_4F_9)_3N$	–	–	1.02

*From Lennard-Jones potential function (Breck, 1974, p. 636).

†Evaluated from bond lengths, bond angles, and van der Waals radii (Pitcher, 1972).

‡Moore and Katzer, 1972.

compounds. Table 7.3 lists channel sizes based on the structure of hydrated zeolites and also the kinetic diameters in the dehydrated form, as estimated from sorption measurements. During dehydration some movement of cations may occur; the cations may also move by interaction with adsorbed species and thus alter the structure. Hence the effective pore size of dehydrated zeolites cannot be directly inferred from that of the hydrated form. The effective pore diameter may also vary with the nature of the cation present, for example, Na, Ca, H, or with the dehydroxylated form, and with other treatments. For use in reactions, such as catalytic cracking, high temperature stability is important, and this is generally increased by minimum sodium content.

A number of other materials also have a fine and more or less regular pore structure, such as some fine-pore silicas, porous glass, montmorillonite and other clays, and porous carbons prepared by controlled pyrolysis of synthetic polymers. These may therefore also exhibit molecular-sieving properties. The term *molecular-sieve zeolites* is sometimes used to distinguish zeolites from this broader group of materials, although the term *molecular sieve* is sometimes used loosely and perhaps misleadingly as a synonym for a zeolite even when the molecular-sieving property of a zeolite is not being utilized.

The pore structures may be designated as one-, two-, or three-dimensional (Breck,

1974, p. 59). Mordenite has a two-dimensional structure in which the larger channels are interconnected by other cross channels. These, however, are sufficiently smaller that they do not generally provide a means for transport of molecules between adjacent passageways. Hence mordenite has effectively a one-dimensional structure that may be regarded as an array of parallel noninterconnecting channels. In the ideal mordenite pore structure these are slightly elliptical, with dimensions of about 0.70 × 0.67 nm. Unlike type Y, sodium mordenite can be treated directly with acid to produce a highly active and stable hydrogen form. This treatment also may increase channel size, up to about 0.8 to 1.0 nm diameter, by selective leaching of alumina. Si/Al ratios in the leached mordenite can range up to 50:1 or more, if desired, without collapse of structure.

The faujasite-type zeolites (X and Y), chabazite, and erionite have three-dimensional intersecting channels in which the minimum free diameter is the same in each direction. In erionite a zigzag channel is formed, and some commercial materials may be an intergrown mixture of erionite and offretite. The X and Y types consist of an array of cavities having internal diameters of about 1.2 nm. Access to each cavity (also termed a *supercage*) is through six equispaced necks having a diameter of about 0.74 nm. The X and Y zeolites have among the largest minimum aperture restrictions of any zeolite, and the highest void fractions. Their structure may be visualized by the line drawing of Fig. 7.8. In the foreground is one of the necks, consisting of a ring of 12 oxygen atoms, through which may be seen slight portions of three other necks. The

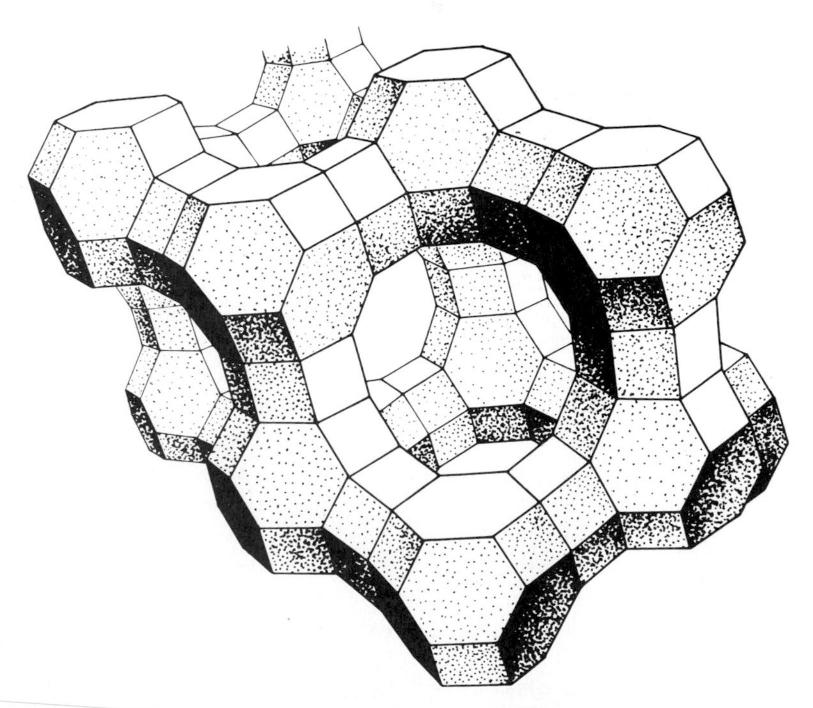

Figure 7.8 Line drawing of faujasite structure. (*Courtesy of P. B. Venuto.*)

rows of necks comprise an array of passageways perpendicular to one another in three dimensions.

A number of zeolites having relatively open structures can be synthesized in the presence of a tetraalkylammonium cation (see below), which can later be removed by heat to leave behind a relatively open pore structure. Examples are types O, Ω, or TMA offretite, which are synthesized in the presence of tetramethylammonium (TMA) cation. The ZSM group of zeolites are similarly synthesized, utilizing various alkyl groups in the quaternary ammonium compound. The basic concept is to replace an inorganic cation such as Na^+ with an organic cation of unit charge whose bulk can be varied by varying the size of the alkyl groups. It appears that the effective size and structure of the passageways can be controlled to a significant extent by this technique.

7.7.2 Synthesis

Zeolites are generally synthesized by a hydrothermal process; sodium hydroxide is added to such reactants as sodium silicate and sodium aluminate, causing gelation to occur (Breck, 1974, p. 245). The freshly coprecipitated gel is allowed to age, for example, at about 25°C, after which it is heated to a temperature of from about 50 to 200°C. The pressure is equal to the saturated vapor pressure of the water present. All zeolites are unstable under synthesis conditions, and those with wide pores and exchangeable cations tend to be formed first. At higher temperatures the most compact type of zeolites dominate (Smith, 1976). The ultimate desired crystal form is controlled by choice of initial composition, reaction conditions, reaction time, and nucleation. Even for a fixed initial bulk composition, the product can vary markedly with the type of reactant, e.g., by changing from sodium silicate to colloidal silica.

Naturally occurring minerals may also be suitable starting materials–kaolin-type clays in particular (Breck, 1974, p. 313). Aluminum may be removed by leaching or additional silica may be added to secure a Si/Al ratio higher than that existing in the clay. A naturally occurring clinoptilolite and one of the commercial mordenites (Huber) have been recrystallized into faujasite (Robson et al., 1977) by treatment with sodium hydroxide at about 100°C. Zeolites prepared directly from minerals may contain foreign constituents and, although satisfactory for some uses, may be unsuitable as catalysts. The final crystalline product is typically about a micrometer or slightly larger in size. This fine powder is usually pelletized or incorporated in a gel to convert it to a form usable in a reactor. The external area of the zeolite crystals is of the order of 1 percent of the internal area. The latter is typically in the region of 500 to 800 m^2/g.

7.7.3 Effective Pore Size

Since the pores in a zeolite are so small, the question of whether reacting molecules have free access to the interior and whether possible product molecules can readily escape under a particular set of reaction conditions assumes considerable importance. Only limited evidence is available. Minimum pore sizes are frequently assigned on the basis of crystallographic measurements, but they are also affected by the nature of the cations present and by heat treatments. Many materials also incorporate stacking faults, impurities, or deformations of various kinds that restrict channels, or they may have been treated in various ways to open up channels. A more useful characterization in practice, especially on a relative scale, stems from measurements with fluids consisting of molecules of progressively increasing size to determine which are absorbed into the pores and which are not under a specified set of conditions (Breck, 1974, p. 633).

Several factors restrict a precise quantitative application of the method. The lattice may be distorted to various degrees by the degree of polarity of the absorbed molecule, and the effective diameter of a probe molecule may be somewhat uncertain. For spherical, nonpolar molecules the kinetic diameter from the Lennard-Jones potential is preferred. For more complex molecules the minimum cross-sectional diameter is preferred, which may be estimated from bond lengths, bond angles, and van der Waals radii. Table 7.3 lists critical diameters as determined by the two methods for a number of substances of interest. Measurements of a separation factor (partition coefficient) from, say, a binary mixture must be interpreted with caution. Considerable separation may be achieved even when both compounds have full access to the pore interiors and even when the two compounds are very similar in structure (Satterfield and Cheng, 1972).

For well-characterized zeolites a ranking of pore sizes by sorption measurements generally agrees with a ranking from crystallographic measurements, but with mordenite two forms, *large port* and *small port*, have been discussed. These have substantially different sorption characteristics but exhibit the same X-ray patterns. The reasons are not clear, but since mordenite has an effectively one-dimensional pore structure, a slight degree of blocking by, e.g., foreign ions, could have a major effect. The characterization of a variety of zeolites by sorption and diffusion measurements is reviewed by Eberly (1976).

7.7.4 Diffusion in Zeolites

When the molecules in the pore are nearly the size of the passageway, the diffusing molecule is never away from the influence of the wall, and the rate of diffusion becomes relatively slow. This regime has been variously termed *restricted diffusion* or *configurational diffusion*. In contrast, *Knudsen diffusion* occurs in pores sufficiently small that the mean free path is much greater than the pore size but molecular motion occurs by free flight interrupted by momentary adsorption and desorption on the wall. Diffusion in zeolites is more complex than Knudsen diffusion or bulk diffusion, and the activation energy is usually substantially greater than that for Knudsen diffusion or bulk diffusion.

Typical reported diffusion coefficients in zeolites range downwards from about 10^{-11} cm^2/s, the highest that can be measured by sorption rate experiments. For comparison, representative values are in the region of 10^{-1} cm^2/s for bulk diffusion of liquids, and 10^{-3} cm^2/s and less for Knudsen diffusion of gases (Knudsen diffusion does not occur with liquids) (Fig. 7.9). By the usual method of measurement, the unsteady-state rate of sorption into or desorption out of zeolite crystals, the lowest

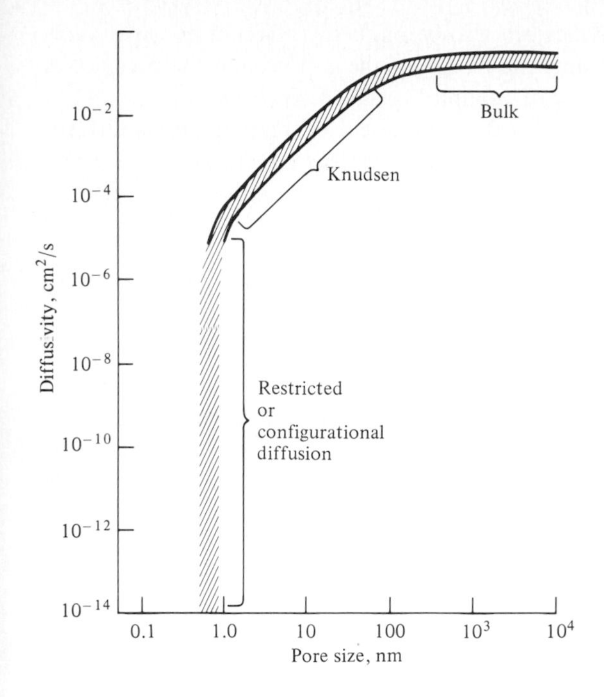

Figure 7.9 Effect of pore size on diffusivity. (*Adapted from Weisz, 1973.*) (*Reprinted with permission from Chemtech. Copyright by the American Chemical Society.*)

values measurable are limited only by the patience of the investigator. The highest are set by the maximum rate that can be accurately followed by available experimental methods. For crystals of the order of 1 μm, this maximum corresponds to diffusivities of about 10^{-11} cm^2/s. Nuclear magnetic resonance (NMR) techniques can also be used to measure diffusivities, and they provide a means of making observations of diffusivities higher than those observable by the sorption methods. Diffusion coefficients for zeolites of the order of 10^{-3} to 10^{-6} cm^2/s have been reported, especially in the older literature. When based on sorption measurements on pressed pellets or beds of powder, they probably represent diffusion in the interstices between zeolite crystals rather than within the crystals themselves. Diffusion in pores of roughly 1 to 5 nm diameter, in which Knudsen diffusion merges into configurational diffusion, is essentially unexplored, but surface diffusion may well play a significant role under these circumstances.

In the configurational diffusion regime, the increase in diffusivity with change in zeolite generally parallels the increase in pore size, as determined by sorption measurements with different-sized molecules. For a specific zeolite, the activation energy generally increases with increasing molecular size of diffusing species, within a series of similar types of molecules. However adsorption phenomena and the interaction energy between diffusing molecule and pore walls are also significant variables.

A number of measurements have been reported of diffusion coefficients for hydrocarbons in zeolites such as mordenite and type Y under ambient and near-ambient conditions (Satterfield et al. 1967–1974; Moore and Katzer, 1972; Eberly, 1976). Unlike Knudsen diffusion, in which the fluxes of oppositely moving molecules are independent of one another, the diffusion in zeolites is such that the flux in one direction is markedly hindered by the opposite flux. Such counterdiffusion is inevitable in catalytic reaction. Diffusivities can also be markedly affected by slight variations in the zeolite structure, including the nature of the cations and the presence of impurities, and by the size and polarity of counterdiffusing molecules.

There is insufficient basis for extrapolating the effect of temperature on diffusivity from presently available data at ambient or near ambient conditions to the generally much higher temperatures of reaction in order to determine the presence or absence of significant diffusion limitations under reaction conditions. The apparent activation energy for zeolite diffusion is greater than that for bulk or Knudsen diffusion. The few values that have been reported are relatively uncertain because of the limited temperature range covered and other difficulties such as deviations from Fick's law, sorption effects, and questions concerning the difference between unidirectional and counterdiffusion. Generally, the smaller the diffusion coefficient, the higher the activation energy. As a measure of the significance of a modest change in activation energy here, an increase in temperature from 25 to 310°C would increase the diffusivity by a factor of 60 with an activation energy of 21 kJ/mol or by a factor of 5000 with an activation energy of 46 kJ/mol.

Diffusion limitations may be inferred by comparing the observed activation energy to that for the same reaction under non-diffusion-limiting conditions, or by the effect of crystal particle size on reaction rate (Sec. 11.2). However, it is difficult to obtain uniform zeolite crystals over a very wide range of crystal sizes. Miale et al. (1966) concluded that diffusion was *not* a rate-limiting process for the cracking of *n*-hexane on synthetic hydrogen mordenite and on several faujasite catalysts at 270 to 500°C, since the observed activation energy was the same as that on silica-alumina. Chutoransky and Dwyer (1973) studied the liquid-phase isomerization of xylenes on a zeolite-containing catalyst described as an AP (aromatics processing) catalyst (but not further characterized), at temperatures of about 150 to 300°C. They reported that the rate was not diffusion-limited utilizing 0.2 to 0.4-μm crystallites, but was significantly limited at the higher temperatures with 2- to 4-μm crystallites.

In their commercial form zeolite catalysts are incorporated into a gel matrix, but diffusivities are much greater in the pores of the matrix than in the zeolite pores. Diffusion limitations inside the zeolite pores may not be significant in commercial catalytic cracking, but it depends on the feedstock, amount of carbon present on the catalyst, and other factors. Evidence is contradictory (Magee and Blazek, 1976). On the basis of their size, three- and four-ring compounds may be expected to be effectively excluded from the pores of a Y-type zeolite. Although some reaction may occur in the matrix, the major activity and selectivity effects are attributed primarily to the properties and content of the zeolites present (Magee, 1977). The matrix may have a beneficial effect on zeolite stability by adsorbing sodium ions from it at the high temperatures of reaction.

A useful method of distinguishing between zeolites of different minimum aperture size is to compare the rate of reaction on a series of zeolites of two compounds, similar chemically but of significantly different minimum molecular diameters. The Mobil Oil Co. in its patent literature refers to a "constraint index" based upon the ratio of the rate of cracking of *n*-hexane and that of 3-methylpentane in a 50:50 mixture of the two. This is used to characterize a class of zeolites exemplified by

ZSM-5 and TEA mordenite that exhibit unusual selectivity in conversion of methanol to aliphatic and aromatic compounds in the boiling-point range of gasoline. The effectiveness of these catalysts appears to be dependent, at least in part, on a shape-selectivity characteristic. By measurements with *n*-hexane, 3-methylpentane, and 2,3-methylbutane, Chen and Garwood (1978) show that ZSM-5 has pore openings intermediate between such small-pore zeolites as chabazite and erionite on the one hand and large-pore zeolites such as the faujasites on the other.

7.7.5 Shape-Selective Catalysis

Weisz and Frilette (1960) were apparently the first to describe shape selectivity, and they, together with Chen, Miale, and coworkers, have published extensively on a variety of applications. The subject was reviewed by Csicsery (1976). The pore size and shape in a zeolite may affect the selectivity of a reaction in two ways: *Reactant selectivity* occurs when the aperture size of the zeolite is such that it admits only certain smaller molecules and excludes larger molecules; hence, in a mixture, effectively only the smaller molecules react. *Product selectivity* occurs when bulkier product molecules cannot diffuse out, and if formed, they are converted to smaller molecules or to carbonaceous deposits within the pore. These eventually may cause pore blockage. It is also possible that pore shape may sterically hinder certain types of reactions and thereby affect selectivity, but the evidence is ambiguous.

For shape selectivity to occur, essentially all the active catalytic sites must be in the interior of the pores. The exterior area of zeolite crystals is only about 1 percent of the total, but if diffusion limitations are significant, it may become necessary to poison or inactivate exterior sites so they do not contribute excessively to reaction. Some zeolites having suitable aperture sizes are not stable under reaction conditions. With some metal-loaded zeolites, metal may migrate out of the pores with time, thus destroying this type of catalyst selectivity.

A commercial application of reactant selectivity catalysis is the Mobil *Selectoforming process* used in several installations for hydrocracking *n*-paraffins that are in a mixture with other paraffins and aromatics. Representative reaction conditions are 1.3 to 4 MPa, 300 to 500°C, and a hydrogen/hydrocarbon mole ratio of 2:4 (Chen et al., 1968). Aromatics are not reacted, and neither are branched or cyclic paraffins. This process is typically carried out on the product obtained from a catalytic reformer or may be carried out in the last reactor of a reformer. The advantage of Selectoforming is that branched paraffins are not destroyed, so an improvement in octane number (Sec. 9.3) is achieved with less yield loss.

The catalyst in Selectoforming is of the erionite type, probably intergrown with some offretite. It has a low potassium content and contains a nonnoble metal, probably nickel, as the hydrogenation component (Bolton, 1976). The hydrogenation component helps minimize the rate of formation of carbonaceous deposits, as with reforming catalysts.

Examples of product selectivity are seen in the distribution of products from hydrocracking in the Selectoforming process. These products are predominantly smaller *n*-paraffins, although isoparaffins dominate in non-shape selective hydrocracking reactions. In conventional hydrocracking, utilizing type Y zeolites, shape selectivity does not seem to be involved. In catalytic isomerization of *n*-paraffins (Sec. 9.6), highly branched products are desired and pore sizes must be sufficiently large to allow free movement of products and reactants.

Zeolite catalysts from Mobil Oil termed AP (aromatics processing) catalysts may make possible product selectivity based on pore structure. In the isomerization of xylenes it is desired to minimize the formation of polymethyl benzenes by disproportionation reactions. The high selectivity of the AP catalysts may rest on aperture sizes that allow xylenes to escape but not substituted benzenes with three or more methyl groups. The AP catalyst is reportedly similar to zeolite Ω (Bolton, 1976), which has a one-dimensional, large-pore structure.

It is gradually becoming clearer that the diffusivity and reaction pattern with zeolites may not change monotonically within the members of a homologous series of reactants. Gorring (1973) reported that the diffusivity of *n*-paraffins in K-exchanged zeolite T (probably offretite intergrown with some erionite) dropped to a minimum with an increase in molecular weight to the C_8 paraffin, rose to a maximum with the C_{12} member of the series, and dropped again with C_{14}. Chen and Garwood (1973) hydrocracked three multicomponent feedstocks comprising *n*-paraffins ranging from C_4 to C_{16} on erionite and reported two maxima in the rate constants, one with the C_6 paraffin and the other with the C_{10} and C_{11} paraffins. Relative reaction rates and product distribution are different in mixtures than in single pure compounds, but the above parallelism with the diffusivity behavior, reported by Gorring, strongly suggests that pore diffusion played an important role.

Under the same reaction conditions the apparent activation energies for C_4, C_5, and C_6 *n*-paraffins, studied individually and with a mixture of C_5 and C_6, suggested the onset of a diffusion limitation for C_5 and C_6 in mixtures at about 400°C (but not for C_5 alone). The C_6 reaction was diffusion-limited at 370°C, the lowest temperature studied. The diffusion and reaction results were interpreted in terms of a unique "cage effect," the relationship between the size of a molecule and the size of a cage or "supercavity" inside a zeolite that is somewhat larger than the neck size leading to it.

Irregularities in a homologous system also appear in liquid sorption equilibria. On NaY zeolite a brief study was made of binary mixtures of *n*-octane with either C_{10}, C_{12}, or C_{14} *n*-paraffins. *n*-Octane is always preferentially adsorbed into the zeolite, but a minimum value of the separation factor was found in the mixture with C_{12} (Satterfield and Smeets, 1974).

7.7.6 Catalytic Cracking with Zeolites

The first significant industrial application of zeolites in catalysis was the incorporation into a silica-alumina matrix of up to 15 percent of a Y-type zeolite to form a new type of cracking catalyst. This was first introduced commercially in about 1962. Although zeolites in the acidic form are much more active than conventional silica-alumina, their behavior under industrial conditions was not readily predictable from laboratory or pilot plant experience. It was not so much high activity as such that led to their use, as the finding that under suitable commercial operating conditions it was possible to achieve a higher yield of gasoline from gas oil. Moreover, the gasoline had a higher octane number. The best performance with zeolite-containing catalysts is achieved with light and paraffinic feedstocks and with good regeneration of the catalyst to produce low coke levels (Magee and Blazek, 1976). This required substantial modification of existing reactors or the design of new reactors (e.g., use of riser-cracking instead of dense fluid-bed reactors) and modification of regenerator operation. It also led to some designs in which fresh virgin feed was processed under different conditions than recycle streams, which are more aromatic in nature (Sec. 9.4).

The activation energy for cracking *n*-hexane is about 125 kJ/mol on each of several zeolites and on silica-alumina. The rate of reaction, which under a specified set of conditions requires 540°C on amorphous silica-alumina, can be achieved at 270°C or less on zeolites in the H form (Miale et al., 1966). Assuming that this activation energy holds over the entire temperature range and on the group of catalysts studied, this would indicate that the zeolites are at least 10^4 times as active as amorphous silica-alumina.

The reasons for this behavior are not clear. The concentration of acid sites on zeolites may be 10 to 100 times greater than on silica-alumina. It also has been suggested that, in addition, in the vicinity of unshielded cations within small pores high electrostatic fields are created that may enhance reactivity. This is somewhat supported by the fact that on NaY the cracking patterns are more typical of those encountered in thermal pyrolysis, which involves free radicals, rather than those encountered in the carbonium-ion mechanisms found on silica-alumina. However the activity of NaY, although comparable to that of silica-alumina, is much less than that of the acidic forms of Y, and a mechanism stemming from electrostatic effects does not appear to be of major significance under commercial cracking conditions. At 500°C and atmospheric pressure, KY, which is probably similar to NaY, showed only about a fivefold increase in rate over that of the purely thermal reaction (Poutsma, 1976, p. 517).

The zeolite must have sufficiently large pores to accommodate most feed molecules, have good stability to the high temperatures encountered in reaction and regeneration, and have good stability to steam, which is formed in regeneration and used as a purge between reactor and regenerator. Stability increases roughly with Si/Al ratio. Type Y, which is more hydrothermally stable and has a higher Si/Al atomic ratio than X (about $2\frac{1}{2}$ versus about $1\frac{1}{4}$), is now generally used in preference to X. The Y zeolite is synthesized in the Na form, but this may be replaced with a wide variety of cations and/or with H^+. An enormous amount of detailed attention has been devoted to the best economical procedures for obtaining an active and highly stable catalyst. In many cases, rare earth ions are incorporated into zeolite Y, replacing from 80 to 97+ percent of the equivalent Na, which provides additional structural strength. These are termed REY or REHY. The principal rare earth elements as found in nature are Ce, La, and Nd, decreasing in this order. The ratios vary somewhat with the mineral source. When used as a mixture, no preferential exchange occurs and the element ratio in the rare earth mixture does not affect catalyst activity or selectivity. However a high La content is preferred since it (or Nd) provides greater stability than Ce.

The hydrogen Y form cannot be obtained by direct treatment of NaY with acid since the structure is attacked. Instead the Na^+ is replaced by NH_4^+. Upon heating, ammonia is driven off and then, or somewhat simultaneously, water, to form what is termed *decationized Y*. Not all Na ions are identical in the crystal structure, and some are more difficult to exchange than others. Various procedures have been utilized to achieve a high degree of removal of Na^+, desired for increased temperature stability. Other stabilization procedures involve extraction of some of the aluminum from the Y framework by chelation followed by heat treatment. This causes slight contraction of the structure. So-called *ultrastable forms* of Y may be prepared by these various means, in which the crystal form may be retained at temperatures as high as 1000°C. In the commercial forms in which the zeolite is incorporated into a silica-alumina matrix, the stability of the zeolite may be enhanced, apparently by migration of Na ions into the matrix. Zeolite stability and ultrastable zeolites are reviewed by McDaniel and Maher (1976).

Detailed commercial procedures for manufacture of zeolite cracking catalysts are described by Magee and Blazek (1976). The zeolite is usually prepared separately and then incorporated into a silica-alumina gel matrix. This is spray-dried to yield microspheroidal particles averaging about 60 μm in diameter, the desired form for use in fluidized-bed reactors. The product is washed, base-exchanged to remove contaminant ions such as Na^+ and SO_4^{2-}, and dried again, as in a rotary drier. It is also possible to manipulate kaolin clay and caustic in such a manner as to produce a final product that contains some zeolite structure, but this method is not widely used in commercial manufacture.

For hydrocarbon transformations in general, HX catalysts are less active than HY (Poutsma, 1976). Hydrogen mordenite is generally more active than HY, but its activity usually drops rapidly, probably because of blockage of the one-dimensional pore structure by carbonaceous deposits. Zeolites in the alkali metal form such as NaY and NaX are relatively inactive, although reactions that are relatively easy to catalyze, such as a double-bond shift in olefins, may be observed. NaY shows no acidity by titration methods or by infrared (ir) methods, but CaY has substantial activity and exhibits considerable acid strength. Poutsma (1976, p. 469) lists the order of reactivity as a function of the exchanged cation for X and Y zeolites for some 11 reaction studies. The order may be affected by such variables as activation temperature and exchange level, but the usual sequence is

$$H^+ > RE^{3+} > \text{group IIA}^{2+} > \text{group IA}^+$$

for both X and Y zeolites. RE^{3+} symbolizes exchange with rare earth elements.

The amount of acidity and acid-strength distribution varies with degree of exchange, nature of the cation present, and heat treatment. HY zeolite (90.4 percent exchanged) has acid strengths in the range of $H_0 = -4$ to -8, as measured by butylamine titration. A LaX zeolite showed an increase in both amount of acid and acid strength as degree of exchange was increased (Otouma et al., 1969). It is clear that very high acid strength, either with X, Y, or silica-alumina, leads to excessive formation of coke and low-molecular-weight gases at the expense of gasoline (Moscou and Moné, 1973). The acid strength of HY is not greater than that in amorphous silica-alumina, but REY may have stronger sites than HY. Acid strength can be decreased by either heat treatment or steaming. Further details concerning acidic and catalytic properties of zeolites are reviewed by Barthomeuf (1977).

Catalytic cracking in the presence of zeolites results in less olefins and naphthenes and more aromatics and paraffins than are found with silica-alumina, attributed to increased activity for hydrogen-transfer reactions. More C_5 to C_{10} products are formed and less C_3 and C_4 products, again attributed to the increasing importance of hydrogen-transfer reactions that stop the cracking at a higher molecular weight [Eq. (7.8)]. Coke is formed by a complex series of condensation, polymerization, and cracking reactions, and the rate of its formation is increased by any type of hydrogen-transfer reaction from the adsorbed molecules comprising coke precursors to reactants in the fluid phase. The lower rate of formation of coke on zeolites is attributed to a higher rate of hydrogen transfer to proton acceptors among product molecules, relative to the rate of hydrogen transfer from coke precursors.

7.7.7 Literature on Zeolites

The preparation and commercial utilization of zeolite cracking catalysts is reviewed by Magee and Blazek (1976) and Magee (1977). The use of zeolites in other catalytic processes is reviewed by Bolton (1976). Minachev and Isakov (1976) review metal-loaded zeolite catalysts and give extensive references. Rudham and Stockwell (1977) review in detail catalysis on faujasite-type zeolites. The use of zeolite catalysts in hydrocracking is summarized in Sec. 9.7.

Much of the information on zeolites in general is published in the proceedings of the international congresses which have been held about every 3 to 4 years. The first was at the University of London in 1967 (Society of Chemical Industry, 1968) followed by conferences at Worcester Polytechnic Institute, Worcester, Massachusetts in 1970 (Flanigen and Sand, 1971), in Zurich in 1973 (Meier and Uytterhoeven, 1973), and at the University of Chicago in 1977 (Katzer, 1977). The fifth international conference was held in Naples, Italy, in 1980. Sorption and reaction were reviewed by Riekert (1970) from a somewhat theoretical approach. A useful, comprehensive, and detailed book on zeolites has been published by Breck (1974); it does not however, treat zeolite catalysis. A detailed monograph edited by Rabo (1976) contains several chapters on reactions, mechanism, and technology. Jacobs (1977) critically reviews catalytic reactions on zeolites and the relationships between activity and physical and acidic properties, focusing mainly on the literature during the period 1970 to 1976. Much of the early work on zeolites was by Barrer, who reviews some aspects of their science and technology (1968). In a recent book (Barrer, 1978) he treats their sorption and diffusion characteristics in detail, including that of related substances such as clays and layered silicates.

7.8 OTHER SOLID ACIDS

A number of solids other than those discussed above show acidic properties and may be used as catalysts. Synthetic layered silicates and aluminosilicates having structures related to natural clays have been prepared. An example is synthetic mica montmorillonite (SMM), a layered aluminum silicate that has been evaluated as a cracking catalyst (Swift, 1977). This material can also be ion-exchanged to incorporate metal into the structure and form a hydrocracking catalyst.

Cation-exchange resins, in the form of small beads, have been used as acid catalysts for some liquid-phase reactions. The most common formulation is styrene cross-linked with divinylbenzene, which is then sulfonated (Dowex, Amberlite). The effective pore structure in these materials is very fine, however, so reactions tend to be highly diffusion-limited. This can be substantially improved by preparing a so-called *macroreticular structure*, which has a more open set of pores.

Certain forms of carbon activated under oxidizing conditions will have oxygenated structures on the surface that are acidic, such as carboxylic acids. Although usually these structures are not deliberately used as acidic catalysts, they may contribute an acid-catalyzed reaction where it is not desired.

REFERENCES

Barrer, R. M.: *Chem. Ind.*, Sept. 7, 1968, p. 1203.

——: *Zeolites and Clay Minerals, as Sorbents and Molecular Sieves*, Academic, New York, 1978.

Barthomeuf, D., in J. R. Katzer (ed.): Molecular Sieves, II, *ACS Symp. Ser. No. 40*, 1977, p. 453.

Benesi, H. A.: *J. Phys. Chem.*, **61**, 970 (1957).

——, and B. H. C. Winquist: *Adv. Catal.*, **27**, 97 (1978).

Bolton, A. P., in J. A. Rabo (ed.): *Zeolite Chemistry and Catalysis, ACS Monogr. No. 171*, 1976, p. 714.

Breck, D. W.: *Zeolite Molecular Sieves: Structure, Chemistry and Use*, Wiley, New York, 1974.

Chen, N. Y., and W. E. Garwood: "Molecular Sieves," *Adv. Chem. Ser. No. 121*, 1973, p. 545.

——, and ——: *J. Catal.*, **52**, 453 (1978).

——, J. Maziuk, A. B. Schwartz, and P. B. Weisz: *Oil Gas J.*, **66** (47), 154 (1968).

Csicsery, S. M., in J. A. Rabo (ed.): "Zeolite Chemistry and Catalysis," *ACS Monogr. No. 171*, 1976, p. 680.

Chutoransky, P., Jr., and F. G. Dwyer: "Molecular Sieves," *Adv. Chem. Ser. No. 121*, 1973, p. 540.

Eberly, P. E., Jr. in J. A. Rabo (ed.): "Zeolite Chemistry and Catalysis," *ACS Monogr. No. 171*, 1976, p. 392.

Emmett, P. H.: *Catalysis Then and Now*, Franklin, Englewood, N.J., 1965.

Flanigen, E. M. and L. B. Sand (eds.): "Molecular Sieve Zeolites, I and II," *Adv. Chem. Ser. Nos. 101 and 102*, 1971.

Forni, L.: *Catal. Rev.*, **8**, 65 (1974).

Gates, B. C., J. R. Katzer, and G. C. A. Schuit: *Chemistry of Catalytic Processes*, McGraw-Hill, New York, 1979.

Goldstein, M. S., in R. B. Anderson (ed.): *Experimental Methods in Catalytic Research*, Academic, New York, 1968, p. 361.

Gorring, R. L.: *J. Catal.*, **31**, 13 (1973).

Greensfelder, B. S., H. H. Voge, and G. M. Good: *Ind. Eng. Chem.*, **41**, 2573 (1949).

Hansford, R. C.: *Adv. Catal.*, **4**, 1 (1952).

Hirschler, A. E.: *J. Catal.*, **2**, 428 (1963).

Jacobs, P. A.: *Carboniogenic Activity of Zeolites*, Elsevier, New York, 1977.

Johnson, O.: *J. Phys. Chem.*, **59**, 827 (1955).

Katzer, J. R. (ed.): "Molecular Sieves, II," *ACS Symp. Ser. No. 40*, 1977.

McDaniel, C. V., and P. K. Maher in J. A. Rabo (ed.): "Zeolite Chemistry and Catalysis," *ACS Monogr. No. 171*, 1976, p. 285.

McMahon, J. F., C. Bednars, and E. Solomon in J. J. McKetta (ed.): *Advances in Petroleum Chemistry and Refining*, vol. 7, Wiley, New York, 1963, p. 284.

Magee, J. S., in J. R. Katzer (ed.): "Molecular Sieves, II," *ACS Symp. Ser. No. 40*, 1977, p. 650.
——, and J. J. Blazek in J. A. Rabo (ed.): "Zeolite Chemistry and Catalysis," *ACS Monogr. No. 171*, 1976, p. 615.
Meier, W. M. and J. B. Uytterhoeven (eds.): "Molecular Sieves," *Adv. Chem. Ser. No. 121.* [Discussion and recent progress reports published as a separate volume: J. B. Uytterhoeven (ed.): *Proceedings of the Third International Conference on Molecular Sieves*, Leuven Univ. Press, Leuven, Belgium, 1973.]
Miale, J. N., N. Y. Chen, and P. B. Weisz: *J. Catal.*, **6**, 278 (1966).
Minachev, Kh. M., and Ya. I. Isokov in J. A. Rabo (ed.): "Zeolite Chemistry and Catalysis," *ACS Monogr. No. 171*, 1976, p. 552.
Moore, R. M., and J. R. Katzer: *AIChEJ.*, **18**, 816 (1972).
Moscou, L., and R. Moné: *J. Catal.*, **30**, 417 (1973).
Oblad, A. G., T. H. Milliken, Jr., and G. A. Mills: *Adv. Catal.*, **3**, 199 (1951).
Otouma, H., Y. Arai, and H. Ukihashi: *Bull. Chem. Soc. Jpn.*, **42**, 2449 (1969).
Pines, H.: *Intra-Sci. Chem. Rep.*, **6**(2), 1 (1972).
——, and W. O. Haag: *J. Am. Chem. Soc.*, **82**, 2471 (1960).
——, and L. A. Schaap: *Adv. Catal.*, **12**, 117 (1960).
——, and W. M. Stalick: *Base-Catalyzed Reactions of Hydrocarbons and Related Compounds*, Academic, New York, 1977.
Pitcher, W. H., Jr.: Sc.D. thesis, M.I.T., Cambridge, Mass., 1972.
Poutsma, M. L., in J. A. Rabo (ed.): "Zeolite Chemistry and Catalysis," *ACS Monogr. No. 171*, 1976, p. 437.
Rabo, J. A. (ed.): "Zeolite Chemistry and Catalysis," *ACS Monogr. No. 171*, 1976.
Riekert, L.: *Adv. Catal.*, **21**, 281 (1970).
Ries, H. E., Jr.: *Adv. Catal.*, **4**, 87 (1952).
Robson, H. E., K. L. Riley, and D. D. Maness in J. R. Katzer (ed.): "Molecular Sieves, II," *ACS Symp. Ser. No. 40*, 1977, p. 233.
Rudham, R., and A. Stockwell: *Catalysis*, The Chemical Society, London, 1977, vol. 1, chap. 3, p. 87.
Ryland, L. B., M. W. Tamele, and J. N. Wilson in P. H. Emmett (ed.): *Catalysis*, vol. 7, Reinhold, New York, 1960, p. 1.
Satterfield, C. N., and C. S. Cheng: *AIChE Symp. Ser. No. 117*, **67**, 43 (1971).
——, and ——: *AIChEJ.*, **18**, 720 (1972).
——, and G. T. Chiu, *AIChEJ.*, **20**, 522 (1974).
——, and A. J. Frabetti, Jr.: *AIChEJ.*: **13**, 731 (1967).
——, and J. R. Katzer in E. M. Flanigen and L. B. Sand (eds.): *Adv. Chem. Ser.* "Molecular Sieve Zeolites, II," *No. 102*, 1971, p. 193.
——, ——, and W. R. Vieth: *Ind. Eng. Chem. Fundam.*, **10**, 478 (1971).
——, and W. G. Margetts: *AIChEJ.*, **17**, 295 (1971).
——, and J. K. Smeets: *AIChEJ.*, **20**, 618 (1974).
Schaad, R. E., in B. T. Brooks et al. (eds.): *The Chemistry of Petroleum Hydrocarbons*, vol. 3, Reinhold, New York, 1955, p. 221.
Sherman, J. D., in J. R. Katzer (ed.): "Molecular Sieves, II," *ACS Symp. Ser. No. 40*, 1977, p. 30.
Smith, J. V., in J. A. Rabo (ed.): "Zeolite Chemistry and Catalysis," *ACS Monogr. No. 171*, 1976, p. 3.
Society of Chemical Industry (London): *Molecular Sieves*, 1968.
Swift, H. E., in J. J. Burton and R. L. Garten (eds.): *Advanced Materials in Catalysis*, Academic, New York, 1977, p. 209.
Takeshita, T., R. Ohnishi, and K. Tanabe: *Catal. Rev.*, **8**, 29 (1974).
Tanabe, K.: *Solid Acids and Bases*, Academic, New York, 1970.
Thomas, C. L., *Ind. Eng. Chem.*, **41**, 2564 (1949).
——, and D. S. Barmby: *J. Catal.*, **12**, 341 (1968).
Villadsen, J., and H. Livbjerg, *Catal. Rev., Sci. Eng.*, **17**, 203 (1978).
Weisz, P. B., *Chemtech*, **3**, 498 (1973).
——, and V. J. Frilette: *J. Phys. Chem.*, **64**, 382 (1960).

CHAPTER

EIGHT

CATALYTIC OXIDATION

Partial oxidation processes using air or oxygen are used to manufacture a variety of chemicals, and complete catalytic oxidation is a practicable method for elimination of organic pollutants in gaseous streams. In the manufacture of organic chemicals, oxygen may be incorporated into the final product, as in the oxidation of propylene to acrolein or *o*-xylene to phthalic anhydride; or the reaction may be an oxidative dehydrogenation in which the oxygen does not appear in the desired product, as in the conversion of butylene to butadiene. The desired reaction may or may not involve C—C bond scission. Closely related are *ammoxidation* reactions (also termed *oxidative ammonolysis*) in which a mixture of air and ammonia is reacted catalytically with an organic compound to form a nitrile, as in the ammoxidation of propylene to acrylonitrile. With the advent of more active and selective catalysts, direct oxidation processes have gradually replaced earlier processes that utilized such oxidizing agents as nitrogen dioxide, chromic acid, and hypochlorous acid.

Catalytic oxidation processes for manufacture of organic chemicals have several features in common.

1. They are highly exothermic. Heat- and mass-transfer effects may be very important, so multitube heat-exchange reactors or fluidized beds are usually used. An important element of reactor design is to prevent catalyst deactivation by excessive temperature or runaway reaction.
2. Certain composition regions may be explosive. The ratio of organic compound to air (or oxygen) in the feed stream may be selected in large part to avoid these regions. For the same reason, air may be introduced at multiple points. Operation may be either "fuel-rich" or "fuel-lean."
3. The desired product must be sufficiently stable relative to the reactant under reaction conditions that it can be removed from product gases in an economic yield,

usually involving rapid quenching, before it decomposes or undergoes further reaction. Most of the compounds currently made industrially have ring structures that are highly stable, for example, phthalic anhydride, maleic anhydride, and ethylene oxide; or a conjugated structure which imparts stability, such as C═C—C═C, C═C—C≡N, and C═C—C═O. In contrast, although there have been many attempts to develop a process for the direct heterogeneous oxidation of methane to formaldehyde, no economic method has been invented since the product is so much less stable than the reactant. Sometimes relatively good selectivity to a desired product can be achieved at low percent conversions. In that case the product is typically removed from the reactor exit gas and the remaining stream sent to a second reactor. Alternately the stream may be recycled, perhaps after purification, to the reactor inlet.

Effective catalysts for oxidation reactions fall into three categories:

1. Transition metal oxides in which oxygen is readily transferred to and from the structure. Most but not all of the industrial catalysts of this type are mixed oxides containing two or more cations, and the compounds are nonstoichiometric. Examples are an iron molybdate catalyst for oxidizing methanol to formaldehyde, bismuth molybdates for oxidizing propylene to acrolein or ammoxidation of propylene to acrylonitrile, and catalysts based on vanadium oxide for converting benzene to maleic anhydride and naphthalene or *o*-xylene to phthalic anhydride.
2. Metals onto which oxygen is chemisorbed. Examples are ethylene to ethylene oxide on a supported silver catalyst, ammonia oxidation to nitric oxide on platinum gauze, and methanol to formaldehyde on bulk silver.
3. Metal oxides in which the active species is chemisorbed oxygen, as molecules or atoms. These may also provide a significant additional mechanism under some conditions with metal oxide catalysts that also contain interstitial oxygen as an active species.

Practical oxide catalysts may be much more difficult to characterize than other types of catalysts. In the case of a supported oxide catalyst, methods for determining active area in contrast to total area—analogous to the use of selective chemisorption for supported metal catalysts or base titration to estimate the number of acid sites on an acidic catalyst—are much more rudimentary. Knözinger (1976) has recently reviewed in detail methods of specific adsorption that may be suitable. Many of the oxide catalysts may be more or less amorphous, and even when an X-ray pattern can be observed, the contribution of the crystalline phase to total catalytic properties may be far from clear. Some useful catalyst compositions may contain as many as four or five metal elements, each of which must be present for optimum performance, but their role remains obscure. Defects in a crystal structure appear to play a role, but the extent to which they exist under reaction conditions is speculative.

In a few cases, at reaction temperatures the catalyst appears to be a liquid held in the pores of a support rather than a solid. This is the case under at least some reaction conditions for the vanadium oxide-potassium sulfate catalyst used for oxidation of sulfur dioxide to sulfur trioxide (Sec. 8.9), and also for some metal chloride catalysts used in oxychlorination (Sec. 8.8).

8.1 REDOX MECHANISM

The behavior of most oxidation catalysts can be interpreted within the framework of a redox mechanism. This postulates that the catalytic reaction comprises two steps:

1. Reaction between catalyst in an oxidized form, Cat–O, and the hydrocarbon R, in which the oxide becomes reduced: Cat–O + R → RO + Cat.
2. The reduced catalyst, Cat, becomes oxidized again by oxygen from the gas phase: 2Cat + O_2 → 2Cat–O.

Under steady-state conditions the rates of the two steps must be the same.

Within this framework more specific models were developed by Mars and van Krevelen (1954) to explain the kinetic behavior of the partial oxidation of several aromatic hydrocarbons and of sulfur dioxide over vanadium oxide catalysts of various formulations. It was noted in particular that the rate was independent of the nature of the hydrocarbon or its partial pressure over a substantial range of concentration.

Strictly, their derivation makes no assumption about the form of the oxygen in the catalyst; that is, it can be either chemisorbed or lattice oxygen. However lattice oxygen systems seem to follow this model in a general way, whereas some chemisorbed oxygen systems exhibit kinetic behavior that may be more clearly understood in terms of the Langmuir-Hinshelwood or Rideal models. In the case of lattice oxygen, the active species is presumably the O^{2-} ion.

For the hydrocarbon oxidation studies, to develop a simple mathematical model Mars and van Krevelen assumed that the rate of oxidation of the reactant is proportional to the fraction of active sites in the oxidized state and to the hydrocarbon partial pressure. The rate of reoxidation of the catalyst is taken to be proportional to the fraction of sites on the catalyst in the reduced (or empty) state and to $P_{O_2}{}^n$. Then

$$r = kP_{HC}(1 - \theta) = \frac{k^*}{\beta} P_{O_2}{}^n \theta \tag{8.1}$$

where, in present nomenclature,

P_{HC} = partial pressure of hydrocarbon
k = reaction rate constant for oxidation of hydrocarbon
k^* = reaction rate constant for surface reoxidation
β = moles O_2 consumed per mole of hydrocarbon reacted
θ = fraction of active sites in reduced state

$$\theta = \frac{kP_{HC}}{kP_{HC} + (k^*/\beta)\, P_{O_2}{}^n} \tag{8.2}$$

Substituting Eq. (8.2) into the second equality of Eq. (8.1) gives

$$r = \frac{1}{(\beta/k^* P_{O_2}{}^n) + (1/kP_{HC})} \tag{8.3}$$

For the hydrocarbon oxidation reactions, their data were best fitted by $n = 1$.

Equation (8.3) leads to several conclusions. If the potential rate of oxidation of the hydrocarbon by the catalyst exceeds that of reoxidation of the catalyst, then

$$kP_{HC} \gg \frac{k^* P_{O_2}{}^n}{\beta} \tag{8.4}$$

and Eq. (8.3) reduces to

$$r = \frac{k^*}{\beta} P_{O_2}{}^n \tag{8.5}$$

The overall observed reaction rate should then be equal to the potential rate of oxidation of the catalyst surface and therefore should be independent of hydrocarbon partial pressure (i.e., zero order). The rate constant k^* at a fixed temperature should be independent of the nature of the hydrocarbon. Mars and van Krevelen indeed found this to be true for several aromatic substances on a catalyst consisting of 9 wt % V_2O_5, 2.9 wt % MoO_3, and 0.03 wt % P_2O_5 on a corundum carrier.

In a later study of oxidation of *o*-xylene to phthalic anhydride on a V_2O_5/SiC catalyst (Calderbank and Caldwell, 1972), the reoxidation was likewise reported to be the rate-limiting step, with an activation energy of about 167 kJ/mol. The activation energy for the hydrocarbon oxidation step was estimated at about 113 kJ/mol. Thus at relatively higher temperatures, the inequality of Eq. (8.4) could conceivably be reversed, whereupon the reaction would become independent of oxygen pressure and proportional to hydrocarbon partial pressure.

For other oxidation reactions this inequality is indeed reversed ($k^*/\beta \gg k$ for $n = 1$), and the rate is then proportional to the hydrocarbon partial pressure and independent of the oxygen pressure. This is the case, for example, for ammoxidation of propylene to acrylonitrile on the bismuth molybdate or antimony uranium oxide catalysts used industrially. Here the rate-limiting process is the reaction of hydrocarbon with the catalyst rather than reoxidation of the catalyst. A further consequence is that for these cases only a slight excess of oxygen should be needed to keep the catalyst in a high oxidation state.

If one considers a broad range of oxide catalysts, it might be expected that, for a specified reaction, the maximum rate would be encountered at some intermediate degree of heat of reaction for reoxidation of the catalyst, Q_0. That is, a plot of reaction rate versus Q_0 would show a volcano curve analogous to that found for some reactions on metals. This concept has been developed by many investigators and was reviewed by Germain (1972) and Haber (1975). In a more precise formulation of this concept, Balandin specified that the maximum rate should occur when $Q_0 \approx Q_R/2$, where Q_R is the overall heat of reaction for conversion of reactant to product. This has been reported by a number of investigators. See, for example, Roiter et al. (1968) for oxidation of hydrogen and total oxidation of propylene.

Applying the Polanyi relationship here (Sec. 3.4), for low values of Q_0 the activation energy for reoxidation should be high. The rate-limiting step should then be that of reoxidation of the catalyst. In the basic Mars–van Krevelen scheme the reaction should then be zero order in reactant and n order with respect to oxygen. At high values of Q_0, reoxidation is fast, and the rate-limiting step would be that of oxidation of the adsorbed reactant. The rate should then be proportional to reactant partial pressure and independent of oxygen pressure. A study by Moro-oka et al. (1967) of oxidation of propylene and other hydrocarbons fitted the pattern of this general prediction. With propylene, as Q_0 was increased, activity declined. The kinetic order for propylene went from -1 to +1, and that for oxygen dropped from 0.5 to 0.

A number of other reports show a general decline in activity with increased value of Q_0, but with less satisfactory correlations with kinetic order and with more scatter of results. For the gas-phase oxidation of hydrogen, ammonia, methane, ethylene, propene, carbon monoxide, or toluene, the order of activity varied somewhat with the reactant, but the general pattern of activity found was (Germain, 1972):

$$\mathrm{Ti} < \mathrm{V} < \mathrm{Cr} < \underline{\mathrm{Mn}} > \mathrm{Fe} < \underline{\mathrm{Co}} > \mathrm{Ni} < \mathrm{Cu} > \mathrm{Zn} \tag{8.6}$$

In almost all cases the most active catalyst was manganese oxide or cobalt oxide; the least active were the oxides of titanium or zinc. The oxides of vanadium, chromium, and nickel were mavericks, being relatively active for some reactants, relatively inactive for others.

This pattern of reactivity corresponds in a general way to the mobility of oxygen. The latter can be expressed in several ways, such as the heat of dissociation of the first oxygen from the oxide to the next lower oxidation state, or the rate of exchange of isotopic oxygen between the bulk-gas phase and oxygen in the catalyst. It can also be characterized by the partial pressure of oxygen above the oxide at a specified temperature.

A rank ordering of the heat of formation of metal oxides per oxygen atom, as used by Moro-oka et al. (1967), is:

$$\mathrm{Ag} < \mathrm{Pt} < \mathrm{Pd} < \mathrm{Cu} < \mathrm{Co} < \mathrm{Ni} < \mathrm{Cd} \approx \mathrm{Mn} < \mathrm{Fe} < \mathrm{V} < \mathrm{Cr} < \mathrm{Ce} < \mathrm{Al} < \mathrm{Th} \tag{8.7}$$

If the heat of dissociation of the first oxygen from the metal oxide is used instead (Simons et al., 1968), the ranking is

$$\mathrm{MnO_2} < \mathrm{V_2O_5} < \mathrm{CuO} < \mathrm{Co_3O_4} < \mathrm{Fe_2O_3} < \mathrm{NiO} < \mathrm{TiO_2} < \mathrm{ZnO} \tag{8.8}$$

Boreskov calculated values of heats of desorption from the effect of temperature on the equilibrium partial pressure of oxygen above the oxides, and obtained the following sequence:

$$\mathrm{Co_3O_4} < \mathrm{CuO} < \mathrm{NiO} < \mathrm{MnO_2} < \mathrm{Cr_2O_3} < \mathrm{Fe_2O_3} < \mathrm{V_2O_5} < \mathrm{ZnO} < \mathrm{TiO_2} \tag{8.9}$$

The activation energy of isotopic exchange gives the following sequence:

$$\mathrm{Co_3O_4} > \mathrm{MnO_2} > \mathrm{NiO} > \mathrm{CuO} > \mathrm{Fe_2O_3} > (\mathrm{TiO_2}) > \mathrm{ZnO} > \mathrm{Cr_2O_3} > \mathrm{V_2O_5} \tag{8.10}$$

It is seen that, although there is considerable variation among the various methods of characterizing oxygen mobility, the oxides rank in very roughly the same order.

The overall conclusion as to the usefulness of this approach, as drawn by Germain in a detailed review (1972), is that for simple metal oxide catalysts there is a limited correlation between catalytic activity and oxygen mobility. Direct methods of estimation of oxygen mobility, such as by oxygen chemisorption and oxygen exchange, give a better correlation with reactivity than the thermodynamic calculation of heat of dissociation of a metal oxide. One difficulty with this approach is that the nature of the reactant oxidation step cannot realistically be expected to remain constant over a broad range of catalysts.

Sachtler (1970) and Sachtler and deBoer (1965) have suggested that maximum selectivity is associated with an optimum degree of oxygen mobility. For an optimum combination of activity and selectivity there should be a matching between the difficulty of oxidizing the reactant and the ease of removal of oxygen from the catalyst. Tightly bound oxygen should result in a low-activity catalyst. Highly mobile oxygen should result in a high-activity catalyst, but one that is nonselective. Indeed Simons et al. (1968) reported a maximum selectivity for dehydrogenation of 1-butene to butadiene at an intermediate degree of oxygen mobility.

There are several problems in pressing oxygen mobility very far as a useful concept. The approach is perhaps most useful in considering a very broad range of catalyst compositions, but many other factors are involved. For high selectivity in a reaction of the type $A \rightarrow B \rightarrow C$ it is necessary that the rate of desorption of a desired intermediate B be high relative to the rate of further oxidation. This is related to the strength of adsorption of B, which is determined by other aspects of the catalyst structure. Another limitation is that most of the oxide catalysts exhibiting high selectivity are not simple oxides but are instead complex structures containing more than one type of metal, whose properties may vary substantially with fairly minor changes in composition. Furthermore the adsorptivity and reactivity of a reactant or intermediate can be markedly affected by the defect nature of the catalyst.

In at least several cases maximum activity of an oxidation catalyst is observed when the catalyst is in a somewhat reduced state rather than completely oxidized. This was shown, for example, by Sleight and Linn (1976) for oxidation of butene over various lead bismuth molybdate catalysts. With vanadium oxide, maximum activity or selectivity has been reported for the composition V_2O_4 or compositions intermediate between V_2O_4 and V_2O_5 (see Sec. 8.6). However, selectivity may also be enhanced by the degree of acidity of a catalyst and by the presence of water vapor, which may assist the desorption of a desired intermediate.

The concept of selectivity as affected by adsorption of an intermediate has been generalized by Germain (1972) in what he terms a *rake mechanism.* Considering a species such as propylene, successive abstraction of H atoms leads to a "rake" of adsorbed species:

$$C_3H_6 \longrightarrow C_3H_5 \longrightarrow C_3H_4 \longrightarrow C_3H_3$$

The third can be desorbed as acrolein in oxidation and the fourth as acrylonitrile in ammoxidation (Sec. 8.2). The usefulness of this concept is that it suggests the possibility of improving reaction selectivity by modifying catalyst composition so as to enhance the desorption of a desired intermediate.

With oxide catalysts, chemisorbed surface oxygen as well as lattice oxygen may play a role. A study by Sancier et al. (1975) of the oxidation of propylene to acrolein on a $BiMo/SiO_2$ catalyst in a pulsed reactor showed by isotopic oxygen studies that surface chemisorbed oxygen as well as lattice oxygen could contribute to the overall reaction. The reaction path with chemisorbed oxygen had an activation energy of about 54 kJ/mol, whereas that with lattice oxygen had an activation energy of about 84 kJ/mol; hence the mechanism involving surface oxygen contributed less to the overall reaction at the higher temperatures. In this particular case the fraction of the total acrolein formed by this mechanism was only 8 percent at about 590°K and 1 percent at about 770°K. It seems plausible that in many other cases chemisorbed oxygen would lead to a different set of products than lattice oxygen and both mechanisms could be significant. On the basis of other studies, Haber (1975) advances the hypothesis that surface-adsorbed oxygen may in general lead to products of complete oxidation and that lattice oxygen is needed for partially oxidized products, but more study is needed to test this proposal.

Most of the metal oxides of interest are nonstoichiometric, and the oxygen/metal ratio in the solid will vary with temperature, oxygen partial pressure, and the environment. Hence there may be a subtle interaction between the effect of reaction conditions on the state of oxidation of the catalyst and, in turn, by catalyst composition on activity and selectivity.

V_2O_x can be readily reduced down to a composition with x nearly as low as 3 and yet be rapidly reoxidized to V_2O_5. This has led to the suggestion that vanadium and other metal oxides capable of rapid and reversible oxidation-reduction could be used as an oxygen carrier to cause a partial oxidation reaction, the carrier being reoxidized in a separate vessel with air. This has the merit of achieving different (and possibly superior) solid compositions than that corresponding to the steady-state condition in the presence of a reactant-oxygen mixture and the possibility of synthesizing products that would be too reactive in the presence of oxygen. However, there is no information that a process of this type has actually been put in operation. One disadvantage is the large quantity of solid that would need to be moved back and forth between reactor and regenerator relative to the quantity of product formed. A second is the necessity of developing a solid composition that, in addition to having the requisite chemical properties, would also be resistant to mechanical attrition and repetitive cycling between different temperatures and gas compositions. Isolation of the fluids in the reactor and regenerator from one another is also vital, so inadvertent mixing, which might lead to an explosion, cannot occur.

Many commercial oxidation catalysts are complex metal oxides containing two or more different cations, and the structure is much different than that of the component oxides. In many cases the structure is relatively complex and uncertain. A number of studies have been made with three important classes of complex metal oxides whose structures are well established; they are stable under reaction conditions and exhibit substantial catalytic activity per unit area for reactions of industrial interest. These are the spinels (Sec. 4.5.1), the scheelites (see Sec. 8.2 and Sleight, 1977) and the perovskites

(Voorhoeve, 1977). The principal disadvantage to their use in practice is that by conventional methods the starting materials must be heated to a high temperature to cause the desired structure to form. The product is then nonporous, and typical surface areas are of the magnitude of 0.1 m^2/g. Some of these crystalline materials are inert and are of interest as supports because of their stability.

8.2 OXIDATION AND AMMOXIDATION OF PROPYLENE

8.2.1 Acrylonitrile

The manufacture of acrylonitrile by the ammoxidation of propylene has become the dominant industrial process, based primarily on the discovery of a series of mixed metal oxide catalysts by Idol and coworkers at the Standard Oil Co. of Ohio (Sohio). The process, which first became commercial in about 1960, utilizes a fluidized-bed reactor, as do closely related processes by Montedison and Nitto. Ammonia, air, and propylene are supplied to the reactor at a pressure of about 2 atm and in nearly stoichiometric ratios, according to the equation

$$NH_3 + C_3H_6 + \tfrac{3}{2}O_2 \longrightarrow CH_2{=}CHCN + 3H_2O \qquad (8.11)$$

$$-\Delta H = -515 \text{ kJ/mol}$$

The reactor temperature is in the range of 400 to 500°C, and the residence time is a few seconds. Several byproducts are also formed, in particular hydrogen cyanide and acetonitrile, which may be recovered as byproducts, and carbon oxides. With some catalyst compositions and under proper operating conditions, essentially all the ammonia and propylene are consumed. With other catalysts a small amount of ammonia remains in the product gas, which may be recovered or scrubbed with sulfuric acid to convert it to ammonium sulfate before the acrylonitrile is separated. Heat released by the reaction is recovered as steam by cooling coils immersed in the reactor. The same reaction is carried out in a multitube fixed-bed reactor in a process developed by British Petroleum (Distillers)/Ugine, which uses a catalyst containing the mixed oxides of antimony and tin, and possibly also some iron. This process is not now widely practiced.

The first of the commercial ammoxidation catalysts for the Sohio fluid-bed process was a bismuth phosphomolybdate, superceded by an antimony oxide-uranium oxide composition or (in Japan) an antimony oxide-iron oxide composition. These catalysts may also contain small amounts of other elements. In 1972 a third-generation catalyst was announced by Sohio. The patents and publications indicate that it has a complicated composition such as $M_8{}^{II}Fe_3{}^{III}Bi^{III}(MoO_4)_{12}O_{12}$. M represents various amounts of Ni^{2+} and Co^{2+} (Gates et al., 1979). This catalyst produces more acrylonitrile and less byproduct acetonitrile than its predecessors, and may be operated for essentially complete consumption of reactants on a once-through basis. All these catalysts are supported on silica for attrition resistance in the fluidized-bed reactor. The Montedison catalyst consists of tellurium, cerium, and molybdenum oxides on silica and operating conditions are similar to that used with the Sohio catalyst, but apparently a small amount of ammonia passes through unreacted.

Bismuth molybdate and other molybdate catalysts have been among the most intensively studied mixed-oxide compositions, and the reactions of propylene have been a popular model reaction, in considerable part because of the industrial importance of molybdenum-based catalysts and of the manufacture of acrylonitrile. The overall features of the oxidation and ammoxidation reactions are now reasonably well established, but many of the details of the mechanisms themselves are still debatable.

The first step in either the oxidation or ammoxidation of propylene is dissociative adsorption with abstraction of a hydrogen atom to form an adsorbed symmetric allyl radical, C_3H_5. This is shown, for example, by tracer studies with ^{14}C (Sachtler, 1970). The catalysts effective for ammoxidation are also usually effective for formation of acrolein in the presence of oxygen but in the absence of ammonia (Sec. 8.2.2). In the presence of ammonia, acrolein in some cases may be found in small amounts as an isolable intermediate. Presumably a surface intermediate is formed that can be either desorbed as acrolein or converted to acrylonitrile, perhaps via an iminelike structure as an intermediate step. Alternatively, this or another surface intermediate is converted to carbon oxides. The manner in which ammonia is incorporated into the product is speculative. A plausible mechanism would involve dissociation to form NH_2 or NH, although there is no direct evidence. These steps are shown in Fig. 8.1.

The first step, formation of adsorbed allyl, is clearly the rate-limiting step, based on a number of observations. The rate of reaction is essentially the same for both the oxidation and ammoxidation reactions, and they both have about the same activation energy (about 80 to 90 kJ/mol) on various bismuth molybdates (Callahan et al., 1970). The selectivity for acrylonitrile probably centers largely about the fate of an adsorbed intermediate closely related in structure to acrolein. If oxygen is too mobile or accessible in the lattice structure, the intermediate is converted to carbon monoxide and carbon dioxide instead of acrylonitrile. It also appears that selectivity may be reduced by a form of heterohomogeneous catalysis (Sec. 1.3.5). Aldehydes, formed on the catalyst surface, may be desorbed into the gas phase, where they may readily decompose, initiating rapid gas-phase oxidation of propylene (Cathala and Germain, 1971).

All catalysts showing high activity and selectivity in the ammoxidation reaction consist of a complex oxide structure incorporating two or more metals. This has quite

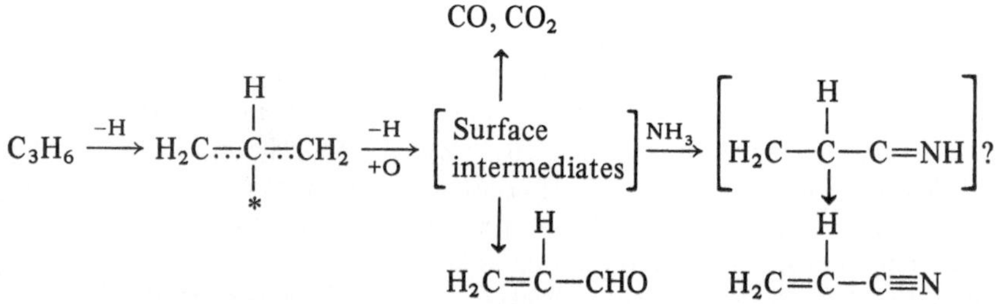

$$C_3H_6 \xrightarrow{-H} H_2C \overset{H}{\underset{*}{\cdots C \cdots}} CH_2 \xrightarrow[+O]{-H} \left[\begin{matrix}\text{Surface} \\ \text{intermediates}\end{matrix}\right] \xrightarrow{NH_3} \left[H_2C{-}\overset{H}{C}{-}C{=}NH\right]?$$

$$\text{Surface intermediates} \rightarrow CO, CO_2$$

$$\text{Surface intermediates} \rightarrow H_2C{=}\overset{H}{C}{-}CHO$$

$$\left[H_2C{-}\overset{H}{C}{-}C{=}NH\right] \rightarrow H_2C{=}\overset{H}{C}{-}C{\equiv}N$$

Figure 8.1 Plausible reaction mechanism for oxidation or ammoxidation of propylene.

different properties in oxidation or ammoxidation than do the single oxides. For oxidation of propylene, Bi_2O_3 alone has low activity and causes essentially complete oxidation. MoO_3 alone shows good selectivity but has even lower activity than Bi_2O_3. Much study has been devoted to Bi_2O_3-MoO_3 systems, but the structures are still not well characterized and the relationships between structure and activity are still obscure. Three phases, α, β, and γ, having mole ratios of MoO_3 to Bi_2O_3 of 3, 2, or 1 are frequently discussed, but there are variations of structure within a single "phase." Possibly more important than the nature of the phase is the degree of nonstoichiometry and the nature of the defect structure, which are in part determined by the method of preparation (Sleight and Linn, 1976).

In the U-Sb mixed-oxide system the active and selective phase for formation of acrolein or acrylonitrile is USb_3O_{10}. Sb_2O_4 by itself is inactive, and UO_3 by itself strongly catalyzes combustion to CO_2 and CO.

The principal structure identified by X ray in the third-generation Sohio catalyst appears to be cobalt nickel molybdate (Gates et al., 1979), but it is not clear what portion of the total this represents or in what form the other elements are present. These various molybdates can be prepared with an excess of molybdenum trioxide. Iron molybdate, used industrially for oxidative dehydrogenation of methanol to formaldehyde, has a maximum selectivity for this reaction when it contains some excess molybdenum trioxide. The detailed structures of these multicomponent ammoxidation catalysts are complex, may vary considerably between the core and outer layers, and as yet are poorly understood.

A variety of multicomponent molybdates such as those containing cobalt, iron, bismuth, tellurium, or other heavy metals have been characterized, and attempts have been made to relate structure to activity and selectivity for a variety of partial oxidation reactions (Higgins and Hayden, 1977). Such patterns that have been proposed are as yet subject to debate. $CoMoO_4$ can exist in two forms, α and β, the α form being the stable form above 420°C. However the α form can be stabilized for an indefinite period at lower temperatures on some supports or by suitable additions. The two forms clearly have quite different catalytic properties. Another poorly understood variable of importance for partial oxidation in general is the acidity of the catalyst.

The catalyst of choice for an industrial process rests upon a complicated interplay of various factors. These include aging characteristics, stability to fluctuations in process conditions, and maximum achievable selectivity. The fact that the reaction to form acrylonitrile is zero order with respect to oxygen (also zero order with respect to ammonia) and first order with respect to propylene indicates that the rate of reoxidation of the catalyst is rapid and not a rate-limiting process. The mechanism in Fig. 8.1 suggests that maximum selectivity may be achieved with the catalyst in a relatively reduced state. However, an excessive degree of reduction can lead to a change in structure and loss of the active phase. Moreover, if a metallic phase appeared, it might cause undesired decomposition of ammonia to the elements. One of the functions of some of the minor components added may be to stabilize the desired phase to fluctuations in temperature and gas composition.

The marked effects of defects on activity and selectivity in the conversion of propylene to acrolein or to acrylonitrile are shown in studies by Aykan et al. (1974)

with scheelite-type structures. These are highly active and selective for this reaction and are easier to characterize than the molybdates. Scheelites are named after the mineral scheelite, $CaWO_4$, and their structure and catalytic properties are reviewed by Sleight (1977). They have the ideal structure AMO_4, in which all M and A cations are equivalent and all the oxygens are equivalent. The A cations are relatively large ($r > 0.08$ nm), and the M cations are relatively small ($r < 0.06$ nm). A-type cation vacancies can be produced in considerable concentrations in a number of scheelite-type structures in which A represents one or more than one type of large cation. The resulting structure can be represented as $A_{1-x}\phi_x MO_4$ where ϕ is the number of vacant A-cation sites, which are randomly distributed. For a group of scheelites as unsupported catalysts in which A consists of lead and bismuth in various ratios, and M is molybdenum, Fig. 8.2 shows the marked change in rate of reaction and selectivity to acrylonitrile as the composition was changed. The value of ϕ was varied by varying the ratio of bismuth to lead in the mixture of starting ingredients.

It is difficult in these kinds of studies to separate the role of the defects from the role of the bismuth. Studies on a variety of scheelite structures led Sleight (1977) to conclude that the most important role of A-type cation defects was to promote allyl formation from an olefin. Bismuth is not required for this step; it promotes high mobility of O^{2-} and may well be the site for adsorption of oxygen. The maximum conversion to acrylonitrile here was greater than that reported for any industrial bismuth molybdate catalyst, but because of other factors, such as the low surface area of the scheelites, they do not appear to have been used commercially.

The details of alternate mechanisms for propylene oxidation and ammoxidation are discussed more fully by Sleight and Linn (1976), Hucknall (1974), and Gates et al. (1979). The last also describe in considerable detail the bulk structures of various

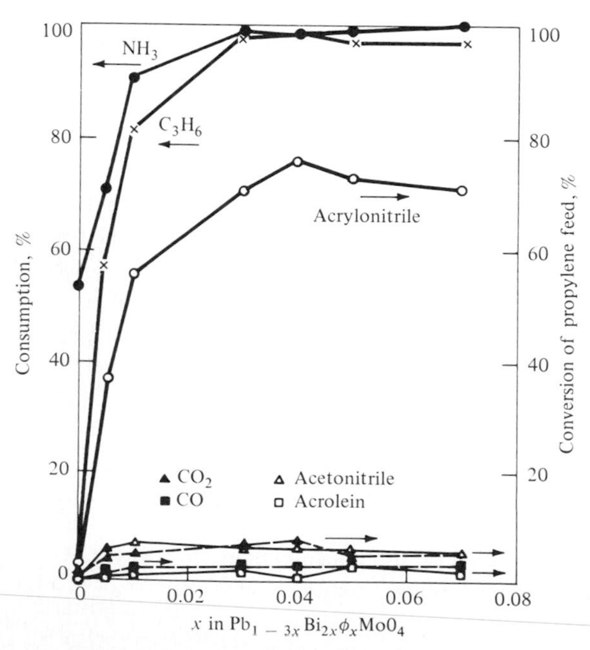

Figure 8.2 Percent consumption and conversion during ammoxidation of propylene versus x in the $Pb_{1-3x}Bi_{2x}\phi_x(MoO_4)$ system. The temperature was 440°C, and the feed was 4.0% C_3H_6, 4.8% NH_3, 47.7% air, and 43.5% N_2 with a 6.0-s contact time. (Aykan et al., 1974.) (*Reprinted with permission from Advances in Catalysis. Copyright by Academic Press.*)

binary oxides that appear to be of catalytic importance and of related single metal oxides. However little is known of the surface structures of these materials and how they may differ from the bulk. Higgins and Hayden (1977) review in detail the literature on mixed-oxide catalysts and selective hydrocarbon oxidation reactions for the period of 1973 to mid-1976.

The various methods of preparing acrylonitrile, including earlier manufacturing processes such as the reaction of acetylene and hydrogen cyanide, are described by Dalin et al. (1971). Some 1400 patents and literature references are cited. The book edited by Hancock (1973) also provides considerable information. A paper by Kolchin (1973) summarizes the performance characteristics of a large number of catalyst compositions and contains numerous references. The catalytic ammoxidation of propane has received considerable study since propane is a lower-cost raw material than propylene, but no commercial process is known to have been put into operation.

About one-half of the acrylonitrile produced is used for manufacture of acrylic fibers. It is also hydrodimerized electrolytically (e.g., Monsanto process) to adiponitrile for manufacture of nylon 66. Some is incorporated into copolymers such as acrylonitrile-butadiene-styrene.

8.2.2 Acrolein and Acrylic Acid

Propylene is partially oxidized with either air or oxygen by the reaction:

$$CH_3CH{=}CH_2 + O_2 \longrightarrow CH_2{=}CHCHO + H_2O \qquad (8.12)$$

$$-\Delta H = 341 \text{ kJ/mol}$$

In the Shell process, propylene is diluted with steam and reacted with air at about 350°C and 200 kPa over a supported cuprous oxide catalyst. About an 85 percent yield is obtained at about 20 percent conversion. In the Sohio acrolein process, propylene, air, and steam are reacted in a fixed-bed reactor at 200 to 300 kPa pressure and about 320 to 370°C, using a bismuth molybdate catalyst. Acetaldehyde and acrylic acid are produced as byproducts.

Acrylic acid is made by several similar processes. In the Distillers process used in Great Britain by Border Chemicals and in the United States by Union Carbide, a mixture of propylene, air, and steam in the ratio 1:5:4 is oxidized to acrylic acid and acrolein in a multitube reactor at slightly above atmospheric pressure. With a conversion of 22.5 percent of feed per pass, 12.5 percent is acrylic acid and 10 percent is acrolein (Hahn, 1970, p. 207). Patents disclose a complex antimony-tin-vanadium mixed-oxide composition for this reaction. Acrolein can be recovered from the first reactor, or the exit stream from the first can be further oxidized to form more acrylic acid in a second reactor, probably with a different catalyst and different operating conditions. The conversion of acrolein to acrylic acid can be written:

$$2CH_2{=}CHCHO + O_2 \longrightarrow 2CH_2{=}CHCOOH \qquad (8.13)$$

$$-\Delta H = 254 \text{ kJ/mol}$$

In a process for acrylic acid licensed by Sohio, a two-reactor, fixed-bed system is used at 200 to 300 kPa and about 290 to 400°C with different catalysts in the two reactors. In the Toyo Soda process reaction temperatures of 330 to 370°C are reported for the first reactor and 260 to 300°C for the second; an overall yield of acrylic acid from propylene of about 67 percent is reported. In all these processes some acetic acid is produced as a byproduct.

The primary use of acrolein is as an intermediate to acrylic acid, which is then converted to various esters such as methyl and ethyl acrylates, typically by liquid-phase reaction using an ion-exchange-resin catalyst.

A closely related reaction is the partial oxidation of isobutylene to methacrolein, which can be converted to methacrylic acid. A number of complex metal oxide catalyst compositions have been patented for the formation of methacrolein, but this route has apparently not yet been commercialized.

8.2.3 Other Ammoxidation Processes

A substantial literature exists on ammoxidation reactions of aromatic compounds to form nitriles. An industrial process, first commercialized in Japan about 1968 to 1970 (Mitsubishi/Badger), converts *m*-xylene to isophthalonitrile in a fluid-bed reactor. A similar process is operated by Showa Denko. Small amounts of benzonitrile and *m*-tolunitrile are also formed, as well as carbon dioxide, carbon monoxide, and hydrogen cyanide. By similar reactions heterocyclic compounds such as an alkyl pyridine may be converted to the corresponding nitrile. A nitrile in turn can be hydrolyzed to the corresponding carboxylic acid or hydrogenated to the corresponding amine. Klink et al. (1976) give some information on the development and use of a catalytic process for ammoxidation of a methylthiazole to the corresponding cyanothiazole. A book by Suvorov (1971), in Russian, is devoted entirely to the oxidative ammonolysis of organic compounds.

8.3 ETHYLENE TO ETHYLENE OXIDE

Ethylene can be oxidized to ethylene oxide with high selectivity over supported silver catalysts, and this direct oxidation process now accounts for essentially all ethylene oxide produced industrially. The process was first developed by Union Carbide, who started the world's first plant in 1937. In the 1950s a direct oxidation process was also developed by Scientific Design Co., and a process using substantially pure oxygen instead of air was developed by Shell Development Co. Both of these have also been widely used. Most of the ethylene oxide produced is hydrolyzed to ethylene glycol, used as an antifreeze in automobile cooling systems and as one of the two monomers in polyethyleneterephthalate (polyester) fibers and polymers. In a newly developed competitive process by Oxirane for making ethylene glycol, utilized in a plant under construction, ethylene is reacted catalytically with acetic acid to form ethyl acetates. These are then hydrolyzed to the glycol, and acetic acid is recycled. No ethylene oxide is formed.

For the ethylene oxide process the optimum reaction temperature is about 260 to 280°C when air oxidation is utilized. The contact time is about 1 s. When oxygen is used, the optimum reaction temperature is about 230°C. Pressure is 1 to 3 MPa.

Industrial reactors are shell-and-tube type (2.5 to 5 cm tube diameter). The heat typically is removed by an organic coolant and it is then recovered as steam, or is removed by a boiling coolant system. The feed composition consists of fresh feed mixed with recycled gas and is typically 5 to 10 vol % ethylene, and 5 to 10% oxygen, the remainder being mostly carbon dioxide and nitrogen. With the use of oxygen, a higher ratio of ethylene to oxygen can be used, resulting in a slightly higher selectivity. For high selectivity (70 to 80 percent) the conversion is held to about 30 percent. With the use of air, after removal of ethylene oxide from the exit gas by scrubbing with water under pressure, a portion of the exit stream is recycled. The remainder is sent to a second reactor, which is used as a way of scavenging remaining ethylene and purging nitrogen and other inert gases from the system. Recycling may also be used in the second reactor. The carbon dioxide content of the mixed feed to the first reactor is controlled by removing carbon dioxide from a portion of the recycle stream to the first reactor with an absorbent such as a potassium carbonate solution. With essentially pure oxygen, the second reactor is unnecessary, and usually only one reactor is used; in this case unabsorbed gases are recycled and a portion of the recycle stream is treated for carbon dioxide removal.

Use of pure oxygen allows a higher reactor productivity (production rate per unit volume of catalyst) and less costly recycle of ethylene. These may more than compensate for the higher cost of pure oxygen. Usually the oxygen-based process is preferred in large plants, and some air-based plants have been converted to use of oxygen. With oxygen a diluent such as methane can be used which narrows the inflammability limits and allows the use of higher concentrations of ethylene and oxygen. This gives higher productivity without reducing yield. A fluid-bed reactor system has been considered for this process, but it does not appear to have been commercialized. More details concerning these processes are given by Kiguchi et al. (1976), Gans and Ozero (1976), and in an earlier review by Voge and Adams (1967).

Several features of the reaction are unique. No other metal is at all comparable to silver in selectivity, and this epoxidation reaction is unique to ethylene. Propylene or butylene do not undergo analogous reactions on silver or any other known heterogeneous catalysts. Possibly they are adsorbed in the form of a π-bonded allyl that is readily oxidized to carbon oxides and water. In practice about 10 to 15 wt % silver is supported on low area ($<1\ m^2/g$) α-alumina, presumably to provide large pores and avoid the diffusion-limiting regime which would reduce selectivity. For the same reason the silver may be deposited in a shell on the outside portion of the catalyst.

In order to obtain maximum selectivity it is necessary to partially poison the catalyst (see later) by adding continuously a chlorinated compound such as ethylene dichloride to the feed. This, however, reduces activity, so that the optimum concentration is a few parts per million. This causes a chloride to be adsorbed onto the surface; but the chloride is not retained, hence the necessity for continuous addition. Calcium or barium salts are widely quoted as promoters, but they have an opposite effect; they increase conversion but decrease selectivity.

The overall kinetics are of the type:

$$\begin{array}{ccc} C_2H_4 + O_2 & \xrightarrow{k_1} & C_2H_4O \\ & \searrow^{k_2} \quad {}^{k_3}\swarrow & \\ & CO_2 + H_2O & \end{array} \tag{8.14}$$

The heats of reaction, per mole of ethylene, are $-\Delta H_1 = 146$ kJ/mol and $-\Delta H_2 = 1320$ kJ/mol. Most of the carbon dioxide is apparently formed by reaction 2 rather than reaction 3 under industrial conditions. The mechanism of the epoxidation reaction has been studied extensively, but much is still in dispute. The arguments center around the various forms in which oxygen is chemisorbed onto the silver and ways in which ethylene reacts with these species and with silver atoms.

The oxygen complexes adsorbed on silver have been characterized by a wide variety of studies, including calorimetric studies, infrared, isotopic exchange, electron diffraction, and kinetic studies of adsorption rates. The interpretation has been ambiguous in many cases because of the marked effects of traces of impurities on the silver. Perhaps the most plausible reaction mechanism is that advanced by Kilty and Sachtler (1974) and recently discussed by Carra' and Forzatti (1977). The key evidence is that diatomic oxygen species are formed on silver and they react readily with ethylene to form ethylene oxide. If it is clean, silver metal itself does not adsorb ethylene, but it may under industrial operating conditions. At sufficiently high ethylene concentrations the reaction is first order with respect to oxygen and zero order with respect to ethylene.

The desired reaction is believed to proceed via a Rideal mechanism in which ethylene is strongly adsorbed onto a previously adsorbed diatomic oxygen ion, O_2^-. This complex splits at the O—O bond to form ethylene oxide, leaving an adsorbed oxygen atom.

```
                   H H            H H
                   | |            | |
H2C=CH2          H-C-C-H        H-C-C-H          (8.15)
   +               \ /            \ /
   O               O+              O
   |      --->      |     --->
   O               O-              O
   |                |              |
   Ag              Ag              Ag
```

The key first step may therefore be written as

$$O_2 + Ag \longrightarrow O_2^-(ads) + Ag^+ \tag{8.16}$$

in which gas-phase oxygen is adsorbed onto a single silver atom. Oxygen can also be dissociatively adsorbed to form monoatomic oxygen by interaction with a cluster of four adjacent silver atoms.

$$O_2 + 4Ag(adj) \longrightarrow 2O^{2-}(ads) + 4Ag^+(adj) \tag{8.17}$$

This is a nonactivated process, but it can be completely inhibited at any temperature by adsorption of chlorine atoms when the surface is about one-quarter covered with chloride.

It is postulated that clusters of silver atoms can also be produced by an activated process involving migration, which increases in importance at higher temperatures.

$$O_2 + 4Ag(nonadj) \longrightarrow 2O^{2-}(ads) + 4Ag^+(adj) \tag{8.18}$$

Reactions (8.17) and (8.18) are not desired since adsorbed oxygen ions (or atoms) are postulated to lead to carbon monoxide and carbon dioxide.

Maximum selectivity is achieved if further reaction of ethylene oxide is negligible, reaction (8.17) is blocked by adsorbed chlorine, and reaction (8.18) is negligible. However by the above mechanism the maximum selectivity achievable cannot be 100 percent, since some of the ethylene must be consumed to reduce the silver with adsorbed oxygen atoms formed in reaction (8.15) back to metallic silver. The minimum quantity of ethylene required for this step would occur if the products were carbon dioxide and water, leading to the following overall stoichiometry:

$$6O(ads) + C_2H_4 \longrightarrow 2CO_2 + 2H_2O \tag{8.19}$$

The maximum selectivity achievable according to this mechanism is $\frac{6}{7}$ or 85.6 percent. This conclusion is consistent with the fact that apparently no selectivities higher than this figure have ever been reported.

The reasons for the uniqueness of silver for this reaction are still unclear. The species O_2^-, which is a key intermediate, is also formed on other metals, so this is a necessary but not sufficient condition. The role of the chloride is to inhibit the dissociative adsorption of oxygen, which is postulated to occur on four-atom clusters of silver. Alloying silver with gold without using chlorine addition also increases the selectivity of the reaction (Flank and Beachell, 1967), which can be interpreted as reducing the concentration of the four-atom clusters of silver by a dilution effect (see also Sec. 6.3). Conceivably a small concentration of carbonaceous residues on the catalyst could also improve selectivity in the absence of chloride by, in effect, preventing clusters of silver atoms from being exposed to reactants.

Propylene oxide cannot be manufactured by a process analogous to that used for ethylene oxide. Instead, a current process, developed by Halcon and Oxirane, involves the liquid-phase oxidation with air of ethylbenzene to form the hydroperoxide. This reacts with propylene in the liquid phase to form propylene oxide and phenylmethylcarbinol (phenylethyl alcohol). The latter is then dehydrated in the vapor phase to styrene, using a TiO_2 catalyst at about 180 to 280°C. Another version of this process was used in the first plant, built in the United States, in which isobutane was used instead of ethylbenzene, isobutylene being formed as the byproduct.

8.4 METHANOL TO FORMALDEHYDE

The commercial production of formaldehyde started in Germany in about 1890, using the catalytic partial oxidation of methanol with air in a fuel-rich mixture. The catalyst was unsupported metallic copper, used at approximately atmospheric pressure, and the product was rapidly quenched by solution in water. Since about 1910 copper was replaced by silver in processes used in Germany and the United States; this gave higher yields. A different process, apparently first put into use in the 1940s and 1950s, uses an iron molybdate catalyst under fuel-lean conditions (Reichhold-Skanska process). These two processes account for essentially all commercial manufacture of formaldehyde, the silver-catalyzed process being the more widely used. A small portion of the total production is obtained from the homogeneous oxidation of butane or propane.

The possibility of direct conversion of methane to formaldehyde has been extensively studied, but it is not commercially practicable, probably because of the much greater reactivity of formaldehyde relative to methane. Methanol will decompose directly to formaldehyde and hydrogen, an endothermic process, but no catalyst appears to have been developed which does not also substantially decompose the formaldehyde further to carbon monoxide and hydrogen. The methanol-air explosive region is from about 6.7 to 36.5 vol % at atmospheric pressure and expands somewhat with increasing temperature. Operations must be outside these limits for safety. Some details concerning the two processes have been published in the works by Walker (1964), Dixon and Longfield (1960, p. 231), and Chauvel et al. (1973).

8.4.1 Methanol-Rich System (Silver Catalyst)

A methanol-air mixture (about 50:50 mol ratio), at marginally above atmospheric pressure, is passed through a fixed, thin catalyst bed consisting of silver wire gauze or silver crystals (0.5 to 3 mm in size). The exit temperature is about 600°C, and the contact time is about 0.01 s or less. The reaction is not equilibrium-limited but is highly mass-transfer-controlled, and operation is essentially adiabatic. The silver catalyst is more sensitive to iron-group impurities than the iron molybdate catalyst described below. The reactor configuration is simple and much less costly than a multitube reactor with external cooling, as used with the iron molybdate catalyst, but the overall yield with the silver catalyst system is less than with the iron molybdate system. The operating temperature is well above the Tammann temperature, $\sim T_m/2 = 615$ K, and the silver undergoes major physical reconstruction during the reaction.

The exit gases always contain considerable hydrogen as well as water, so the reaction may be regarded as a mixture of partial oxidation and dehydrogenation.

$$CH_3OH + \tfrac{1}{2}O_2 \longrightarrow HCHO + H_2O \qquad (8.20)$$

$$-\Delta H = 158 \text{ kJ/mol}$$

$$CH_3OH \longrightarrow HCHO + H_2 \qquad (8.21)$$

$$-\Delta H = -84 \text{ kJ/mol}$$

Metallic silver as such has little catalytic activity for the decomposition of methanol even at high temperatures. To be active, oxygen must be chemisorbed onto it. Methanol is not readily adsorbed on silver metal but is readily adsorbed onto the chemisorbed oxygen. The situation here is somewhat analogous to ethylene oxide formation, where ethylene is adsorbed onto oxygen, which is, in turn, chemisorbed onto silver. At the much higher temperatures here, however, the chemisorbed oxygen is postulated to be monatomic. Since the feed ratio is oxygen-lean, the above implies that the downstream portion of a packed bed may have little activity once the oxygen has been substantially consumed. Adsorptivity for oxygen decreases in the order: copper, silver, gold. The intermediate adsorptivity of silver makes it the preferred catalyst over the other two.

From studies on metallic copper and silver, Wachs and Madix (1978) propose that the principal mechanism is:

$$CH_3OH(g) + O(ads) \longrightarrow CH_3O(ads) + OH(ads) \qquad (8.22)$$

$$CH_3OH(g) + OH(ads) \longrightarrow CH_3O(ads) + H_2O(g) \tag{8.23}$$

$$2CH_3O(ads) \longrightarrow 2CHOH(ads) + 2H(ads) \tag{8.24}$$

$$2H(ads) \longrightarrow H_2(g) \tag{8.25}$$

This implies that the oxidation and dehydrogenation reactions do not occur as separate processes on different sites but are intimately related. They further suggest that at a high degree of oxygen coverage, formation of a formate becomes more important, which decreases selectivity. Therefore a relatively low coverage of chemisorbed oxygen is optimum. The formate would rapidly decompose to form carbon dioxide and hydrogen or carbon monoxide and water. A low concentration of steam is also frequently added to the feed to help eliminate carbon deposits on the catalyst, and this may allow a somewhat higher degree of conversion to be achieved before selectivity becomes unacceptably low. The product must be rapidly quenched to minimize decomposition of formaldehyde into carbon monoxide and hydrogen. Considerable carbon dioxide is found in the byproducts. This is probably not formed by oxidation of methanol directly and possibly is formed by catalytic oxidation of carbon monoxide.

If a single reactor is used, conversion is limited to about 70 percent of the methanol. The product is then distilled, and unconverted methanol is recycled. To achieve essentially complete reaction without recycling, it is necessary to use a second reactor with additional air introduced interstage, to avoid forming an explosive mixture initially. This eliminates the cost of the distillation step, but the overall yield with respect to methanol is less.

8.4.2 Methanol-Lean System (Iron Molybdate Catalyst)

The concentration of CH_3OH in air here is about 6 to 9 mol %, and the catalyst is $Fe(MoO_4)_3$ plus some excess MoO_3. The operating temperature is about 350 to 400°C, with a maximum of about 425°C, set largely to minimize the volatilization of molybdenum oxides at higher temperatures. Such volatilization can cause the formation of "molybdenum-deficient" iron molybdates, which reduces the selectivity. Excess MoO_3 added in preparation of the catalyst helps prevent formation of the molybdenum-deficient species. Nevertheless the Mo/Fe ratio gradually drops during use, and typical lifetime is about 6 to 12 months. Small amounts of a chromium oxide or a cobalt oxide are also sometimes added, perhaps to help stabilize the iron molybdate structure. Representative forms are 3.5 × 3.5 mm pellets or rings of OD × ID × h = 4 × 2 × 3.5 mm.

This catalyst becomes inert in the presence of excess methanol and requires a fairly high oxygen partial pressure to maintain its activity. Considerable excess air must thus be fed with the methanol. The reaction is more exothermic than that occurring on silver and requires cooling during reaction to keep the temperature down to the required level. A multitube reactor externally cooled with Dowtherm or a molten salt is conventionally used, with about 0.5-s superficial contact time at reactor conditions. Nearly complete reaction is obtained in one pass, and reported yields are about 91 to 94 percent. Air-compression costs are substantial, so operation is with as high a methanol/air ratio as is compatible with safety and maintenance of an active

Figure 8.3 Formaldehyde plant in Castellanza (Varese), Italy. (*Courtesy of Montedison, S.p.A.*)

catalyst. Figure 8.3 is a photograph of a plant having a capacity of 90,000 tons per year of formaldehyde, expressed in terms of the 37 wt % solution. The reactor is on the right, and the (large) absorption tower for product gases is on the left.

A redox-type mechanism presumably operates here; it is of the form

$$CH_3OH + Cat\text{-}O \longrightarrow CH_2O(g) + H_2O(g) + Cat \tag{8.26}$$

$$2Cat + O_2 \longrightarrow 2Cat\text{-}O \tag{8.27}$$

where Cat-O represents the catalyst in the oxidized form, and Cat in the reduced form. The active oxidizing agent is lattice oxygen, but the activity of molybdate-type catalysts for formation of formaldehyde appears to be closely related to their acidity (Ai, 1978). This in turn may be affected by heat treatment and activation procedures.

The formaldehyde yield is higher with the iron molybdate catalyst than with the silver catalyst, but the reactor is more complex and expensive, the catalyst must be replaced more frequently, and air-blowing costs are greater.

8.5 BUTADIENE BY OXIDATIVE DEHYDROGENATION

Normal butylenes may be converted to butadiene by oxidative dehydrogenation as an alternate to direct dehydrogenation (Sec. 9.11).

$$C_4H_8 + \tfrac{1}{2}O_2 \longrightarrow C_4H_6 + H_2O \tag{8.28}$$

$$-\Delta H = 121 \text{ kJ/mol}$$

Unlike direct dehydrogenation, the reaction is not equilibrium-limited. Higher olefins, for example, pentene, react similarly, but give a variety of products. Isobutylene is converted mostly to combustion products; paraffins are relatively unreactive.

This reaction has been carried out industrially by Petro-Tex since 1965 and about 75 percent of United States primary butadiene capacity as of 1978 is reportedly based on their technology (Welch et al., 1978). Patents suggest that the catalyst is a manganese iron spinel whose structure may be written as $MnFe_2O_4$ or $Fe[MnFe]O_4$, or a magnesium chromium ferrite. A zinc chromium ferrite is also reported to give high conversion and selectivity [Rennard and Kehl (1971) and Massoth and Scarpiello (1971)]. These structures may be written as $Zn(Cr_{2-x}Fe_x)O_4$ or $Mg(Cr_{2-x}Fe_x)O_4$. Maximum selectivity occurs with values of x of about 1. Sterrett and McIlvried (1974) report a kinetic study on an unsupported $ZnCrFeO_4$ spinel having approximately a 1:1:1 ratio of the three metals and review earlier literature on oxidative dehydrogenation of butene. The surface area of this catalyst is in the range of 1 to 10 m^2/g, depending on the calcination temperature.

Studies of these and ferrite spinels containing various other metals such as copper and cobalt have also been reported in a series of papers by Hightower and coworkers (Gibson and Hightower, 1976; Gibson et al., 1977). Bismuth molybdate catalysts are also active and selective for this reaction (Batist et al., 1967, 1968), as is a phosphorus-tin oxide composition (Pitzer, 1972). In a process described by British Petroleum Chemicals, reaction in the presence of steam and air is carried out at 400 to 450°C with a catalyst that may be a mixed oxide of antimony and tin.

In the Petro-Tex process a mixture of air and butenes in the presence of a high concentration of steam is reacted in a fixed-bed adiabatic reactor using a moderate stoichiometric excess of butenes at a pressure slightly above atmospheric. The kindling temperature is about 345 to 360°C, and the maximum allowable exit temperature is about 575 to 595°C to prevent catalyst deactivation. Essentially all the oxygen is consumed, and under representative operating conditions about 65 percent of the butylenes are converted to butadiene per pass, in about 93 percent selectivity, using a steam/butylenes feed mol ratio of about 12 (Welch et al., 1978). This ratio is set largely as the minimum needed to limit the adiabatic temperature rise to an acceptable level, which in turn is determined by the hydrocarbon/oxygen ratio and the extent of side reactions. The principal byproducts are vinylacetylene, formaldehyde, acetaldehyde, and acrolein, plus carbon oxides. Some information about a similar process operated by Phillips was published by Hutson et al. (1974).

The oxidative dehydrogenation of hydrocarbons in general was reviewed by Sharchenko (1969), who gives extensive references. The oxidative dehydrogenation of paraffins catalyzed by gaseous halogens such as bromine or iodine, either with or without a solid catalyst, has been studied extensively. No process based on this chemistry appears to have been commercialized, in part because of corrosion problems and the necessity for high halogen recovery.

8.6 PHTHALIC ANHYDRIDE AND MALEIC ANHYDRIDE

The reactions of commercial importance are: the oxidation of naphthalene or *o*-xylene to phthalic anhydride and the oxidation of benzene, *n*-butane, or *n*-butene to maleic anhydride.

$$\text{naphthalene} + (7+x)/2\,O_2 \longrightarrow \text{phthalic anhydride} + 2H_2O + xCO_2 + (2-x)\,CO \qquad (8.29)$$

$$-\Delta H = 1790 \text{ kJ/mol for } x = 2$$

$$\text{o-xylene} + 3O_2 \longrightarrow \text{phthalic anhydride} + 3H_2O \qquad (8.30)$$

$$-\Delta H = 1200 \text{ kJ/mol } o\text{-xylene}$$

$$\text{benzene} + (7+x)/2\,O_2 \longrightarrow \text{maleic anhydride} + 2H_2O + xCO_2 + (2-x)\,CO \qquad (8.31)$$

$$-\Delta H = 1870 \text{ kJ/mol for } x = 2$$

$$nC_4H_8 + 3O_2 \longrightarrow \text{maleic anhydride} + 3H_2O \qquad (8.32)$$

$$-\Delta H = 1315 \text{ kJ/mol}$$

Some maleic anhydride may also be produced as a byproduct in the oxidation of *o*-xylene, especially if it is carried out at relatively high temperatures. The main catalyst ingredient in each case is vanadia, but the support and other additives vary with the process and reactant. A very substantial literature exists in the form of patents and publications. The application of vanadium pentoxide for these types of reactions goes back to at least 1917, when Gibbs and Conover obtained a patent covering some of these uses.

8.6.1 Oxidation of Aromatics

The reactions of interest are the conversion of benzene to maleic anhydride and naphthalene or *o*-xylene to phthalic anhydride.

Maleic anhydride The catalytic oxidation of benzene in air to maleic anhydride was described by Weiss and Downs in 1920, and first industrial production was in 1928. A representative catalyst (Montedison MAT5) comprises V_2O_4 and MoO_3 in a molar ratio of about 2 plus a small amount of Na_2O, on an alumina support. The alumina makes up 88 to 89 wt % of the total. The catalyst is typically available as 5×5 mm pellets and is used in a multitube fixed-bed reactor. One composition (Becker and Barker, 1961) contains a very small amount of phosphate and nickel in addition to vanadia and molybdena. These fast fixed-bed reactions use low-area supports (for example, $\sim$1 m^2/g) and sometimes a coated-type catalyst in order to remain outside the pore diffusion–limiting regime that would otherwise reduce the selectivity. Typically the air/benzene molar ratio is about 65 to 85, pressure is slightly above atmospheric, and contact time is of the order of 0.5 to 1 s.

The reaction is highly exothermic, and considerable temperature gradients are typically encountered along a packed tube. Typically the array of tubes is cooled with a molten inorganic salt (for example, a mixture of 7% $NaNO_3$, 40% $NaNO_2$, and 53% KNO_3, which is a eutectic) held in this case at about 375°C. The molten salt in turn is used to generate high-pressure steam. (In early processes mercury was used.) The benzene-air mixture is typically fed at about 200°C and is rapidly preheated in the first portion of the packed tubes (for example, about the first 30 cm) to a "kindling temperature" of about 350°C. The reaction rate then becomes appreciable, and heat is transferred out rather than in. Even with the relatively small tubes, about 2 cm in inside diameter, the rate of heat transfer is insufficient to hold the reacting gas nearly isothermal, and a hot zone develops part way down the tube that may have a peak temperature as much as 100°C above that of the coolant. The reacting gas then cools down towards the end of the reactor tube as the rate diminishes with depletion of reactant. Typically 97 to 98 percent conversion is achieved with an initial selectivity of over 74 percent, which gradually drops with time. Under normal conditions catalyst life is of the order of 2 to 3 years.

Phthalic anhydride The first industrial process for making phthalic anhydride by catalytic air oxidation appeared in the 1920s, using refined naphthalene from coal tar in a fixed-bed reactor. In 1946 Chevron initiated production by the vapor-phase oxidation of *o*-xylene. The use of naphthalene received an impetus by the development of a fluidized-bed process (Sherwin-Williams and Badger), first commercialized in the early 1950s. This gave good yields with high-purity naphthalene but uneconomic yields with *o*-xylene. Coupled with this was the development in the early 1960s of processes for production of naphthalene from petroleum by hydrodealkylation of alkyl naphthalenes present in certain refinery product streams (Sec. 9.12). Alkyl naphthalenes such as methylnaphthalene give considerably lower yields in the partial oxidation reaction than does naphthalene itself.

o-Xylene has become the preferred feed, and formation of phthalic anhydride from *o*-xylene in a multitube fixed-bed process has become the process of choice in most new installations. In the Von Heyden, Badische Anilin und Soda Fabrik (BASF) process the reaction temperature is about 350 to 360°C, pressure is slightly above atmospheric, contact time is of the order of 5 s, and few byproducts other than carbon oxides are formed. In another version (e.g., Chevron), temperatures of about 400 to 475°C and contact time of about 0.5 s are used, and an appreciable quantity of maleic anhydride and other byproducts are formed. The higher-temperature process utilizes a smaller reactor and smaller catalyst load, but the yield is slightly less.

Either process can also be used with naphthalene, either from petroleum or unrefined naphthalene from coal tar. In the Von Heyden process the yield on a weight basis is about 82 percent with a naphthalene feed, and about 92 percent with *o*-xylene. (However, the stoichiometric yield is greater for naphthalene than for *o*-xylene.) The maximum concentration of aromatic in the feed is limited by the necessity to avoid an explosive mixture, and in the case of *o*-xylene this is about 1 vol %. The fluid-bed process (Sherwin-Williams and Badger) is now obsolescent.

Since feedstock and product are purchased and sold on the basis of weight, a considerable advantage of *o*-xylene over naphthalene as a feedstock is that no carbon is lost in forming the desired product. Hence, with 100 percent selectivity a theoretical yield of 139.6 kg of product is possible per 100 kg of *o*-xylene fed. (Yields up to about 80 percent of theoretical have been reported.) The theoretical heat of reaction for naphthalene is substantially greater than for *o*-xylene, which exacerbates the heat-transfer problem in multitube fixed-bed reactors.

o-Xylene is also preferred because xylenes are readily available from catalytic reforming, and suitable processes have been developed for separation of *o*-xylene from its isomers. Commercially, material of about 95 percent purity is used as a feedstock. As a liquid, it is easier to transport and store than naphthalene. In Sec. 11.9 representative axial-temperature profiles are given for the partial oxidation of *o*-xylene to phthalic anhydride to show the effect of such variables as tube diameter and the wall temperature, based on a simplified kinetic model and data published by Froment (1967). Industrial reactors are generally operated so as to obtain essentially complete disappearance of the reactant, in order to reduce separation costs.

In the reactions to form phthalic anhydride and maleic anhydride, there are a number of subtleties in the relationships between catalyst composition, reactor configuration and operation, and feed composition on the one hand, and activity and selectivity on the other. A considerable variety of catalyst compositions, all based on vanadia, appear to have been used. The most active form of the catalyst may be one in which the vanadium has a state of oxidation less than +5. Simard et al. (1955) reported that a lower oxide, $V_{12}O_{27}$ or V_6O_{13}, was the active species in *o*-xylene oxidation; Volf'son et al. (1965) and Schaefer (1967) showed that in the oxidation of naphthalene and benzene, respectively, V_2O_4 was the most active of a number of vanadium oxides studied. It also has been suggested that the role of MoO_3 when added to vanadia catalysts is to stabilize the V^{4+} state. A number of other catalyst compositions containing vanadia seem likewise to be directed at various means of obtaining and stabilizing the V^{4+} form. However again it may be in part an effect caused by defects.

With naphthalene as a feedstock, a catalyst composition used commercially consists of vanadium oxide and potassium sulfate on a silica support, similar to that used in sulfur dioxide oxidation. Optimum catalyst selectivity, activity, and stability is achieved when a considerable amount of pyrosulfate is present. This is obtained by adding a little sulfur dioxide to a pure naphthalene feed or by pretreatment of coal-tar naphthalene to reduce its sulfur content to an optimum value. The optimum sulfur level depends on the type of reactor. In a fluid-bed system this was a SO_3/K_2O ratio of about 2.0, achieved with a naphthalene feed containing about 0.1 wt % sulfur. The optimum in a tubular fixed-bed reactor is different, which may be ascribed to the existence of a hot spot in the fixed-bed case, so the composition of the catalyst at this elevated temperature would differ from that in contact with the same gas composition at a lower temperature. The effects of sulfur in naphthalene on the performance of fixed- and fluidized-bed reactors are also discussed by Saffer (1963).

With an *o*-xylene feed some of the highest yields reported in the patent literature are for a vanadium oxide–titanium dioxide catalyst. The catalytically active material is applied as a thin layer on the outside of an inert core, presumably to minimize diffusion limitations. The principal reaction pathway is the partial oxidation of *o*-xylene to tolualdehyde, which in turn is converted to phthalide and then phthalic anhydride in a series of reactions. Some *o*-xylene may also be converted directly to carbon monoxide and carbon dioxide, and some maleic anhydride may also be formed.

The relative rates of the different steps are affected by catalyst composition and by the particular mixtures of species present. Alkali metal ions inhibit the further oxidation of *o*-tolualdehyde, but this may be counteracted by addition of borax. Overoxidation of the phthalic anhydride product may be inhibited by a small P_2O_5 content (Calderbank et al., 1977). A general reaction scheme and kinetic data were presented by Herten and Froment (1968), and more recently Froment (1976) with Vanhove has developed it further. A similar scheme has been developed in several studies by Calderbank and coworkers. The 1977 paper gives quantitative information on a model for a packed-bed catalytic reactor as well as references to earlier work. Wainwright and Hoffman (1974) also summarize information on reaction mechanisms for *o*-xylene oxidation.

Dicarboxylic acids such as terephthalic acid (1,4 benzenedicarboxylic acid) are too unstable to be manufactured by vapor-phase catalytic oxidation. Instead, the process of choice appears to be liquid-phase oxidation with air of *o*-xylene dissolved in acetic acid, using a homogeneous catalyst system (Mid-Century, Amoco process).

8.6.2 Oxidation of Paraffins and Olefins

Newer processes for manufacture of maleic anhydride are being introduced based on vapor-phase catalytic oxidation of *n*-butane or *n*-butylene (butene-1 and butene-2). In analogous fashion to a comparison of *o*-xylene and naphthalene as alternate feedstocks for manufacture of phthalic anhydride, the C_4 feedstocks offer the potential advantage over benzene in that in the desired reaction no carbon is lost in forming the desired product. Patents for butane oxidation describe complex vanadia-phosphate catalysts, in some cases containing also iron, titania, or other ingredients. A commercial plant for converting *n*-butane to maleic anhydride was recently put into operation by Amoco.

Normal butylenes are becoming increasingly available as a coproduct of thermal "steam" pyrolysis of naphthas, and processes to oxidize the C_4 fraction to maleic anhydride have been developed by several companies, including Mitsubishi in Japan (fluid-bed reactor) and Bayer and BASF in Germany (tubular fixed-bed reactor with molten salt cooling). Mitsubishi put a commercial plant in operation in about 1971. The preferred catalyst appears to be vanadium pentoxide or vanadium and molybdenum oxides, and an alkali phosphate. Patents also describe the incorporation of iron, tin, and/or tungsten as promoters. Either type of process operates at about 350 to 450°C and slightly above atmospheric pressure with a fuel-lean mixture of air and hydrocarbon. More severe conditions are needed with a saturated feedstock. The C_4 product from a steam cracker will consist primarily of butenes and butadiene, plus smaller amounts of butanes. Under optimum operating conditions, of these constituents in the feed stream to a maleic anhydride plant, butanes are unreacted, isobutylene burns to carbon oxides and water, and the *n*-butenes and butadiene are converted to maleic anhydride. Conversions and selectivities for butane or butene to maleic anhydride obtained with a number of catalysts are summarized in a review by Voge and Adams (1967). Varma and Saraf (1979), in a review with extensive references, summarize a large number of patented catalyst compositions and discuss possible kinetics and reaction mechanism.

8.7 VINYL ACETATE

A mixture of ethylene, oxygen, and acetic acid in the vapor phase, with ethylene in considerable excess, is passed over a fixed-bed catalyst at about 150 to 175°C and about 0.5 MPa pressure. The reaction is

$$C_2H_4 + CH_3COOH + \tfrac{1}{2}O_2 \longrightarrow CH_2{=}CH{-}O{-}C(O)CH_3 + H_2O \qquad (8.33)$$

$$-\Delta H = 176 \text{ kJ/mol}$$

The catalyst contains palladium on an acid-resistant support such as silica, in the form of 4- to 5-mm spheres. Several percent of potassium acetate is also added to the catalyst to increase catalyst activity and to reduce the oxidation to carbon dioxide. The potassium acetate slowly migrates from the carrier and must be replaced by injection into the feed stream.

The reaction is highly exothermic, and a multitube reactor using, for example, 2.5-cm-ID tubes is used, the heat being removed by boiling water in the shell. About 10 to 20 percent conversion of ethylene is obtained per pass, with 60 to 70 percent consumption of oxygen. The overall selectivity is 91 to 94 percent based on ethylene reacted. About 1 percent of the ethylene forms acetaldehyde, and the remainder forms carbon dioxide. Reaction of acetic acid with the oxygen is apparently negligible; it is converted only to vinyl acetate, in amounts corresponding to 15 to 30 percent of the acetic acid fed. The product is rapidly cooled, vinyl acetate is removed by scrubbing with an organic solvent such as propylene glycol, and carbon dioxide is removed by scrubbing with a base such as aqueous, hot carbonate. The remaining stream is mixed with fresh feed and recycled. It appears to be desirable to allow the concentration of carbon dioxide and other inert gases, such as nitrogen and argon brought in with the oxygen, to build up to a certain extent to help keep the oxygen concentration below the explosive limit.

The process, first put into commercial operation in about 1968, was developed almost simultaneously by Bayer and Hoechst in Germany and National Distillers (U.S. Industrial Chemicals) in the United States, and appears to be the process of choice for new plants. It replaces an earlier liquid-phase process utilizing a homogeneous palladium and copper chloride redox catalyst with ethylene, similar to that in the Wacker process for manufacture of acetaldehyde from ethylene. Corrosion problems in the liquid-phase ethylene process were especially severe, largely because of HCl which is formed as an intermediate. It is doubtful that the process is still operated. An earlier process based on acetylene was the process of choice in the 1960s but is now becoming obsolete. Acetylene and acetic acid are reacted at about 180 to 230°C over zinc acetate deposited on carbon, in a multitube reactor. Further details on this and competitive processes are given by Stobaugh et al. (1972) and by Krekeler and Kronig (1967).

A mechanistic study by Nakamura and Yasui (1970) of this reaction on a palladium catalyst revealed that no vinyl acetate is formed from ethylene and acetic acid in the absence of oxygen, and they concluded that acetic acid is activated by abstraction of hydrogen by palladium in the presence of oxygen. Ethylene is adsorbed with abstraction of hydrogen by palladium even in the absence of oxygen. They concluded that vinyl acetate is formed from combination of dissociatively adsorbed acetic acid and dissociatively adsorbed ethylene, the surface reaction being rate-determining under industrial reaction conditions.

In their work at 120°C the addition of potassium acetate increased the rate of reaction on a 1 wt % Pd/Al_2O_3 catalyst as much as tenfold. The maximum activity was achieved with about 2 to 3 wt % potassium acetate, after which it decreased. Since the compound is slowly lost by volatilization, maintenance of the proper level of potassium acetate under industrial reaction conditions is important.

Of other alkali metal acetates, those of cesium and rubidium are comparable to potassium in activity, sodium and lithium are less active, and alkaline earth acetates are generally even less active. Of other potassium salts, those in which the anion can be readily replaced by acetate ion are about as active as potassium acetate. A compound such as potassium chloride showed no activity, presumably because the chloride ion is not easily replaced by acetate ion. In addition to promoting hydrogen abstraction from adsorbed acetic acid, the potassium is postulated to weaken Pd—O bonds in a palladium acetate surface compound.

Nakamura and Yasui developed a rate expression from their data that is a starting point for modeling this reaction. However, the kinetics are probably very complicated.

Judging from the reaction conditions, it is highly possible that a liquid phase is present in at least some of the pores of the catalyst under at least some portion of the reaction conditions. (The boiling point of acetic acid at atmospheric pressure is 119°C.)

From propylene, acetic acid and oxygen allyl acetate is formed, in much higher selectivity than in the equivalent liquid-phase reaction using a palladium salt and cocatalysts.

Most vinyl acetate is polymerized to polyvinyl acetate, used in emulsion paints, adhesives, coatings, etc., or it is processed into polyvinyl alcohol (used in adhesives, sizings, etc.) or polyvinyl butyral. The latter is used as a laminate in safety glass. Some vinyl acetate is also copolymerized with vinyl chloride or other monomers.

8.8 OXYCHLORINATION

A wide variety of chlorinated compounds are manufactured by the chlorination of a hydrocarbon, hydrogen chloride being produced as a byproduct. Hydrogen chloride is difficult to transport and overall is produced in quantities greater than those needed in industry. Consequently there has long been interest in processes to convert hydrogen chloride directly back to chlorine, either in a separate process or simultaneously during the chlorination process. In an old procedure, the *Deacon process*, the direct reaction

$$4HCl + O_2 \longrightarrow 2Cl_2 + 2H_2O \tag{8.34}$$

was catalyzed by a copper chloride catalyst at about 600°C. Particular difficulties are volatilization of the metal chloride at this reaction temperature and corrosion, which is especially severe for a mixture of water and hydrogen chloride at an elevated temperature. The conversion for this exothermic reaction may also be limited by equilibrium.

After World War II a commercial fluidized-bed process (Shell) was developed using a catalyst containing potassium chloride and a rare earth chloride, for example, lanthanum chloride, in addition to copper chlorides. This exhibited suitable activity at about 420°C. The potassium salt appears to reduce volatility and the lanthanum salt markedly increases activity. A competitive process (Kel-Chlor) for reaction (8.34) uses nitrogen oxide compounds in the presence of about 70 to 80% sulfuric acid as reaction intermediates rather than a heterogeneous catalyst. The process may operate at 0.1 to 1.5 MPa and temperatures of about 130 to 180°C (Schreiner et al., 1974).

Chlorination and oxidation of the hydrogen chloride product can also be caused to occur simultaneously, termed *oxychlorination.* The most important process is the conversion of ethylene to 1,2-dichloroethane (ethylene dichloride), which in turn is converted to vinyl chloride by pyrolysis:

$$CH_2{=}CH_2 + 2HCl + \tfrac{1}{2}O_2 \longrightarrow CH_2ClCH_2Cl + H_2O \tag{8.35}$$

$$-\Delta H = 119 \text{ kJ/mol}$$

$$CH_2ClCH_2Cl \longrightarrow CH_2{=}CHCl + HCl \tag{8.36}$$

Because of the overall stoichiometry of the process, some ethylene is also reacted directly with chlorine to form ethylene dichloride.

$$CH_2{=}CH_2 + Cl_2 \longrightarrow CH_2ClCH_2Cl \tag{8.37}$$

Reaction (8.37) may be carried out with a catalyst such as ferric chloride, at 50 to 60°C, using liquid dichloroethane as a reaction medium. Homogeneous reaction in the vapor phase may be carried out at 370 to 500°C. In the latter case some vinyl chloride and hydrogen chloride may also be formed. More than 90 percent of United States production capacity for vinyl chloride utilizes this so-called *balanced technology*, which came into practice in the 1960s. This replaces earlier technology in which acetylene was reacted with hydrogen chloride to form vinyl chloride using a mercury(II) chloride/carbon catalyst, in a multitube reactor cooled by boiling water. In this system tube-wall temperature can be controlled within limits by the pressure on the water.

For the oxychlorination reaction the catalyst is typically cupric chloride on alumina, modified with potassium chloride to reduce the volatility of the copper chloride. The latter may be held onto the catalyst by bonding with OH groups on alumina to form a structure such as —Al—O—CuCl. The reaction temperature is 250 to 315°C, and the pressure is atmospheric or higher. The conversion of hydrogen chloride decreases at higher temperatures, possibly because of decreased adsorption of hydrogen chloride onto the catalyst. The overall reaction may be regarded as a combination of the Deacon reaction and chlorination, in which the Deacon reaction is probably the rate-controlling step. There are some indications that the dichlorethane product is more strongly adsorbed than the reactants and thereby inhibits the rate. Carrubba and Spencer (1970) report a kinetic study of the reaction using a copper chloride/alumina catalyst at 180°C.

The feed stream typically contains an excess of ethylene and oxygen to obtain essentially complete conversion of hydrogen chloride. Steam may be added to bring the composition below the explosive limit. The reaction is highly exothermic, and either a multitube fixed-bed or a fluid-bed reactor may be used, with air or oxygen. In a process utilized by Pechiney-St. Gobain, ethylene, hydrogen chloride, and air are reacted at 280 to 480°C and 0.2 to 0.8 MPa pressure in a fluidized-bed reactor using a copper salt/oxide as the catalyst.

It would be clearly desirable to combine reactions (8.35) and (8.36), and a number of catalyst compositions have been patented. However, no commercial process to do so has apparently yet been put in operation. Typical process flow sheets and further details are given by Wimer and Feathers (1976) and Reich (1976). The catalyst in the Shell version of the Deacon process is a molten salt under reaction conditions held in the pores of a support; in oxychlorination the catalyst may also be molten under some conditions.

A process to oxychlorinate methane (Transcat, C-E Lummus), recently put into commercial operation (*Chem. Eng.*, 1974), uses a molten salt mixture of cupric chloride, cuprous chloride, and potassium chloride (which depresses the melting point) at a pressure below 0.7 MPa and temperatures of about 370 to 450°C. The molten salt is circulated by gas lifts between two reactors, an oxidizing reactor and a chlorination/oxychlorination reactor. In the first reactor the molten salt flows downward in a packed bed against a stream of air. Waste chlorocarbons are pyrolyzed elsewhere, and their products are also fed to this reactor. CuCl is oxidized to $CuO \cdot Cl_2$, some CuCl reacts with Cl_2 to form $CuCl_2$ and some reacts with HCl and O_2 in air to form $CuCl_2$ and H_2O. In the second reactor methane is converted to various chloromethanes by chlorination, and some of the HCl formed is converted to water and Cl_2 in situ by oxygen released from the molten salt. The process can also be used for chlorination of ethane. The catalyst composition and reaction mechanisms are reviewed by Kenney (1975) and Villadsen and Livbjerg (1978).

Although 90 percent or more of the HCl consumed in the United States is produced as a byproduct, it is shipped primarily as an aqueous solution and transportation costs are an important factor. Hence a small amount of HCl is produced by direct reaction of Cl_2 and H_2, "chlorine burning," where regional economics supports this method.

8.9 SULFURIC ACID

Sulfuric acid is one of the most widely used and important inorganic chemicals. In the nineteenth century it was manufactured by the lead chamber process in which dilute sulfur dioxide, usually produced by burning iron pyrites or elemental sulfur in air, was contacted with nitrogen oxides and passed slowly through a series of chambers (usually lined with lead inside for inertness). This provided time for a complex series of reactions to take place in both the gas and the liquid phases, leading to the formation of dilute sulfuric acid. The nitrogen oxides and an intermediate unstable compound $HNOSO_4$, nitrosyl sulfuric acid, acted as a homogeneous catalyst and were recovered for reuse. An advantage of the process was that relatively impure gases, as from smelters, could be reacted without extensive prepurification; but concentrated acid could not be produced directly, the equipment required was huge and cumbersome, and the process is now obsolete.

The use of a heterogeneous catalyst to catalyze the reaction

$$SO_2 + \tfrac{1}{2}O_2 \longrightarrow SO_3 \qquad (8.38)$$

$$-\Delta H^\circ_{900^\circ K} = 95.8 \text{ kJ/mol}$$

became commercial in Europe at about the turn of the century, spurred by the fact that this process permitted the direct production of fuming sulfuric acid, "oleums," consisting of SO_3 dissolved in H_2SO_4. These were particularly desired for sulfonation reactions in the growing dye industry. The SO_3 formed is dissolved in 98% H_2SO_4 since if it is attempted to dissolve the SO_3 in water directly or into a weaker acid, the water vapor pressure causes the formation of an acid mist that is very difficult to remove. The fortified H_2SO_4 formed may then be diluted to the desired strength.

In the early plants the catalyst was platinum supported on an acid-resistant material, usually asbestos, magnesium sulfate, or silica gel. This gradually became replaced between 1920 and 1940 by a catalyst comprising vanadium oxide and potassium sulfate on a silica support that, although slightly less active, is cheaper and less suscep-

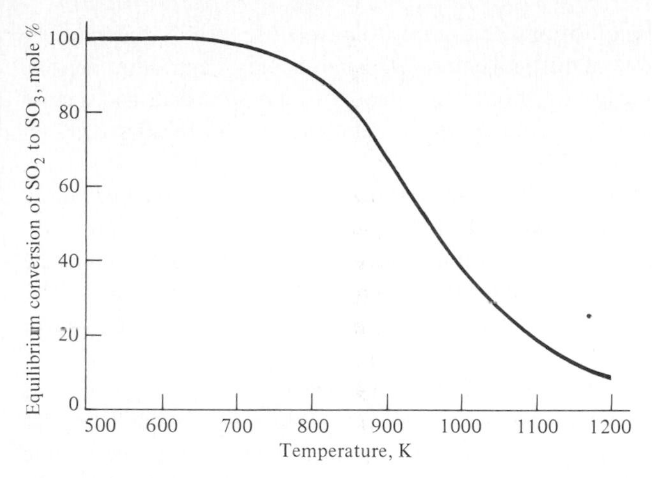

Figure 8.4 Effect of temperature on the equilibrium conversion of sulfur dioxide to sulfur trioxide. (Initial sulfur dioxide concentration 8.0 percent by volume in air.)

tible to poisoning. All plants constructed since World War II are believed to use a vanadium-type catalyst.

The equilibrium for reaction (8.38) as a function of temperature is shown in Fig. 8.4 for an initial composition of 8.0 mol % SO_2, 13% O_2, and 79% N_2 at atmospheric pressure. Since the reaction reverses at higher temperatures, the reactor usually consists of four trays in series which operate adiabatically at slightly above atmospheric pressure. The reacting gas is cooled between the trays. The inlet SO_2 concentration is typically about 7% (from a metallurgical smelter operation) to 10.5 wt % (from elemental sulfur combustion).

With an inlet gas containing about 9% SO_2 the temperature entering the first tray is adjusted to about 420°C. It leaves at about 600°C, and the heat in this stream is usually recovered in a waste-heat steam boiler. The temperature drop across subsequent trays is typically 50 to 60°C, 10 to 15°C, and 5°C, respectively, and the gas is cooled by heat exchangers between trays. Typically about 75 percent of the total conversion occurs on the first tray, and only 2 percent or so on the last. The inlet temperature to each succeeding tray is adjusted to an optimum value that varies with the details of the process and the inlet SO_2 concentration, but in each case it is typically between 420 and 440°C. In a sulfur-burning plant additional air may be supplied after the second and third passes to provide additional oxygen for the reaction. This results in a somewhat lower required catalyst loading but slightly higher SO_2 emissions from the final absorption tower.

The catalyst layers are from about 45 to 75 cm deep, and the cost of the catalyst is a large portion of the cost of the loaded reactor. A plant that produces 1000 tons of acid per day will contain 150,000 to 200,000 liters of catalyst. An optimum distribution of catalyst over the trays assumes some importance, and it is apparent that an unequal distribution, with the latter trays loaded more heavily, is most economic. Fariss (1963) calculated optimum catalyst loadings for a Monsanto commercial reactor for representative gas compositions and percent conversion, results that must be tempered by the varying degree of dust collection on different trays, and other factors. The catalyst is extremely long-lived, 10 or more years not being unusual, and a common procedure is to occasionally screen or replace the catalyst on the first tray on which dust, scale, and poisons slowly accumulate.

The overall conversion is typically 99.7 percent or so, and is determined primarily by air-pollution control limitations rather than by internal economics. To meet air-pollution requirements limiting SO_2 concentrations to about 300 to 1000 ppm in the discharge to the atmosphere, SO_3 is substantially removed from product gases after the third tray by absorption, and the remaining SO_2 is reacted further on the fourth tray.

Figure 8.5 is a schematic diagram of a representative converter design. Only a portion of the gas from the second tray is removed for cooling, conditions being adjusted so that the mixture of the cooled gas and the uncooled gas is at the desired temperature for entering the third tray. That leaving the third tray is cooled against gas returning to the fourth tray by a heat exchanger, passes to an absorber, and returns through the heat exchanger to the fourth tray. Heat is then recovered, and the gas passes to a second absorber. The use of two absorbers increases the overall conversion and minimizes SO_2 discharge to the atmosphere. Figure 8.6 is a photograph of a sulfuric acid plant having a capacity of 850 tons of acid per day. The converter is on the left, and one of the heat exchangers is on the right foreground.

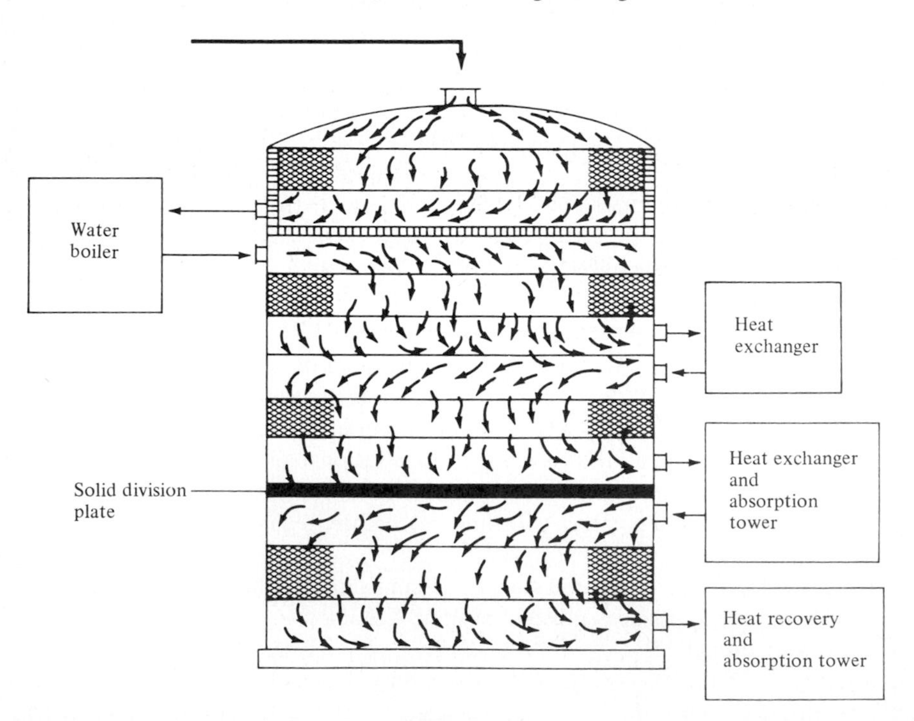

Figure 8.5 Representative sulfur dioxide converter design. (*Courtesy of Monsanto Enviro Chem Systems, Inc.*)

Figure 8.6 Sulfuric acid plant at Scarlino/Follonica (Grosseto), Italy. (*Courtesy of Montedison S.p.A.*)

The usual catalyst comprises a mixture of vanadia and potassium sulfate on a silica support, the K/V atomic ratio being in the range of 2 to 3.5. For example, Haldor Topsøe VK38 contains 8% V_2O_5, has a K/V atomic ratio of 3.0, and is supported on kieselguhr (diatomaceous earth). Such a catalyst is much more active than vanadia alone. In a typical manufacturing process kieselguhr that has been acid-washed (to remove iron and other impurities) and calcined is mixed with an ammonium vanadate and potassium hydroxide, plus some potassium silicate that acts as a bonding agent. The thick dough is extruded, is air-dried, and is calcined in contact with SO_2 or an SO_2-SO_3 mixture to convert it to the final chemical form. The reaction of SO_2 with alkali and its oxidation is relatively exothermic, and, desirably, it is carried out before the catalyst is placed into service in the reactor.

These catalysts are unusual in that the potassium and vanadium compounds react with one another and with SO_2 and SO_3 to form a mixture of complex potassium vanadia pyrosulfates, which is partly or completely liquid under reaction conditions (Tandy, 1956). The pyrosulfates have lower melting points than the corresponding sulfates. This melt forms a film on the surface of the pores in the inert silica support, which can migrate from particle to particle under operating conditions. The composition of the melt changes with temperature and gas composition and with a change in operating conditions. This influences its activity. If the temperature is changed, many hours may be required before a new steady-state rate is reached, especially at lower temperatures. Various models have been proposed for ways in which the liquid may be distributed on the support, and optimal liquid loadings have been suggested; but these depend on the model assumed. These theoretical aspects have been reviewed in detail by Livbjerg et al. (1974) and Villadsen and Livbjerg (1978).

Commercial catalysts are typically extrudates about 5.5 mm in diameter by about 8 mm in length. A larger size, for example, 8 × 11 mm, may be placed on the first tray, in order to lower pressure drop, especially since dirt, scale, etc. build up in this location. In some cases a vanadium oxide catalyst of slightly different composition is placed on the final trays of the converter, the catalyst having higher activity for the partially converted gases. The average pore size of these catalysts is in the micrometer range, and the area is about 1 m^2/g, a value which does not seem to change much with the melt conditions.

If one considers the various alkalis as promoters, according to Tandy (1956) maintenance of the V^{5+} state is enhanced in the order Ce $>$ Rb $>$ K $>$ Na. The melting point and viscosity also increase in this order. Hence the greater activity of compositions containing potassium over those containing sodium may be in part a physical effect and in part a stabilization of the +5 valence form. It is reasonably clear that the +4 form is less active for this reaction than the +5 form. The ratio of the +4 to the +5 form is also enhanced at lower temperatures and at lower values of the ratio P_{SO_2}/P_{SO_3}. All commercially used catalysts appear to contain potassium, and many of them some sodium as well, which enhances low-temperature activity. The cost of cesium and rubidium outweighs their increased activity.

Sulfur dioxide in air can be oxidized to sulfur trioxide by bubbling it through a potassium sulfate-vanadia melt (Haldor-Topsøe and Nielsen, 1947). Such a melt will also catalyze other oxidation reactions for which solid vanadium oxide is a catalyst, e.g., the partial oxidation of *o*-xylene to phthalic anhydride (Satterfield and Loftus, 1965). However in the latter case the rate on a volume basis is low compared to that observed with a solid catalyst.

The apparent activation energy of the forward reaction of oxidation of SO_2 to SO_3 is much greater at temperatures below about 440 to 460°C than at higher temperatures, the reasons for which are not clear. There is not just one published kinetic expression that can represent the rate over the whole range of industrial conditions, and a large number have been reported. Livbjerg and Villadsen (1972) list 12 such expressions. In part they represent studies under different experimental conditions, but they also conflict with one another in some degree. It is evident that none should be extrapolated for use beyond the specific range of conditions for which the basic data were obtained. In general the reaction is about first order with respect to oxygen and somewhat inhibited by SO_3, even far from equilibrium. The effect of SO_2 concentration is complex. On 6 × 6 mm pellets effectiveness factors in the range of 0.3 to 0.8 for representative industrial conditions have been estimated (Livbjerg and Villadsen, 1972), although it is difficult to do so quantitatively because of the complex kinetics. Diffusion effects in an SVD (Russian) catalyst and an Imperial Chemical Industries (I.C.I.) 6 × 4 mm catalyst have also been reported (Kadlec et al., 1970).

Fariss (1963) has published rate data, on a relative basis, for a commercial K_2SO_4-V_2O_5 Monsanto catalyst, shown in Figs. 8.7 and 8.8. These were obtained in a reactor operated under differential conditions. The catalyst is probably very similar to those used as of 1980, and the figures illustrate the unusual kinetic behavior. At 420°C the rate of reaction decreased unevenly with percent conversion. At 480°C the rate of reaction *increased* with conversion up to about 15 percent, and then decreased. The catalyst melt composition is significantly altered by the composition of the gas in contact with it, which evidently has a significant effect on catalyst activity.

The apparent activation energy for the reaction was much greater at temperatures up to about 460°C than at higher temperatures, in general agreement with the results of other studies.

Details on catalyst compositions and processes are given in the book by Duecker and West (1959). Dixon and Longfield (1960, p. 322) review kinetics and possible mechanisms for both platinum and vanadium oxide-based catalysts. The reaction has been intensively studied by Boreskov and his coworkers for many years, and a report

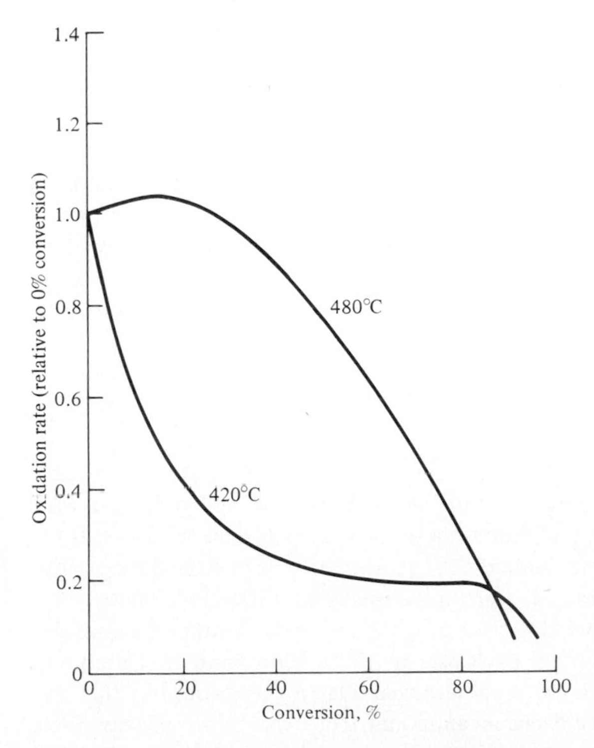

Figure 8.7 Oxidation of sulfur dioxide. Effect of percentage conversion on oxidation rate. Inlet concentration, 13% SO_2, 8% O_2. (*Fariss, 1963.*)

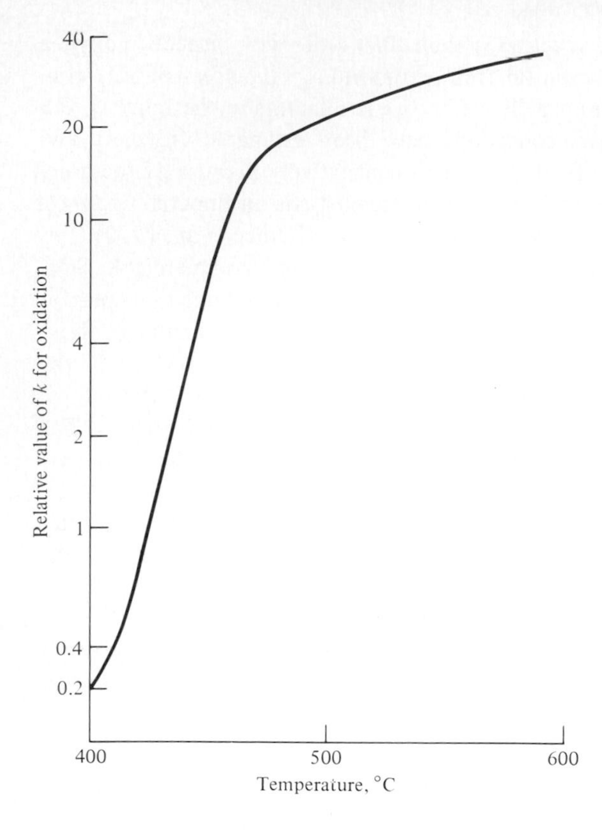

Figure 8.8 Oxidation of sulfur dioxide. Effect of temperature on rate constant. (*Fariss, 1963.*)

on a kinetic study in 1967 (Boreskov et al., 1967) reviews earlier work. Herce et al. (1977) report kinetic data at 0.1, 0.5, and 1.0 MPa for a vanadia catalyst containing barium oxide instead of a potassium compound, although commercially there seems to be little economic incentive to operate a catalytic reactor at much above atmospheric pressure. In a review of the use of molten salts as catalysts Kenney (1975) discusses the kinetics of the reaction. Villadsen and Livbjerg (1978) cover the same area in great detail in a later review, with particular attention to composition of these melts as a function of the reaction conditions.

8.10 AMMONIA OXIDATION

The basis for present-day nitric acid manufacture stems from work by Kuhlmann, who filed a patent in 1838 for oxidation of ammonia in air over platinum sponge to form nitric oxide. Ostwald and coworkers studied this reaction on a pilot scale during 1901 to 1904, leading to the first commercial plant in Germany in 1906. This plant had a capacity of 300 kg of nitric acid per day. Use of platinum in the form of a gauze, the present-day configuration, was patented by Kaiser in 1909. Early industrial ammonia was primarily a byproduct of coking of coal and was relatively impure, but this was soon replaced by relatively pure and cheap ammonia from the Haber process (Sec.

10.5). This stimulated the advent of large-scale nitric acid production from ammonia oxidation in the 1920s. Worldwide, some 75 to 80 percent of the nitric acid produced today is used for manufacture of fertilizer. Other uses are for explosives and oxidation of chemical intermediates.

In present-day practice a mixture of ammonia and air is passed downward through a pad of fine platinum alloy gauzes at a pressure ranging from slightly above atmospheric to about 0.9 MPa (Fig. 8.9). The desired reaction is:

$$4NH_3 + 5O_2 \longrightarrow 4NO + 6H_2O \tag{8.39}$$

$$-\Delta H_{298} = 227 \text{ kJ/mol } NH_3$$

The stoichiometric ratio for the reaction corresponds to 14.2% NH_3, but the explosive limit of NH_3 in air is 12.4 mol % at 0.8 MPa and slightly more at 0.1 MPa. Hence the NH_3 concentration used does not exceed about 11% in a high-pressure plant and about 13.5% at atmospheric pressure. The highly exothermic reaction is extremely rapid, highly mass-transfer-controlled and is complete in about 1 ms or less. Intermediates such as NH, HNO, and NH_2OH have been postulated, but little is actually known about the reaction mechanism.

To initiate the reaction it is necessary to preheat the catalyst, as by use of a flame, to a temperature at which the reaction becomes self-starting. After this the reactor heats itself up to a steady-state reaction temperature. The latter is always much greater than the ignition temperature. Conditions for starting an exothermic reaction such as this and for maintaining the reactor stable have been extensively treated in chemical engineering and other literature for many years. An analysis of the factors determining the starting and stability phenomena in ammonia oxidation was published in 1918 by Liljenroth and further developed by Wagner in 1945.

The exit gases are rapidly cooled by placing the gauze pad on a ceramic support that, in turn, is positioned on top of a heat exchanger. NO in the cooled gases is allowed to oxidize to NO_2 homogeneously, and the NO_2 is then absorbed into water, accompanied by further oxidation, to form HNO_3. The oxidation and absorption are always carried out at elevated pressure, from about 0.5 MPa to 0.8 or 0.9 MPa. The catalytic oxidation of NH_3 may be carried out at approximately atmospheric pressure, followed by compression of product gas, or it may be carried out at a pressure dictated by the downstream operations without intermediate compression. The latter is the more common practice.

Increased pressure increases the rate of mass transfer of NH_3 and O_2 to the catalyst surface, and hence increases the rate of reaction. Thus the catalyst inventory can be less, and a smaller reactor can be used. However, at the optimum reaction conditions the selectivity is slightly less and Pt losses are greater than with atmospheric-pressure operation. At atmospheric pressure, the optimum gauze temperature is about 810 to 850°C; at 0.8 MPa, about 920 to 940°C. The yield is about 97 to 98 percent for atmospheric-pressure processes and about 95 percent for operation at 0.7 to 0.9 MPa, the remainder of the NH_3 appearing as N_2. Platinum is lost from the gauze by volatilization of platinum oxide (see later) and by mechanical losses, and this typically amounts to 200 to 400 mg per ton of 100% HNO_3 produced, at the higher-temperature operation. Higher losses are also encountered with higher flow rates per unit cross-

Figure 8.9 Ammonia oxidation reactors. (*Courtesy of Johnson Matthey & Co., Ltd. and I.C.I. Agricultural Division.*)

sectional area of gauze. At atmospheric pressure and 800°C, a platinum loss rate of about 50 mg per ton HNO_3 is representative.

NH_3 will also react homogeneously with NO to form N_2 by the reaction

$$4NH_3 + 6NO \longrightarrow 5N_2 + 6H_2O \tag{8.40}$$

$$-\Delta H = 1810 \text{ kJ/mol}$$

This undesirable reaction may occur if the flow through the gauze pad is uneven, allowing some NH_3 to escape downstream. For this reason the pad is built up of individual flat gauzes laid very carefully on one another to ensure that no free spaces exist between them (Fig. 8.10). At excessively low flow rates some NO may also be lost by the catalytic decomposition of NO

$$2NO \longrightarrow N_2 + O_2 \tag{8.41}$$

The oxidation of NH_3 to elemental nitrogen may also be significant:

$$4NH_3 + 3O_2 \longrightarrow 2N_2 + 6H_2O \tag{8.42}$$

$$-\Delta H = 1265 \text{ kJ}$$

A representative pad is about 5 mm thick, and it may be as much as 4 m or more in diameter. Since the pressure drop is so low, careful engineering is required to obtain even distribution of gas flow. A long cone-type entrance section is frequently placed above the pad for this purpose (Fig. 8.9). The pad may contain as many as 40 gauzes, but reaction will be essentially complete on the first two or three gauzes. The additional gauzes may increase the overall efficiency by increasing the pressure drop and

Figure 8.10 Installation of new catalyst gauzes in an ammonia oxidation plant. (*Courtesy of Johnson Matthey & Co., Ltd. and Thames Nitrogen Co., Ltd.*)

thereby causing more uniform gas distribution and more uniform temperature. Some are also needed to take over the reaction as the first gauzes disintegrate or become inactivated (see below). However the overall rate of precious metal loss increases with the number of gauzes in the pad. In some designs the lower portion of the catalyst pack is replaced with a porous pad made of a nickel-chrome alloy, which reduces the platinum inventory required while providing the minimum pressure drop needed.

Fresh gauze is not very active, but during the first few days of reaction reconstruction of the metal occurs with complex outgrowths and deep fissuring, so gauzes become rough and the activity is greatly enhanced, as shown in Fig. 8.11. The restructuring occurs primarily on those gauzes on which the major portion of the reaction is

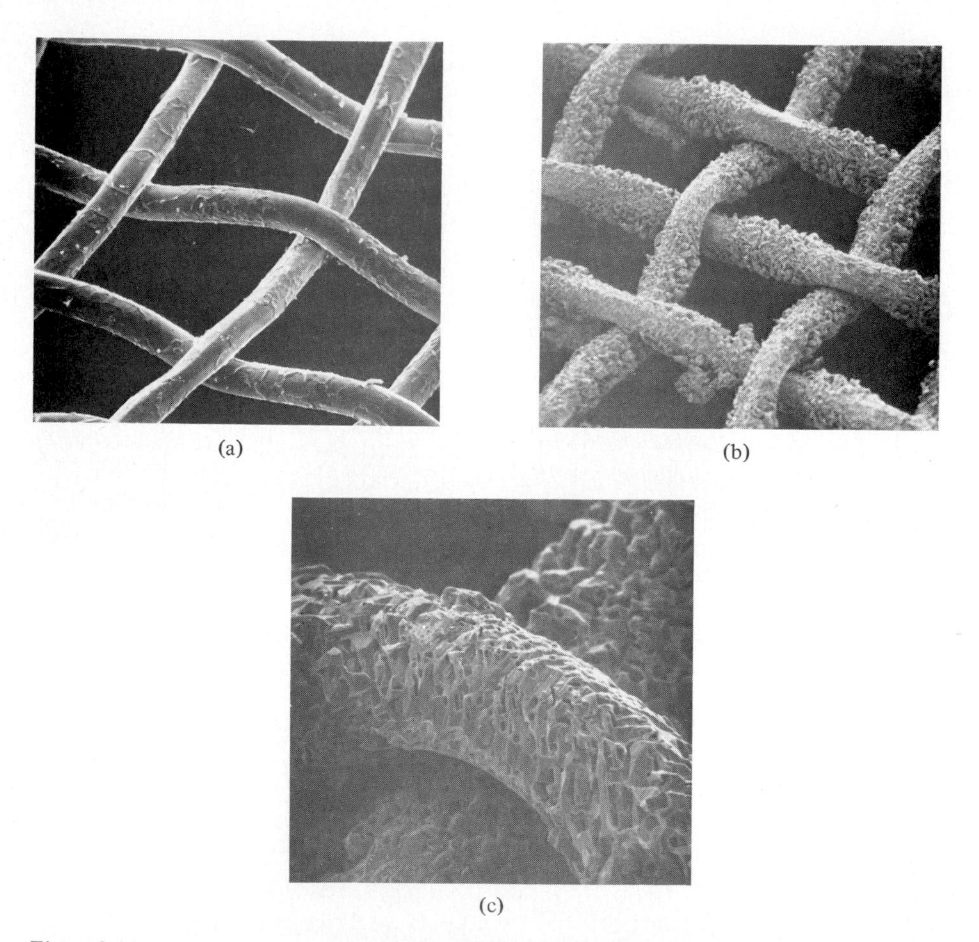

(a) (b) (c)

Figure 8.11 Scanning electron micrographs of 10% Rh–90% Pt gauze wire 0.075 mm (0.003 in) in diameter. (*a*) Original gauze, as drawn, showing uniform grain structure and absence of nodules on the wires. (*b*) The gauze of Fig. 8.11*a* after one-half the normal gauze life installed in a nitric acid plant, showing the nodular pattern typical of an active gauze. (*c*) An enlarged view of the crossover point of two gauze wires in Fig. 8.11*b* showing the characteristic octagonal crystal structure and the nodules, which are concentrated on the wire surfaces between crossover points. [*Courtesy of Platinum Metals Review, 15(2), 52 (1971).*]

occurring, so major differences are seen between the first, the second, and the third gauzes, and between the top and bottom surfaces of the active gauzes. Continuous electron microscope studies during reaction indicate that reconstruction of active catalyst surfaces probably occurs constantly throughout the working life of the catalyst. This reconstruction is associated with the reaction itself since it does not occur at reaction temperature in the presence of either ammonia alone or oxygen alone.

During operation platinum is lost at a fairly steady rate, caused primarily by the reaction

$$Pt(s) + O_2 \rightleftharpoons PtO_2(g) \tag{8.43}$$

The vapor pressure of platinum metal itself (for example, 8.9×10^{-7} Pa at 1570°C) is too low for platinum volatilization itself to be a significant mechanism. The equilibrium of reaction (8.43) is shifted to the right at higher temperatures. The rate of loss is governed by the mass transport of slightly volatile PtO_2 from the surface of the catalyst (Nowak, 1969), so it increases at higher temperatures and higher flow rates. The PtO_2 is reconverted to the metallic form downstream from the reactor by the reverse of reaction (8.43). Some platinum may be recovered by filters, some may be deposited on downstream heat-exchange surfaces, and some may also settle out in a variety of other locations. A "getter" system (Degussa) (see Holzmann, 1968) is used in some plants in which gauzes of a 20% gold–80% palladium alloy are placed directly below the platinum-alloy catalyst pad. This immediately alloys with volatilized platinum oxide to form a solid solution. However in turn some palladium is lost from the "getter gauzes," and the incremental recovery of platinum drops with increased thickness of the "getter" pad. These gold-palladium gauzes are removed and replaced from time to time for Pt recovery. The optimum economic recovery of noble metal is about 55 to 70 percent. The cost of refining and manufacture of getter pads does not usually justify use of thicker pads for higher recovery.

The platinum loss occurs predominantly from the top pads, and some of it occurs mechanically as the gauze disintegrates. Consequently from time to time new gauzes are added to the pack, usually being placed on the bottom, while used ones are gradually moved to the top. This eliminates the activation period.

The addition of a few percent of rhodium to the platinum increases conversion efficiency and reduces catalyst loss (Handforth and Tilley, 1934). Hence the platinum alloy in the gauzes usually consists of 95% platinum–5% rhodium or 90% platinum–10% rhodium, the latter generally being preferred for high-pressure, high-temperature catalytic reaction. Rhodium in small amounts contributes mechanical strength to the alloy, but concentrations higher than about 10 percent result in a more brittle material that is more difficult to fabricate. Rhodium is also more expensive than platinum. A 90% platinum–5% palladium–5% rhodium may also be used, in which half of the rhodium is replaced by much less expensive palladium, apparently without loss of yield. The standard gauze is 80 mesh per inch (31.5 wires per centimeter) woven from wire 0.003 in (0.075 mm) in diameter. This represents a reasonable optimum between the desire for fine wire and high area on the one hand and mechanical requirements on the other.

At atmospheric pressure platinum is preferentially lost by volatilization, so the surface of the alloy becomes enriched with rhodium, which is nearly as active as platinum. However operating conditions are near the boundary separating rhodium and a rhodium oxide as the stable phase. Rhodium oxides are favored by lower temperatures, high oxygen partial pressures, and higher rhodium content of the alloy. Some rhodium may be lost by formation of RhO_2, which, however, is less volatile than PtO_2. Solid Rh_2O_3, which is relatively unreactive, may also be formed and may blanket some of the platinum, thereby inactivating it (Harbord, 1974). Since formation of Rh_2O_3 is favored by lower temperatures, a drop of the order of 50°C below normal operating temperature could cause a platinum-rhodium gauze to become inactive (Sperner and Hohmann, 1976). Segregation of rhodium at the surface is readily shown by electron microprobe studies (Sec. 5.5.4 and Fig. 8.12).

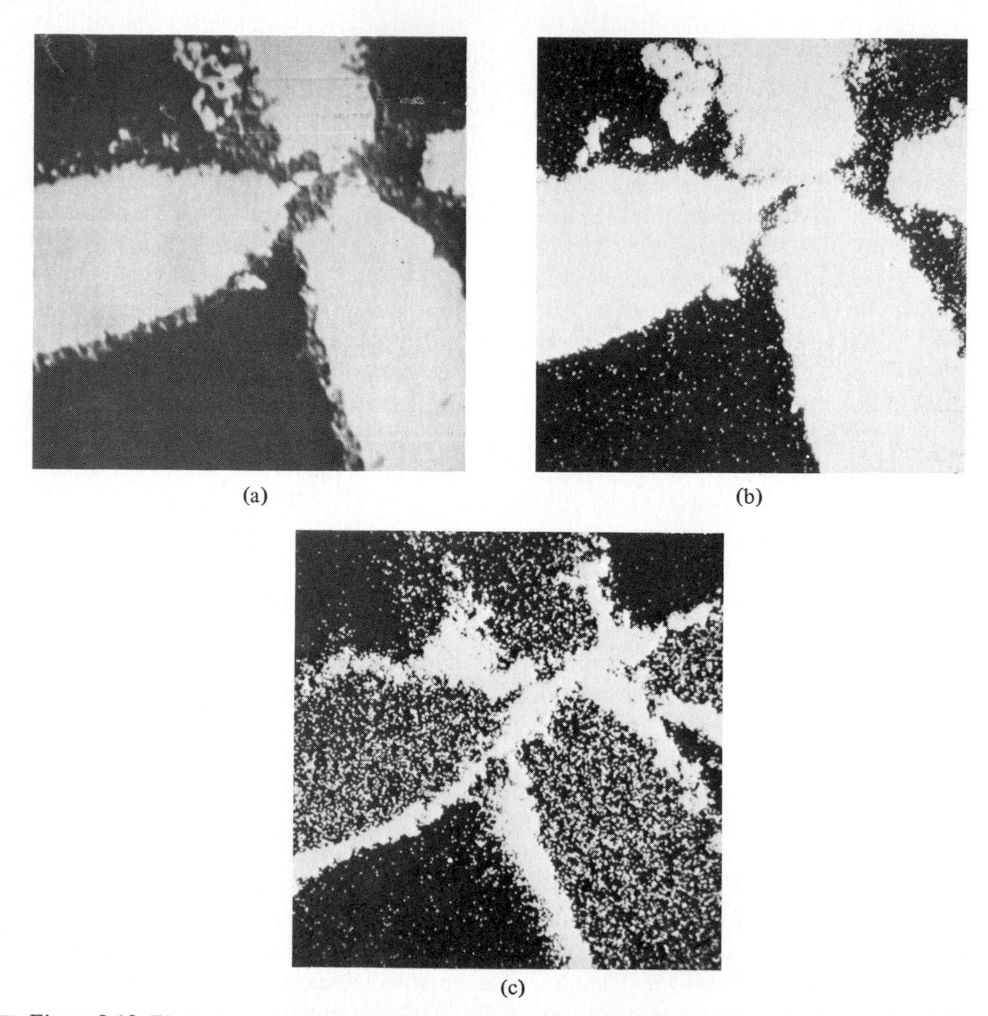

(a) (b) (c)

Figure 8.12 Electron microprobe and X-ray images of spent inactive gauze showing segregation of rhodium at the surface. (*a*) Electron image of cross section of gauze. (*b*) Pt X-ray image. (*c*) Rh X-ray image. [*Courtesy of Platinum Metals Review, 15(2), 52 (1971).*]

A loss in yield can also be produced by iron contamination of the gauzes from adventitious impurities in the plant (iron is a good ammonia-decomposition catalyst), or from surface iron acquired during fabrication. Catalyst activity can be markedly affected by small variations in surface-iron concentration.

Various base metal catalysts have long been studied as possible replacements for platinum-alloy catalysts, and a composition of 90% Fe_2O_3, 5% Mn_2O_3, and 5% Bi_2O_3 was used during World War I as a substitute. Cobalt oxide alone and combinations of cobalt oxide with other oxides such as iron oxide and chromium oxide, supported or unsupported, in fixed or fluidized beds, have also been studied. However the base metal catalysts are not economically competitive with the platinum-alloy system. The active species seems to be a Co(II, III) oxide which is reduced to an inactive Co(II) form with ammonia concentrations in air above about 7 percent.

Ammonia oxidation is reviewed by Dixon and Longfield (1960, p. 281) who discuss base metal oxide catalysts as well as use of platinum. The high-pressure nitric acid process was developed by the duPont Co. in the 1920s and is described by Chilton (1968). Details of 10 current nitric acid processes with emphasis on flow sheets and economics were published in a European Chemical News Supplement (*Eur. Chem. News*, 1970).

8.11 SYNTHESIS OF HYDROGEN CYANIDE

The principal present-day process for manufacture of hydrogen cyanide was developed by Andrussow from studies in the early 1930s. A mixture of methane, ammonia, and air is reacted by passing it through a platinum-alloy gauze pad in a manner closely analogous to that used in ammonia oxidation. The desired reaction may be written as

$$CH_4 + NH_3 + \tfrac{3}{2}O_2 \longrightarrow HCN + 3H_2O \qquad (8.44)$$

$$-\Delta H = 482 \text{ kJ/mol}$$

The product gas also contains appreciable amounts of H_2 and CO as well as unreacted NH_3 and a small amount of CH_4 in addition to N_2 from the air. Natural gas and NH_3 are each usually fed in slight stoichiometric excess to air as expressed by reaction (8.44), and some degree of preheat may or may not be used.

The catalyst is a pad of 90% Pt-10% Rh woven screens like that used for NH_3 oxidation and is supported on a ceramic structure. This in turn is placed directly atop a heat exchanger for rapid quenching of reaction products, since HCN will react rapidly with the water formed by the homogeneous reaction,

$$HCN + H_2O \longrightarrow NH_3 + CO \qquad (8.45)$$

Under equilibrium conditions the HCN concentration in reaction (8.45) is negligible.

The reaction pressure is about 0.2 MPa, the temperature about 1100 to 1150°C, and contact time is of the order of a millisecond. Lower temperatures may lead to carbon formation, which reduces reactivity and can cause mechanical breakup of the gauze. The yield based on NH_3 fed is typically about 70 percent, and NH_3 is recovered from the product gas for reuse. As with NH_3 oxidation, traces of Fe contamination on the Pt catalyst reduce yield and here promote coke deposition.

Platinum losses here are small, presumably because formation of volatile platinum oxides is less likely in the overall reducing environment. A new gauze pack does not reach full activity and selectivity until after a period of 2 to 3 days, during which the gauze undergoes a major rearrangement with a large increase in roughness and surface area (Fig. 8.13). The activity and selectivity then slowly decline with time over a period of months. The addition of a trace of sulfur-containing compounds during initial operation reportedly decreases the activation period. The reason for the effect is unknown, although electron photomicrographs show that it causes the catalyst structure to rearrange more rapidly. Schmidt and Luss (1971) report on the detailed examination by various instruments of Pt-Rh gauze catalysts after being used for NH_3 oxidation or the HCN reaction, the latter with and without sulfur treatment. Pan (1971) reports on morphological changes of a Pt-Rh gauze catalyst after use in a

(a)

(b)

(c)

Figure 8.13 Appearance of leading gauze in the pack after prolonged exposure in the Andrussow process. (*a*) Magnification about ×40. Note the matt appearance and considerable reduction in apertures of the gauze. (*b*) Scanning electron micrograph of Fig. 8.13*a*, ×150. Note transformation of formerly solid wire into a mass of crystallites, many with well-developed pyramidal faces. (*c*) Same as Fig. 8.13*b*, ×375. [*Courtesy of Platinum Metals Review, 22(4), 131 (1978)*.]

10-cm-ID pilot plant reactor. He also calculated that the reaction was not significantly mass-transfer-controlled, although this seems unlikely.

Earlier studies of the reaction are reported, e.g., by Andrussow (1951) and Pfeil and Hoffman (1963). The most comprehensive rate and selectivity data appear to be those of Pan and Roth (1968). They report information on conversion and selectivity of NH_3 to HCN as a function of CH_4/NH_3 mol ratio and the ratio of air to reactants (CH_4 and NH_3), obtained in a 10-cm-ID reactor. The NH_3 may react to form HCN, it may pass through unreacted, or it may decompose to the elements. At air/reactant mole ratios lower than about 2.8, the HCN yield dropped off substantially. Also the catalyst can become inactivated by carbon deposits if insufficient air is provided. Above a ratio of about 3.25 the mixture becomes explosive. Representative values for the maximum yield of HCN based on NH_3 were about 85 percent at a CH_4/NH_3 mol ratio of about 1.0 and an air/(CH_4 + NH_3) ratio of 2.8. The maximum yield was about 88 percent at a CH_4/NH_3 mol ratio of about 1.6 and air/(CH_4 + NH_3) ratio of 3.25. An increasing fraction of the NH_3 is found to pass through the gauze pad unreacted as the CH_4/NH_3 ratio is increased (e.g., above 0.8), termed *leakage*. The fraction of NH_3 decomposed and therefore unrecoverable increases as the CH_4/NH_3 ratio is decreased below about 1.2.

The ratio of NH_3/CH_4 chosen varies somewhat depending upon whether emphasis is on yield or on production. For a fixed feed-gas temperature to the reactor, the reaction temperature changes with reactant ratios and mass flow rate, and this may also affect yield and conversion. In an industrial reactor it may not be easy to obtain as high yields as those reported by Pan and Roth, because of the difficulty of achieving uniform flow over a large-diameter, thin pad of gauze and the possible occurrence of downstream reactions.

The mechanism of the reaction is speculative. At sufficiently high temperatures NH_3 and CH_4 will react directly in the absence of oxygen to form HCN (see below). Possibly an exothermic reaction occurs on the first one or two screens to generate heat for the subsequent endothermic reaction. In an earlier commercial process, apparently not now used, NO from NH_3 oxidation was reacted with CH_4 at about 600°C on Pt to form HCN. Thus NO may be formed as an intermediate in the Andrussow process, although it is not found in the exit gases. The effect of the increase in roughness on activity is probably at least in part a physical effect; it increases surface area and may increase mass transfer by decreasing the effective boundary layer.

In an alternate process (Degussa), which is not widely used, a mixture of methane and NH_3 is passed through an array of ceramic tubes about 2 m in length and 2 cm in diameter, coated internally with a Pt-containing catalyst layer and heated externally. Reaction occurs at 1200 to 1300°C, and H_2 is formed as a byproduct. The reaction is highly endothermic.

$$CH_4 + NH_3 \longrightarrow HCN + 3H_2 \tag{8.46}$$

$$-\Delta H = -250 \text{ kJ/mol}$$

Studies of this reaction are reported by Koberstein (1973), who used a reactor tube nearly identical to those used industrially. No consecutive reactions reportedly occur, and with this particular catalyst ammonia decomposition reportedly is negligible. Most of the reaction occurs in a short portion of the tube and largely under mass-transfer-controlling conditions; the first portion of the tube is required for heat-up and the last for cooling. Because of the large difference in molecular weight between H_2 and the other species present and the large temperature gradient between wall and gas at the exit, Koberstein calculates that considerable separation of the products occurs by thermal diffusion.

A higher concentration of HCN is produced in this process than in the Andrussow process, which lowers subsequent costs of purification; but the process is sensitive to operate, and it seems to be economic only in a small installation and where natural gas is expensive.

Some HCN is also produced as a byproduct of manufacture of acrylonitrile. HCN is used primarily to make adiponitrile by reaction with butadiene (for nylon 66) and methyl methacrylate via reaction with acetone and esterification with methanol. Methyl methacrylate is polymerized to form transparent plastics such as Lucite and Plexiglas and protective coatings. Some HCN is also used to make chelating agents.

8.12 CONTROL OF AUTOMOBILE-ENGINE EMISSIONS

The exhaust from the internal-combustion engine of an automobile contains small concentrations of hydrocarbons and CO from incomplete combustion of fuel, and of nitrogen oxides, NO_x, from nitrogen fixation at the high temperature of combustion. These contaminants are a major contributor to air pollution and one of particular concern in urban areas.

Federal legislation in the United States, which went into effect for 1975-model cars, established maximum permissible emission levels for automobiles in terms of the emissions of three pollutants, expressed as grams of hydrocarbons, CO, and NO_x per mile traveled. The concentrations are averaged in a detailed test procedure that simulates representative modes of operation of an automobile during warm-up and driving after the engine reaches normal operating temperature. Pollutant emissions can be reduced to some extent by internal-combustion engine modifications and by precise control of fuel/air ratio. However with a few exceptions, mostly small cars, the specified emission levels for CO and hydrocarbons can be reached only by use of a catalyst for oxidation of these species in the exhaust to CO_2 and H_2O. The situation has been complicated in the United States by the fact that a schedule of gradually lowering maximum allowable pollutant emission levels has been established by law and then modified and/or postponed, as the technical problems and costs to be borne by the public became clearer. For the model years of 1981 and some time into the future the maximum allowable levels (in grams per mile) have been set as 0.41 hydrocarbons, 3.4 CO, and 1.0 NO_x. A level of 0.40 NO_x called for in the original legislation has subsequently been recognized as being very difficult to meet within reasonable economic constraints and remains as a research goal.

Individual states may set more stringent levels than those called for by federal law. For a period those in California were more stringent than the federal standards, but that is not the case for 1981-model cars. Legally, the catalyst system in the new vehicle must continue to meet specified standards for 50,000 mi or 5 years, whichever comes first. The federal standard for NO_x was 2.0 g per mile for cars of model years through 1980, which was generally met by engine modifications. However, the level of 1.0 g per mile set for 1981-model cars necessitates the use of a catalyst system of a different type than that introduced for 1975-model cars, which was an oxidation catalyst only. (See Sec. 8.12.1.)

The catalyst application to automobiles is unique. It was the first large-scale use of a catalyst in a product for consumers, supposed to perform for at least 5 years without attention by technically trained people, or indeed by anyone at all. Nearly complete oxidation is required of an intermittent gaseous stream whose temperature, composition, and flow rate varies in an erratic pattern and which may contain traces of any of a wide variety of catalyst poisons from fuel additives, lubricating oil, and corrosion.[1] Further the unit must be compact because of stringent volume limitations, and the thermal mass must be as low as possible so the catalyst can reach ignition temperature rapidly, whereupon the emission control starts. The catalyst bed must operate at an extremely low pressure drop in order to minimize power loss from the engine. The sums of money expended for research and development of this single product stretch into the hundreds of millions of dollars and probably exceed that for any other single catalyst application. To date catalyst units in automobiles have been remarkably free of consumer complaint, probably in part because as yet their performance is rarely evaluated after the car has been acquired by the user. This situation will probably change in the future.

In spite of intensive study no catalyst composition has yet been developed that exhibits the requisite activity and durability when the fuel contains lead alkyls, commonly added to raise the octane number of gasoline (Chap. 9), or phosphorus compounds. Hence nonleaded gasoline must be used. Various base metal oxides exhibit good catalytic activity for oxidation, two of the most active being cobalt oxide and copper chromite. However they may be insufficiently resistant to sintering at high temperatures to meet durability standards, and they are less active than noble metal catalysts at the low temperatures of the engine warm-up period. They also may be more susceptible to poisoning from SO_x in the exhaust gases. Some base oxide compositions may also deteriorate when subjected alternately to a net oxidizing and net reducing environment.

The catalysts used for oxidation of CO and hydrocarbons as of 1980 are in all cases platinum or platinum plus a little palladium. These are more active and more durable than base oxide systems. Either of two catalyst configurations is used. In one the noble metal is supported on alumina beads held in a thin layer in a rather flat pancakelike alloy steel container. The gases flow down through the layer. In the second a

[1]The potential cumulative effect of even traces of poisons in the reacting gases in contact with a catalyst can be vividly seen in this application. On average an automobile burns its weight in fuel every year. With a 1000-kg car, over a 5-year period about 5000 kg of fuel would be consumed at an air/fuel weight ratio of about 14.5/1. About 72,500 kg of air with the products of combustion will contact the catalyst. If the usual leaded gasoline containing 3 g of lead per gallon were to be used, 5 kg of lead would pass through a catalyst on which about 1 to 2 g of precious metal is supported. Even with "lead-free" gasoline, which is specified as containing no more than 0.05 g of lead per gallon, over a 5-year period this amounts to about 80 g. With a sulfur content of about 100 ppm in gasoline, which is set as the desirable standard for fuel to be used with a catalyst, about 9 kg of sulfur dioxide will contact the catalyst.

honeycomb monolith is used consisting of a block of parallel, nonintersecting channels. (See Figs. 4.13, 4.14, and 4.15.) This is composed of cordierite, a magnesium aluminum silicate, $Mg_2Al_3(AlSi_5O_{18})$, which has a low coefficient of thermal expansion and is mechanically strong. The monolith surfaces are covered with a wash coat of alumina, and the noble metal is impregnated onto this. The open cross-sectional area is typically 70 percent or so of the total, and the number of cells per square inch ranges from 200 to 400 (31 to 62 cells per square centimeter), the corresponding wall thicknesses being, for example, 10 or 6 mils (0.25 or 0.15 mm), respectively. The material of higher cell count is somewhat more expensive to fabricate but allows a smaller unit to be utilized. The less the mass, the more rapidly it becomes heated up to the ignition temperature. For the same reason, specially fabricated alumina pellets of low mass are used in the packed-bed configuration.

Platinum is more active for oxidation of paraffin hydrocarbons, palladium for oxidation of CO and possibly unsaturated hydrocarbons. In an oxygen atmosphere, platinum on alumina sinters more readily than does palladium (Klimisch et al., 1975), but platinum is more resistant to lead poisoning than is palladium. Under most oxidizing conditions palladium probably exists in the form of PdO, which is more active for oxidation than the metallic form. Platinum is more active as the metal, on which oxygen is chemisorbed.

Platinum and palladium are found together in ores located predominantly in the USSR and South Africa. Their individual price in the marketplace is dictated by the relative demand for the two metals compared to the ratio in which they occur in nature. The demand for platinum in the Western world was about 2,500,000 troy oz per year as of 1978,[2] of which about one-half was for use in jewelry, especially in Japan. Use in automobile catalysts was the next largest demand, making up about one-third of the total. Over the past several years about two-thirds of the supply has come from South Africa, about 5 to 10 percent as a byproduct of nickel refining from the Inco operations at Sudbury, Canada, and almost all of the remainder from the USSR. The supply from the USSR is irregular but has averaged in the neighborhood of 600,000 troy oz per year over the last several years. This is probably a by-product of nickel operations, and the ratio of palladium to platinum appears to be much higher than that in South African ores. The amount from the Inco operations depends on the primary demand for nickel, whereas the South African ores are mined for primary production of platinum and palladium; other precious metals and nickel are produced as by-products. The largest demand for palladium is in the electronics industry for use largely in conductive inks and protective coatings.

The world production of the *platinum-group* metals (Ru, Rh, Pd, Os, Ir, Pt) was about 6,000,000 troy oz per year as of 1978, of which about one-third was used in the United States. In South Africa the ores are mined from what is known as the Merensky Reef and the platinum-group metal content is about 0.2 troy oz per ton of ore. Their relative weight proportions are about as follows: Pt, 1; Pd, 0.6 to 0.4; Ru, 0.15 to 0.07; Rh, 0.06 to 0.04; Ir, 0.02 to 0.01; Os ~ 0.002. Some osmium and

[2] Precious metals are sold by the troy ounce (1 troy oz = 31.1 g), but the price per gram-atom may be a more meaningful unit in catalysis in cases in which one element may be substituted for another. Historically, platinum has sold at a higher price per unit weight than palladium, but since palladium has an atomic weight of 106 versus 195 for platinum, the economics in favor of palladium are greater than a casual observation might imply.

iridium are also recovered as byproducts of refining gold from certain South African ores. This concentrate is termed *osmiridium* and typically is one-third Os, one-third Ir, and about one-sixth Pt.

In the United States little Pt is used in jewelry. About one-third of the total Pt plus Pd is used for motor-vehicle catalysts, and about one-quarter in chemical and petroleum processing. The primary producers usually set a *producer price* for long-term customers under contract, which is changed with supply and demand. Spot prices may fluctuate greatly and relatively quickly both above and below the producer price, especially for the precious metals other than platinum and palladium. The production of these is essentially inflexible, so slight changes in demand can markedly affect price. The market prices for platinum and palladium also tend to move with current prices for gold since they are also purchased as an investment or speculative vehicle. The producer price for platinum in the United States during 1977 and 1978 varied between \$162 and \$300 per troy ounce and spot prices on the free market varied between about \$144 and \$392 per troy ounce, although market fluctuations during this period were unusually great. The average producer price in 1978 was \$237 per troy ounce for platinum and \$71 per troy ounce for palladium. Approximate prices in 1978, in dollars per troy ounce, for the other platinum-group metals were about: Ru, 60; Rh, 500; Ir, 250; Os, 150. In comparison, in recent years market prices for such base metal catalysts as Ni, Cr, Mo, and Co have been in the range of \$2 to \$20 per pound of metal. The average 1978 price for rhenium, used in reforming catalysts, was about \$350 per pound.

Since platinum-group metals are expensive, the loading used in auto catalysts is the minimum necessary to meet standards. This works out to an amount of platinum plus palladium in the range typically of about 1 to 2 g per catalyst unit, to meet federal standards for 1980 models. The catalysts may be somewhat poisoned by SO_x in the exhaust gases, but the poisoning appears to be reversible; that is, SO_x is adsorbed at lower temperatures and desorbed at higher temperatures.

The catalyst temperature is in the range of 400 to 600°C during normal operation, but may rise several hundred degrees above this under extreme driving conditions or with engine malfunction. The reaction is highly mass-transfer–limited after the engine reaches operating temperature, and the most effective use of the noble metal catalyst is to deposit it in a relatively thin layer near the outside catalyst surface. Poisons will generally accumulate on the outermost surface of the porous support, so optimally the noble metal layer should probably be displaced slightly into the interior to prolong catalyst life. This is sometimes termed an *eggshell*-type deposit. A further complication is that the rate of CO oxidation on platinum is an inverse function of CO concentration. In a diffusion-limited reaction on a catalyst with uniform Pt distribution, the local rate will *increase* with depth of penetration up to a certain percent conversion (Wei and Becker, 1975). This is unlike the usual case in which reaction rate is decreased by a diffusion limitation.

8.12.1 NO_x Removal Catalyst

The more stringent limitations set for NO_x emissions (NO and NO_2) from automobiles for 1981 and later years requires in most cases a suitable catalyst for this purpose. No catalysts of sufficient activity for decomposition of NO_x are known, and NO_x must therefore be removed by reduction to N_2. More than one type of catalyst configuration may be utilized, depending on engine characteristics and other design factors. The central problem is that a vehicle does not run at steady-state conditions.

For many cases it appears that the concentrations of hydrocarbons, CO, and NO_x can be reduced to the desired level simultaneously in a single catalyst unit operated in a narrow range around the stoichiometric air/fuel ratio. This requires precise control of the fuel-air mixture fed to the automobile engine at all operating conditions, so as to operate the catalyst in the narrow "window" of composition in which this degree of pollution control can be achieved. Generally an oxygen sensor in the engine exhaust stream with a feedback control system, is required. This type of unit is termed a *three-way catalyst* because all three pollutants are removed simultaneouly.

The most active catalyst is platinum, but some of the NO_x is reduced to NH_3 instead of N_2, which is not desired. The most effective catalyst for conversion of NO_x to N_2 appears to be one in which some rhodium is added to the platinum. However there is no primary production of rhodium anywhere in the world; it is produced as a byproduct of production of platinum and palladium, the ratio of rhodium to platinum in South African ores, for example, being about 0.06 to 0.04. Therefore the optimum catalyst composition will be markedly influenced by market economics which set the price of a byproduct of strictly limited availability. Catalyst compositions containing ruthenium in various forms appear to be effective for conversion of NO_x to N_2. However these may be converted to ruthenium oxides under oxidizing conditions, as might occur occasionally during engine operation. These oxides are volatile and poisonous. Ruthenium is likewise produced only as a byproduct, but in South African ores mined primarily for platinum it is present in greater amounts than rhodium.

In a number of car applications it may be necessary to provide a second oxidation catalyst following the three-way catalyst, additional air being supplied between the two catalyst beds.

Information on automobile catalysts is extensive, widely scattered, heavily patented, and highly proprietary, and much of it is of rather short half-life because of the rapidly moving state of affairs. The papers in a symposium edited by McEvoy (1975) show some of the complexities involved in attempting to reach an overall optimum solution. A review by Dwyer (1972) treats some of the earlier catalysts considered. A review by Shelef (1975) considers NO_x-removal reactions in great detail. Wei (1975) reviews automobile catalysts with particular attention to kinetics and reactor engineering. Resistance to sintering and to agglomeration of precious metal crystallites is of prime importance, a subject discussed in Chaps. 4 and 6. Poisoning of automobile catalysts is reviewed by Shelef et al. (1978).

8.13 CATALYTIC COMBUSTION

There are a large variety of uses of catalytic combustion beyond that in automobiles, and some have been applied for many years. Low concentrations of hydrocarbon and organic pollutants in air, as from painting operations, can thus be eliminated. In some cases a gaseous stream may contain CO and hydrocarbons in sufficiently high concentration to represent energy that is recoverable, but the fuel is of insufficient concentration to support normal combustion. Of growing interest is catalytic combustion for primary energy production, as in stationary gas turbines. Here it shows promise of producing significantly lower emissions, especially of NO_x, than do conventional combustors. The reactions taking place are probably partly catalytic and partly thermal, and in some cases the catalyst may primarily play the role of a flame holder. In all cases the catalyst is a supported noble metal. For some applications a principal limitation is to develop a rugged support that is stable at temperatures that may be as high as 1500°C.

Closely related are catalytic abators, which have been used for 20 years or more to reduce NO_x emissions from nitric acid plants or chemical operations emitting NO_x fumes (Fig. 8.14). A representative composition of the tail gas from a nitric acid plant

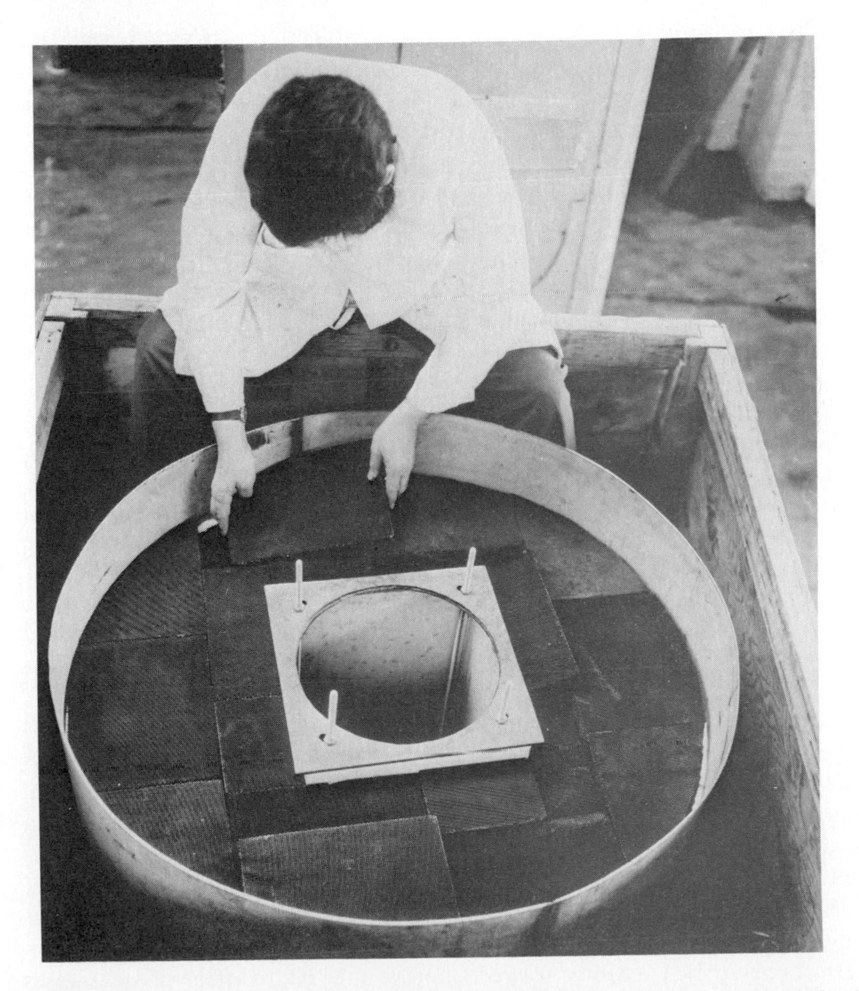

Figure 8.14 A cylindrical tail gas catalytic abator unit for a nitric acid plant being loaded with platinum on a ceramic "honeycomb" catalyst. [*Courtesy of Platinum Metals Review, 17(2)*, 57 *(1973)*.]

(dry basis) is 2 to 3% O_2, 0.3 to 0.4% NO_x, and the remainder N_2. Fuel is added to this, and the mixture is passed over the catalyst, consisting of platinum and/or palladium on a support. The gas must be preheated to the *ignition temperature*, i.e., the temperature at which the reaction first becomes economically rapid. Because of the exponential effect of temperature on rate, the ignition temperature for a given system can often be specified within fairly close limits. The addition of some rhodium to platinum is reported to reduce the ignition temperature with natural gas below that achievable with platinum alone, which is desirable. Early units were operated primarily to reduce NO_2 (brown color) to NO (colorless), but with little reduction of the fixed N content. This is easy to accomplish. Under the more recent impact of legislation for control of air pollution, design and operation has been altered by which the total NO_x concentration can be reduced to a few hundred parts per million (Adlhart et al., 1971; Newman, 1971).

The ignition temperature (for a platinum and/or palladium catalyst) depends on the fuel used; some approximate representative values are CH_4, 450 to 500°C; naphtha, 400°C; synthesis gas, 200°C; hydrogen, 150 to 200°C. These will slowly rise as the catalyst deactivates with use. Significant conversion of NO to N_2 can be obtained only after the oxygen present is substantially consumed. The reactor operates adiabatically, and if methane or a light naphtha is the fuel, the temperature increase through the reactor is about 130°C for each percent of O_2 initially present; it is about 160°C if hydrogen is used. The maximum O_2 concentration in the tail gas at which a single-stage abator can operate is thus limited by the thermal stability of the catalyst. This is about 750 to 800°C for representative substances. The maximum inlet O_2 concentration that can be handled is thus about 2 to 2.5% with methane and about 3.5 to 3.7% with hydrogen. For higher O_2 concentrations a two-stage abator may be used, some fuel being introduced between the two stages. The tail gas from a HNO_3 high-pressure plant is at a pressure of 0.8 to 0.9 MPa. Abators are usually run at this pressure, and energy is recovered as power or steam.

An alternate method which has been developed recently is the selective reduction of nitrogen oxides to N_2 by addition of ammonia, a reaction which occurs in the presence of oxygen.

$$6NO_2 + 8NH_3 \longrightarrow 7N_2 + 12H_2O \tag{8.47}$$

$$6NO + 4NH_3 \longrightarrow 5N_2 + 6H_2O \tag{8.48}$$

This requires precise control to maintain the rate of NH_3 addition to the tail gas in the proper ratio to NO_x. Good mixing is also needed, as is the maintenance of the temperature at the proper level, about 250 to 300°C. At lower temperatures ammonium nitrate and nitrites are formed, and at higher temperatures some ammonia may be oxidized to nitrogen oxides. Anderson et al. (1961) report studies of this reaction, as well as results with a variety of fuels, on supported platinum or palladium.

The desired reactions will proceed homogeneously at much higher temperatures, but a catalyst appears to be needed for a high degree of NO_x removal. The same reaction has also been applied recently for reduction of the NO_x content of stack gases from furnaces, boilers, etc., especially in Japan. In this application the inevitable presence of sulfur oxides from sulfur in the fuel must be taken into consideration, and a low-pressure-drop system is important. The catalysts generally are supported, nonprecious metal compositions.

8.14 LITERATURE

A considerable number of books and reviews treat catalysis of partial oxidation reactions in considerable detail. The book by Hucknall (1974) discusses the selective oxidation of olefins and alkanes, but it does not cover aromatic or heterocyclic compounds. The review by Sampson and Shooter (1965) treats ethylene, propylene, butane, butylene, benzene, naphthalene, and *o*-xylene. Voge and Adams (1967) cover the catalytic oxidation of olefins, from ethylene to pentenes. The book by Germain (1969) treats hydrocarbon reactions in general, with particular emphasis on mechanism, and includes well over 1000 references. Sharchenko (1969) reviews oxidative dehydrogenation of hydrocarbons with extensive references, many on work in the USSR. Carra' and Forzatti (1977) summarize mechanisms and kinetic expressions for a variety of reactions, and some of the factors involved in reactor design. A review by Schuit (1969) focuses on oxide catalysts, especially molybdates, but this is essentially superseded by a chapter in the book by Gates et al. (1979). Information on approximate commercial operating conditions and catalyst compositions is given for a variety of processes by Waddams (1973).

In the earlier literature, Margolis (1963) gives a general review of hydrocarbon oxidation, focusing primarily on kinetics and mechanism. Dixon and Longfield (1960, p. 183) present a broad review which covers aromatics, olefins, and heterocyclic compounds. Companion reviews cover oxidation of NH_3, SO_2, and CO (1960, p. 281) and include a discussion of reaction mechanisms and a list of references to other miscellaneous catalytic oxidations (1960, pp. 347, 363).

REFERENCES

Adams, C. R.: *Am. Chem. Soc., Div. Pet. Chem., Prepr.*, **14**(3), C6 (1969).
Adlhart, O. J., S. G. Hindin, and R. E. Kenson: *Chem. Eng. Progr.*, **67**(2), 73 (1971).
Ai, M.: *J. Catal.*, **54**, 426 (1978).
Anderson, H. C., W. J. Green, and D. R. Steele: *Ind. Eng. Chem.*, **53**(3), 199 (1961).
Andrussow, L.: *Angew. Chemie*, **63**, 21 (1951).
Aykan, K., D. Halvorson, A. W. Sleight, and D. B. Rogers: *J. Catal.*, **35**, 401 (1974).
Batist, Ph. A., A. H. W. M. Der Kinderen, Y. Leeuwenburgh, F. A. M. G. Metz, and G. C. A. Schuit: *J. Catal.*, **12**, 45 (1968).
——, C. J. Kapteijns, B. C. Lippens, and G. C. A. Schuit: *J. Catal.*, **7**, 33 (1967).
Becker, M., and R. S. Barker: U.S. Patent 2,967,185 (to Scientific Design Co.), January 3, 1961.
Boreskov, G. K., R. A. Buyanov, and A. A. Ivanov: *Kinet. Catal.*, **8**, 126 (1967).
Calderbank, P. H.: *Adv. Chem. Ser., No. 133*, 1974, p. 646.
——, and A. D. Caldwell: *Adv. Chem. Ser. No. 109*, 1972, p. 38.
——, K. Chandrasekháran, C. Fumagalli, *Chem. Eng. Sci.*, **32**, 1435 (1977).
Callahan, J. L., R. K. Grasselli, E. C. Milberger, and H. A. Strecker: *Ind. Eng. Chem., Prod. Res. Dev.*, **9**, 134 (1970).
Carra', S., and P. Forzatti: *Catal. Rev., Sci. Eng.*, **15**(1), 1, (1977).
Carrubba, R. V., and J. L. Spencer: *Ind. Eng. Chem., Process Des. Dev.*, **9**, 414 (1970).

Cathala, M., and J. E. Germain: *Bull. Soc. chim. Fr.*, **38,** 2174 (1971).
Chauvel, A. R., P. R. Courty, R. Maux, and C. Petitpas: *Hydrocarbon Process.*, September 1973, p. 179.
Chem. Eng., June 24, 1974, p. 114.
Chilton, T. H.: *Strong Water. Nitric Acid: Sources, Methods of Manufacture and Uses*, M.I.T., Cambridge, Mass., 1968. [See also Chem. Eng. Prog. Monogr. Ser. **56,** (3), 1960.]
Dalin, M. A., I. K. Kolchin, and B. R. Serebryakov: *Acrylonitrile*, Technomic Pub. Co., Westport, Conn., 1971.
Dixon, J. K., and J. E. Longfield in P. H. Emmett (ed.): *Catalysis*, vol. 7, Reinhold, New York, 1960.
Duecker, W. W., and J. R. West: *The Manufacture of Sulfuric Acid*, Reinhold, New York, 1959.
Dwyer, F. G.: *Catal. Rev.*, **6,** 261 (1972).
Eur. Chem. News, Nitric Acid Supplement, Jan. 30, 1970.
Fariss, R. H., in "Catalysis in Practice," *Symposium, Institution of Chemical Engineers*, London, 1963, p. 51.
Flank, W. H., and H. C. Beachell: *J. Catal.*, **8,** 316 (1967).
Froment, G. F.: *Ind. Eng. Chem.*, **59**(2), 23 (1967).
——: "Chemical Reaction Engineering," *Fourth International/Sixth European Symposium*, DECHEMA, 1976, p. 421.
Gans, M., and B. J. Ozero: *Hydrocarbon Process.*, March 1976, p. 73.
Gates, B. C., J. R. Katzer, and G. C. A. Schuit: *Chemistry of Catalytic Processes*, McGraw-Hill, New York, 1979.
Germain, J. E.: *Catalytic Conversion of Hydrocarbons*, Academic, New York, 1969.
——: *Intra-Sci. Chem. Rep.*, **6,** 101 (1972).
Gibson, M. A., W. R. Cares, and J. W. Hightower: *Am. Chem. Soc., Div. Pet. Chem., Prepr.*, **22,** 475 (1977).
——, and J. W. Hightower: *J. Catal.*, **41,** 420, 431 (1976) (and earlier studies).
Gillespie, G. R., and R. E. Kenson: *Chemtech*, October 1971, p. 627.
Haber, J.: *Int. Chem. Eng.*, **15,** 21 (1975). [See also *Z. für Chem.*, **13**(7), 241 (1973).]
Hahn, A. V. G., *The Petrochemical Industry: Market and Economics*, McGraw-Hill, New York, 1970.
Haldor-Topsøe, F. A., and A. Nielsen: *Trans. Dan. Acad. Tech. Sci.*, **1,** 18 (1947).
Hancock, E. G. (ed.): *Propylene and its Industrial Derivatives*, Benn, London, 1973.
Handforth, S. L., and J. N. Tilley: *Ind. Eng. Chem.*, **26,** 1288 (1934).
Harbord, N. H.: *Platinum Met. Rev.*, **18,** 97 (1974).
Herce, J. L., J. B. Gros, and R. Bugarel: *Chem. Eng. Sci.*, **32,** 729 (1977).
Herten, J., and G. F. Froment: *Ind. Eng. Chem., Process Des. Dev.*, **7,** 516 (1968).
Higgins, R., and P. Hayden: *Catalysis*, vol. 1, The Chemical Society, London, 1977, chap. 5, p. 168.
Holzmann, H.: *Chemie Ing. Tech.*, **40,** 1229 (1968).
Hucknall, D. J.: *Selective Oxidation of Hydrocarbons*, Academic, New York, 1974.
Hutson, T., R. D. Skinner, and R. S. Logan: *Hydrocarbon Process.*, June 1974, p. 133.
Kadlec, B., J. Michálek, and A. Šimeček: *Chem. Eng. Sci.*, **25,** 319 (1970).
Kenney, C. N.: *Catal. Rev., Sci. Eng.*, **11,** 197 (1975).
Kiguchi, I., T. Kumazawa, and T. Nakai: *Hydrocarbon Process.*, March 1976, p. 69.
Kilty, P. A., and W. M. H. Sachtler: *Catal. Rev., Sci. Eng.*, **10,** 1 (1974).
Klimisch, R. L., J. C. Summers, and J. C. Schlatter: *Adv. Chem. Ser. No. 143*, 1975, p. 103.
Klink, A., E. Paul, J. Gillin, and W. Sklarz: *Proceedings of the Fourth International/Sixth European Symposium on Chemical Reaction Engineering*, DECHEMA, 1976, p. 327.
Knözinger, H.: *Adv. Catal.*, **25,** 184 (1976).
Koberstein, E.: *Ind. Eng. Chem., Process Des. Dev.*, **12,** 444 (1973).
Kolchin, I. K.: *Khim. Prom.*, **49**(11), 815 (1973).
Krekeler, H., and W. Kronig: *Seventh World Petroleum Congress*, vol. 5, Elsevier, 1967, p. 41.
Liljenroth, F. G.: *Chem. Metall. Eng.*, **19**(6), 287 (1918).
Livbjerg, H. B., and J. Villadsen: *Chem. Eng. Sci.*, **27,** 21 (1972).
——, B. Sorenson, and J. Villadsen: *Adv. Chem. Ser., No. 133*, 1974, p. 242.

McEvoy, J. E. (chairman): "Catalysts for the Control of Automotive Pollutants," *Symposium, Adv. Chem. Ser. No. 143*, 1975.
Margolis, L. Ya.: *Adv. Catal.*, **14,** 429 (1973).
Mars, P., and D. W. van Krevelen: *Chem. Eng. Sci.*, **3** (special supplement), 41 (1954).
Massoth, F. E., and D. A. Scarpiello: *J. Catal.*, **21,** 294 (1971).
Moro-oka, Y., Y. Morikawa, and A. Ozaki: *J. Catal.*, **7,** 23 (1967).
Nakamura, S., and T. Yasui: *J. Catal.*, **17,** 366 (1970).
Newman, D. J.: *Chem. Eng. Prog.*, **67**(2), 79 (1971).
Nowak, E. J.: *Chem. Eng. Sci.*, **21,** 19 (1966); **24,** 421 (1969).
Pan, B. Y. K., *J. Catal.*, **21,** 27 (1971).
——, and R. G. Roth: *Ind. Eng. Chem., Process Des. Dev.*, **7,** 53 (1968). [See also B. Y. K. Pan, *Ind. Eng. Chem., Process Des. Dev.*, **8,** 262 (1969).]
Pfeil, E., and P. Hoffmann: *Ber. Bunsenges. physik. Chem.*, **67,** 229 (1963).
Pitzer, E. W.: *Ind. Eng. Chem., Prod. Res. Dev.*, **11,** 299 (1972).
Reich, P.: *Hydrocarbon Process.*, March 1976, p. 85.
Rennard, R. J. and W. L. Kehl: *J. Catal.*, **21,** 282 (1971).
Roiter, V. A., G. I. Golodets, Yu. I. Pyatnitzky: *Proceedings of the Fourth International Congress on Catalysis, Moscow*, Akadémíai Kiadó, Budapest, 1972, paper no. 35. [See also *Scientific Selection of Catalysts*, A. A. Balandin et al. (eds.), English translation, Keter Pub. House, Israel, 1968, p. 59.]
Sachtler, W. M. H. et al.: *Rec. trav. chim.*, **89,** 460 (1970).
—— and N. H. deBoer: *Third Congress on Catalysis*, North-Holland Publishing Company, Amsterdam, 1965, p. 252.
Saffer, A.: in "Catalysis in Practice," *Symposium*, Institution of Chemical Engineers, London, 1963, p. 54.
Sampson, R. J., and D. Shooter: *Oxid. Combust. Rev.*, **1,** 223 (1965).
Sancier, K. M., P. R. Wentrcek, and H. Wise: *J. Catal.*, **39,** 141 (1975).
Satterfield, C. N., and J. Loftus: *AIChEJ.*, **11,** 1103 (1965).
Schaefer, von. H.: *Ber. Bunsenges. physik. Chem.* **71** (8), 222 (1967).
Schmidt, L. D., and D. Luss: *J. Catal.*, **22,** 269 (1971).
Schreiner, W. C., A. E. Cover, W. D. Hunter, C. P. van Dijk, and H. S. Jongenburger: *Hydrocarbon Process.*, November 1974, p. 151.
Schuit, G. A.: *Chim. Ind.*, **51,** 1307 (1969) (in English).
Sharchenko, V. K.: *Int. Chem. Eng.*, **9,** 1 (1969).
Shelef, M.: *Catal. Rev., Sci. Eng.*, **11,** 1 (1975).
——, K. Otto, and N. C. Otto, *Adv. Catal.*, **27,** 311 (1978).
Simard, G. L., J. F. Steger, R. J. Arnott, and L. A. Siegel: *Ind. Eng. Chem.*, **47,** 1424 (1955).
Simons, Th. G. J., E. J. M. Verheijen, Ph. A. Batist, and G. C. A. Schuit: *Adv. Chem. Ser. No. 76*, 1968, p. 261.
Sleight, A. W., in J. J. Burton and R. L. Garten (eds.): *Advanced Materials in Catalysis*, Academic, New York, 1977, p. 181.
——, and W. J. Linn: *Ann. N.Y. Acad. Sci.*, **272,** 22 (1976).
Sperner, F., and W. Hohmann: *Platinum Met. Rev.*, **20,** 12 (1976).
Sterrett, J. S., and H. G. McIlvried: *Ind. Eng. Chem., Process Des. Dev.*, **13,** 54 (1974).
Stobaugh, R. B., W. C. Allen, Jr., and Van R. H. Sternberg: *Hydrocarbon Process.*, May 1972, p. 153.
Suvorov, B. V.: "Oxidative Ammonolysis of Organic Compounds," Academy of Sciences, KazakhSSR, Nauka ("Science") Publishing House, Alma-Ata, 1971. (In Russian.)
Tandy, G. H.: *J. Appl. Chem.*, **6,** 68 (1956).
Varma, R. L., and D. N. Saraf: *Ind. Eng. Chem., Prod. Res. Dev.*, **18,** 7 (1979).
Villadsen, J., and H. Livbjerg: *Catal. Rev., Sci. Eng.*, **17,** 203 (1978).
Voge, H. H., and C. R. Adams: *Adv. Catal.*, **17,** 151 (1967).
Volf'son, V. Ya., Ya. V. Zhigailo, E. F. Totskaya, and V. V. Raksha: *Kinet. Catal.*, **6,** 138 (1965).
Voorhoeve, R. J. H., in J. J. Burton and R. L. Garten (eds.): *Advanced Materials in Catalysis*, Academic, New York, 1977, p. 129.

——, D. W. Johnson, Jr., J. P. Remeika, and P. K. Gallagher: *Science*, **195**, 827 (1977).
Wachs, I. E., and R. J. Madix: *J. Catal.*, **53**, 208 (1978); *Surf. Sci.*, 1978, in press.
Waddams, A. L.: *Chemicals from Petroleum*, 3d ed., Wiley, New York, 1973.
Wagner, C., *Chem. Tech.* (*Chem. Fabr., Neue Folge*), **18**, 1, 28 (1945). (In German.)
Wainwright, M. S., and T. W. Hoffman: *Adv. Chem. Ser. No. 133*, 1974, p. 669.
Walker, J. F.: *Formaldehyde*, 3d ed., Reinhold, New York, 1964.
Wei, J.: *Adv. Catal.*, **24**, 57 (1975).
——, and E. R. Becker: *Adv. Chem. Ser., No. 143*, 1975, p. 116.
Welch, L. M., L. J. Croce, and H. F. Christmann: *Hydrocarbon Process.*, November 1978, p. 131.
Wimer, W. E., and R. E. Feathers: *Hydrocarbon Process.*, March 1976, p. 81.

CHAPTER

NINE

PROCESSING OF PETROLEUM AND HYDROCARBONS

The largest-volume catalytic processes are to be found in the refining of petroleum. Catalytic reforming, catalytic cracking, hydrocracking, and hydrodesulfurization are carried out on a very large scale and catalytic hydrotreating is used to improve the quality of various products and intermediate feedstocks. The primary products are a range of fuels, but a small yet growing fraction of the crude oil processed (currently about 5 percent in the United States) is converted to chemicals. The most important, in terms of quantity produced, are ethylene, propylene, butylenes, 1, 3-butadiene, and benzene, toluene, and xylene, (BTX) compounds. The BTX aromatic compounds are produced primarily by catalytic reforming and are in demand as high-octane components in gasoline as well as for chemical manufacture.

Some low-molecular-weight olefins are produced as a byproduct of refinery operations such as catalytic cracking. A mixture of olefins and a wide variety of other compounds is also produced by thermal cracking of various hydrocarbon feedstocks in the presence of steam, so-called *steam cracking*. The composition of the product varies substantially, depending upon the feedstock (e.g., ethane in contrast to various distillate fractions) and the severity of the cracking reaction. Steam cracking is operated primarily to produce ethylene, but with the present trends towards heavier feedstocks in the United States it is anticipated that most or all of the United States domestic requirements for butadiene (1, 3-butadiene) will in the future be supplied as a coproduct from steam cracking. An increasing portion of United States benzene requirements also will be extracted from the *pyrolysis gasoline* fraction of the product.

Acetylene has also been an important chemical, produced by subjecting natural gas or other hydrocarbons to very high temperatures. For ethane cracking to produce ethylene typical temperatures are 750 to 900°C, and for formation of acetylene about 1200°C. These are set by equilibrium considerations. The use of acetylene as the starting point for chemical manufacture has been steadily decreasing over the past 10 to 15 years as processes based on lower-cost olefins have been developed. The use of acetylene for production of vinyl chloride has been replaced by ethylene, and use of acetylene for the manufacture of acrylonitrile has been replaced by propylene. However use of acetylene may increase again if, as predicted, steam cracking of heavier feedstocks becomes more important, since larger amounts of acetylene are then produced as a byproduct.

9.1 COMPOSITION OF PETROLEUM

Petroleum consists predominantly of various hydrocarbons but also contains lesser amounts of sulfur, nitrogen, and oxygen in the form of a variety of organic compounds. Small amounts of vanadium and nickel are also present, mostly in the form of porphyrins and related structures. Elemental sulfur is also found in small amounts in some crude oils. The distribution of type of compound and of molecular weights varies greatly with location and the particular formation (depth) from which the crude oil was obtained. In primary recovery methods, the crude oil is forced to the surface by the natural energy of the reservoir. This may be from natural gas under pressure, either dissolved in the oil or present in a cap over the oil, or from the surrounding water.

Some portion of the petroleum always remains in the formation after primary recovery, and a variety of "secondary" and "tertiary" methods are being increasingly applied to recover additional petroleum from depleted fields or to increase the recovery rate from partially exhausted fields. These include injection of energy from the surface, as by pressurization with a gas or water, injection of steam to lower viscosity, or use of surface-active agents or low-viscosity solvents to assist the flow of oil.

Petroleum recovered in primary production is accompanied by natural gas, which typically contains 80 to 90% methane. The remainder of the hydrocarbons is primarily ethane, propane, and butane in quantities decreasing in this order. Gas containing appreciable amounts of these higher hydrocarbons is termed *wet gas.* Natural gas recovered in the absence of petroleum, as in Western Europe, frequently contains no more than 3 or 4 percent of hydrocarbons other than methane and is termed *dry gas.* Most of the natural gas in the United States is wet gas. In either case considerable quantities of hydrogen sulfide, carbon dioxide and/or nitrogen may also be present. In some United States production the gas contains helium in concentrations that can be economically recovered. C_2 to C_4 hydrocarbons are removed from wet gas by absorption in an oil or, in some newer plants, by a cryogenic process. Ethane is an important feedstock for manufacture of ethylene by thermal (steam) cracking. The C_3 and C_4 gases may be sold as *liquefied petroleum gas* or used as a feedstock for thermal cracking. As of 1980, in the United States the amount of *n*- and *iso*-C_4 separated from natural gas was about twice the amount separated from refinery streams.

The hydrocarbons in petroleum are primarily paraffins, aromatic ring compounds, and *naphthenes* (saturated rings with five or six C atoms in the ring). No olefins or acetylenes are present. Petroleum may be classified by composition loosely into *paraffinic-base* and *asphaltic-base* crudes, some being intermediate in character. Paraffinic-base crudes contain considerable quantities of paraffins (alkanes), which may be linear or branched. They are more highly concentrated in the lower-boiling fractions. The major portion of the crude, however, consists of ring compounds, either aromatic or naphthenic. Asphaltic-base crudes contain varying quantities of high-molecular-weight, nonvolatile matter that can be separated by solvents into fractions termed *resins* and *asphaltenes.*[1] The asphaltenes have molecular weights ranging typically from 1000 to 100,000 and are of complex composition. They appear to consist of clusters of polyaromatics that may disperse to various degrees in different solvents. The intermediate-boiling fractions of asphaltic crudes are usually more aromatic than naphthenic in character.

The concentration of paraffins decreases in the higher-boiling fractions of all crude oils. They are mostly branched. The single-ring compounds (both aromatic and naphthenic) in light fractions of many crudes usually have one, two, or three short alkyl chains. Molecular complexity increases with molecular weight, and in the higher-boiling fractions there occur multiple-ring structures. These may be linked by alkyl groups, be conjugated (as in bicyclohexyl) or fused [as in decahydronaphthalene (Decalin) or naphthalene]. In a multiple-ring structure some rings may be aromatic and others saturated.

The most important impurities in petroleum from the point of view of catalytic processing of distillate fractions are sulfur and nitrogen compounds. Organic sulfur compounds occur in all crude oils, in amounts varying from as little as 0.05 wt % sulfur to as high as 5 wt %. Hydrogen sulfide is also often present, and sometimes there are traces of elemental sulfur. The sulfur content (expressed as a weight fraction) increases with overall molecular weight, so a residual fraction will have a considerably higher sulfur content than the crude from which it is derived. In the low-boiling fractions, the principal sulfur compounds are organic sulfides or disulfides, mercaptans (thiols; R—S—H, in which R is an aliphatic group or a saturated cyclic ring), and thiophenes. In higher-boiling fractions sulfur is present largely in the form of thiophene derivatives such as benzo- and dibenzothiophenes. The nonheterocyclic sulfur compounds are the less stable thermally, and they may react or be substantially decomposed during distillation.

Most crude oils contain 0.1 wt % nitrogen or less, but some heavy asphaltic-base crudes may contain up to about 0.9 wt %. Shale oil derived from retorting United States western shales contains up to 2 wt % nitrogen, and so-called synthetic crudes derived from coal or tar sands may contain up to about 1 wt % nitrogen. As with sulfur compounds, the nitrogen content increases with molecular weight of the crude-oil fraction. The structures of the various nitrogen compounds have not been as well characterized as those of the sulfur compounds. Those identified appear to be mostly heterocyclic, and they are typically grouped into basic compounds (e.g., pyridines and quinolines, which may have alkyl substituents) and nonbasic compounds (derivatives of indole, pyrrole, and carbazole).

[1] The usual separation scheme involves extraction with *n*-pentane. The insoluble fraction is termed *asphaltenes.* The soluble fraction is termed *maltenes* and may be further separated into *resins* and *oils.*

Oxygen is present in low- and medium-boiling range fractions of petroleum primarily as aliphatic and cyclic carboxylic acids, and sometimes, in small amounts, as phenols. Vanadium and nickel are found in crude oil in concentrations from as low as 3 ppm to 500 to 600 ppm, the weight ratio of vanadium to nickel typically being 5:1 to 10:1. These are found mostly in the asphaltenes, which are in the residual fraction. Hence high metal content is typically associated with asphaltic-base crudes of high asphaltene content.

9.2 FRACTIONATION

The first step in processing is to separate the crude oil by fractional distillation into cuts characterized by their boiling point range. These cuts may vary considerably from one refiner to another, depending upon the subsequent use to which they are put; they may be "wide" or "narrow," and they may overlap one another somewhat. Table 9.1 lists those that are typically obtained, in order of decreasing volatility. A range limit may be described by a carbon number, instead of a boiling point, for the more volatile fractions. The last column describes typical uses for the fraction.

Different crude oils vary greatly in the boiling-point distribution of their components. A light crude from Algeria, Libya, or Nigeria may contain as much as 30 percent in the boiling range up to about 200°C, whereas a heavy crude, as found in Venezuela,

Table 9.1 Representative fractions from distillation of petroleum

Fraction	Component range and/or boiling point range, °C (°F)	Use
1. Gas	Up to C_4	Burned as fuel. Ethane may be thermally cracked to produce ethylene. Propane or a mixture of propane and butane may be sold as liquified petroleum gas (LPG).
2. Straight-run gasoline	C_4–C_5	Blended into gasoline.
3. Virgin naphtha (light distillate)	C_5–150 (300)	Used as a feed to catalytic reformer or blended into gasoline.
4. Heavy naphtha (kerosene)	120–200 (250–400) (Up to ~C_{15})	Jet fuel, kerosene.
5. Light gas oil	200–310 (400–600) (Up to ~C_{20})	Used as No. 2 distillate fuel oil, or blending stock for jet fuel and/or diesel fuel.
6. Gas oil (heavy distillate)	Up to ~350 (650) (~C_{25})	Used as a feed to catalytic cracker or sold as heavy fuel oil.
7. Atmospheric residual	~350+ (650+)	Various uses. May be distilled under vacuum to produce vacuum-gas oil, coked, or burned as fuel.
8. Vacuum residual	~560+ (1050+) equivalent boiling point	

may have hardly any material boiling below 200°C but may contain 35 to 50 percent with an equivalent boiling point above 560°C (1050°F).

The highest temperature to which the liquid can be subjected at atmospheric pressure without appreciable decomposition is about 430°C (800°F). However the least volatile distilled fraction typically has an end boiling point of about 350°C (650°F), in order to avoid excessive viscosity of the product at ambient temperatures (high *pour point*) and to achieve other desirable properties such as suitable combustion characteristics. The residue can be further separated by vacuum distillation into one or more *heavy gas oils*. A final nondistillable fraction is left having an equivalent boiling point at a pressure of 1 atm of about 560°C+ (1050°F), so-called *vacuum residual*. The heavy gas oils may be sent to a catalytic cracker or hydrocracker, or if they are of appropriate composition, they may be further processed into lubricating oils or waxes.

The relative demand for various petroleum products varies considerably in different parts of the world. Historically, in the United States the refining of petroleum has centered on the manufacture of gasoline for motor cars. This has accounted for nearly 50 percent of the crude oil processed.[2] This is in part because of the past emphasis on private car transportation and the abundant supply of natural gas that provided a considerable portion of the overall fuel demands. In contrast, in Western Europe and Japan emphasis has historically been more on production of fuel oils, and gasoline accounts for only about 15 to 20 percent of the crude oil processed. Consequently, large United States refineries have tended to include a substantial variety of chemical processes that are integrated with one another, and directed at production of gasoline, whereas a large refinery in another part of the world may do little more than distill crude oil into fractions.

9.3 GASOLINE

The two principle requirements for gasoline to be used as fuel in an internal-combustion engine are:

1. The fuel must have a suitable volatility range. Some portion must vaporize readily in the carburetor and manifold of the engine when cold so that the engine may be started, yet the mixture must not be so volatile that it causes *vapor lock* when the engine is warm. This is the formation of bubbles of gasoline vapor in critical locations in the fuel system, such as the fuel pump or carburetor, which interferes with engine performance. The heaviest components must be vaporizable, and the fuel as a whole must be sufficiently nonviscous that it will flow freely in cold weather. The composition to provide this is modified somewhat in different climates and from summer to winter to provide approximately the same volatility under these different conditions. At a given temperature, volatility is determined primarily by molecular weight. The suitable hydrocarbons are primarily those in the C_5 to C_8 range, to which some C_4 may be added to achieve additional volatility.

[2] As of 1980 gasoline was a somewhat smaller fraction of the total petroleum products consumed in the United States since substantial quantities of fuel oils, but relatively little gasoline, were imported.

2. The fuel must have an octane number equal to or exceeding some minimum value. The higher the compression ratio of an internal-combustion engine, the higher the octane number required of the fuel, but the more efficiently the engine can be operated.

The octane number in essence is a quantitative but imprecise measure of the maximum compression ratio at which a particular fuel can be utilized in an internal-combustion engine without some of the fuel-air mixture undergoing premature self-ignition. The fuel-air mixture in the cylinder is ignited by a spark, but the advancing flame front compresses the remaining unburned mixture until it may reach the point of self-ignition before being ignited by the flame. This causes an excessive rate of pressure increase, termed *knocking*, which reduces the engine power. Under severe conditions it can cause engine damage. The octane number of a fuel is obtained by comparing its knocking characteristics with various blends of isooctane (2, 2, 4-trimethylpentane), arbitrarily assigned a value of 100, and *n*-heptane, assigned a value of 0. The blends of those two compounds are termed the *primary reference fuels.* A fuel which matches the knocking characteristics of a mixture of 90 parts isooctane and 10 parts *n*-heptane by volume is assigned an octane number of 90.

The octane number is measured in a standardized single-cylinder variable compression ratio engine in which conditions of incipient knock can be determined. The engine can be operated in either of two ways. One simulates mild operating conditions and is termed the *research method* or *F-1 method*; the other simulates severe operating conditions such as high-speed, high-load conditions and is termed the *motor method* or *F-2 method.* The research octane number (RON) is always higher than the motor octane number (MON) (for octane numbers of 100 and below), the difference between the two being termed the *sensitivity.* The research octane number is that most frequently cited in the literature.

Operation of the actual engine on the road is, of course, the ultimate test, and this is also affected by variables such as engine design, spark timing, age, and carbon deposits in the cylinders. *Road octane numbers* (RdON) may be obtained by operation of a car on the road or on a chassis dynanometer, although these data are more expensive to obtain. Empirical relationships may thus be developed between RdON on the one hand and RON and MON on the other for each significant variety of automobile, and these formulas are used by the refiner as a guide in the blending of gasoline.

Mixtures of hydrocarbons do not behave by the additive rule defined by isooctane and *n*-heptane mixtures. Some compounds such as toluene and xylene contribute effectively to the antiknock characteristics of a paraffinic-base or other base stock and are assigned a high *blending octane number*, whereas a compound such as benzene is less effective and is assigned a much lower blending number. Table 9.2 gives the research octane numbers of some representative pure hydrocarbons and, for comparison, the blending octane numbers. The blending number is a function of the base stock into which the compound is blended and the concentration of a specified compound in the final mixture.

Although the use of octane numbers in practice is a complex but highly developed

Table 9.2 Octane numbers of selected compounds (research method)

Compound	Octane number		Boiling point, °C
	Actual	Blending*	
Paraffins			
Methane	>120		-161
Ethane	118		-88
n-Propane	112		-42
n-Butane	93	113	0
n-Pentane	62	62	36
n-Hexane	25	19	69
n-Heptane	0	0	98
2,2,4-isooctane	100	100	
Other trimethyl pentanes (2,2,3-; 2,3,3-; 2,3,4-)	103–106		99–119 (Other trimethyl pentanes, Dimethyl hexanes, Methyl heptanes)
Dimethyl hexanes	55–75		
Methyl heptanes	22–33		
Naphthenes			
Cyclopentane	101	141	49
Methylcyclopentane	91	107	72
Cyclohexane	83	110	81
Higher alkyl cyclopentanes and alkyl cyclohexanes	80–70 and less		88–132 and up
Olefins compared to paraffins			
2-Methyl-1-hexene	91		91
2-Methylhexane	42		90
3-Methyl-2-pentene	97		68
3-Methylpentane	74		63
Aromatics			
Benzene	>100	99	80
Toluene	120	124	111
m-Xylene	117	145	139
Alcohols			
Methanol	106 (92)†	134–138§	65
Ethanol	99	128–135§	78.5
2-Propanol	90		82
Methyl *tert*-butyl ether (MTBE)	117 (101)†	118‡	55
Tert-amyl methyl ether (TAME)		112‡	

*Based on 20 vol % of the compound in 80 vol % of a 60:40 mixture of isooctane and *n*-heptane, using a linear scale.

†Motor octane number.

‡Based on 5–20 vol % of the compound blended with unleaded regular gasoline of a base RON of 89.

§Based on 10 vol % of the compound blended with an unleaded gasoline of a base RON of 90 to 93.

procedure, some general trends are clearly shown by Table 9.2. For a given molecular weight, branched paraffins have much higher octane numbers than linear (normal) paraffins, and the highest octane numbers are obtained when the side groups are bunched together in the center of the chain (2, 2, 3-trimethylpentane has the highest octane number of the octanes). The octane number of paraffins increases markedly with decreasing carbon number. An olefin generally has a higher octane number than the corresponding paraffin, and the octane number is greater if the double bond is in the center portion of the molecule. Diolefins tend to form gums on standing and are not desired.

Aromatic compounds have among the highest octane numbers, but they are also much in demand as feedstocks for chemical manufacture. They are produced primarily by catalytic reforming in which a major reaction is dehydrogenation of naphthenes. The product from a reformer varies with *severity* (Sec. 9.5) and with the nature of the feedstock. An operation to form a product with a research octane number of 95 typically yields about 60 percent aromatics in the product fraction consisting of C_5 and higher hydrocarbons. A typical aromatic composition is: benzene, 10%; toluene, 47%; xylenes, 32%; remainder (e.g., ethylbenzene), 11%. There is much greater demand for benzene than for toluene as a chemical feedstock, hence considerable benzene is manufactured by hydrodealkylation of toluene (Sec. 9.12).

Toluene is prized for its combination of high octane number and suitable volatility. In effect benzene by itself does not knock, but it does not blend nearly as effectively with other gasoline components as do most other compounds. Since it may also be a health hazard, its concentration is typically limited to not more than about 2 percent in United States gasolines, but considerably higher concentrations may be present in some gasolines elsewhere.

Xylenes and ethylbenzene have considerably higher boiling points than C_8 paraffins and olefins. Although their octane numbers are high, they are not sufficiently volatile to be allowed to make up a large fraction of the total. The total aromatic content of gasoline is determined in part by an economic balance among various sources of octane number blending stocks, but high concentrations may lead to smoke in the exhaust and carbon buildup in the engine. Most of the fuel mixture must also be more volatile than toluene, and it is desirable to have high-octane compounds distributed throughout the volatility range. The maximum aromatic content of gasoline is typically about 35 percent.

Some compounds and mixtures have octane numbers greater than 100, as calculated by an extrapolation procedure, and in these cases the F-2 number may exceed F-1. Various alcohols and ethers also have high octane numbers as such or high blending octane numbers. Methanol and ethanol are good blending components, but trace amounts of water in blends with gasoline can cause phase separation. Ethanol is of interest in that it can be readily produced from biomass, and it has been added to gasoline at various times during the past half-century or more in various countries, in concentrations of up to about 10 percent. Ethers of suitable boiling point can be manufactured from olefins and methanol. With ethers, phase separation with traces of water is less of a problem, and they are effective additives for increasing octane number. Methyl *tert*-butyl ether (MTBE), for example, can be manufactured from isobutylene and methanol utilizing an ion-exchange resin catalyst (Pecci and Floris, 1977). Its use provides a method of upgrading C_4 streams, as from steam cracking plants, which otherwise could be utilized only as plant fuel. As of 1980 two commercial plants, in Italy and Germany, were making MTBE as a gasoline blending component and several plants were being completed in the United States.

The octane number of gasoline may be increased by addition of alkyl lead compounds. A chemical or physical blend of tetraethyllead and tetramethyllead is usually used to achieve a range of volatility. It is also necessary to add ethylene dichloride and ethylene dibromide to avoid the undesired buildup of nonvolatile lead deposits in the engine cylinders and cylinder head. Instead, with these compounds volatile lead oxyhalides are formed and are emitted in the exhaust.

Until about 1970 almost all gasoline in the United States contained up to about 3 g of lead per gallon (0.8 g per liter). Beginning with the 1975-model automobile manufacture in the United States, most cars have been equipped with catalyst units to reduce automobile pollution by oxidizing most of the CO and hydrocarbons in the exhaust gas to CO_2 and H_2O. In spite of extensive study no catalyst system has yet been developed having acceptable activity and durability with gasoline containing the lead alkyls and accompanying organohalogen compounds. Catalysts are gradually inactivated by lead compounds, and alumina, the usual catalyst support, is gradually attacked by halogen acids formed from the ethylene dichloride and ethylene dibromide. Cars equipped with catalysts of present design must use gasoline in which the requisite octane number is obtained without the addition of lead compounds. Concern with the health hazards of emitting lead compounds into the environment has also led to a program in the United States of gradually reducing the maximum allowable concentration of lead in gasoline in general.

For a short time very small amounts of methyl cyclopentadienyl manganese tricarbonyl (MMT), of the order of $\frac{1}{16}$ to $\frac{1}{32}$ g Mn per gallon, were added to much of the unleaded gasoline manufactured in the United States to improve its octane rating. However the long-term effects of this on catalytic converters and on the environment are unknown, and as of 1980 use of the MMT additive in gasoline in the United States was on a regular basis discontinued.

Gasoline is blended from streams from several sources in the refinery. The principal ones are the portion of the product of suitable volatility from five sources:

1. Catalytic cracking
2. Catalytic reforming
3. Hydrocracking
4. Alkylation
5. Straight-run gasoline

Light isoparaffins from various refinery systems and *n*-butane may also be added. *Alkylation* in the fuels industry refers to the reaction of C_3 to C_5 olefins with isobutane to form C_7 to C_9 isoparaffins, catalyzed by sulfuric acid or nearly anhydrous hydrogen fluoride. A large excess of isobutane is used to minimize polymerization of the olefins with themselves. Two liquid phases are present. The reaction can also be catalyzed by acidic zeolites, but no commercial process based on zeolites or other solid acids appears to have been put into practice.

Very few refineries operate all four catalytic processes above, so the extent to which an individual refiner draws upon various sources for gasoline components will vary substantially from case to case and will depend upon such factors as the fraction of the crude oil processed to be converted to gasoline, the scale of operations, the demand for aromatics for sale as chemicals, and the outside supply of high-octane components.

9.4 CATALYTIC CRACKING

The demand for gasoline relative to other liquid fuels in the United States has exceeded the fraction of crude petroleum having a suitable range of volatility ever since the advent of the mass-produced automobile. It has long been known that large molecules could be thermally (noncatalytically) decomposed to form smaller ones. With the growth in demand for gasoline just prior to World War I, there appeared commercial processes utilizing thermal cracking to increase the fraction of petroleum that could be utilized for this product.

Clays, especially after various treatments, were also long known to be active for hydrocarbon reactions. It remained for Eugene Houdry to recognize the improved gasoline quality obtainable from decomposition reactions utilizing clay catalysts and to develop a commercial process, first put in operation in 1936. This utilized an acid-treated montmorillonite in a fixed-bed reactor operated cyclically. After about 10-min contact time the bed was purged with steam and carbonaceous deposits were burned off, thus restoring the catalyst activity. The commercialization of catalytic cracking was further spurred shortly before World War II by the fact that the product had an octane number greater than that of thermal cracking, and a large demand developed for high-octane gasoline for propeller-driven aircraft.

Acid-treated montmorillonite catalysts are sensitive to high-temperature regeneration. Iron, present in the crystal lattice, becomes active in the presence of sulfur compounds. It is then oxidized during regeneration and catalyzes the undesired formation of coke and hydrogen. Other acid-treated clays were also used, but these were superseded by synthetic silica-alumina, a homogeneous xerogel containing 10 to 20% Al_2O_3 and free of iron. Various semisynthetic materials have also been used. In the 1960s these in turn were generally replaced by compositions containing several percent of zeolites (crystalline silica-aluminas; Sec. 7.7) dispersed in an amorphous silica-alumina matrix. These further increased the yield of gasoline from a specified amount of crude oil. Other acidic compositions such as silica-magnesia also cause cracking, but none has been used commercially to a significant extent.

The activity of virgin cracking catalysts drops rapidly during the first few seconds of contact time and then more slowly thereafter. For industrial purposes reaction measurements are generally made on *equilibrium catalyst*, which is material that has been subjected to a sufficient number of cycles of reaction and regeneration that steady-state properties have been achieved.

Catalytic cracking is the largest-volume catalytic process practiced industrially. In the United States about 35 percent of the crude oil processed is catalytically cracked, but since a considerable amount of product is recycled, the capacity of catalytic cracking units in absolute terms is higher. The early development of cracking catalysts and their properties and characterization are described by Ries (1952), by Ryland et al. (1960), and by Oblad et al. (1951).

Carbonaceous deposits (coke) are rapidly formed on the catalyst, and the reactor configuration must provide a means whereby a catalyst can be regenerated by burning off the coke with air after a short contact time. In the first reactor configuration, a fixed bed was subjected to alternate cycles of cracking and regeneration. This was superseded by a moving-bed process in which reaction and regeneration occurred in different portions of the vessel. A fluidized-bed process was developed in the United States just prior to World War II and became the dominating process in the 1950s and 1960s. Finely divided catalyst was rapidly recirculated back and forth between the reactor and a regenerator in which deposited coke was removed by combustion with air. The heat required for the endothermic cracking reactions is supplied by the sensible heat of the catalyst from the regenerator, which is mixed with fresh feedstock. Close coupling between the two units of the process is required.

With the advent of zeolite fluid cracking catalysts, reactor configurations have been modified to exploit most effectively the properties of these newer materials. Optimum operation corresponds to higher temperatures and shorter contact times, so the fluid-bed reactor has been modified or replaced by a *riser cracker*. The feed is mixed with and vaporized by contact with hot, regenerated catalyst and passes upwards concurrently with the catalyst in a vertical pipe in essentially plug flow and at velocities that are an order of magnitude greater than that in conventional dense fluidized beds. Contact times of the order to 2 to 4 s are thus achieved. Regeneration continues to be carried out in a fluid bed. A number of modified and new reactor configurations to achieve riser cracking are described by Magee and Blazek (1976).

It is desired to obtain as high a ratio of CO_2 to CO as possible in the gases leaving the regenerator in order to maximize heat utilization in the form of hot catalyst (the heat of combustion of carbon at 800 K to form CO_2 is 395 kJ/mol but is only 123 kJ/mol to form CO). It is also desired to minimize afterburning of CO in the dilute phase above the regenerator fluid bed, which may damage the catalyst, and CO discharged to the atmosphere must also be limited to meet air pollution control standards. A high ratio of CO_2 to CO may be achieved by operation of the regenerator at high temperatures, but this may shorten catalyst life and special materials of construction may be needed.

Some years ago small amounts of chromia were sometimes added to a silica-alumina catalyst to promote CO oxidation in the regenerator, but this lowered catalyst selectivity in the cracking reactor. A recent development, put into widespread commercial practice, has been to incorporate minute amounts of platinum (of the order of 1 to 10 ppm) into conventional cracking catalyst (Chester et al., 1978; Upson, 1978; Schwartz, 1978; Hemler and Stine, 1979). This concentration is sufficiently small that the reactions in the cracking reactor are not noticeably altered, but sufficiently great to significantly increase the combustion of CO in the regenerator. The platinum may also be furnished in the form of a separate solid additive which allows the platinum level in the circulating system to be increased rapidly when desired, so as to adjust to variations in feed composition to the cracking reactor. By designing the solid to have an attrition rate greater than that of the cracking catalyst it is also possible to cause it to be preferentially removed from the system, so the platinum level could also be rapidly decreased if desired. Although platinum is expensive and it is lost with the catalyst fines, the amounts here are so small that this application is economical.

Use of either a high temperature regenerator operation or a special catalyst system to obtain a high CO_2/CO ratio also makes possible adjustment of the catalyst/oil ratio and lowering of residual carbon on regenerated catalyst so as to improve performance in the cracking reactor (Magee, 1977).

Reaction temperatures were about 460°C with silica-alumina in fluidized beds and are in the neighborhood of 520 to 540°C with the zeolite catalysts. The research octane number is increased by 1.5 to 3.0 by the use of riser cracking, and higher conversions are obtainable without excessive secondary cracking. There has been a gradual trend towards higher pressure in the regenerator, to the neighborhood of 0.3 MPa. This improves regeneration kinetics and cyclone efficiency. The reactor must then also be operated at the higher pressure, although this may affect product distribution adversely.

Thermal cracking for the production of gasoline is no longer practiced, but thermal cracking of a variety of feedstocks in the presence of steam (steam cracking) is utilized to produce ethylene, propylene, etc., for chemical manufacture. The viscosity of the residue from distillation may be decreased by a mild thermal cracking termed *visbreaking*, or the residue may be converted to petroleum coke and lighter products by a severe thermal cracking operation.

The feed to the catalytic cracking reactor may be any distilled fraction, atmospheric- or vacuum-distilled, that is to be reduced in molecular weight. Usually it is a fraction with an initial boiling point above about 200°C since more volatile material can be processed into gasoline. The feed thus may contain a wide range of types of hydrocarbons and of molecular weights. If considerable organosulfur is present, this may be lowered by catalytic desulfurization before being fed to the catalytic cracker. The reactions that occur are complex, and the products formed are considerably different than those from thermal cracking. An extensive literature exists on catalysts and reactions and has been reviewed by Haensel (1951), Voge (1958), and Ryland et al. (1960). All cracking catalysts are highly acidic, and the mechanisms of these various reactions involve the formation of carbonium-ion intermediates, discussed in more detail in Chap. 7.

About one-half of the product from catalytic cracking is of the volatility range suitable for gasoline. The hydrocarbon composition is, of course, quite different than that of uncracked material (straight run) of the same boiling-point range. The sulfur and nitrogen content is usually higher since the concentration of these impurities increases with molecular weight of the crude fraction. The product from the catalytic cracker is typically separated by distillation into streams that may be further processed as follows:

1. *C_4 and lighter fraction.* This may be fed to an alkylation unit.

2. *Light catalytic naphtha*, bp 18 to 93°C (65 to 200°F). This goes to the gasoline pool.
3. *Intermediate catalytic naphtha*, bp 93 to 165°C (200 to 330°F). It may be necessary to give this fraction a light hydrogen treatment to lower the diolefin content. (The latter tends to form gums on storage.) In so doing it is important to avoid hydrogenation of monoolefins, which would lower the octane number. The octane number of this fraction is lower than that of the light catalytic naphtha, and it may be fed to a catalytic reformer to improve the octane rating. (The light and intermediate naphtha may also be treated as a single stream.)
4. *Heavy catalytic naphtha*, bp 150 to 235°C (300 to 450°F). The aromatics may be extracted from this fraction; the paraffins may then be used for jet fuel. Otherwise, this fraction may be fed to a hydrocracker, or it may be blended into heating oil.
5. *Fraction above 235°C* (450°F). This is highly aromatic. It may be hydrodesulfurized and used as fuel, or recycled to the catalytic cracker (sometimes it is termed *cycle oil*). Aromatic rings do not crack appreciably and must be presaturated (hydrogenated) to cause significant ring breakage.

9.5 CATALYTIC REFORMING

The principal objective of catalytic reforming is to process a hydrocarbon fraction having a volatility range suitable for use as gasoline so as to increase its octane number, but without significantly changing the molecular weight. Catalytic reforming is also the principal source of aromatic chemicals (benzene, toluene, and xylenes). It is utilized on a major scale in petroleum refining; the quantity of feedstock so processed in the United States is nearly one-quarter of the total crude oil processed. A wide variety of reactions are involved, consisting primarily of dehydrogenation and isomerization reactions, plus some hydrocracking. The rates decrease in the approximate order shown in Table 9.3, which lists for each type of reaction a specific example for a C_6 hydrocarbon. The equilibrium constant K (partial pressures in atmospheres) and heat of reaction ΔH are for 500°C (Edmonds in Tedder et al., 1975, p. 58). The hydrogen produced by reforming is an important byproduct since it is the source of hydrogen used for hydrodesulfurization and hydrotreating in other portions of a refinery. The catalyst consists of platinum or platinum plus another metal, such as rhenium, supported on acidified alumina.

The first three reactions listed in Table 9.3 occur very rapidly, essentially to equilibrium. The olefins concentration in the final product is low since they are thermodynamically unfavored, but olefin formation is an intermediate step in paraffin isomerization (see below). The equilibrium of reaction (3) is likewise unfavorable, but the reaction is driven to the right by the rapid dehydrogenation of the cycloparaffin formed by isomerization. The combined reaction is termed *dehydroisomerization.* The dehydroisomerization of alkyl cyclopentanes is an important reaction since the ratio of their concentration to that of alkyl cyclohexanes varies between about 0.5 to 1.5 in representative straight-run naphthas. The isomerization of n-C_6H_{14} to the isoparaffins does not normally proceed to the equilibrium distribution of dimethyl butanes. The *dehydrocyclization* of an n-paraffin may also occur, but the rate is relatively

Table 9.3 Representative catalytic reforming reactions

Reaction	K	ΔH, kJ/mol
1. Dehydrogenation of naphthenes to aromatics	6×10^5	+221
(cyclohexane) $\longrightarrow$ (benzene) $+ 3H_2$		
2. Dehydrogenation of paraffins to olefins	$\sim 4 \times 10^{-2}$	~+130
$C_6H_{14} \longrightarrow C_6H_{12} + H_2$		
3. Isomerization of alkyl cyclopentanes	8.6×10^{-2}	-15.9
(methylcyclopentane) $\longrightarrow$ (cyclohexane)		
4. Isomerization of n-paraffins to isoparaffins	~1	~5
$n\text{-}C_6H_{14} \longrightarrow \text{iso-}C_6H_{14}$		
5. Dehydrocyclization of paraffins to aromatics		
$n\text{-}C_6H_{14} \longrightarrow C_6H_6 + 4H_2$		
6. Hydrocracking of paraffins		
$C_6H_{14} + H_2 \longrightarrow$ olefins and paraffins		

low, increasing with molecular weight of the paraffin. The mechanism from n-hexane is believed to involve the intermediate formation of methylcyclopentane and methylcyclopentene.

The catalyst is the classic example of *dual functionality* in that in order for isomerization, dehydrocyclization, or hydrocracking to occur it is necessary that both metallic sites (e.g., platinum) and acidic sites be present on the catalyst surface. The isomerization of, e.g., an n-paraffin to an isoparaffin does not proceed directly, but via the dehydrogenation to an n-olefin. This is isomerized to an isoolefin, which is then hydrogenated to the corresponding isoparaffin. The hydrogenation-dehydrogenation steps proceed on the platinum sites, the isomerization on acidic sites.

This concept of a dual-function catalyst with two distinctly different kinds of

Table 9.4 Hexane isomerization on mechanical catalyst mixtures*

Catalyst charge to reactor	Wt % conversion to isohexanes
10 cm^3 of platinum/silica	0.9
10 cm^3 of silica-alumina	0.3
Mixture of 10 cm^3 of each of above	6.8

*Weisz (1962). (Reprinted with permission from *Advances in Catalysis*. Copyright by Academic Press.)

sites was introduced by Mills et al. (1953) by work with catalysts containing only an acidic function, only a dehydrogenation function, or both. Later Weisz (1962) showed by studies with mechanical mixtures that the intermediates are true gas-phase species that are transported by gaseous diffusion between the different sites. For example, under comparable conditions little isomerization of hexane occurs on either a platinum/silica catalyst alone or silica-alumina catalyst (acidic) alone, but a mechanical mixture is quite active (Table 9.4). These data were obtained at 373°C with a 5 : 1 molar ratio of hydrogen to n-hexane. Studies with n-heptane isomerization show that with platinum/silica or platinum/carbon little reaction occurred (Fig. 9.1), but a mechanical mixture of either of these with silica-alumina gave much more reaction. The three upper curves show that the smaller the component particle sizes used, the more the isomerization rate increased. The finest particle mixture gave results comparable to platinum impregnated on silica-alumina. Weisz and others (see, for example, Thomas and Thomas, 1967, p. 338) have calculated so-called intimacy requirements concerning the maximum distance between sites that can be allowed without diffusion

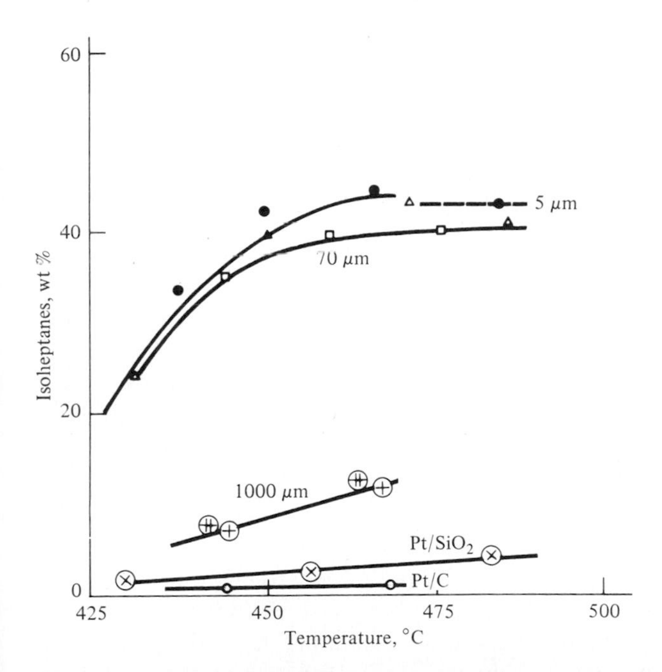

Figure 9.1 Isomerization of n-heptane, conversion versus temperature.

Pt/C	1000 μm	○
Pt/SiO_2	1000 μm	⊗
$Pt/SiO_2 + SiO_2 \cdot Al_2O_3$	1000 μm	⊞
$Pt/C + SiO_2 \cdot Al_2O_3$	1000 μm	⊕
Supported Pt + $SiO_2 \cdot Al_2O_3$	70 μm	□
Supported Pt + $SiO_2 \cdot Al_2O_3$	5 μm	△
Pt impregnated on $SiO_2 \cdot Al_2O_3$		●

(Weisz, 1962.) (Reprinted with permission from Advances in Catalysis. Copyright by Academic Press.)

Table 9.5 Hydrocracking of high-molecular-weight paraffins on mechanical catalyst mixtures*

Catalyst	Conversion, %	
	n-$C_{12}H_{26}$ (to 175°C EP†)	n-$C_{16}H_{34}$ (to 275°C EP†)
Silica-alumina	2.7	2.1
Platinum/carbon	4.0	2.5
Platinum/carbon + silica-alumina	13.2	36.5

*Weisz (1962). (Reprinted with permission from *Advances in Catalysis.* Copyright by Academic Press.)

†EP = distillation endpoint to define products; particle size = 0.8–1.4 mm.

becoming a significant rate-limiting step. This is a function of other variables such as rates of individual reactions and thermodynamic equilibria.

Hydrocracking[3] likewise requires a dual-function catalyst, as shown by the data in Table 9.5 for n-$C_{12}H_{26}$ or n-$C_{16}H_{34}$ obtained in similar fashion at 370°C and 75 kPa of hydrogen. The product distribution starting with 1-dodecene was essentially the same as that from dodecane, demonstrating that the olefin must be an intermediate. For some of the other reactions occurring in reforming, such as dehydrocyclization of an n-paraffin such as n-hexane, mechanical mixtures are not very effective (Sinfelt, 1964). Presumably some key intermediate, perhaps a diolefin, is formed that is not easily desorbed. In that event the intermediate would have to migrate by surface diffusion to a nearby acid site or interact with a nearby acidic site.

Comparisons of a mechanical mixture of different monofunctional catalysts with a single multifunctional catalyst must be made carefully to ensure that the results are truly significant, especially if the kinetics are complicated. Even if the same total loadings of active components are used, differences in contact time and concentration profiles between the two sets of experiments may contribute their own subtle effects. Figure 9.2 generalizes the concept of dual functionality to a larger variety of reactions of importance in catalytic reforming (Mills et al., 1953), taking the C_6 hydrocarbons as an example. A number of the reactions observed involve several steps in sequence. In view of the distribution of products obtained in reforming reactions over a molybdena catalyst, this may also be termed bifunctional, in which both functions are shown by a single compound. This can be rationalized by attributing the acidic property to the metal oxide and the hydrogenation-dehydrogenation function to the metal ions. Multifunctional catalysts are reviewed by Weisz (1962) and by Sinfelt (1964).

In addition to its role in providing sites for hydrogenation-dehydrogenation reactions to occur, platinum also acts to minimize the rate of formation of carbonaceous deposits. Hydrogen chemisorbs onto the platinum with dissociation and appears to diffuse along the surface of the catalyst, perhaps along carbonaceous deposits. (This phenomenon of adsorption followed by surface diffusion is sometimes termed *spill-*

[3]Selective hydrocracking in which the molecule fissions in the center, requires an acidic site. Hydrocracking at the end C—C bond to form CH_4 will proceed on platinum by itself (see Chap. 6).

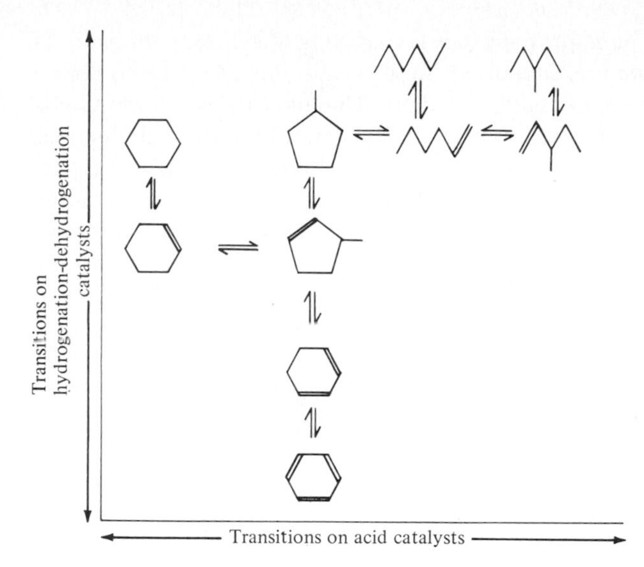

Figure 9.2 Sequence of reactions on a dual-function catalyst. C_6 hydrocarbons used as an example. (*Mills et al., 1953.*) (*Reprinted with permission from Industrial and Engineering Chemistry. Copyright by the American Chemical Society.*)

over.) Coke precursors are hydrogenated, which improves their desorption from the catalyst. The rate of coke formation is a function of hydrogen pressure. The minimum concentration of platinum that can be practicably used on the catalyst is set more by that needed to keep the catalyst "clean" rather than the amounts needed to provide an adequate number of metallic sites. This amount in turn is determined in part by the nature of the feedstock and operating conditions.

For straight-run naphtha from distillation a practicable minimum platinum content is in the neighborhood of 0.2 to 0.3 wt %, depending in part on the frequency of regeneration for which the reactors and process are designed. There is increasing interest in feeding cracked stocks (e.g., naphtha from a catalytic cracker or from steam cracking) to a reformer, but the presence of olefins and especially diolefins causes a much more rapid increase in the rate of coking. This can be overcome to some extent by use of a higher platinum content. Under industrial operating conditions and with industrial catalysts, the hydrogenation-dehydrogenation reactions appear to be essentially in equilibrium, and the sites are so close to one another that diffusion is not rate-limiting. The rate-limiting processes are the isomerization reactions on the acid sites. Some isomerization occurs on metallic sites, but this is not a major contribution under industrial reaction conditions.

Optimum operating conditions are determined by an economic balance among several factors. The equilibrium for dehydrogenation reactions is favored by high temperatures and low hydrogen partial pressure, but this increases the rate at which the catalyst becomes inactivated by coke deposits. Higher platinum loadings decrease the rate of inactivation, but the catalyst is then more costly.

At least three reactor configurations have been used. In each case in effect multiple adiabatic reactors are used in series, the product being reheated between reactors. The naphthene dehydrogenation is so rapid and so endothermic that the temperature drops rapidly in the first bed to the point that the rate is uneconomically slow.

Beginning in about 1967 an improved catalyst consisting of a mixture of platinum and rhenium, a *bimetallic catalyst*, supported on an acidic alumina came to replace much of the commercial use of the earlier platinum/alumina catalyst. A mixture of platinum and iridium is also used to some extent. These newer catalysts maintain their activity at lower hydrogen pressures over long periods of time. The lower hydrogen pressure also allows an increased degree of dehydrocyclization of paraffins to aromatics to occur and generally a higher yield of C_5+ product and of hydrogen. However they are much more readily poisoned by sulfur compounds. The reasons for the effectiveness of these bimetallic catalysts are not yet clear. Some of the rhenium may not be reduced to the metal and may modify the support. When platinum is alloyed with another metal, geometrical considerations, i.e., the particular grouping of atoms at the site of reaction, may be important (Sec. 6.3). Hydrogenation-dehydrogenation reactions are relatively insensitive to geometry, but hydrogenolysis and hydrocyclization require more complex types of sites.

The bimetallic catalysts are typically 3 to 4 times more active than the all-platinum catalyst they replace, and the refiner can use this enhanced activity in existing equipment in any one or a combination of several ways. The refiner can:

1. Decrease the catalyst charge
2. Maintain the same catalyst charge but increase the aromatic concentration in the reformate product by operating under more severe conditions, keeping the same schedule of regeneration
3. Maintain the same catalyst charge but increase the cycle time between regenerations
4. Maintain the same catalyst charge but decrease the pressure and thus improve gasoline yield

Operation at lower pressure increases the degree of isomerization and dehydrocyclization and reduces hydrocracking, all of which increase the octane number of the reformate. Lower-pressure processes using bimetallic catalysts reportedly accounted for about 40 percent of United States reforming capacity as of 1974.

The earlier conventional reforming process using a Pt/Al_2O_3 catalyst consisted of three (occasionally four) fixed-bed reactors with furnace reheaters between the units. In a representative four-reactor system the temperature drop along each reactor would be about 55, 40, 20, and 5°C respectively. In this configuration the catalyst is distributed nonuniformly, the first reactor containing the least amount and the last reactor the most. Approximately every 6 months or so the entire group of units is shut down, carbonaceous and sulfur compounds accumulated on the catalyst burned off, and then the entire unit put back on stream. These reformers are termed *semiregenerative*.

More recently the trend has been to *cyclic* reactors in which four to six fixed-bed units are operated in series, one being shut down for regeneration at all times. In this case a temperature drop of about 25 to 30°C will be taken through each reactor, and the catalyst is distributed uniformly among all reactors. One or two reactors may be regenerated each day, and since carbonaceous deposits are removed quite frequently, a higher rate of formation is allowable. These reactors can be operated satisfactorily at lower pressures and lower recycle gas ratios than the semiregenerative units, which allows an increased yield to form gasoline. The cyclic units also provide a higher degree of flexibility to accommodate varying feedstocks, changing octane-number requirements, etc. They are especially favored where a high-octane product is required and for operation with a variety of feedstocks.

A third variation in equipment consists of a continuous moving-bed design in which in effect three units are stacked on top of one another. Addition and withdrawal of catalyst is continuous, although the average residence time of the catalyst in such a unit is so long—of the order of 50 to 60 days—that for some conditions truly continuous regeneration is not actually required.

Representative operating conditions are about 480°C and 2 MPa with an all-platinum catalyst or about 1 MPa with a bimetallic catalyst. In the semiregenerative reformers the temperature is gradually raised as the catalyst deactivates so as to keep the octane number of the product nearly constant. An upper limit of about 520 to 550°C is set by materials of construction or furnace capacity. When this temperature is reached, the reactor is shut down and the catalyst is regenerated. Instead, the run may be terminated by excessive pressure drop, caused by coke buildup, or by decline in yield.

Operating conditions are also determined by the nature of the feedstock and the uses to which the product is to be put. With *low-severity* operation, only dehydrogenation reactions and dehydroisomerization of alkyl cyclopentanes occur. A high yield (volume basis) of material in the suitable boiling range for gasoline but of lower octane number is obtained. High severity increases the aromatics content (via, in part, dehydrocyclization reactions), but hydrocracking also increases. The yield of C_5+ products drops, but its aromatics content and octane number increases. If aromatic chemicals are to be produced, a reformer is usually run at high severity to produce a more concentrated aromatics product and thereby reduce the cost of subsequent processing (extraction, fractionation). However an increased portion of the feed becomes hydrocracked material that is too volatile to use in gasoline. Beyond operating conditions, the percentage of the feedstock converted to aromatics depends largely on the naphthenic content of the feed. Much Middle East crude is high in paraffins and produces lower yields of aromatics than can be obtained with most United States or South American crudes.

Hundreds of patents have been issued on catalyst compositions. All commercially used catalysts contain platinum, either alone or in a combination with another metal. In the monometallic forms the platinum content is typically 0.35 to 0.6 wt %. The bimetallic catalysts with rhenium are typically about 0.3 wt % platinum and 0.3 wt % rhenium. Other metal compositions also typically contain about 0.3 wt % platinum. The minimum content is set by an otherwise excessive rate of coking and is also affected by the nature of the feedstock. Diolefins, for example, cause more rapid coking, which can be counteracted by a higher platinum content. For maximum utilization of a very expensive material the platinum is spread out in as finely divided form as possible. A large fraction of the total platinum atoms present are on the surface in a commercial catalyst. The support is an alumina, in the γ or η form, which may have been acidified by the incorporation into the structure of about 1% chloride or fluoride during the catalyst preparation. The use of chloroplatinic acid for impregnation leads to the natural incorporation of chloride into the alumina during preparation. Acidity is gradually lost during reaction and may be restored by adding small amounts of an organic chloride or fluoride to the feedstock. Presumably this decomposes to release a halogen acid, which is adsorbed by the alumina.

The bimetallic catalysts are sometimes referred to as *alloy catalysts*, but the actual form under reaction conditions is unknown. The structure may also gradually change upon repeated regenerations.

All these catalysts are gradually poisoned by sulfur compounds in the feed that, being predominantly in the divalent state, adsorb on and coordinate with metal sites. A small amount of sulfur may be beneficial, and some of the adsorption may be reversible and some irreversible. The platinum monometallic catalysts may be operated satisfactorily with sulfur levels in the feedstock up to the range of 10 to 50 ppm, depending partly on the platinum loading on the catalyst and the frequency of regeneration that is practicable. If sulfur compounds are completely removed from the feedstock while reaction is continued, at least some of the sulfur, adsorbed on the catalyst, will slowly be hydrogenated to hydrogen sulfide and desorbed. The Re-Pt/Al_2O_3 catalysts are more sensitive to sulfur, and a level not exceeding the order of 1 ppm in the feedstock is typically required. This will usually necessitate prior hydrodesulfurization of the feed. Consequently a refinery not possessing such equipment may be able to use only the all-platinum catalyst. Rhenium sulfide appears to be more difficult to convert back to the metal than is platinum sulfide.

A new platinum or platinum-rhenium catalyst may cause an excessive degree of hydrocracking, which, being exothermic, may result in an excessive temperature rise. Hence a low concentration of a sulfur compound may be introduced into the feed initially to avoid this behavior. This treatment method is sometimes referred to as *tempering* or "taking the edge off" a catalyst. Avoiding excessive initial activity may reduce the subsequent rate of buildup of coke deposits and prolong catalyst life. Apparently some very active sites that cause hydrocracking, polymerization, and condensation reactions are poisoned by this procedure. It also appears that low sulfur concentrations may be beneficial under steady-state conditions with some catalysts. This may be related to an enhancement of acidity (Pfefferle, 1970) or to suppression of hydrocracking (Menon and Prased, 1977). In any event too high a sulfur concentration leads to excessive coking and rapid deactivation. Basic nitrogen compounds adsorb on acid sites and reduce isomerization and cracking activity, but they have little effect on dehydrogenation activity. Tiny amounts of arsenic or of other heavy metals can produce irreversible poisoning.

Regeneration is accomplished by carefully burning coke off the catalyst, avoiding overheating that might damage it. In the regeneration of a platinum catalyst, typically an inert gas is circulated through the catalyst bed and air is introduced to provide initially an oxygen content of about 1 percent. After the more reactive portion of the deposits has been removed, the oxygen content may be raised, for example, to 2 percent and the inlet temperature raised to 480 to 500°C, or a substantially higher oxygen content may be used. These conditions are held until all carbon is removed. During this process a variety of processes occur that may cause the platinum either to sinter or to disperse. This seems to involve the intermediate formation of oxides, such as platinum dioxide, that have appreciable volatility or surface mobility. Under commercial conditions possibly other compounds are also formed by reaction of platinum with carbon monoxide or chlorine. The final stage in regeneration is sometimes termed an *oxygen soak* in which conditions are adjusted to maximize redispersion (see Sec. 6.5).

The first commercial catalytic reforming process in the United States was introduced in 1939, stimulated by the wartime demand for high-octane gasoline for propeller-driven aircraft and for aromatic chemicals such as toluene for use in making trinitrotoluene (TNT). A molybdena/alumina catalyst was utilized in a fixed-bed reactor operated under a high pressure of hydrogen. It was soon followed by processes utilizing a similar catalyst in a moving-bed or fluidized-bed reactor. The catalyst became coked rapidly, thus requiring frequent regeneration, and these processes gradually became uneconomic during the 1950s upon the development of a catalyst consisting of platinum on an acidic alumina.

The platinum catalyst was utilized in a fixed-bed reactor and was introduced in 1949 by the Universal Oil Products Co. (UOP) based on work by Haensel and his coworkers. Hydrogen, formed in the process, was recycled through the reactor to maintain a sufficiently high hydrogen partial pressure so that coke-forming reactions could be minimized. Thus the reactor could be operated for periods of many months before regeneration was needed. This was the first use of a precious metal catalyst in the petroleum industry, and it was followed rapidly by a number of related processes utilizing platinum that differed primarily in the nature of the support.

It was recognized that the support needed to be acidic, but there are a number of subtleties about the optimum kind of acidity that were gradually revealed through intensive research. In the first UOP catalyst the alumina contained combined halogen, as fluoride or chloride. This had poor tolerance to low concentrations of water vapor as formed in regeneration, which removed halogen from the support. The halogen could be replaced and its effectiveness substantially restored by adding to the feed a small amount of an organohalogen compound that decomposed in the reactor. Better tolerance to water was achieved by using a silica-alumina base that contained no halogen. It was necessary to treat the silica-alumina to achieve a controlled degree of acidity, but silica-alumina is not now used since it causes too much hydrocracking. The next development was the use of η-alumina as a base, which has an inherently stronger acidity than the more common γ form (see Sec. 4.5.1 and Fig. 7.5). It could thus be used without the necessity for added halogen. It could also be more easily regenerated, and it provided a high ratio of paraffin cyclization relative to hydrocracking.

The development of catalytic reforming processes and understanding of the chemical reactions involved is reviewed by Ciapetta et al. (1958). Reactions on chromia, which is effective for dehydrocyclization of paraffins to aromatics, and molybdena are described in considerable detail. A later review by Ciapetta and Wallace (1971) discusses Pt/Al_2O_3 and $Pt\text{-}Re/Al_2O_3$ catalysts and the reforming reactions they catalyze. An engaging personal account of the early development of the platinum reforming catalyst and the process based on it has been published by Sterba and Haensel (1976). The great industrial importance of this process has led to much fundamental research on methods of obtaining a high degree of dispersion of platinum and other metals on alumina, and on characterization of such dispersions, including chemisorption methods for measuring the surface area of supported metal. It has also stimulated studies of the effect of particle size on reactivity and studies of highly dispersed alloy catalysts (Chap. 6).

9.6 ISOMERIZATION

9.6.1 Paraffin Isomerization

Paraffins in the C_5 and C_6 range may be isomerized in order to increase the octane number of a light blending stock for gasoline. (*n*-Pentane has a research octane number of 62 compared with a value of 92 for isopentane.) The process is closely related to catalytic reforming, but aims at achieving a closer approach to equilibrium with respect to paraffin isomerization reactions. Thus isomerization increases the quantity of blending stocks that can be used for high-octane gasoline. This is of particular importance as the use of lead additives is gradually phased out, but paraffin isomerization is used to a much lesser extent than reforming.

Under equilibrium conditions the highest concentrations of the highly branched paraffins, which have the highest octane numbers, occur at the lowest temperature (Fig. 9.3). Hence a highly active catalyst is needed. Such a catalyst is similar to that

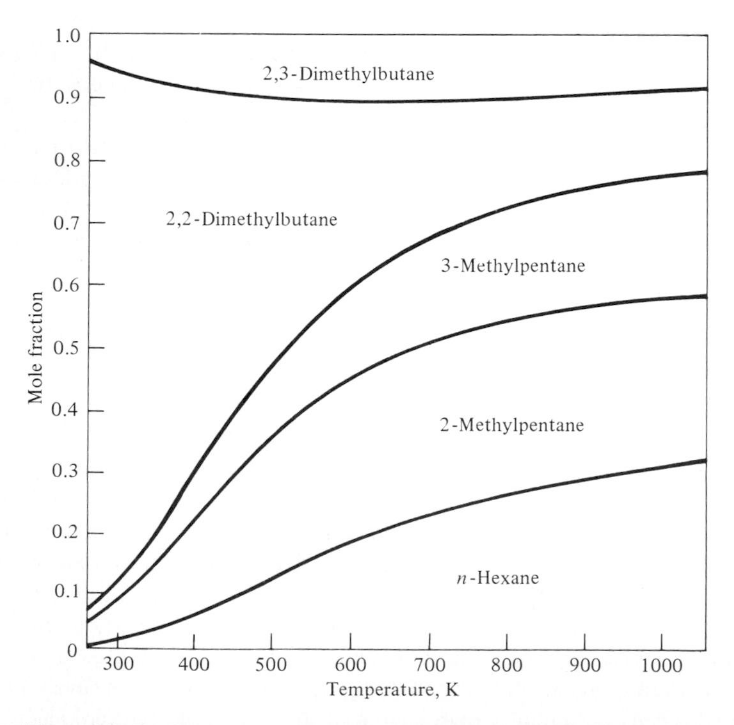

Figure 9.3 Equilibrium concentrations of five hexanes in the gaseous phase (width of band measures mole fraction of indicated isomer at equilibrium). (*Rossini et al., 1941.*)

used in reforming but usually has somewhat stronger acidity; thus platinum on fluorided alumina or platinum on silica-alumina may be used. Representative reaction conditions are 200 to 400°C (considerably lower than those in reforming) and pressures of 2 to 3.5 MPa, using a high partial pressure of hydrogen to minimize coke formation. An acidic zeolite loaded with noble metal may also be used, at somewhat lower temperatures and pressures. In the Hysomer process (Shell), introduced in 1970, the catalyst is a low-sodium hydrogen mordenite containing a noble metal (Minachev and Isakov, 1976; Bolton, 1976). The process is not used on C_7 or higher paraffins, because excessive cracking (undesired) occurs on these highly acidic catalysts. The isomerization of *n*-butane to isobutane is also carried out to produce a feed for alkylation.

During World War II the demand for high-octane aviation fuel led to a number of plants being built utilizing an aluminum chloride catalyst for isomerization. The reaction occurs at low temperatures and thus produces a high-octane product; but the process is expensive, and disposal of the sludge formed is a severe problem. These processes were superceded in the 1950s in the United States by the use of dual-function catalysts similar to those developed for reforming, although in some cases a noble metal other than platinum was used or even a nonnoble metal. Asselin et al. (1972) review catalysts, mechanisms, and processes and refer to earlier reviews for more details.

9.6.2 Xylene Isomerization

This process is used in conjunction with the separation of the C_8 aromatic fraction from catalytic reforming into the three xylene isomers and ethylbenzene. Of the xylenes, the greatest demand is for the ortho and para isomers, but typically the meta isomer is present in the greatest concentration in the product from a reformer. *o*-Xylene and ethylbenzene may be separated from each other and from the remaining xylenes by fractionation. *p*-Xylene may be separated from the mixture by a zeolite adsorption process (UOP Parex process) or by low-temperature fractional crystallization. In either case it may be desirable to react the remaining stream to cause conversion to the equilibrium composition of the three xylenes. Disproportionation (transalkylation) may also occur to form toluene, trimethylbenzene, and other products.

A number of processes are used commercially. Unlike paraffin isomerization, equilibrium is not appreciably affected by temperature. An essentially equilibrium composition can be achieved at about 400°C using platinum/silica-alumina in the presence of hydrogen to minimize coke formation (Bolton, 1976). One of the commercial processes using this type of catalyst is Octafining (Uhlig and Pfefferle, 1970) of the Engelhard Co. Representative conditions are about 430 to 480°C and 1 to 2 MPa. One of its advantages is that a high conversion of ethylbenzene to xylenes is obtained, a reaction which seems to involve the formation of ethylcyclohexane as an intermediate. Silica-alumina alone will cause isomerization of the xylenes to occur, but not conversion of ethylbenzene. In the absence of platinum and hydrogen, carbonaceous deposits build up rapidly.

A liquid-phase process using a zeolite catalyst at considerably lower temperatures has been announced by Mobil Oil. (See also Sec. 7.7.5.) The reaction conditions are about 200 to 260°C and 2 MPa. (At 230°C the equilibrium composition is 21% *o*- 55%

m- and 24% *p*-xylene.) In this process the feed is diluted with 10 to 20% toluene, which inhibits disproportionation losses. The rate of catalytic deactivation in the liquid phase is less than in the vapor phase, probably because the liquid helps act as a solvent to remove deposits. Ethylbenzene is not converted to xylenes over this catalyst, so it is removed from the feed before reaction.

9.7 HYDROCRACKING

Hydrocracking is a combination of catalytic cracking and hydrogenation under substantial pressure. The hydrogen hydrogenates coke precursors on the catalyst and thereby minimizes the rate of coke formation such that a long catalyst life can be achieved before regeneration is required. Hydrocracking processes were developed in Germany and in England in the 1930s to supply aviation gasoline for military purposes. The catalyst consisted of various metal sulfides such as those of nickel or tungsten, supported on HF-treated montmorillonite or silica-alumina. High pressures were required with these catalysts, e.g., over 20 MPa, and the processes were uneconomic under subsequent peacetime conditions.

In 1959 the first modern type of hydrocracking process was announced by the Standard Oil Co. of California, which uses a dual-function catalyst. The balance between metallic function and acid function may be substantially varied, and a variety of feedstocks processed, to produce a considerable variety of products. In the early 1960s the focus of attention was particularly on conversion of distillates in the 200 to 350°C range to gasoline. More recently, emphasis has been placed on the versatility of a particular process to be operated so as to maximize gasoline production in the summer and fuel oils in the winter. Modern hydrocracking processes were developed primarily to process feeds having a high content of polycyclic aromatic compounds, which are relatively unreactive in catalytic cracking. This includes heavy vacuum-gas oils, deasphalted vacuum residuum, heavy gas oil from petroleum coking, and catalytic cycle oil.

The reaction temperature is lower than that in catalytic cracking, in the range of 200 to 400°C, and the hydrogen pressure may vary from 1 to 10 MPa, depending on the feedstock and type of reactions desired. The minimum hydrogen pressure is usually set by that required to minimize the rate of coke formation, but too high a hydrogen pressure causes excessive hydrogen consumption. This may also lower product quality, e.g., by converting aromatics to naphthenes and thus lowering the octane number of a gasoline fraction.

The early metal sulfide/acidic support catalysts were strongly acidic but had mild hydrogenation activity. The product distribution indicates that the mechanism was essentially like that observed with catalytic cracking, accompanied by some hydrogenation and hydroisomerization. Little isomerization of an *n*-paraffin occurred without cracking. The product from these catalysts in general is highly aromatic and highly branched, which provides excellent qualities for a gasoline but poor qualities for a distillate fuel.

Catalysts of high hydrogenation activity relative to acidity give more highly saturated products, which are desired in distillate fuels. Examples are unsulfided platinum, palladium, or nickel, each on silica-alumina. If the support is nonacidic, substantially no isomerization occurs and hydrogenolysis is the only cracking reaction of paraffins found. With an acidic support, isomerization occurs (without cracking) to essentially thermodynamic equilibrium and, especially with noble metals, cyclization. The thermodynamic iso/normal ratio is less than that found from catalytic cracking or with a mild hydrogenation catalyst (Sec. 9.4), which is undesirable. The ratio of gas (up to C_4) to liquids (C_5+) is less, indicating less secondary splitting, perhaps because the intermediate species are converted more rapidly to stable paraffins. The balance between hydrogenation activity and acidity on the one hand and severity on the other can be adjusted to secure an optimum group of reactions. In general a high ratio of hydrogenation activity to acidity favors high liquid yields (C_5+), a low ratio favors a higher-octane product.

A highly aromatic feed is generally needed to make a high-octane gasoline fraction directly from the hydrocracker. With feeds of lower aromaticity the naphtha fraction of the product must generally be fed to a reformer to increase the octane number. In the usual hydrocracker operation, recycling is used to extinction; that is, all the feed is eventually converted to a product boiling below a specified maximum temperature.

Several types of zeolite catalysts, loaded with noble or nonnoble metals, are reportedly used commercially. Mordenite-based catalysts are satisfactory for producing C_3 to C_4 hydrocarbons from naphtha but are insufficiently stable for use with gas-oil feeds (Bolton, 1976). Y-type zeolites in low-sodium decationized or rare earth forms (Sec. 7.7) are highly active, but they must be back-exchanged with a divalent cation to achieve the desired stability. Palladium on Y zeolite reportedly is a suitable catalyst, used in the Union Oil/Exxon process, but palladium on types L and Ω is unsuitable.

These catalysts are poisoned by low concentrations of organosulfur and organonitrogen compounds, so the feedstocks are usually first fed to a hydrodesulfurizer (HDS) reactor in which these substances are converted to H_2S and NH_3, respectively. The zeolite catalysts can tolerate greater concentrations of H_2S and NH_3 than can silica-alumina, hence with their use it is usually unnecessary to separate the NH_3 and H_2S formed in the desulfurizer from the remainder of the product before it is passed to the hydrocracker. This results in considerable cost savings.

The first hydrocracking process utilizing zeolites was commercialized by Union Oil in 1964 and offered to the industry jointly by Union Oil and Exxon. Operating characteristics of a number of installations of this process are detailed by Bolton (1976). The chemical reactions occurring in hydrocracking are discussed in more detail by Langlois and Sullivan (1969). Reactions and processes are treated in an ACS Petroleum Division symposium chaired by Watkins and Hutchings (1972) and in a symposium organized by Ward and Qader (1975). Vlugter and Van'T Spijker (1971) review catalyst compositions for the hydrocracking of residua and distillates to produce a variety of products.

9.8 HYDRODESULFURIZATION (HDS)

The sulfur content of a crude oil varies greatly with its origin. The highly paraffinic crudes from North Africa (Libya, Algeria) and from Nigeria and Indonesia may have as

Table 9.6 Variation of sulfur content of a Kuwait crude oil with boiling point*

Fraction; boiling point range, °C (°F)	Sulfur content, wt %
Naphtha; C_4-150 (300)	0.02
Kerosene; 150–230 (300–450)	0.175
Furnace oil; 230–345 (450–650)	1.23
Heavy furnace oil; 345–370 (650–700)	2.37
Heavy gas oil; 350–550 (660–1020)	2.91
Residue; >370 (700)	4.22
Residue; >550 (1020)	5.12

*F. W. B. Porter et al., British Patent 710,342, in Schuman and Shalit, 1970, p. 293.

little as 0.2 wt % sulfur and a very low metal content (for example, ~3 ppm). In the Near East a light Arabian crude may contain typically 1.5 wt % sulfur, and a Kuwait crude 2.5 to 4 wt % sulfur. The metal contents are of the order of 10 to 30 ppm. Midcontinent United States crudes may vary from 0.2 to 2.5 wt % sulfur. Heavy crudes, having a high nonvolatile fraction, are usually also high in sulfur content and may also have a relatively high metal content. Thus Venezuelan crude oils typically contain 2 to 4 wt % sulfur, and as much as 400 to 550 ppm of metals.

The sulfur content increases progressively with the boiling point of the fraction, as illustrated in Table 9.6 for Kuwait crude (Schuman and Shalit, 1970, p. 293). The metals are found almost entirely in the residuum and greatly complicate HDS processes applied to a residuum. They form deposits inside catalyst pores or on the outside surfaces of catalyst pellets, which ultimately cause plugging and catalyst deactivation.

It is necessary to reduce the sulfur content of petroleum fractions for a variety of reasons:

1. The maximum acceptable sulfur content in the feed to a catalytic reformer is determined by the nature of the catalyst. Bimetallic reforming catalysts are especially sensitive, and the sulfur content must be limited to the vicinity of 1 ppm or less.
2. Air-pollution control standards require removal of, in some cases, 80 percent or more of the sulfur otherwise present in various fuel oils.
3. Much of the sulfur in a gas oil fed to a catalytic cracker may be deposited in the form of coke, the sulfur content of which is converted to sulfur dioxide in the regenerator and emitted to the atmosphere in the combustion gases. To limit air pollution from this source the sulfur content of the gas oil may be reduced before being fed to the catalytic cracking unit.
4. The organosulfur content of the feed to a hydrocracker must be reduced to avoid poisoning of the hydrocracking catalyst.
5. Reduction of sulfur content reduces corrosion during refining and handling and improves the odor of the product.

The process used for this purpose is catalytic treatment with hydrogen to convert the various sulfur compounds present to hydrogen sulfide.[4] The hydrogen sulfide is readily separated and converted to elemental sulfur, a convenient form for handling, by the Claus process. Here some of the hydrogen sulfide is oxidized to sulfur dioxide by air, and sulfur is formed by the overall reaction:

$$2H_2S + SO_2 \longrightarrow 3S\downarrow + 2H_2O \tag{9.1}$$

The catalyst is a form of alumina. Grancher (1978) discusses reaction kinetics and the process.

Interest in hydrodesulfurization was initially stimulated by the availability of hydrogen from catalytic reformers. However, the demand for hydrogen for hydrodesulfurization and hydrotreating now often outstrips that available in a given refinery, and it must then be generated specifically for these purposes. Under a suitable set of circumstances sufficient hydrogen may be formed from dehydrogenation side reactions in the hydrodesulfurization reactor, and this can be continuously recycled back to the inlet. An example occurs with a charge stock that is high in naphthenes, low in olefins, and operated under conditions such that considerable dehydrogenation of naphthenes to aromatics also occurs (e.g., Autofining, a British Petroleum process).

During hydrodesulfurization a number of other hydrogenation reactions occur—some desired and some not desired. Diolefins are readily hydrogenated, which improves the stability of the product and reduces gum formation. The nitrogen content may also be somewhat reduced by hydrodenitrogenation, which occurs simultaneously with HDS (see Sec. 9.9). Basic nitrogen compounds inhibit the catalytic cracking reaction by adsorption on acid sites, and they accelerate the formation of coke. This is an additional reason for hydrogen pretreatment of the gas-oil feed to a catalytic cracker. Multiring aromatic compounds will not crack significantly, so it is necessary to hydrogenate one or more of these rings before the feedstock goes to a catalytic cracker. One thus obtains in the product the single-ring aromatics which are particularly desired for gasoline. The term *hydrofining* is sometimes used for a process in which hydrodesulfurization and various hydrogenation reactions are carried out simultaneously on a distillate or fraction from a catalytic cracker. (See also Sec. 9.10.) For a product to be blended into gasoline it is desirable to minimize the hydrogenation of monoolefins and of aromatics in order not to reduce octane number. This can usually be accomplished by careful control of temperature and pressure.

Hydrogen is expensive to manufacture, so it is desirable to operate all hydrodesulfurization and hydrotreating processes to optimize the desired reactions yet minimize hydrogen consumption. Operating conditions may vary substantially with the particular distillate fraction to be treated. A *straight-run* or *virgin fraction* (that separated from crude oil by distillation only) may have a substantially different chemical com-

[4]Sulfur compounds may also be removed by adsorption on a bed of zinc oxide, which is thereby converted to zinc sulfide. The maximum reaction temperature is about 400°C, limited by side reactions of the feedstock being treated. Only those sulfur compounds that are readily decomposed thermally to H_2S can be removed. This includes mercaptans and disulfides, but most sulfides and all thiophenes are relatively stable at the maximum temperature of operation. The spent zinc oxide is not regenerated, and the process is not practicable if the sulfur content is very great.

position than a fraction of the same boiling-point range that has been subjected previously to chemical reaction, such as catalytic cracking.

For low-boiling-point and middle-boiling-point distillates (up to about 400°C) representative HDS reaction conditions are about 300 to 400°C, and 1 to 7 MPa hydrogen pressure. The higher the boiling point of the feedstock the higher is the sulfur content (Table 9.6) and the less reactive are the sulfur compounds present. More severe conditions (higher temperature and pressure and longer contact time) are then needed. Combinations of high pressure (for example, 7 MPa) and low temperature (for example, 300°C) are generally avoided, if possible, to minimize excessive consumption of hydrogen, as by conversion of aromatics to naphthenes. Usually a considerable excess of hydrogen over that needed stoichiometrically is fed to the reactor inlet, and the hydrogen remaining in the exit gases is recycled to the inlet (with some purge to remove product gases and impurities in the hydrogen). Recycling provides a high partial pressure of hydrogen at a minimum total pressure and high overall utilization of hydrogen.

Hydrodesulfurization and hydrogenation reactions are exothermic. Reactors are adiabatic fixed beds and may be multistage. Cooling is usually provided by injecting additional hydrogen between the stages, termed *cold-shot cooling.* The feed at reaction conditions may be all vapor, mixed vapor and liquid, or essentially all liquid. In the latter two cases the liquid is usually caused to flow concurrently downward with the gas through a fixed catalyst bed, termed a *trickle-bed reactor.* Alternately, for reaction of a heavy or residual oil, the "H-oil" process has been used in a few installations. Liquid and hydrogen gas are passed upwards through an ebullient bed of small catalyst particles. This process minimizes accumulation of solid deposits on the outside surface of catalyst particles, as of vanadium and nickel compounds from residual oils, provides good temperature control, and provides a convenient method of continuously adding new catalyst and removing working catalyst.

The sulfur is present largely in the form of thiols (mercaptans), sulfides, disulfides, and various thiophenes and thiophene derivatives. Mercaptans and sulfides react to form hydrogen sulfide and hydrocarbons.

$$RSSR' + H_2 \longrightarrow RH + R'H + H_2S \tag{9.2}$$

$$RSH + H_2 \longrightarrow RH + H_2S \tag{9.3}$$

$$RSR' + 2H_2 \longrightarrow RH + R'H + H_2S \tag{9.4}$$

where R and R$'$ are various hydrocarbon groups.

The order of reactivity is about RSH $>$ R—S—S—R$'$ $>$ R—S—R$'$ $>$ thiophenes. Reactivity decreases with increased molecular size and varies depending upon whether R is an aliphatic or aromatic group. Among thiophene derivatives, reactivity under industrial conditions decreases in about the following order: thiophene $>$ benzothiophene $\approx$ dibenzothiophene $>$ methyl-substituted benzothiophenes. The alkyl thiophenes are in general less reactive than thiophene, but there is considerable variation with location of the alkyl group. Methyl groups in the 2 and 5 position cause the greatest inhibition, presumably by steric effects. With dibenzothiophene, the presence of methyl groups in the 4 or 4 and 6 positions likewise greatly decreases the reactivity of dibenzothiophene, but the reactivity is increased when they are in the 3 and 7 or 2 and 8 positions. Presumably in the latter cases inductive (accelerating) effects predominate, and in the former case steric (inhibiting) effects are the more important (Houalla et al., 1977).

The main reaction pathway for thiophene is:

$$\text{S} + 3H_2 \longrightarrow H_2S + C_4H_8 \quad \text{(mixed isomers)} \tag{9.5}$$

Small amounts of butadiene are formed, possibly as an intermediate, but this is rapidly hydrogenated to a butene. The butenes in turn are more slowly hydrogenated to butane. The thiophene ring is not hydrogenated before sulfur is removed, although the first step may involve an essentially simultaneous removal of a sulfur atom and donation of two hydrogen atoms to the structure. Thiophane (the completely hydrogenated analogue of thiophene) is found in small amounts in the absence of basic nitrogen compounds. In their presence, little or no thiophane is observed (Satterfield et al., 1980). Thiophane may be formed by a side reaction, but, in any event, of itself it is much more reactive than thiophene.

A kinetic study of thiophene hydrodesulfurization at atmospheric pressure and 235 to 265°C on a commercial Co-Mo/Al_2O_3 catalyst in a differential reactor (Satterfield and Roberts, 1968) showed that the reaction is inhibited by hydrogen sulfide. The rate could be expressed by a Langmuir-Hinshelwood type of expression as follows:

$$-r = \frac{kP_T P_H^n}{(1 + K_T P_T + K_S P_S)^2} \tag{9.6}$$

(The subscript T refers to thiophene, S to hydrogen sulfide, and H to hydrogen.) This and other studies indicate that the hydrodesulfurization and subsequent hydrogenation reactions occur on separate sites. The HDS reaction appears to be between half and first order with respect to hydrogen at pressures above atmospheric, and is severely inhibited by basic nitrogen compounds.

In the case of benzothiophene, substituted or unsubstituted, the thiophene ring is hydrogenated to the thiophane derivative before the sulfur atom is removed, in contrast to the behavior of thiophene.

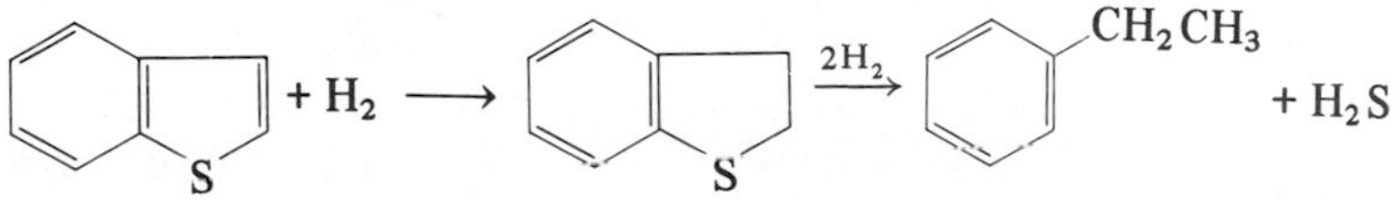

$$\text{S} + H_2 \longrightarrow \text{S} \xrightarrow{2H_2} \text{-}CH_2CH_3 + H_2S \tag{9.7}$$

As with thiophene, methyl substitution reduces reactivity. Dibenzothiophene reacts to form predominantly biphenyl, plus smaller amounts of phenylcyclohexane.

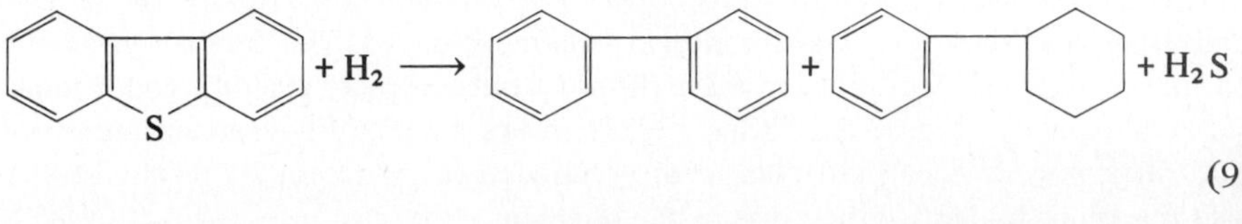

$$\text{S} + H_2 \longrightarrow \quad + \quad + H_2S \tag{9.8}$$

Biphenyl may be hydrogenated further to phenylcyclohexane, but there is good evidence that some phenylcyclohexane is formed as an initial reaction product. (See also Sec. 3.3.1, Case 5.)

All the hydrodesulfurization reactions are essentially irreversible under most industrial conditions, although if extremely low sulfur content (for example, less than 1 ppm) is desired, this may be limited by thermodynamic considerations. Thus a minimum sulfur content may be set by a reaction of the type:

$$RCH{=}R'CH + H_2S \rightleftharpoons \text{mercaptans} + H_2 \tag{9.9}$$

Such a reaction is exothermic and is shifted to the right by lower temperatures.

The most commonly used catalyst is a mixture of cobalt and molybdenum oxides on a γ-alumina support, which is sulfided before use. The ratio of Mo/Co is always considerably greater than 1, a representative composition being 3 wt % CoO and 12 wt % MoO_3. An alternate catalyst composition is Ni-Mo/Al_2O_3, used especially if hydrodenitrogenation or hydrogenation reactions are to be emphasized. As with Co-Mo, the ratio Mo/Ni exceeds 1. Less widely used are Ni-W/Al_2O_3 catalysts, which are quite active but more costly. The method of presulfiding the catalyst is considerably more important in the case of Ni-Mo/Al_2O_3 than with Co-Mo/Al_2O_3 catalysts. If not presulfided before use, NiO may be reduced to metallic Ni by the reducing environment in the reactor, and this may then be difficult to convert to the sulfide. Metallic Ni may cause undesirable reactions and probably sinters more rapidly than a nickel sulfide.

The structure of the active catalysts has been subject to much discussion, but there is no consensus as yet. With a Co-Mo catalyst the sulfided forms may be represented as MoS_2 and Co_9S_8, but their compositions actually are complex. MoS_2 by itself is considerably more active than Co_9S_8, but a mixture of the two is more active than either alone. The mechanism of their interaction is still speculative. These species may also react with the alumina support, but the main weight of evidence is that such compounds formed with alumina contribute little, if any, catalytic activity. Their role, as well as that of the alumina itself, is essentially that of a support. The sulfided catalyst is quite different in structure from the oxide precursors. MoS_2 and WS_2 form layered-type structures in which layers of S atoms alternate with layers of metal atoms. It has been suggested that at the edges of such a structure, atoms of Ni (or Co) might *intercalate* into the MoS_2 or WS_2 structure, forming the active sites. There seems to be no quantitative method presently available for distinguishing between the effective area of metal sulfides and the total area, analogous to the gas chemisorption methods that distinguish between metal area and total area of a supported metal catalyst.

In residual oils quantities of vanadium and nickel compounds are present in concentrations of up to thousands of parts per million. Under hydrodesulfurization conditions these are deposited in the pores of the catalyst, eventually plugging and deactivating it. The reactor itself can also become plugged. The deposition reaction does not occur significantly on an inert support under typical reaction conditions; it requires a catalyst. Figure 5.10 (Sec. 5.5.4) shows a series of electron microprobe studies of a plug from a pilot plant reactor utilized for processing a residual oil and their utilization to identify the cause of the plugging.

More details on catalyst structure, reaction mechanisms, and kinetic models are given in a review by Schuit and Gates (1973), a book by Gates et al. (1979), and in earlier reviews by Schuman and Shalit (1970) and McKinley (1957). The book by Weisser and Landa (1973) gives a detailed treatment of sulfide catalysts and reactions catalyzed by them, with copious references. The nature of sulfur compounds in petroleum is summarized by Drushel (1970) and work over many years on characterization of sulfur in petroleum at the U.S. Bureau of Mines is summarized by Coleman et al. (1970). Trickle-bed reactors as used for hydrodesulfurization and other processes are reviewed by Satterfield (1975).

The term *sweetening process* is applied to any of a variety of methods for removing objectionable odors from various petroleum products. The odors are caused primarily by mercaptans. These may be removed by extractive or adsorptive processes or by catalytic air oxidation in the liquid phase to disulfides, which are innocuous. The patent literature indicates that supported metal phthalocyanines are effective, such as cobalt phthalocyanine on a carbon support. Other methods of sulfur removal are discussed in Sec. 10.1.4.

9.9 HYDRODENITROGENATION (HDN)

Organonitrogen compounds are present in smaller concentrations in crude oil than are sulfur compounds. The weight ratio N/S varies from about 1 : 2 in some high-nitrogen crudes to 1 : 5 to 1 : 10 in other crudes. A high-nitrogen California crude may contain 0.9 wt % N and a Caribbean crude 0.4 wt % N (together with 2.5 to 4.0 wt % S), but in many crude oils the nitrogen content is in the vicinity of 0.1 wt %. Some of the organonitrogen compounds are converted to ammonia during hydrodesulfurization, but the nitrogen compounds are in general less reactive. Thus desulfurization of a Caribbean heavy fraction to reduce the sulfur content from 4 to 1 wt % might typically lower the nitrogen content from 0.4 wt % to about 0.25 wt %. As with sulfur compounds, the nitrogen content of distillate fractions increases with boiling point.

The nature of nitrogen and oxygen compounds in petroleum is summarized by Snyder (1970). In crude oils, nitrogen is present largely in the form of heterocyclic compounds having five or six-membered rings, mostly unsaturated. The nonheterocyclic compounds include anilines, aliphatic amines, and nitriles. Some of these are easier to denitrogenate than the heterocyclic nitrogen compounds.

Crude shale oil is produced from oil shale by thermal retorting, which decomposes the less stable nitrogen compounds and yields a product typically containing about 1 wt % nitrogen in the naphtha fraction and about 2 wt % nitrogen total. Oxygen compounds are present in amounts of 1 wt % oxygen or more and sulfur compounds are about 1 wt %. Both are distributed more or less uniformly through the boiling-point range, in contrast to the distribution found in natural crude oils. The heterocyclic and other nitrogen compounds are often grouped into strong bases (quinolines, pyridines, acridines) versus weak bases or nonbasic compounds (indoles, pyrroles, carbazoles). (See Table 9.7.) The strong basic compounds are of particular concern in acid-catalyzed reactions, such as catalytic cracking, since they will poison acidic sites.

Table 9.7 Representative heterocyclic nitrogen compounds

Name	Formula	Structure
Pyrrole	C_4H_5N	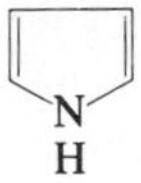
Indole	C_8H_7N	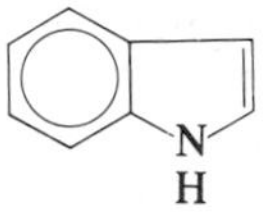
Carbazole	$C_{12}H_9N$	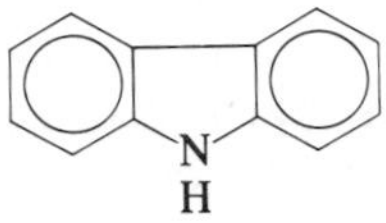
Pyridine	C_5H_5N	
Quinoline	C_9H_7N	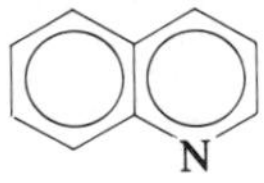
Isoquinoline	C_9H_7N	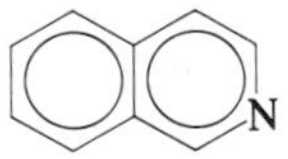
Acridine	$C_{13}H_9N$	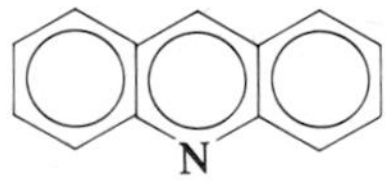

Less is known about the hydrodenitrogenation of heterocyclic ring compounds than of the analogous reactions with heterocyclic sulfur compounds. In general the heterocyclic ring is first saturated, followed by ring fracture at a C—N bond. Nitrogen is removed from the resulting amine or aniline as ammonia (Cocchetto and Satterfield, 1976). Aliphatic amines react readily, but aromatic amines are less reactive. With pyridine, the ring is first hydrogenated to piperidine, which forms, in turn, pentylamine and pentane plus ammonia (McIlvried, 1971).

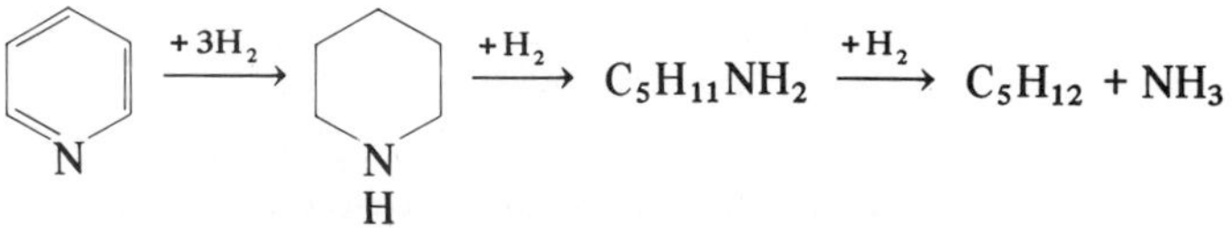

$$\text{Pyridine} \xrightarrow{+3H_2} \text{Piperidine} \xrightarrow{+H_2} C_5H_{11}NH_2 \xrightarrow{+H_2} C_5H_{12} + NH_3 \qquad (9.10)$$

On a Ni-Mo catalyst, which accelerates hydrogenation reactions, the first step can proceed to a nearly equilibrium concentration of piperidine under a wide range of reaction conditions of industrial interest (Satterfield and Cocchetto, 1975; Satterfield et al., 1980).

With quinoline, analogous reactions occur but a greater variety of intermediate compounds are formed. These include 1,2,3,4-tetrahydroquinoline, 5,6,7,8-tetrahydroquinoline, and decahydroquinoline, followed by hydrocracking to form aromatic and saturated cyclic amines, and then cyclic hydrocarbons. The overall reaction network is relatively complex (Satterfield et al., 1978; Satterfield and Cocchetto, 1980).

Hydrodenitrogenation is not now conventionally carried out as a separate process but is one in the overall set of complex reactions that occur in catalytic hydrotreating. Hydrodesulfurization and hydrodenitrogenation reactions interact with each other in a complicated manner. Under some circumstances one inhibits the other, but under other circumstances hydrogen sulfide formed by the HDS reactions enhances the HDN reactions (Satterfield et al., 1975, Satterfield et al., 1980).

9.10 HYDROTREATING (HYDROPROCESSING)

This term is used in a general way to include a variety of catalytic hydrogenation processes used in fuels refining or for purification of products such as industrial solvents. These may be grouped into processes for improving the quality of a final product or processes in which the hydrotreated stream becomes the feedstock for a subsequent process that benefits from the pretreatment. The term may include processes in which emphasis is on hydrodesulfurization or hydrocracking as well as those in which emphasis is on saturation of some or all of various unsaturated species present in a feedstock containing little or essentially no sulfur or nitrogen compounds.

As examples of the first group of processes, hydrotreating may be used as a finishing process in the manufacture of lubricants and various special oils, to improve their color and stability. The catalyst is typically Co-Mo/Al_2O_3 or Ni-Mo/Al_2O_3, it is sulfided before use, and a minimum concentration of hydrogen sulfide must be present in the reactor environment to maintain the catalyst in the sulfide form. Sufficient hydrogen sulfide is usually formed from the sulfur compounds present. In its absence the catalyst ages faster, perhaps because nickel sulfide becomes reduced to the metal, which then sinters and loses area. If necessary, a small amount of hydrogen sulfide may be deliberately added to prevent this effect.

Another application as a finishing process is the partial hydrogenation of *pyrolysis gasoline*, the product fraction of the volatility range of gasoline produced from a steam cracking process. This is too unstable to be incorporated directly into gasoline. Mild hydrogenation of diolefins and other gum-formers stabilizes it. As another example, jet-fuel specifications often include a *luminometer number*. This is a measure of the carbon-forming propensity of the fuel and hence of smoke emissions and of radiation heat-transfer characteristics in the aircraft-engine combustion chamber. These are related essentially to the aromatics content of the fuel. A portion of the feed to the jet-fuel "pool" may be hydrogenated to limit the aromatics content to an acceptable level. Likewise, for diesel fuel paraffins are needed rather than aromatics to achieve the desired self-ignition properties, and these are characterized by a *cetane number*.

As an intermediate processing step, hydrotreating is used to improve the quality of a feedstock. Thus hydrodesulfurization is often necessary before a feedstock can be catalytically reformed. Before being fed to a catalytic cracker, a gas-oil fraction may be hydrotreated to accomplish several objectives simultaneously.

1. Prior hydrodesulfurization provides a means for control of air pollution because some of the sulfur present in a feedstock will be deposited in the form of coke on the cracking catalyst. This will be emitted to the air in the cracking catalyst regenerator.
2. HDN removes nitrogen compounds that otherwise deactivate acidic sites on cracking catalysts and also contribute to coke formation.
3. Saturation of an aromatic ring is required to cause it to crack. Without such prior saturation, multiring aromatic compounds will pass through a catalytic cracking reactor and undergo little or no reaction.

Weekman (1976) has published a comprehensive review of the reaction-engineering aspects of hydroprocessing in general. Some of the processes for liquefaction of coal involve suspending finely divided coal in a liquid in the presence of a catalyst such as Co-Mo/Al_2O_3 and treating it with hydrogen at elevated pressure and temperature. The reactions occurring are complicated, and diffusion effects and clogging of catalyst pores by finely divided solids in the coal play an important role (Oblad, 1976).

9.11 DEHYDROGENATION

9.11.1 Butadiene from Butane and Butene

Butadiene is one of the basic petrochemicals. It is used to make butadiene-styrene and other synthetic rubbers and is a starting material for a variety of chemical syntheses. Basically, two types of processes are operated: a two-stage process from *n*-butane (Phillips), in which butene is converted to butadiene in the second stage, or a single-stage process from *n*-butane and *n*-butylenes (the Houdry Catadiene process). Butenes from refinery or other sources may also be converted in a one-stage process to butadiene. These processes were hastily put into commercial application in the United States at the beginning of World War II for synthetic rubber production when sources of natural rubber were suddenly cut off. The reactions are endothermic and thermodynamically limited.

$$C_4H_{10} \longrightarrow C_4H_8 + H_2 \tag{9.11}$$

$$C_4H_8 \longrightarrow C_4H_6 + H_2 \tag{9.12}$$

The ΔH for each reaction is about 134 kJ/mol. Since the number of moles increases on reaction, improved conversion at equilibrium is obtained by operation at as low a partial pressure of hydrocarbon as feasible. For favorable equilibrium a temperature of 500 to 600°C is needed for reaction (9.11) and 600 to 700°C for reaction (9.12). Contact times of a fraction of a second are used for maximum selectivity.

In the two-stage process butane is converted to butenes at about 1 atm over a chromia-alumina catalyst held in tubes heated to about 600°C in a furnace. Operation is cyclic. About once an hour the feed is stopped and coke deposits are removed by passing flue gas containing 2 to 3% O_2 over the catalyst.

Butenes separated from the product are mixed with steam in a mole ratio of steam to butenes of about 10:1 and reacted at 600 to 700°C over an unsupported iron oxide–chromia catalyst promoted with a potassium compound. The usual composition, known as Shell 105 catalyst, is about 90% Fe_2O_3–4% Cr_2O_3–6% K_2CO_3. The stable form of iron under reaction conditions is probably Fe_3O_4, which is stabilized by the chromia. The catalyst is calcined at high temperatures for strength, and this results in a low area, for example, of the order of 2 m^2/g. The potassium compound helps promote the reaction of carbon with steam and hence helps keep coke deposits from forming. At the reaction temperature K_2CO_3 is appreciably volatile and is slowly lost. Hence it must be replenished by addition to the feed. A chromia-alumina catalyst as used, e.g., in the Houdry process (see below) is not stable in the presence of steam.

In an alternate process a mixture of butene with superheated steam at about 660°C is passed continuously through an adiabatic reactor containing a calcium nickel phosphate catalyst having the composition $Ca_8Ni(PO_4)_6$ plus about 2% Cr_2O_3, at a total pressure slightly above atmospheric. A mole ratio of steam/butylene of about 16 to 17 is used to control the temperature and minimize the rate of formation of carbonaceous deposits. This also lowers the partial pressure of the hydrocarbon, which is desirable. The catalyst is periodically regenerated with a mixture of air and steam. Although this catalyst is more expensive than the Shell 105 catalyst, a higher yield of butadiene is apparently obtained. Even with the heat capacity supplied by the great excess of steam, the reaction temperature drops about 50°C through the reactor.

A problem in operation is the possible formation of mounds of carbon in the bed that grow to the point that the catalyst must be discharged and replaced. These mounds are initiated by metallic nickel, which may be formed by breakdown of the $Ca_8Ni(PO_4)_6$ lattice or from uncombined nickel oxide in the catalyst. Nickel oxide is readily reduced to metallic nickel by hydrocarbons at the reaction temperature, and nickel readily promotes the deposit of coke. Swift et al. (1976) report that addition of hydrogen sulfide to the feed to make up about 5 to 12 ppm effectively controlled carbon mound formation. In other nickel-catalyzed reactions addition of a small amount of hydrogen sulfide has likewise been found effective for control of carbon deposits, although in some cases the catalyst activity may be significantly reduced. Presumably the metallic nickel is covered with a surface nickel sulfide.

In the manufacture of this catalyst it is important that all the nickel be bound up in the crystal lattice and also that this particular orthophosphate be formed. The desired structure is achieved by control of the ratios of starting materials and by roasting at a sufficiently high temperature. The surface area of the final catalyst is in the range of 5 to 10 m^2/g.

The process may be improved by adding a small amount of oxygen to the feed (for example, 0.1 to 0.2 moles/mole butene). This reduces carbon deposition and also reacts with the hydrogen product, making the overall process more thermally neutral and improving the equilibrium limitation (Alexander and Firko, 1967). In this case the calcium nickel phosphate catalyzes the direct hydrogen-oxygen reaction. In newer oxidative dehydrogenation processes hydrogen may not be formed as a separate intermediate (Sec. 8.5).

In the single-stage (Houdry) process, *n*-butane or *n*-butylenes are preheated to about 620°C and passed through a bed of 18 to 20 wt % chromia on alumina in the form of 3- to 4-mm pellets, to which considerable inert material is added to provide additional heat capacity. The beds are wide, horizontal, and only about 1 m thick, to minimize pressure drop. The total pressure is about 10 to 20 kPa (vacuum operation), and the contact time is about 0.2 s. Several reactors are used in parallel. Each reactor operates essentially adiabatically, and after about 5 to 10 min the feed is switched to another reactor. The first reactor is purged with steam and then evacuated, and carbon deposits are then burned off by blowing with air, which reheats the bed. The bed is then evacuated, the catalyst is reduced, and the cycle is repeated. The reaction temperature is about 625 to 675°C. Above about 700°C excessive polymerization and coking occur and the undesirable formation of acetylenes is increased. Below about 600°C the rate becomes uneconomically low. The total cycle time is about 15 to 30 min, so a minimum of three reactors is needed to achieve continuous operation. Commercially as many as eight have been used.

A mixture of butadiene and butenes is produced, from which butadiene is separated and butenes and unreacted butane are recycled. The overall yield of butadiene from butane is about 60 percent. Information on kinetics, catalyst coking, and reactor design and a guide to the earlier literature is provided by Dumez and Froment (1976).

n-Butenes are available in considerable quantities from catalytic cracking in refineries and as a coproduct of steam cracking as well as from catalytic dehydrogenation of *n*-butane. Considerable butadiene is produced by thermal (steam) cracking of naphtha and gas oil, and the trend in recent years to use of these heavier feedstocks instead of ethane or natural gas liquids (propane and butane) is making thermal cracking an increasingly important source of butadiene. Almost all butadiene consumed in Western Europe comes from steam cracking, but catalytic dehydrogenation is important in the United States. However, oxidative dehydrogenation (Sec. 8.5) is preferred over direct catalytic dehydrogenation for new plants.

The historical development of these catalysts and processes is reviewed by Kearby (1955) and Hornaday et al. (1961). Chromia-alumina has a complex structure that may be markedly affected by the method of preparation and has received much study. If prepared by impregnation of alumina, the chromia is present as clusters of tiny crystallites that only partially cover the alumina. The distribution of chromia through the pellet can vary considerably depending on the impregnating solution and techniques used. Chromia-alumina can be regarded for some purposes as a dual-function catalyst that has both acidic properties and a dehydrogenation function. Its physical and chemical properties are reviewed by Poole and MacIver (1967). Carrá and Forni (1972) discuss kinetics and possible reaction mechanisms.

9.11.2 Styrene from Ethylbenzene

Traditionally, ethylbenzene has been made by alkylation of benzene with ethylene using an aluminum chloride catalyst. In new processes developed by UOP and Mobil, a fixed-bed heterogeneous catalyst is used at high temperature and moderate pressure. After separation of the product, polyalkylated aromatics are recycled to the reactor, where transalkylation reaction with benzene occurs to form more ethylbenzene.

Ethylbenzene may be dehydrogenated to styrene using catalysts similar to those used for dehydrogenating butene to butadiene, although styrene dehydrogenation is easier to carry out.

$$C_6H_5CH_2CH_3 \longrightarrow C_6H_5CH{=}CH_2 + H_2 \tag{9.13}$$

$$-\Delta H = -121 \text{ kJ/mol}$$

Chromia is poisoned in the presence of steam, and the preferred catalyst for commercial use is an unsupported iron oxide promoted with a potassium compound such as Shell 105 (see above). Even though this is a low-area catalyst with large pores, the reaction is substantially diffusion-limited on 6-mm particles. Selectivity is improved with smaller sizes, but these cause an increased pressure drop, which is not desired. Commercially 4- to 6-mm-diameter extrudates are used.

The process is operated typically with a mole ratio of steam/ethylbenzene of 10 to 15, at about atmospheric pressure, about 600 to 650°C, and with a contact time of the order of 1 s. Two adiabatic fixed-bed reactors with reheat in between are used. Styrene yields of about 90 percent are achieved at conversions of 40 to 60 percent. Some toluene is produced as a byproduct.

For maximum conversion and yield such reactors and ancillary equipment are especially designed to provide a minimum pressure drop, so a radial fixed-bed reactor may be utilized. Catalyst life is of the order of 1 to 2 years. Kearby (1955) gives equilibrium data and describes early catalyst compositions, and Lee (1974) describes detailed studies with iron oxide catalysts and with commercial Shell 105 catalyst.

The potassium promoter migrates with time towards the center of the pellets (Lee, 1974) because the center is slightly cooler than the outside, caused by the endothermic reaction. Potassium also migrates downstream. Maximum activity occurs at a level of a few percent potassium; hence the potassium migration causes the outside of the catalyst pellet to deactivate, from loss of potassium, and causes the center also to deactivate, from accumulation of excess potassium.

9.12 HYDRODEALKYLATION

Hydrodealkylation is the hydrocracking of an alkyl aromatic to form a paraffin and an unsubstituted aromatic ring. Commercially, toluene and other alkyl benzenes are converted to benzene, or alkyl naphthalenes are converted to naphthalene.

$$C_6H_5CH_3 + H_2 \longrightarrow C_6H_6 + CH_4 \tag{9.14}$$

$$\text{Methylnaphthalene} + H_2 \longrightarrow \text{naphthalene} + CH_4 \tag{9.15}$$

$$-\Delta H \sim 50 \text{ kJ/mol } CH_4$$

Typical operating conditions are 500 to 600°C, 3 to 7 MPa, and a H_2/feed mole ratio of 5 to 15, utilizing an adiabatic reactor. Operating conditions are nearly the same for the two reactions and are chosen so as to achieve good yields and reaction rates without ring hydrogenation. The feedstock is usually a highly aromatic stream such as a reformate or "cycle oil" from catalytic gas-oil cracking or a light oil from coking of coal. Hence a variety of types of compounds are present, and a variety of other reactions can occur in addition to the desired ones. The exothermic nature of the reaction means that it can potentially run away. This requires careful attention to the design and operation of the reactor.

Under reaction conditions equilibrium amounts of saturated ring compounds, e.g., cyclohexane, are very small. The desired reactions are essentially irreversible, and mole yields in the range of 95 to 97 percent are achieved. The large excess of hydrogen minimizes side reactions and coke formation on the catalyst. One or two adiabatic reactors are used in series. The reactions may also be carried out thermally at slightly higher temperature and pressure, and both catalytic and noncatalytic processes (Hydeal process of UOP and Detol process of Houdry Division of Air Products) are practiced commercially. The thermal processes are more flexible with respect to feedstock, but the catalytic processes permit higher yields and conversion per pass and lower hydrogen consumption. Little information has been published about catalyst compositions, but the UOP Hydeal process is reported to use a chromia-alumina catalyst of low sodium content. Hydrodealkylation in general is reviewed by Asselin (1964), and naphthalene from petroleum by Ballard (1965).

Most naphthalene is used as a charge stock for manufacture of phthalic anhydride (Sec. 8.6.1). Alkyl naphthalenes, if used as the feedstock for this partial oxidation reaction, provide an unsatisfactory yield. The economic incentive to operate naphthalene hydrodealkylation is therefore closely related to the price and availability of *o*-xylene, the alternate feedstock for phthalic anhydride manufacture.

Benzene is a widely used starting material for chemical synthesis, but there are few uses for toluene. The economic incentive for dealkylation of toluene to benzene varies greatly with time and location and is influenced by many factors. Most benzene and toluene is produced by catalytic reforming in which the quantities of toluene produced are generally greater than benzene. However toluene is a better high-octane blending component for gasoline than is benzene. Some benzene is also produced as a byproduct of thermal steam cracking.

9.13 REGENERATION OF COKED CATALYSTS BY COMBUSTION

In many hydrocarbon and related reactions, such as catalytic cracking, reforming, hydrodesulfurization, and various dehydrogenations, carbonaceous deposits (so-called coke) are gradually formed on the surface of the catalyst. The continuing accumulation of these deposits reduces the activity of the catalyst to the point that it must be regenerated. This buildup of coke may become significant in less than a minute, as in catalytic cracking, or only over a period of months, as in some catalytic reforming processes.

The character of these deposits is complex and varies substantially with the conditions under which they were formed, and with subsequent treatments. In some cases, especially at relatively low temperatures, they are essentially an ill-defined high-molecular-weight polymer, much of which can be removed by washing with a solvent. Appleby et al. (1962) made a detailed characterization of the coke formed in catalytic cracking and showed that it consisted of large aggregates of polynuclear aromatic molecules comprising essentially condensed systems of fused aromatic rings, plus strongly adsorbed portions of products of the reaction. X-ray diffraction studies likewise show that, when deposited at a high temperature (for example, 400 to 500°C), a substantial portion of the coke is in a pseudographite form. Even so the coke may contain considerable hydrogen, a representative empirical formula being between CH_1 and $CH_{0.5}$, provided that heavy metals are not present on the catalyst in an elemental form. If elemental heavy metals are present, the deposits usually contain little hydrogen (see also Sec. 10.1).

The usual technique of regenerating a coked catalyst is by combustion with air. A mixture of air with flue gas or with other inert gas, to lower the oxygen concentration, is used when an excessive temperature would damage the catalyst. A mixture of air and steam may also be used, in which case the steam-carbon reaction may be significant. If regeneration is required often, a continuous process must provide a facility separate from the reactor for continuous burnoff, as in the catalytic cracking processes. Alternately, a cyclic fixed-bed process may be designed in which reaction and coke-burnoff steps follow in sequence, separated by purge steps to avoid the formation of explosive gas mixtures. An example is the Houdry process for converting butane to butadiene. If regeneration is only required at intervals of several months or longer, a fixed-bed unit can be shut down for the period of time required for regeneration and then put back on stream. If this interruption is too lengthy, or regeneration is difficult to achieve in the reactor vessel, the coked catalyst may be quickly replaced with fresh catalyst and the regeneration carried out elsewhere.

The method of regeneration depends upon many factors. These include the relative availability of steam or inert gas; the sensitivity of the catalyst to elevated temperature, to steam, or to carbon dioxide; the design of the reactor; and the ability to remove heat during the regeneration. Almost all the fundamental information on regeneration is concerned with the reaction of coke and oxygen. In fixed-bed reactors, the burning takes place largely within a zone that moves slowly through the reactor, the temperature along the bed varying greatly with position and with time. The following discussion will focus primarily on the processes occurring in a single pellet or in a differential reactor.

If the effectiveness factor for the primary catalytic reaction is essentially unity (Chap. 11) and the catalyst is uniform in intrinsic activity, the coke usually will be distributed uniformly through the catalyst. If the effectiveness factor for the regeneration reaction likewise approaches unity, the oxygen gradient through the pellet will be negligible, and the coke concentration decreases uniformly throughout the pellet during the combustion reaction.

At a sufficiently high temperature, the regeneration reaction may become diffusion-limited, and oxygen reacts as fast as it is transported to the carbon. In a sphere, reaction then occurs solely at a spherical interface that moves progressively to the center, and the rate is limited by the rate of diffusion of oxygen through a shell of carbon-free porous solid. These facts may be visually demonstrated by cutting a pellet in half after partial removal of coke. With some catalysts the effect may be demonstrated by submerging a pellet in a liquid of the same refractive index, e.g., silica-alumina beads in benzene. Under highly diffusion-controlled regeneration conditions, one sees a black sphere, surrounded concentrically by a white or light-gray spherical shell from which all the coke has been removed. Under intermediate conditions, a progressive darkening toward the center is visible.

9.13.1 Intrinsic Kinetics: Nature of Coke Deposits

A substantial number of reports are available on the rate of combustion in air of carbonaceous deposits in porous catalysts, both of the intrinsic kinetics and under diffusion-controlled conditions. In many of the studies, particularly earlier ones, operating conditions were poorly controlled, the reactor was not truly differential, or measurements were made in the transition zone between intrinsic kinetics and diffusion control. All these factors make interpretation difficult.

An atom of carbon in graphite occupies about 0.04 nm^2 of cross-sectional area; if each atom of carbon in coke occupies the same area, a monoatomic layer of carbon on an oxide-type catalyst or support would comprise about 5 wt % carbon for each 100 m^2/g of surface area. Studies by Weisz and Goodwin (1966) on silica-alumina catalysts of about 250 m^2/g surface area indicated that up to about 6 wt % carbon the rate of combustion was proportional to the amount of carbon present; that is, all atoms of the coke were presumably equally accessible to oxygen. At higher coke contents (for example, 7 to 20% carbon on silica-alumina) the rate for the first 50 percent or so of reaction was less than proportional to the amount of carbon present, indicating that some of the carbon atoms were initially inaccessible, but in the latter part of the reaction the rate again became first order with respect to total carbon present. The same type of behavior at high coke contents has been reported by others.

Considerable hydrogen may be present in the coke, and if so it is removed preferentially in the early stages of the regeneration. If some of the deposit is essentially a hydrocarbon polymer, this may be decomposed and volatilized in the pores, and the products then burn homogeneously outside the catalyst pores. In the best controlled studies, investigators of the regeneration reactions have pretreated their coked catalyst at elevated temperatures in an inert atmosphere before combustion in order to eliminate most of the volatile matter present or to further decompose the coke and thereby obtain more reproducible results. In most laboratory studies the coke has been deposited from a hydrocarbon vapor under nondiffusion-limiting conditions, which at moderate carbon loadings seems to provide a true monolayer.

9.13.2 Carbon Gasification Kinetics

Close agreement has been reported of first-order reaction rate constants for combustion of monolayer coke deposits when they are formed on inert supports or acidic cracking catalysts not containing transition metals. Data were reported by Weisz and

Goodwin (1966) for coke burnoff from silica-alumina, silica-alumina Durabead, silica-magnesia, fuller's earth (an adsorbent clay), or Filtrol 110 (a cracking catalyst derived from natural clay). There was no effect of support on the rate, and the same results were obtained on silica-alumina with deposits from laboratory cracking of gas oil, naphtha, cumene, or propylene. Johnson and Mayland (1955) also reported in an earlier study that the specific burning rates of carbon on silica-alumina, silica-magnesia, clay, silica gel, and cracking catalysts were approximately the same, although detailed information about the kinetics on the various supports was not developed.

The equation for the first-order rate constant is shown in Fig. 9.4 and is given by (Weisz and Goodwin, 1966)

$$k = 1.9 \times 10^8 \, e^{-157/RT} \qquad \text{sec}^{-1} \cdot \text{atm}^{-1} \tag{9.16}$$

where the expression for the rate of reaction per unit volume of pellet is

$$-r = k \cdot P_{O_2} \cdot c_C \tag{9.17}$$

and c_C is the moles of carbon present per cubic centimeter of pellet.

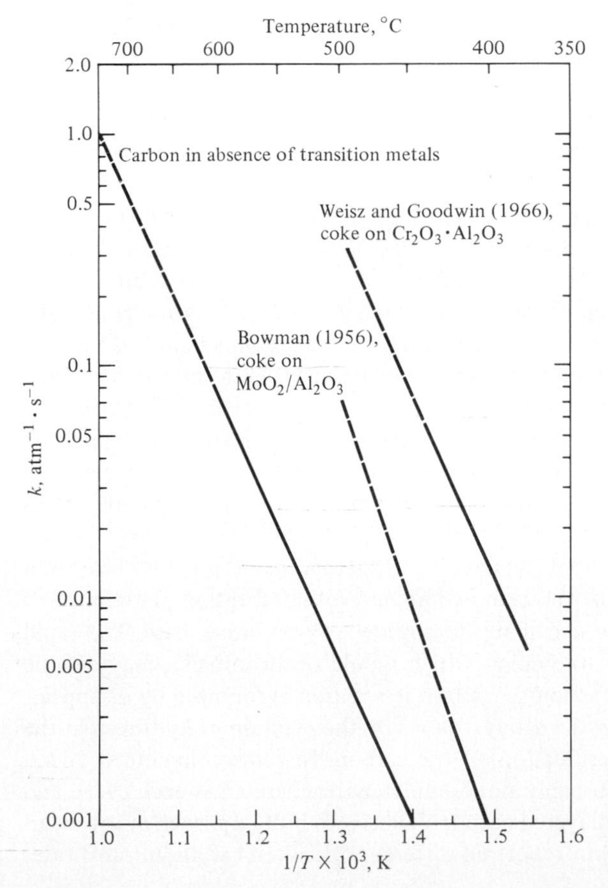

Figure 9.4 First-order reaction rate constants for oxidation of coke deposits. Dashed lines indicate extrapolated data.

Several other investigators show results that agree closely with Eq. (9.16). Hagerbaumer and Lee (1947) reported on the combustion of coke from a synthetic silica-alumina bead catalyst about 3 mm in diameter and having an average pore diameter of 4.7 nm. After removal of the first 40 percent of the carbon, the reaction was first order with respect to the remaining carbon content. The activation energy was 146 kJ/mol, and above about 480°C it fell off, indicating the onset of diffusion control. Hagerbaumer and Lee's data essentially agree with those of Weisz and Goodwin, even though a considerable amount of empirical correlations seemed to have been required in analyzing the data. Pansing (1956) likewise reported data on regeneration of a cracking catalyst and an activation energy of 146 kJ/mol, which agree with the above two studies. Massoth's study (1967) (425 to 480°C and oyxgen partial pressure of 4 to 21 kPa) showed an activation energy of about 168 kJ/mol and rate data, interpreted as burning of small clumps of carbon, that also essentially agree with the above. Adel'son and Zaitova reported studies of coke burning from 0.33-cm-diameter silica-alumina cracking catalyst spheres at 450 to 500°C and 21 kPa oxygen. They also reported the reaction to be first order with respect to carbon with an activation energy of 146 kJ/mol, but their rates were about two-thirds of those reported by Weisz and Goodwin.

Equation (9.16) may thus be taken as representative of the intrinsic combustion rate of monolayer carbon on a noncatalytic surface. There is some controversy concerning the order of the reaction with respect to oxygen under various conditions. Most investigators indicate that it is nearly proportional to the first power in the range from about 10 to 100 kPa partial pressure of oxygen and at the temperatures of interest in catalyst regeneration. These rates are quite close to those reported for oxidation of graphite.

The presence in the porous structure of transition metals or other catalysts greatly accelerates the rate of carbon oxidation. Weisz and Goodwin reported that a chromia-alumina catalyst shows rates two to three orders of magnitude greater than that observed on silica-alumina, although the activation energy remains the same (Fig. 9.4). Even a trace of chromia (for example, 0.1 percent) in silica-alumina increases the rate in fresh catalyst by a factor of about 4, although the rate drops with time under reducing conditions as the chromium is converted from the Cr^{6+} form to Cr^{3+}. Bowman's studies (1956) were on a molybdena-alumina catalyst that evidently had catalytic activity for carbon oxidation. Platinum/alumina reforming catalysts also are catalytic for carbon oxidation and can be rapidly regenerated at relatively lower temperatures and lower oxygen concentration.

The initially rapid rates reported by some investigators may represent desorption of volatile hydrocarbons or further cracking combined with desorption, with much of the combustion of the products occurring homogeneously in either case. The rapid reaction of hydrogen present in the coke, which would occur initially, may also be erroneously interpreted as combustion of carbon if reaction is followed by disappearance of oxygen. Steam formed in the catalyst pores by the reaction of hydrogen in the coke may also accelerate the gasification of the carbon. In some commercial operations, all catalyst particles are probably not completely regenerated at each cycle, and there is an opportunity for multilayered carbon deposits to build up, even at low overall coke levels. Weisz and Goodwin report no difference between carbon-burnoff rates for carbon deposited in the laboratory and that produced in commercial refinery cracking on a moving-bed reactor. Freshly deposited carbon on aged catalysts behaved like freshly deposited carbon on virgin catalysts; that is, alterations in pore size and physical structure of the catalyst upon aging apparently do not affect the intrinsic kinetics as such. However the regeneration reaction may become limited by the diffusion of oxygen into the pores at a temperature in the range of 450°C or higher, depending on catalyst size, pore structure, and other variables (Satterfield, 1970, Chap. 5).

9.13.3 Poisoning

The deposit of coke causes a reduction in catalytic activity by blocking active sites. The distribution of the deposit can vary greatly with the situation, and mathematical analyses are similar to those used to describe catalyst poisoning. At one extreme a strongly adsorbed poison will deposit initially at the pore mouths (near the external surface of catalyst pellets) and on the topmost layers of a fixed bed of catalyst. The outer zone of the catalyst thus becomes an effectively inert annulus through which reactants must diffuse to reach active surface. The thickness of this inert annulus in a pellet will grow with time, and the thickness of the top layers of catalyst containing deposits will likewise increase. Simultaneously pores may become choked if deposits significantly reduce pore size, leading to various diffusion-limiting effects. The resulting effects on activity and selectivity of a fixed bed of catalyst can evidently be highly complex, and their consideration is outside the scope of the present treatment. The subject was analyzed in considerable detail in an early analysis by Wheeler (1951, 1955) and is more recently covered in an extensive review by Butt (1972).

REFERENCES

Adel'son, S. V., and A. Ya. A. Zaitova: *Khim Tekhnol. Topl. Masel*, no. 1, 1962; *Int. Chem. Eng.*, **2**, 360 (1962); **3**, 16 (1963).

Alexander, D. S., and J. Firko: *Seventh World Petroleum Congress*, vol. 5, Elsevier, 1967, p. 115.

Appleby, W. G., J. W. Gibson, and G. M. Good: *Ind. Eng. Chem., Process Des. Dev.*, **1**, 102 (1962).

Asselin, G. F., in J. J. McKetta, Jr. (ed.): *Advances in Petroleum Chemistry and Refining*, vol. 9, Wiley, New York, 1964, p. 284.

——, H. S. Bloch, G. R. Donaldson, V. Haensel, and E. L. Pollitzer: *Am. Chem. Soc., Div. Pet. Chem., Prepr.*, **17**(3), B4 (1972).

Ballard, H. D., Jr., in J. J. McKetta, Jr. (ed.): *Advances in Petroleum Chemistry and Refining*, vol. 10, Wiley, New York, 1965, p. 219.

Bolton, A. P., in J. A. Rabo (ed.): "Zeolite Chemistry and Catalysis," *ACS Monogr. No. 171*, 1976, p. 714.

Bowman, W. H.: Sc.D. thesis, M.I.T., Cambridge, Mass., 1956.

Butt, J. B.: *Adv. Chem. Ser., No. 109*, 1972, p. 259.

Carrà, S., and L. Forni, *Catal. Rev.*, **5**, 159 (1972).

Chester, A. W., A. B. Schwartz, W. A. Stover, and J. P. McWilliams, *Am. Chem. Soc., Div. Pet. Chem., Prepr.*, **24**(2), 624 (1978).

Ciapetta, F. G., R. M. Dobres, and R. W. Baker in P. H. Emmett (ed.): *Catalysis*, vol. 6, Reinhold, New York, 1958, p. 495.

——, and D. N. Wallace, *Catal. Rev.*, **5**, 67 (1971).
Cocchetto, J. F., and C. N. Satterfield: *Ind. Eng. Chem., Process Des. Dev.*, **15**, 272 (1976).
Coleman, H. J., R. L. Hopkins, and C. J. Thompson: *Am. Chem. Soc., Div. Pet. Chem., Prepr.*, **15**(3), A17 (1970).
Drushel, H. V.: *Am. Chem. Soc., Div. Pet. Chem., Prepr.*, **15**(2), C13 (1970).
Dumez, F. J., and G. F. Froment: *Ind. Eng. Chem., Process Des. Dev.*, **15**, 291 (1976).
Gates, B. C., J. R. Katzer, and G. C. A. Schuit: *Chemistry of Catalytic Processes*, McGraw-Hill, New York, 1979.
Grancher, P.: *Hydrocarbon Process.*, July 1978, p. 155; September 1978, p. 257.
Haensel, V.: *Adv. Catal.*, **3**, 179 (1951).
Hagerbaumer, W. A., and R. Lee: *Oil Gas J.*, **45**, March 15, 1947, p. 76.
Hemler, C. L., Jr., and L. O. Stine, U.S. Patent 4,148,751 (April 10, 1979) to UOP, Inc.
Hornaday, G. F., F. M. Ferrell, and G. A. Mills in J. J. McKetta (ed.): *Advances in Petroleum Chemistry and Refining*, vol. 4, Wiley, New York, 1961, p. 451.
Houalla, M., D. Broderick, V. H. J. deBoer, B. C. Gates, and H. Kwart: *Am. Chem. Soc., Div. Pet. Chem., Prepr.*, **22**(3), 941 (1977).
Johnson, M. F. L., and H. C. Mayland: *Ind. Eng. Chem.*, **47**, 127 (1955).
Kearby, K., in B. T. Brooks et al. (eds.): *The Chemistry of Petroleum Hydrocarbons*, vol. 2, Reinhold, New York, 1955, p. 221. See also P. H. Emmett (ed.): *Catalysis*, vol. 3, Reinhold, New York, 1955, p. 453.
Langlois, G. E., and R. F. Sullivan: *Am. Chem. Soc., Div. Pet. Chem., Prepr.*, **14**(4), D18 (1969).
Lee, E. H.: *Catal. Rev.*, **8**, 285 (1974).
McIlvried, H. G., *Ind. Eng. Chem., Process Des. Dev.*, **10**, 125 (1971).
McKinley, J. B., in P. H. Emmett (ed.): *Catalysis*, vol. 5, Reinhold, New York, 1957, p. 405.
Magee, J. S., in J. R. Katzer (ed.): Molecular Sieves II, *ACS Symp. Ser. No.* 40, 1977, p. 650.
Magee, J. S., and J. J. Blazek in J. A. Rabo (ed.): "Zeolite Chemistry and Catalysis," *ACS Monogr. No. 171*, 1976, p. 615.
Massoth, F. E.: *Ind. Eng. Chem., Process Des. Dev.*, **6**, 200 (1967).
Menon, P. G., and J. Prasad: *Proceedings of the Sixth International Congress on Catalysis*, The Chemical Society, London, 1977, p. 1061.
Minachev, Kh. M., and Ya. I. Isakov in J. A. Rabo (ed.): "Zeolite Chemistry and Catalysis," *ACS Monogr. No. 171*, 1976, p. 552.
Mills, G. A., H. Heinemann, T. H. Milliken, and A. G. Oblad: *Ind. Eng. Chem.*, **45**, 134 (1953).
Oblad, A. G.: *Catal. Rev., Sci. Eng.*, **14**, 83 (1976).
——, T. H. Milliken, Jr., and G. A. Mills: *Adv. Catal.* **3**, 199 (1951).
Pansing, W. F.: *AIChE J.*, **2**, 71 (1956).
Pecci, G., and T. Floris: *Hydrocarbon Process.*, December 1977, p. 98.
Pfefferle, W. C.: *Am. Chem. Soc., Div. Pet. Chem., Prepr.*, **15**(1), A21, A27 (1970).
Poole, C. P., Jr., and D. S. MacIver: *Adv. Catal.*, **17**, 223 (1967).
Ries, H. E.: *Adv. Catal.*, **4**, 88 (1952).
Rossini, F. D., E. J. Prosen, and K. S. Pitzer: *J. Res. Nat. Bur. Stand.*, **27**, 529 (1941).
Ryland, L. B., M. W. Tamele, and J. N. Wilson in P. H. Emmett (ed.): *Catalysis*, vol. 7, Reinhold, New York, 1960, p. 1.
Satterfield, C. N.: *AIChE J.*, **21**, 209 (1975).
——: *Mass Transfer in Heterogeneous Catalysis*, M.I.T., Cambridge, Mass., 1970.
——, and J. F. Cocchetto: *AIChE J.*, **21**, 1107 (1975).
——, ——; to be published, 1980.
——, M. Modell, R. Hites, and C. J. DeClerck, *Ind. Eng. Chem., Process Des. Dev.*, **17**, 141 (1978).
——, ——, and J. F. Mayer, *AIChE J.*, **21**, 1100 (1975).
——, and G. W. Roberts: *AIChE J.*, **14**, 159 (1968).
——, ——, and J. A. Wilkens, *Ind. Eng. Chem., Process Des. Dev.*, **19**, 154 (1980).
Schuit, G. C. A., and B. C. Gates: *AIChE J.*, **19**, 417 (1973).
Schuman, S. C., and H. Shalit: *Catal. Rev.*, **4**(2), 245 (1970).
Schwartz, A. B., U.S. Patent 4,072,600 (February 7, 1978), to Mobil Oil Corp.
Sinfelt, J. H.: *Adv. Chem. Eng.*, **5**, 37 (1964).

Snyder, L. R.: *Am. Chem. Soc., Div. Pet. Chem., Prepr.*, **15**(2), C44 (1970).
Sterba, M. J., and V. Haensel: *Ind. Eng. Chem., Prod. Res. Dev.*, **15**, 2 (1976).
Swift, H. E., H. Beuther, and R. J. Rennard, Jr.: *Ind. Eng. Chem., Prod. Res. Dev.*, **15**, 131 (1976).
Tedder, J. M., A Nechvatal, and A. H. Jubb: *Basic Organic Chemistry, Part 5: Industrial Products*, Wiley, New York, 1975.
Thomas, J. M., and W. J. Thomas: *Introduction to the Principles of Heterogeneous Catalysis*, Academic, New York, 1967.
Uhlig, H. F., and W. C. Pfefferele: *Adv. Chem. Ser., No. 97*, 1970, p. 204.
Upson, L. L., *Am. Chem. Soc., Div. Pet. Chem., Prepr.*, **24**(2), 632 (1978).
Vlugter, J. C., and P. Van'T Spijker: *Eighth World Petroleum Congress*, Applied Science Publishers, London, 1971, vol. 4, p. 159.
Voge, H. H., in P. H. Emmett (ed.): *Catalysis*, vol. 6, Reinhold, New York, 1958, p. 407.
Ward, J. W., and S. A. Qader: "Hydrocracking and Hydrotreating," *ACS Symp. Ser. No. 20*, 1975.
Watkins, C. H., and L. E. Hutchings (chairmen): *Symposium on Advances in Distillate and Residual Oil Technology, Am. Chem. Soc., Div. Pet. Chem., Prepr.*, **17**(4), G3 (1972).
Weekman, V. W.: *Fourth International/Sixth European Chemical Reaction Engineering Symposium*, DECHEMA, Frankfurt am Main, 1976, p. 615.
Weisser, O., and S. Landa: *Sulphide Catalysts, Their Properties and Applications*, English translation, Pergamon, New York, 1973.
Weisz, P.: *Adv. Catal.*, **13**, 137 (1962).
——, and R. D. Goodwin: *J. Catal.*, **6**, 227 (1966).
Wheeler, A.: *Adv. Catal.*, **3**, 249 (1951).
——, in P. H. Emmett (ed.): *Catalysis*, vol. 2, Reinhold, New York, 1955.

CHAPTER

TEN

SYNTHESIS GAS AND ASSOCIATED PROCESSES

This chapter is devoted to a closely related group of industrial catalytic processes, many of which are carried out on a large scale and have been of industrial significance for many decades. These center around synthesis gas, a mixture of carbon monoxide and hydrogen (it may also contain nitrogen), which may be made by a variety of processes, either catalytic or noncatalytic. The catalytic processes are steam reforming of a variety of hydrocarbon feedstocks ranging from natural gas to naphtha or a light gas oil. The synthesis gas may then be reacted over a heterogeneous catalyst to form methanol, paraffins (Fischer-Tropsch synthesis), or methane. With a homogeneous catalyst containing cobalt or rhodium, synthesis gas may be reacted with an olefin to form an aldehyde (oxosynthesis). Extensive current research and development work also suggests the future possibility of economic conversion of synthesis gas into chemical feedstocks such as olefins and aromatics and other chemical products.

For the manufacture of hydrogen for hydrogenation reactions or for ammonia synthesis, all the carbon monoxide must be substantially removed. This is accomplished primarily by the water-gas shift reaction, which converts carbon monoxide by reaction with water into carbon dioxide and additional hydrogen. The carbon dioxide is then removed by absorption. The catalytic synthesis of methane has been used as a method of purification of the traces of carbon oxides then remaining. In recent years it has become of interest also as a method of converting low-grade carbonaceous fuels or coal to substitute natural gas (SNG) via the formation of synthesis gas. Certain selected hydrogenation processes are discussed in Sec. 6.7.

Table 10.1 lists the standard free-energy change as a function of temperature for a number of synthesis reactions of interest. Alcohols higher than methanol are not deliberately manufactured from synthesis gas but may be produced as a byproduct of other processes. All these reactions, being exothermic, have the most favorable prod-

Table 10.1 Free energy change (ΔG°) of selected reactions of synthesis gas, kJ/mol

	Temperature, °C				
Reaction	27	127	227	327	427
$CO + 2H_2 \rightarrow CH_3OH$	-26.4	-3.35	+20.9	+45.2	+69.9
$2CO + 4H_2 \rightarrow C_2H_5OH + H_2O$	–	-74.6	-27.1	+21.2	–
$4CO + 8H_2 \rightarrow C_4H_9OH + 3H_2O$	–	-199.3	-102.5	-3.95	–
$2CO \rightarrow CO_2 + C$	-119.6	-101.8	-83.8	-65.9	-47.9
$CO + 3H_2 \rightarrow CH_4 + H_2O$	-141.8	-119.6	-96.3	-72.4	-47.9
$2CO + 2H_2 \rightarrow CH_4 + CO_2$	-170.3	-143.9	-116.7	-88.9	-60.0
$nCO + 2nH_2 \rightarrow C_nH_{2n} + nH_2O$ ($n = 2$)	-114.0	-81.0	-46.5	-11.18	+24.7
$nCO + (2n + 1)H_2 \rightarrow C_nH_{2n+2} + nH_2O$ ($n = 2$)	-214.9	-169.4	-122.2	-73.8	-24.6

uct equilibrium at the lowest temperature. Some reactions, such as synthesis of paraffins, are favorable over the whole temperature range of interest. In each case there is a volume contraction upon reaction, so increased pressure is helpful. The formation of methanol is the most unfavorable, and it requires high pressure, a relatively low temperature, and an active catalyst to cause it to proceed even to a modest degree. Considerably lower pressures suffice for methanation and Fischer-Tropsch synthesis. With the monohydric alcohols the equilibrium becomes progressively more favorable with increase in molecular weight. Which of all these various species will be formed in any given case is largely determined by kinetics rather than thermodynamics, and hence is set by the nature of the catalyst.

The historical development of catalyzed reactions of synthesis gas is too extensive to be treated here. The formation of methane from synthesis gas, apparently first reported by Sabatier and Senderens in 1902, was accomplished using a nickel catalyst at atmospheric pressure and at about 250°C. The commercialization of methanol synthesis by BASF, using a $ZnO\text{-}Cr_2O_3$ catalyst, occurred about 1923.

10.1 STEAM REFORMING

Steam is reacted with natural gas (primarily methane) or with hydrocarbon feedstocks such as naphthas to form a mixture of H_2, CO, CO_2, and CH_4. The term *steam reforming* is something of a misnomer and should not be confused with catalytic reforming or with steam cracking, the latter being a noncatalyzed process. The commercial catalyst is in all cases a supported nickel, but there are considerable variations in the nature of the support and in promoters present. Cobalt and iron are less effective than nickel. The platinum-group metals are highly active but are relatively too expensive.

Starting with hydrocarbons higher than methane, the main course of the reaction appears to be the conversion of the hydrocarbons to carbon monoxide and hydrogen, followed by their reaction to form methane. The water-gas shift reaction accompanies these reactions.

$$C_nH_m + nH_2O \longrightarrow nCO + \left(n + \frac{m}{2}\right)H_2 \tag{10.1}$$

$$-\Delta H_{298} < 0$$

$$CO + 3H_2 \rightleftharpoons CH_4 + H_2O \tag{10.2}$$

$$-\Delta H_{298} = 206 \text{ kJ/mol}$$

$$CO + H_2O \rightleftharpoons CO_2 + H_2 \tag{10.3}$$

$$-\Delta H_{298} = 41.2 \text{ kJ/mol}$$

Under reaction conditions of industrial interest, the amounts of hydrocarbons other than methane present at equilibrium are vanishingly small and reactions (10.2) and (10.3) nearly approach equilibrium.

10.1.1 Formation and Reactions of Carbon

It is particularly important to avoid conditions leading to carbon deposition since this can cause blockage of catalyst pores and catalyst deterioration, leading to premature reactor shutdown. A necessary but not sufficient condition is that operating conditions lie outside of those in which solid carbon can be formed under equilibrium, as from carbon monoxide and methane:

$$2CO \rightleftharpoons C + CO_2 \tag{10.4}$$

$$-\Delta H_{298} = 172.5 \text{ kJ/mol}$$

$$CH_4 \rightleftharpoons C + 2H_2 \tag{10.5}$$

$$-\Delta H_{298} = -75 \text{ kJ/mol}$$

Carbon can be present in a form thermodynamically more active than graphite, so that, for example, the equilibrium ratio CO/CO_2 or CH_4/H_2 can be higher than that calculated for β-graphite. This form of carbon is sometimes termed *Dent carbon*, especially in the earlier literature, from the early work of Dent (1929, 1945). The free energy of this carbon may exceed that of β-graphite by as much as 21 kJ/mol when formed at about 350°C, and up to 15 kJ/mol at 500°C. The properties of the carbon also seem to be influenced by the nature of the support. Rostrup-Nielsen (1975) has reported that the limiting conditions at which coke formation occurred on nickel by reactions (10.4) and (10.5) seemed to be a function of the nickel crystallite size. The smallest crystallites showed the greatest deviation from graphite. (See also Sec. 6.5.)

It is also possible for solid carbon to be formed and to accumulate under conditions in which it would not be present under equilibrium conditions, when reactions leading to its formation are intrinsically faster than those leading to its disappearance. These are kinetic matters, which can be greatly influenced by catalyst composition and by the nature of the hydrocarbons present.

Carbon formation can be eliminated in practice by use of a sufficiently high ratio of steam to hydrocarbon, but that required may be considerably above that called for by thermodynamic equilibrium. For economical operation, it is desirable to reduce this ratio to the minimum consistent with adequate catalyst life. Homogeneous thermal cracking of paraffins can occur simultaneously with the catalytic reaction of the paraffin with steam, and this also forms carbon. The homogeneous reaction has a higher activation energy than the catalytic steam reaction and becomes significant above about 650 to 700°C (Bridger and Chinchen, 1970). To minimize carbon deposits formed by homogeneous processes it is therefore undesirable to subject a hydrocarbon-steam mixture to excessively high temperatures before the hydrocarbon has been substantially reacted catalytically.

The relative rates of carbon formation on nickel catalysts, from various hydrocarbons (Bridger and Chinchen, 1970, p. 74; Rostrup-Nielsen, 1977) is about as follows:

$$\text{Ethylene} \gg \text{benzene, toluene} > n\text{-heptane} > n\text{-hexane} > \text{cyclohexane} > \text{trimethylbutane} \approx n\text{-butane} \approx \text{carbon monoxide} > \text{methane}$$

Carbon formation from natural gas is generally not a significant problem, but it can be much more serious with other hydrocarbons. Aromatics such as benzene and toluene are more conducive to coke formation than paraffins and are reformed more slowly. They typically may be present in the order of several percent in a light naphtha, and a sensitive method of monitoring the slow loss of catalyst activity with a naphtha feedstock is by following the gradual increase in aromatics concentration in the product.

The nature of the carbon deposit varies considerably with operating conditions and causes catalyst deactivation in different ways. Rostrup-Nielsen and Tøttrup (1979) distinguish three forms. Under steam reforming conditions, the carbon deposit dominating at temperatures below 500°C is the result of slow polymerization of CH_x structures to form an *encapsulating film*. That dominating above 450°C is *carbon-filament* growth (Sec. 6.5). Above 600°C thermal cracking of hydrocarbons causes deposition of carbon precursors on the catalyst, which forms *pyrolytic carbon*.

The rate of the carbon-steam reaction is about 2 to 3 times faster than the rate of the carbon-carbon dioxide reaction, but the reaction rate of carbon with hydrogen is insignificant. One of the objectives of catalyst formulation is to catalyze the carbon-steam reaction. This is achieved by addition of an alkaline substance. This may act in part by neutralizing acid sites that catalyze cracking reactions. It may also adsorb and dissociate water, providing a mechanism for reacting deposited carbon with an oxygen-containing species.

10.1.2 Applications of Steam Reforming

Steam reforming may be aimed at any one of several applications, with, consequently, substantially different operating conditions.

The most important ones are:

1. Conversion of a naphtha to *substitute natural gas*, i.e., a gas with a high CH_4 content.
2. Conversion of a naphtha to a mixture of CH_4, CO, and H_2, so-called *town gas*,

which has a heating value of about 500 Btu/ft³ (1 Btu/ft³ = 27,680 J/m³). The product can also be converted to gas of a higher heating value by subsequent methanation.

3. Conversion of a naphtha or natural gas in a several-step process to a 3:1 mole ratio of H_2/N_2 for synthesis of NH_3.
4. Formation of a CO/H_2 mixture for synthesis of CH_3OH or use in oxosynthesis.
5. Manufacture of H_2.

Figure 10.1 shows the calculated equilibrium composition of the product gas from *n*-heptane under representative reaction conditions. Figure 10.2 shows typical reforming conditions for the various applications above (Rostrup-Nielsen, 1975, pp. 23 and 25). The formation of methane is favored at lower temperatures, so in applications where high conversion of methane is desired, high reactor exit temperatures are required. Other figures showing gas composition as a function of temperature, pressure, and steam ratio for methane and naphtha feedstocks are given by Bridger and Chinchen (1970, p. 64). Slack and James (1974, part 1) give tables relating the percent of methane converted at equilibrium as a function of pressure and steam/methane ratio.

For conversion of naphtha to a high methane product using a low H_2O/C ratio in the feed-gas mixture, the overall heat of reaction may be slightly exothermic. For all other cases the reaction is endothermic, and highly so when the hydrocarbon is converted essentially completely to carbon oxides and hydrogen. For conversion of natural

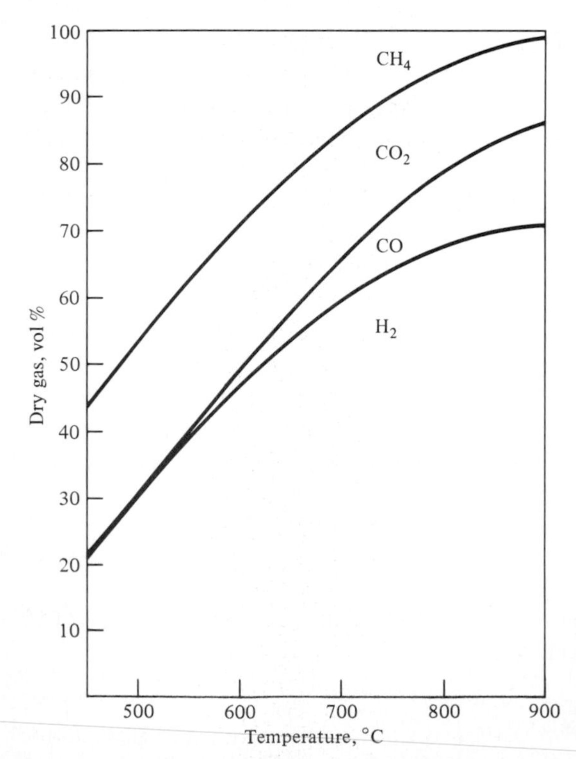

Figure 10.1 Equilibrium composition (dry basis) from steam reforming of *n*-heptane. Pressure: 3 MPa. H_2O/C = 4.0 mol/atom. Width of band indicates percentage of specified component present. (*Rostrup-Nielsen, 1975, p, 23.*)

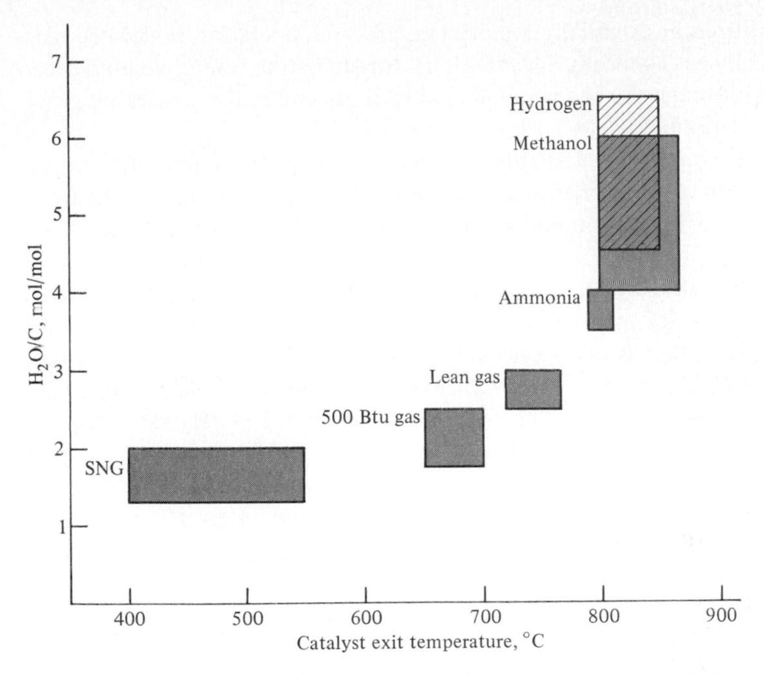

Figure 10.2 Typical reforming conditions for various applications. (*Rostrup-Nielsen, 1975, p. 25.*)

gas to synthesis gas, the natural gas is mixed with steam, in a mole ratio of about $\frac{1}{3}$ to $\frac{1}{4}$, and passed over the catalyst. This is typically held in one or two banks of vertical high-alloy steel (for example, 20% Ni and 25% Cr) tubes, 9 to 16 cm in internal diameter and 6 to 12 m long, heated in a box-type furnace. Feed enters at about 500°C and leaves at a temperature of about 800 to 850°C, the external wall temperature being about 900°C (Rostrup-Nielsen, 1975, p. 31). Pressure is in the range of about 1.5 to 3 MPa or possibly higher, in part because often the product gas is to be used at elevated pressure. The maximum exit temperature allowable at the desired pressure is set largely by metallurgical limitations on the reactor tubes. Operating conditions may vary somewhat with the properties of the feedstock such as the nature and amounts of higher hydrocarbons present and impurities such as sulfur compounds. The methane content in the exit gas should be as low as practicable. This condition is favored by high temperature, but the exit temperature is determined by an economic balance on the entire process. Methane as such is generally inert in ammonia synthesis or in hydrogenation reactions. It is detrimental only because it reduces the partial pressure of the reacting gases and because in its purge from the overall system some valuable synthesis gas or hydrogen is also lost.

In the generation of hydrogen for use in hydrogenation reactors or hydroprocessing in refineries, the methane content in the exit gas is typically in the range of 1 to 3 percent (dry basis). For generation of synthesis gas for ammonia synthesis it is typically in the range of 10 percent (dry basis) (see below). Reactor temperatures are some 100 to 200°C lower for naphtha conversion than for methane conversion. Even when product will be utilized at essentially atmosphere pressure, operation at elevated reactor pressure is usually economically advantageous. It reduces the reactor volume necessary, pressure drop through the bed is less, and natural gas and refinery gases are generally available under pressure.

For ammonia synthesis it is desirable to reduce the methane content in the product gases from the primary reformer to a very low level in order to minimize the purge of methane in the subsequent ammonia synthesis step. To do this and to introduce nitrogen the exit gas from the primary reformer, which contains up to about 10 % methane (dry basis) is mixed with air and fed to a secondary, adiabatic reformer. Here the exothermic reaction of methane and air supplies heat to raise the temperature another 150 to 200°C, that is, to the neighborhood of 1000°C or higher. Such temperatures would be difficult to achieve with externally fired metal alloy tubes. The amount of air introduced is adjusted to form the ultimate 3:1 N_2/H_2 ratio required for ammonia-synthesis gas. The primary and secondary reformer operations clearly must be closely coupled to provide the necessary overall mass and thermal balance. Typically, the secondary reactor is refractory-lined, and the catalyst is nickel supported on a form of alumina that is thermally and mechanically resistant at these very high temperatures. Sometimes a nickel supported on chromia-alumina, which has high thermal resistance, is placed in the top portion of the bed in the secondary reformer.

Product gas is then cooled, additional steam is added, and additional H_2 is produced by passing the mixture through a water-gas shift reactor (Sec. 10.3). CO_2 is then removed by absorption, as by use of an amine-promoted aqueous K_2CO_3 solution. Remaining traces of CO and CO_2 are removed by applying the methanation reaction [Eq. (10.2); Sec. 10.6]. For making feed gas for synthesis of CH_3OH or for the oxo process (CO/H_2 ratios of $\frac{1}{2}$ or 1, respectively), the shift reaction is not used. The higher CO content can be obtained by adding CO_2 to the reformer feed gas.

10.1.3 Reforming Catalysts

The high temperature of the steam reforming reaction and the presence of steam set stringent requirements on the formulation of the supported nickel catalyst. Silica, unless suitably combined, may be slowly volatilized by steam at these temperatures, leading to catalyst degradation and deposit of silica downstream, an effect which is accelerated at higher pressures. Transition aluminas such as γ and η are not stable. The support is therefore a low-area refractory-type material. In early processes operating at pressures up to about 0.3 MPa, the most common support was a calcium aluminate silicate, which was inexpensive and gradually became stronger with use. This catalyst support was unsatisfactory at higher pressures (up to about 3 MPa) because of migration of silica at the higher steam partial pressures, and it had other disadvantages.

In the early 1960s this support was replaced with a low-silica calcium aluminate; it however, had a lower crush strength, gradually lost strength on use, and lost activity by gradual formation of nickel aluminate. This, in turn, was replaced with a low-area α-alumina that, although more expensive, was mechanically stronger and longer lasting. Potassium compounds or other alkaline substances are frequently added to accelerate carbon-removal reactions. However they may be slowly lost from the catalyst by volatilization and be deposited downstream on heat-exchange surfaces or on other catalyst beds. It is also important to minimize sintering of nickel, and this is related to support composition and method of catalyst preparation.

The catalyst formulation may vary somewhat depending upon whether the feedstock is to be methane, light hydrocarbons, or a naphtha, since the propensity to coke formation varies with feedstock composition. Table 10.2 (Bridger and Chinchen, 1970, p. 78) lists the composition of three catalysts supplied by Imperial Chemical Industries (I.C.I.) that are stated to be satisfactory up to at least 3 MPa. Catalyst 57-1 is suitable up to 850°C for a feed of methane or light saturated hydrocarbons, catalyst 46-1 is formulated for use with higher hydrocarbons (naphthas) and catalyst 54-2 is formulated for secondary reforming up to 1300°C. The presence of alkali (as in catalyst 46-1) accelerates carbon-removal reactions but reduces catalyst activity. For a natural-gas feed the use of alkali is usually unnecessary, and a higher activity catalyst may be used instead. With a naphtha, control of carbon-forming reactions is more important, and alkali or alkaline earth substances are often added for this purpose. With an alkali it may be necessary to operate at lower temperatures than with methane to avoid volatilization. The nickel content of catalysts for naphtha reforming may be less than that for methane reforming. The highest degree of activity is not necessary, and a greater tendency to carbon formation with the higher nickel content may require a higher steam/hydrocarbon ratio for control.

Before the early 1960s a catalyst was not available for reforming naphthas without excessive deactivation from carbon formation. In about 1962 a catalyst was developed by the British North Western Gas Board for converting naphtha to town gas. The catalyst consists of nickel and urania supported on corundum (a naturally occurring alumina). Urania reportedly is resistant to sulfur poisoning, has low surface acidity, and promotes dissociation of steam. A representative composition is 13.0 wt % Ni, 12.1 wt % U, 0.3 wt % K. The patent literature (Nicklin and Farrington, 1974) indicates that the maximum resistance to coke deposition occurs with use of a U/Ni ratio of about 0.7. For the reforming of naphthas to produce hydrogen or synthesis gas, a more severe operation than production of town gas, this catalyst does not appear to be used to any significant extent. Compositions such as those of I.C.I. 46-1 and Topsøe RKNR are preferred, although the urania catalyst is used commercially for reforming of natural gas and light hydrocarbons.

Andrew (1969) discusses catalyst formulation from the point of view of eliminat-

Table 10.2 I.C.I. reforming catalysts, typical analyses (loss-free basis)*

	Catalyst designation		
Component	57-1	46-1	54-2
NiO	32%	21%	18%
CaO	14	11	15
SiO_2	0.1	16	0.1
Al_2O_3	54	32	67
MgO	–	13	–
K_2O	–	7	–

*Bridger and Chinchen, 1970, p. 78.

ing carbon deposition in naphtha reforming. He advocated the use of group I alkalies for this purpose, although effective commercial catalysts that do not contain alkali are in use for naphtha reforming. In I.C.I. catalyst 46-1 (Table 10.2) potassium is in the form of kalsilite, $KAlSiO_4$, which reacts slowly with CO_2 to form K_2CO_3, thereby avoiding an excessive rate of loss of volatile potassium compound. Topsøe RKNR contains about 25 wt % Ni on an active magnesia containing about 6 wt % Al. (An electron photomicrograph is shown in Fig. 5.7.) In both formulations an ingredient, potassium or magnesia, is present to enhance steam adsorption. Some catalyst compositions may contain sulfate, which during the initial reduction and first hours of operation will be converted to H_2S and pass downstream. Unless this gas is vented or the H_2S otherwise controlled, it may poison downstream catalysts, such as those in the water-gas shift reactors.

The total surface of these catalysts is about 2 to 3 m^2/g, and that of the nickel metal after being subject to operating conditions is about 0.5 m^2/g. The corresponding crystallite size is about 1 μm. Catalyst life is typically 1 to 3 years. As catalyst activity gradually drops, the furnace temperature is gradually raised to maintain the reaction rate constant. After a period, one or more "hot bands" may appear on the outside of the tubes, caused by a local drop in the rate of the endothermal reaction that keeps the tube walls below the firing temperature. This can be brought about by carbon deposition within the pores of the catalyst, which can ultimately lead to disintegration of the pellet to powder. The process can be self-accelerating since the increased internal reactor temperature accelerates carbon deposition by reactions such as $CH_4 \rightarrow C + 2H_2$. The tube-wall temperature may thus be increased to a level at which a metallurgical limitation necessitates shutdown. The hot spots typically occur at some position intermediate between inlet and exit, where the combination of hydrocarbon concentration and temperature maximizes the rate of carbon deposition.

Shutdown may also be decided upon when the pressure drop reaches an uneconomic level, caused by accumulation of fines. This in turn may be caused by carbon formation and enhanced by plant upsets. Carbon can sometimes be removed satisfactorily by steaming, after which the catalyst is re-reduced and the reactor put back on stream. Even when it is operating satisfactorily, catalyst may be replaced during a scheduled general plant shutdown for overall maintenance, as a precaution against a forced shutdown that might be required if it were to be extended to its full life. Shutdown and startup procedures are rather specific since at elevated temperatures too high a hydrocarbon concentration can cause coke deposits and too high a steam concentration can lead to overoxidation.

Under industrial reaction conditions the reaction is probably highly pore diffusion-limited, so there has been a trend towards use of smaller catalyst particles and/or rings. There are large gradients of temperature and composition between the bulk gas and the outside of catalyst pellets, as well as both axially and radially in an industrial packed tube. It is thus difficult to obtain intrinsic kinetic information for practicable catalysts under industrial operating conditions. The rate of reforming appears to be approximately first order in methane, possibly inhibited to some extent by carbon monoxide and/or hydrogen (Bridger and Chinchen, 1970, p. 91). Allen et al. (1975) give conversion data for the reaction of methane and steam at 640°C and 0.1 to 1.8 MPa on a commercial nickel catalyst (Girdler G-56B) and references to earlier studies of this reaction. Rostrup-Nielsen (1975) presents a kinetic expression for ethane reforming, gives data on other reactions on representative catalysts, and analyzes the possible degree of transport limitations under representative reaction conditions.

Nickel catalysts are very sensitive to poisoning by sulfur compounds, and the sulfur content of the feed gas must be previously reduced to a value below about 0.2 to 0.5 ppm in commercial operations. The coverage of adsorbed sulfur is determined by the ratio P_{H_2S}/P_{H_2}. The nickel becomes saturated at a value of this ratio of the order of 10^{-3} of that at which bulk Ni_3S_2 is formed at equilibrium. This illustrates the fact that surface chemisorbed species can exist under equilibrium conditions where the equivalent bulk solid does not exist. The poisoning is reversible; if sulfur is completely eliminated from the feed gas, the surface nickel sufide is reduced back to elemental nickel.

10.1.4 Reforming Processes

Historically the first steam reformers were started up in Germany by BASF about 1926 to 1928 and by the Standard Oil Co. of New Jersey (now Exxon) in the United States in 1930. For the production of hydrogen, these were usually designed for operation at slightly above atmospheric pressure; beginning in the 1950s, pressures were increased in new designs up to levels in the range of 0.4 to 1 MPa. In the 1960s pressures in the range of 3 MPa came into use, particularly for integrated steam reforming and ammonia-synthesis plants. Increased pressure is unfavorable for conversion of methane to carbon monoxide and hydrogen, but this is compensated for in secondary reforming by the high temperatures obtained. Other industrial developments and many further details on the process are provided in the concise review by Bridger and Chinchen (1970) and in the book by Rostrup-Nielsen (1975), references from which much of the above discussion has been drawn.

For the production of hydrogen for refinery use, typically feed gas or naphtha will be desulfurized, mixed with steam, and reformed, and the product gas will be passed through high- and low-temperature water-gas shift converters. Carbon dioxide will be removed by absorption, and the remaining low concentration of carbon oxides removed by conversion to methane.

With a natural gas or refinery gas consisting of light hydrocarbons, the sulfur compounds and other acid gases may be reduced to a level below 60 to 80 vol ppm by scrubbing with a solution of an organic base such as monoethanolamine. The sulfur level may then be reduced to less than 1 ppm by adsorption at 20 to 50°C on activated carbon impregnated with a metal. This is regenerated by steaming at about 160°C or higher. The effectiveness of scrubbing for removal of sulfur compounds varies greatly with their nature. Carbon adsorption is less effective with sulfur compounds of low boiling points, such as H_2S and COS; it is also less effective in the presence of higher-boiling-point hydrocarbons.

A variety of other processes may be used for removing mercaptans and other sulfur compounds from a feed stream. Mercaptans may be oxidized with air to disulfides, which are then removed by caustic scrubbing, or sulfur compounds may be removed by adsorption on zeolites. Catalytic hydrodesulfurization is usually used with feed streams consisting of naphthas and higher hydrocarbons in which the sulfur compounds present are not removable by scrubbing. Various sulfur compounds may also be removed by adsorption on hot ZnO, which is then discarded (Sec. 9.8). ZnO is effective with a much wider range of substances than is activated carbon, but the fact that it is nonregenerable makes its use costly except with low sulfur concentrations.

Synthesis gas may be prepared by a variety of other methods. The first large-scale industrial uses were for the synthesis of ammonia and methanol, the synthesis gas being made by reacting coke with steam and air in a cyclic process. Steam was injected into a bed of hot coke to form carbon monoxide and hydrogen. The temperature dropped because of the endothermic reaction; after a period the bed was reheated by injection of air, and the cycle was repeated. Coke is used instead of coal to avoid problems from the formation of tar and other volatiles and impurities. From the period of about 1910 to 1940 processes based on coke were predominant. After World War II they were gradually replaced by catalytic processes utilizing natural gas (especially in the United States, where supplies were plentiful and cheap) and naphtha (in Europe, where petroleum became a low-cost source of fuel, replacing coal). Today partial oxidation (noncatalytic) processes are of practical interest for conversion to synthesis gas of carbonaceous substances not readily reacted catalytically, such as heavy hydrocarbon feedstocks, coal, and other solids. Typically, oxygen, steam, and the feedstock are passed through an unpacked reactor at substantial pressure and 1400 to 1600°C, or a fluidized-bed reactor may be used. The synthesis gas is then purified.

10.2 FISCHER-TROPSCH SYNTHESIS

Beginning in the early 1920s Fischer and his coworkers studied a variety of catalyst compositions for the conversion of synthesis gas to fuels, which led to development of forms of iron and cobalt. In Germany, which has no significant indigenous petroleum sources, just before World War II nine plants had been built to convert synthesis gas from coal to liquid fuels, with a total rated output of about 16,000 bbl per day (1 bbl, for petroleum, = 42 gal = 0.159 m^3). These all used a cobalt catalyst in fixed-bed reactors. In the United States a fluidized-bed process utilizing an iron catalyst was installed in the 1950s to convert synthesis gas from then inexpensive natural gas to gasoline. This never operated satisfactorily, and the Fischer-Tropsch process became generally uneconomic because of plentiful supplies of petroleum.

In 1955 there began operation of an installation in Coalbrook in South Africa using an iron catalyst in fixed-bed and fluid-bed reactors with a capacity of about 5000 bbl/day, producing a broad spectrum of chemicals as well as fuels. The plant is state-owned and is known as SASOL (South African Synthetic Oil Limited). A large-scale plant, known as SASOL-II, designed to produce primarily gasoline, is scheduled to come into operation in 1980, with an ultimate capacity of about 50,000 bbl/day. A third plant, of similar capacity to the second, was authorized in 1979. These are the only operating Fischer-Tropsch processes in the world as of 1980. South Africa has large coal reserves but essentially no petroleum, and these plants will supply more than half of South Africa's consumption needs of liquid fuels.

The developments in Fischer-Tropsch synthesis through World War II are detailed in the book by Storch, Golumbic, and Anderson (1951). H. Pichler was a long-time collaborator with Fischer, and he reviews historical developments up to about 1950 (Pichler, 1952). The relatively recent book by Pichler and Kruger (1973) provides more recent information, including some on the operations at SASOL. Recent reviews have been written by Vannice (1976), Henrici-Olivé and Olivé (1976), Shah and Perrotta (1976), and Madon and Shaw (1977), the last focusing on the effect of sulfur on the synthesis.

Fischer-Tropsch synthesis is generally taken to mean the formation of a product consisting predominantly of hydrocarbons higher than methane. By varying catalyst composition, pressure, temperature, and type of reactor, the average molecular weight of products can be markedly affected. Cobalt catalysts give predominantly linear olefins and paraffins. Iron catalysts yield in addition some alcohols, ketones, and other oxygenated species. Nickel forms predominantly methane (see Sec. 10.6). At pressures of 10 MPa and above, oxides that are difficult to reduce, such as thoria, form saturated branched hydrocarbons.

At SASOL, an iron catalyst is used at about 2.5 MPa and in either a fixed-bed (220 to 240°C, H_2/CO ratio of 1.8) or an entrained-bed[1] (310 to 340°C, H_2/CO ratio of 6) reactor. The fixed-bed catalyst is prepared by precipitation of iron hydroxide onto silica gel. The entrained-bed catalyst is made by melting iron oxide and additives in an electric furnace at 1500°C and grinding to form a powder (Henrici-Olivé and Olivé, 1976). The fusion is similar to the process used to prepare commercial ammonia-synthesis catalyst, and presumably a fused catalyst makes an attrition-resistant material. The product from the entrained-bed reactor is predominantly in the gasoline boiling-point range (C_5 to C_{10}), whereas that from the fixed-bed reactor has a much broader molecular weight distribution, ranging from C_3 through paraffin wax. Table 10.3 gives a comparison of the types of compounds present in representative molecular weight fractions from the two types of reactors (Henrici-Olivé and Olivé, 1976). The paraffins from the fixed-bed process are predominantly (79 percent) normal; of those from the entrained-bed process, 55 to 60 percent are normal.

The Fischer-Tropsch synthesis is a polymerization (oligomerization) process, although the mechanism and the nature of intermediates are still in dispute. A species such as M—CHOH may be formed as the first step. This can grow by condensation reactions with other oxygenated species, or by insertion of a CO molecule into the carbon-metal bond, followed by desorption, which terminates the process. It is fairly clear that with respect to hydrocarbon formation the first product is an α-olefin, plus water.

$$n\mathrm{CO} + 2n\mathrm{H}_2 \longrightarrow (\mathrm{CH}_2)_n + \mathrm{H}_2\mathrm{O} \qquad (10.6)$$

The α-olefin may be isomerized to an inner olefin, and either type of olefin may be hydrogenated to a paraffin. The higher hydrocarbons may undergo further polymerization and cracking reactions. CO_2 in the products is apparently formed by the water-gas

[1] In an entrained-bed reactor finely divided catalyst and reacting gas pass concurrently upwards through a vertical pipe at velocities considerably higher than those used in fluid-bed reactors. The principal use of this type of reactor is for the catalytic cracking of gas oils (Sec. 9.4).

Table 10.3 Composition of liquid fractions from the SASOL plant*

	Fixed bed		Entrained bed	
Fraction	C_5–C_{10}	C_{11}–C_{18}	C_5–C_{10}	C_{11}–C_{18}
Olefins	50	40	70	60
Paraffins	45	55	13	15
Oxygenated compounds	5	5	12	10
Aromatics	–	–	5	15

*Henrici-Olivé and Olivé, 1976.

shift reaction and not by the primary synthesis. This however changes the consumption ratio of H_2/CO. For H_2/CO ratios between 1 and 3, the reaction is about first order with respect to H_2 and zero in CO.

The selectivity of iron catalysts can also be strongly affected by promoters. The addition of a small amount of potassium (e.g., up to 1 percent) causes a considerable increase in the formation of higher-molecular-weight products. Dry et al. (1972) conclude that the base donates electrons to the iron, strengthening the carbon-metal bond, thus enhancing CO adsorption and increasing the probability of the growing chain remaining on the catalyst surface. The iron catalyst is very sensitive to poisoning by traces of sulfur.

10.3 WATER-GAS SHIFT REACTION

The catalysis of this reaction [Eq. (10.3)] was developed industrially in conjunction with ammonia synthesis, since it provides a way of increasing the yield of hydrogen from synthesis gas and simultaneously decreasing the carbon monoxide content. Carbon oxides are a poison for ammonia-synthesis catalysts and for most metallic hydrogenation catalysts, and must be reduced to a very low level. Since the shift reaction [Eq. (10.3)] is mildly exothermic to the right, the maximum conversion at equilibrium is attained at the lowest temperatures. Pressure has no effect on the equilibrium. The equilibrium constant as a function of temperature is shown in Fig. 10.3 (Bridger and Chinchen, 1970, p. 64).

10.3.1 High-Temperature Shift Catalyst

The basic catalyst, used for over 50 years, contains Fe_3O_4, the stable phase under reaction conditions, plus some chromia, which acts as a textural promoter. A typical composition contains about 55 wt % Fe and 6% Cr. The catalyst is supplied with a low sulfur content (for example, <0.07 percent) when the high-temperature converter is to be followed by a low-temperature reactor (see below) in which the catalyst used is highly sensitive to low concentrations of sulfur compounds. The catalyst is unsup-

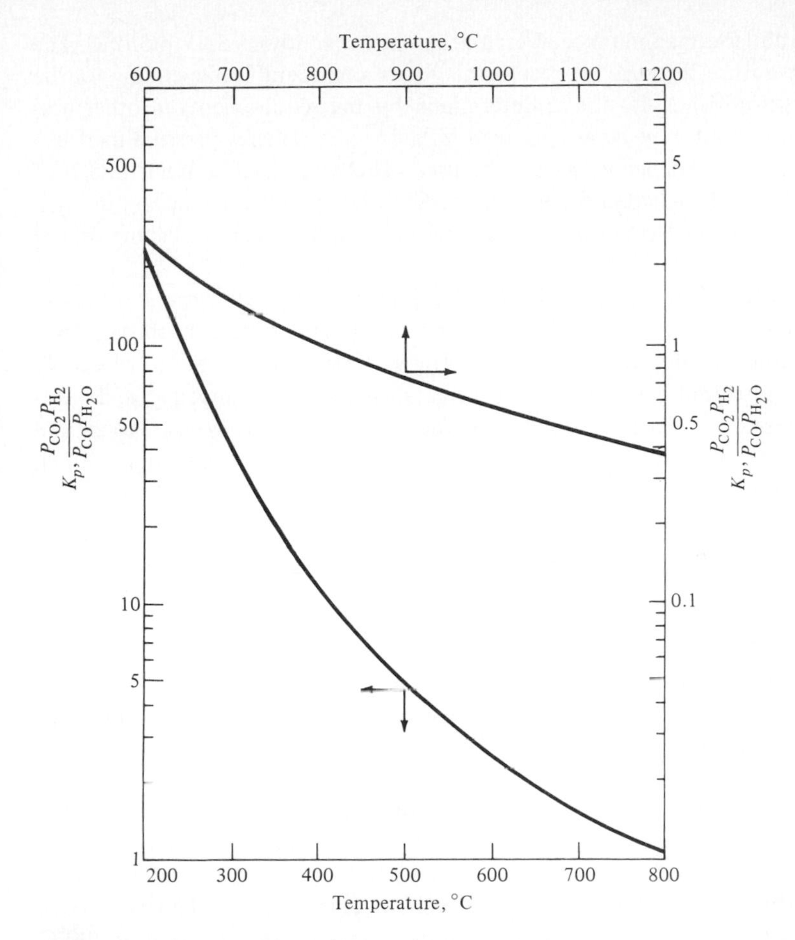

Figure 10.3 Water-gas shift equilibrium. K_p as a function of temperature. (*Bridger and Chinchen, 1970, p. 94.*)

ported, usually prepared by a precipitation process, and is available in the form of tablets (6 X 6 mm in size, for example) or rings (OD X ID X h = 10 X 4 X 8 mm, for example). As prepared, the iron is in the form of Fe_2O_3, and this is reduced to the Fe_3O_4 form before use, usually in situ. This exothermic reaction must be carefully controlled, by the use of a high concentration of steam diluent, so as not to damage the catalyst. The catalyst is reversibly poisoned by sulfur, but does not catalyze the formation of methane.

The gas composition fed to the water-gas shift reactor may vary considerably, depending upon the feedstock and process used to generate the synthesis gas. Typically, after removal of carbon dioxide by scrubbing, a large excess of steam is added to the synthesis gas. The mixture is then passed through an adiabatic converter containing the catalyst. Sufficient steam must be added to prevent reduction of the Fe_3O_4 to metallic iron, which could catalyze the formation of carbon from carbon monoxide. A temperature of about 325 to 350°C must be attained for adequate rate to be achieved, but the maximum temperature is limited to about 530 to 550°C. The increase in temperature through the reactor may vary considerably depending on the feed-gas composition. The pressure is determined by the requirements in other portions of the process, but may be as high as 3 MPa or more. Higher pressure increases the reaction rate, so a smaller vessel can be used. The superficial contact time is of the order of 3 to 9 s (calculated for wet gas at STP) but may be 1 s or less in high-pressure operation. Normal catalyst life is several years, but is decreased somewhat at higher pressures.

The book by Bohlbro (1966) gives detailed rate data and rate expressions from studies on the high-temperature catalyst at atmospheric and superatmospheric pressure. An addendum by Mogenson et al. (1969) gives further information, especially with respect to industrial procedures such as catalyst reduction, loading, and shutdown. The activated catalyst is pyrophoric. If the catalyst is to be exposed to air and is to be re-reduced and used again, it must be stabilized by surface oxidation with an inert gas containing a low concentration of oxygen.

10.3.2 Low-Temperature Shift Catalyst

Considerable carbon monoxide remains in equilibrium (for example, about 3 mol %, dry basis) at the exit temperature with use of the above catalyst. In earlier processes the carbon monoxide was removed together with carbon dioxide by scrubbing with a copper liquor (e.g., an aqueous solution of copper ammonium formate). A substantial improvement came about by the development of a low-temperature shift catalyst, first introduced commercially in 1962. With its use the exit gas from the first converter is cooled, usually by quenching with water, which supplies additional steam, and passed into a second converter. Here additional reaction is achieved over a catalyst containing copper, zinc oxide, and alumina–very similar to one of the catalysts used in methanol synthesis (Sec. 10.4.2). Copper metal and zinc oxide are the stable forms under reaction conditions, and the catalyst is active at temperatures as low as 200°C. Copper metal is the active species, and the principal role of the zinc oxide is to protect the copper from poisoning by adsorbing traces of sulfur compounds by reaction with them. This is analogous to the use of a zinc oxide "guard" bed to remove low concentrations of sulfur compounds from various processing streams. The zinc oxide also acts as a support for the copper. This catalyst is more sensitive to deactivation by thermal sintering than is the high-temperature shift catalyst, since copper has a relatively low melting point. A maximum operating temperature of about 250°C is typical. The temperature rise through the reactor is about 15°C, and the superficial contact time (calculated for wet gas at STP) is about 1 s.

As a representative catalyst, I.C.I. 52-1 has a composition before reduction of 30% CuO, 45% ZnO, and 13% Al_2O_3. Other commercial catalysts contain chromia instead of alumina. If poisons are under control, catalyst life is typically 1 to 2 years. On a commercial Cu-ZnO-Al_2O_3 catalyst operated at 1.3 MPa, the reaction becomes diffusion-limited above about 200°C.

The activity and stability to aging may be markedly affected by the method of precipitation of the catalyst (Campbell, 1970). If precipitation is carried out under alkaline conditions, a zinc compound is substantially precipitated before the copper compound, much smaller particles are formed, and consequently the crystallite size in the ultimate catalyst is smaller. With precipitation carried out under acid conditions, zinc is precipitated last, the catalyst contains larger crystallites, and it is of lower activity and shorter useful life. For maximum activity and stability the copper crystallites must be as small as possible and separated from one another by the zinc oxide and alumina. These must also be finely divided for maximum effectiveness. In a commercial catalyst the copper crystallites may typically be about 4 nm initially and about 8 nm after 6 months of operation (Young and Clark, 1973). The reduction of the copper oxide to metallic copper is highly exothermic and must be carried out carefully, at a temperature not exceeding 220 to 230°C, to avoid sintering. The used catalyst is pyrophoric and is therefore discharged under an inert gas such as nitrogen and then typically doused with water.

The catalyst is poisoned by sulfur and chlorine compounds at concentrations in the range of 1 ppm. However, sulfur compounds react preferentially with the zinc oxide to form zinc sulfide, leaving the copper active. Chlorine causes deactivation of the catalyst by accelerating the sintering of copper, caused by the formation of volatile copper and zinc chlorides. A detailed investigation of catalyst formulation and poisoning effects is given by Campbell (1970), who also includes kinetic information. Use of the low-temperature shift catalyst makes it possible to lower the carbon monoxide content in the exit gas to about 0.2 to 0.4 mol %. This is then further reduced to the parts per million range by catalytic methanation (Sec. 10.6). Copper liquor scrubbing is not generally practiced in current industrial processes.

More details, including kinetic expressions, are given by Campbell et al. (1970), Allen (1974), and Podolski and Kim (1974). In some applications it may be desirable to carry out the shift reaction on a gas stream containing concentrations of sulfur compounds that would rapidly poison the above catalysts. In that event a Co-Mo catalyst in the sulfided state may be used, on a support stable to water vapor at reaction temperature. In this type of application various organosulfur compounds may be converted simultaneously to H_2S. This then may be removed together with CO_2 in a subsequent scrubbing operation.

10.4 METHANOL SYNTHESIS

Methanol is synthesized from carbon monoxide and hydrogen by the reaction:

$$CO + 2H_2 \rightleftharpoons CH_3OH \tag{10.7}$$

$$-\Delta H_{298} = 91 \text{ kJ/mol}$$

First commercialization was in Germany in about 1923 by BASF, using a ZnO-Cr_2O_3 catalyst. The early work was closely associated with catalyst studies directed at conversion of carbon monoxide and hydrogen to synthetic fuels (e.g., Fischer-Tropsch synthesis) and with the technology developed for ammonia synthesis. In the synthesis of either ammonia or methanol, specialized and rather similar equipment operable at high temperatures and pressures is required. Methanol synthesis is like the oxidation of sulfur dioxide to sulfur trioxide, the water-gas shift reaction, and other reactions in that the conversion achievable may be greatly limited by thermodynamic equilibrium. Since the reactions are all exothermic, maximum equilibrium conversion is achievable at the lowest temperature. With methanol and ammonia, the thermodynamic limitation is particularly severe. Reactor design is aimed at operation within a fairly narrow temperature range, set by too low activity at the lower temperatures and a thermodynamic limitation at the higher temperatures. As with ammonia synthesis, high pressures are required in methanol synthesis to achieve reasonable conversions. No byproducts are capable of being formed in significant amount in ammonia synthesis, but with a mixture of carbon monoxide and hydrogen a large variety of products are thermodynamically more stable than methanol (Table 10.1). The methanol-synthesis catalyst must therefore be highly selective as well as active.

To produce a relatively pure methanol product directly requires care in catalyst manufacture, and requires procedures to avoid catalyst contamination. Iron or nickel in the metallic form are good catalysts for formation of methane, and these metals cannot be allowed to come in contact with synthesis gas under reaction conditions. The steel reactor shell is typically copper-lined, although internals may be constructed of 18-8 stainless steel (steel containing 18 wt % Cr and 8 wt % Ni). Iron can also be transported from one portion of the process to another by the formation of iron carbonyl by the reversible reaction:

$$Fe + 5CO \rightleftharpoons Fe(CO)_5 \tag{10.8}$$

The reaction is exothermic to the right, and therefore the formation of iron carbonyl is favored at lower temperatures. Iron carbonyl could thus be formed in an upstream heat exchanger and then decompose at a higher temperature in the reactor to deposit metallic iron on the catalyst.

10.4.1 High-Pressure Process

The catalyst for the original high-pressure process is made by combining a chromium trioxide with an excess of zinc oxide in the presence of water. Zinc hydroxychromates are formed, which are then decomposed to zinc dichromate by heating in air. The resulting complex structure is reduced with hydrogen or synthesis gas. The active species in the catalyst appears to be zinc oxide. Rather than being an active catalyst ingredient, chromia primarily forms a textural promoter that reduces the sintering of zinc oxide. Zinc oxide by itself is an active catalyst initially, but it loses its activity rapidly because of crystal growth. This is shown by X-ray studies. In a commercial catalyst all the chromia is found to be in the form of zinc chromite, the remainder of the catalyst being excess ZnO (Kotera et al., 1976). Williams and Cunningham (1974) state that the form is a disordered zinc chromite spinel (which may be written as $Zn^{II}Cr_2{}^{III}O_4$), and it may contain some interstitial zinc oxide. Zinc chromite as such is apparently not a very active catalyst. This conclusion is supported by the observation that the activation energy for reaction on a $ZnO\text{-}Cr_2O_3$ catalyst is about the same as that observed on ZnO alone.

Even with high-quality commercial ZnO-Cr_2O_3 catalysts, small amounts of methane and dimethyl ether are formed, the latter presumably by dehydration of methanol. Traces of higher alcohols and other species also appear. The reduction stage in catalyst preparation is highly exothermic, and to carry this out in situ requires close temperature control. Hence, as supplied, commercial catalyst is frequently prereduced and stabilized. Williams and Cunningham report on the structural changes that occur in the catalyst during reduction and provide kinetic information on the reduction reaction over the temperature range of 309 to 338°C, as well as a guide to earlier literature on catalyst preparation.

The most widely used type of reactor consists of several (e.g., four or five) adiabatic beds in series, between which product gas is cooled indirectly by water or directly by injection of additional synthesis gas, termed *cold-shot cooling*. The pressure is typically about 24 to 30 MPa. The operating temperature is in the range of 350 to 400°C, and an adiabatic temperature rise of 20 to 25°C is taken along each bed. The rate of reaction decreases markedly with conversion, because the reaction is strongly inhibited both by methanol adsorption on the catalyst (Sec. 10.4.3) and by approach to equilibrium. Typically therefore the gas leaving the last bed will contain only about 3% CH_3OH. The CH_3OH is condensed out, and unreacted gases are recycled. Inert gases are allowed to build up, and for a range of operating conditions used in one type of commercial reactor (Cappelli et al., 1972) the recycle gas contains about 8 to 14% CH_4 and about 6 to 8% N_2.

A representative reactor is shown in Fig. 10.4 (Cappelli et al., 1972). In the mixture of fresh feed and recycle gas the H_2/CO mole ratio is usually much greater than stoichiometric, for example, 6 to 10. Some CO_2 (for example, 0.6 to 3.7%) is sometimes desired in the feed gas. Since it is desirable to operate the reactor within as narrow a temperature range as possible, the excess H_2 and CO_2 serves to allow a closer approach to isothermality by the additional heat capacity they provide. CO_2 may also react by the reverse water-gas shift, which, being endothermic, consumes some of the heat developed by the exothermic CH_3OH synthesis. CO_2 may also minimize poisoning by sulfur compounds and, by the reverse water-gas shift, may be used to adjust for a feed-gas ratio of H_2/CO greater than stoichiometric. CO_2 (and H_2O) may also be able to convert chemisorbed ZnS back to ZnO. A high H_2 content will increase the thermal conductivity of the gas, and thus help even out hot spots. The temperature control provided by the CO_2 and excess H_2 also serves to control the formation of CH_4, since this reaction is highly exothermic and could get out of hand. A locally high temperature could essentially prevent the formation of CH_3OH by the equilibrium limitation and accelerate CH_4 formation, particularly if metallic iron or nickel were present.

The catalyst is typically in the form of pellets about 6 × 6 mm, or rings of OD × ID × h of 10 × 4 × 8 mm, and the reaction is moderately diffusion-limited. Pressure drop through the beds is substantial, and an economic balance is involved. Smaller pellets allow a higher rate of reaction per unit volume to be achieved, but pressure drop and thus compressing costs are then higher. The superficial contact time is 0.1 to 0.2 s, calculated at STP. Normal catalyst life is about 2 years.

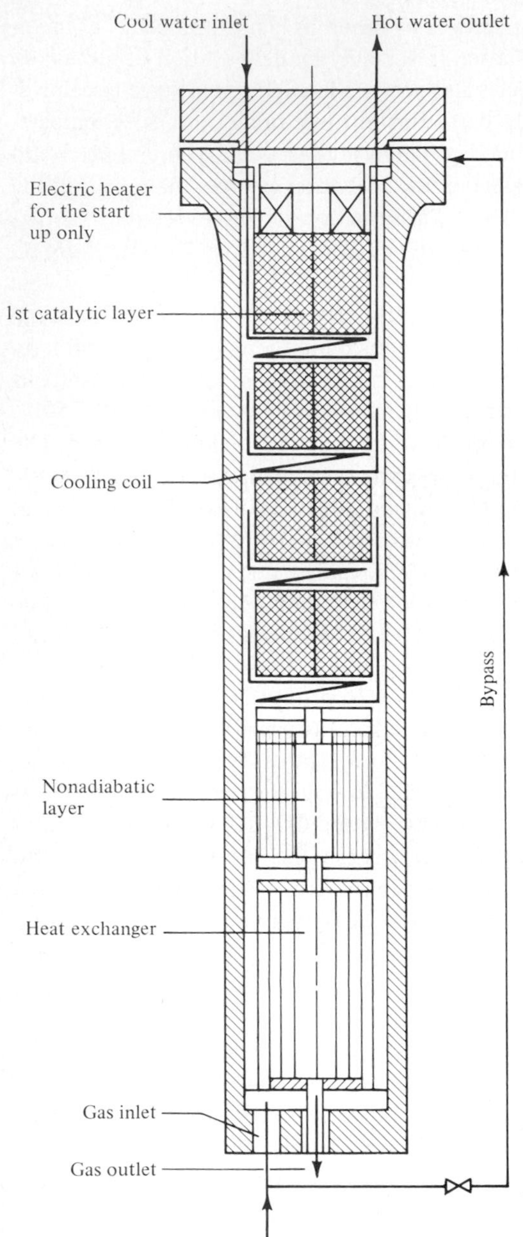

Figure 10.4 Methanol-synthesis reactor, high-pressure process. (*Cappelli et al., 1972.*) (*Reprinted with permission from Industrial and Engineering Chemistry, Process Design and Development. Copyright by the American Chemical Society.*)

10.4.2 Low-Pressure Process

A more active catalyst than the above can be made from a combination of copper and zinc together with a textural promoter such as chromia or alumina. These permit the use of a lower pressure, in the range of about 5 to 10 MPa, and a temperature of about 240 to 260°C. Recent laboratory studies (Herman et al., 1978) indicate that the active phase is a solution of Cu^+ in ZnO and that methanol yields are increased by the presence of CO_2, H_2O, or O_2 in the synthesis gas. If none of these is present, the catalyst gradually loses activity, since the Cu^+-ZnO phase apparently may be gradually reduced to inactive copper metal. This process is irreversible once the crystallites of copper metal have grown. The fact that the copper produces a chemical effect rather than a physical effect is also shown by the fact that this catalyst exhibits a considerably lower apparent activation energy than the ZnO-Cr_2O_3 catalyst. Herman et al. summarize the catalyst compositions and industrial operating conditions as reported in 14 patents on CuO-ZnO/Al_2O_3 and CuO-ZnO/Cr_2O_3 catalysts. The compositions are similar to that of the low-temperature water-gas shift catalyst (Sec. 10.3.2).

The Cu^+-ZnO catalyst is more readily poisoned by impurities such as sulfur (copper sulfide is much more thermodynamically stable than zinc sulfide), it sinters more readily, and it has a shorter life than the ZnO-Cr_2O_3 catalyst. A reactor therefore may be designed specifically to permit rapid discharge and change of catalyst. The Lurgi low-pressure process operates at about 250 to 265°C, and 4 to 5.5 MPa, utilizing a shell-and-tube reactor cooled by boiling water at 4 to 4.5 MPa to produce steam (Supp, 1973). By this design the temperature variation along the tube reactor is held to less than 10°C. The ratio of recycle gas to fresh synthesis gas is 5:1, and the gas leaving the reactor contains 4 to 6.5 vol % MeOH, depending on the gas composition.

The I.C.I. low-pressure process utilizes a single bed of catalyst and quench cooling, obtained by "lozenge" distributors especially designed to obtain good gas distribution and gas mixing and to permit rapid loading and unloading of catalyst. A schematic diagram of such a reactor is shown in Fig. 10.5. A low-pressure methanol synthesis process is advantageously combined with production of synthesis gas by partial oxidation since the latter can be carried out at the methanol-synthesis pressure, thus avoiding the necessity of intermediate gas compression. These low-pressure processes are usually the process of choice in new installations.

10.4.3 Kinetics

The kinetics of the methanol-synthesis reaction, on either a ZnO-Cr_2O_3 (ratio of 89:11) catalyst or a ZnO-CuO/Cr_2O_3 (ratio of 50:25:25) catalyst, were studied by Natta (1955) and his coworkers in a carefully designed laboratory apparatus under industrial synthesis conditions. They developed a kinetic expression for the temperature range of about 330 to 390°C that is used as a basis for design of commercial methanol-synthesis reactors. A kinetic study of this type is difficult to make since it involves a rapid, exothermic reaction at high pressure.

Natta's kinetic expression is:

$$r = \frac{f_{CO} f_{H_2}^2 - (f_{CH_3OH}/K_{eq})}{(A + Bf_{CO} + Cf_{H_2} + Df_{CH_3OH})^3} \tag{10.9}$$

where driving forces are expressed in terms of fugacities and A, B, C, and D are parameters that are a function of temperature. This equation is derivable in terms of Langmuir-Hinshelwood kinetics in which the surface reaction is taken to be a tri-

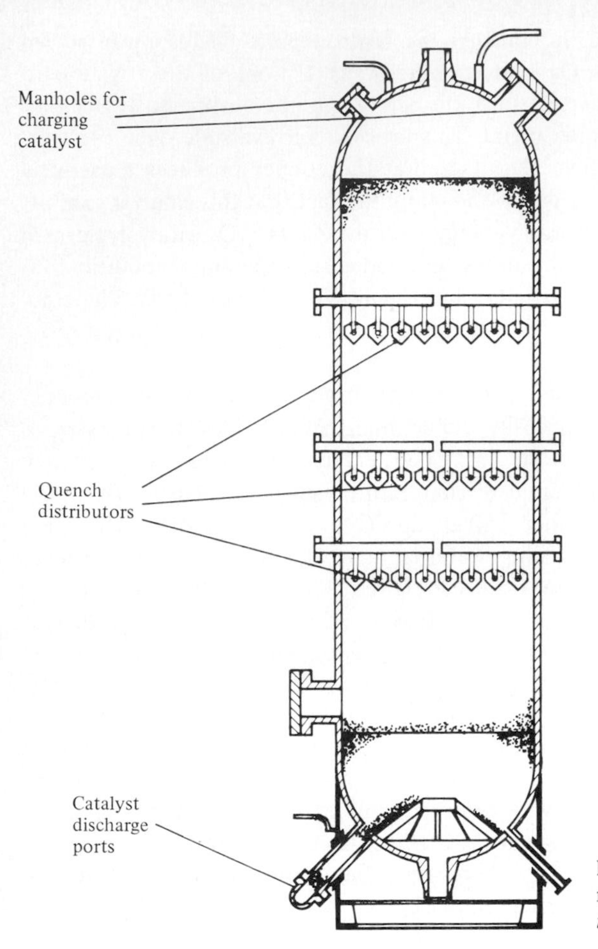

Figure 10.5 I.C.I. internal-quench methanol-synthesis reactor. (*Courtesy of I.C.I. Agricultrual Division.*)

molecular process between two molecules of adsorbed H_2 and one of CO. The assumption of a bimolecular surface process, which a priori might seem more probable, did not lead to kinetic expressions that would fit the data as well.

The constants B, C, and D are proportional to the adsorption equilibrium constants K_{CO}, K_{H_2}, and K_{CH_3OH}, respectively, in the Langmuir-Hinshelwood formulation and, in accordance with theory, decrease with increased temperature. The values decrease in the order $K_{CH_3OH} > K_{CO} > K_{H_2}$, which emphasizes the strong inhibiting effect of the CH_3OH product on the rate of reaction. Also, the maximum rate of reaction will be obtained at a ratio of H_2/CO greater than stoichiometric.

Cappelli et al. (1972) modify the above expression for a ZnO-Cr_2O_3 catalyst to include a term for CO_2. By comparison of the results of their calculations with the performance of an industrial reactor, they conclude that the water-gas shift reaction, consuming CO_2, goes rapidly to equilibrium. The temperature coefficients are somewhat different for the different adsorption constants, and hence the apparent activation energy for the reaction will vary with gas composition. For ZnO (which by itself is an active but short-lived catalyst) or $ZnO\text{-}Cr_2O_3$ the apparent activation energy is about 125 kJ/mol. For $CuO\text{-}ZnO/Cr_2O_3$ it is about 75 kJ/mol. Natta (1955) gives numerical values for Eq. (10.9) for the two types of catalysts and discusses kinetics and effects of catalyst composition in detail. A review by Stiles (1977) focuses particularly on industrial processing. A review by Denny and Whan (1978) discusses and summarizes the considerable variety of kinetic expressions that have been reported.

10.5 AMMONIA SYNTHESIS

The success of ammonia synthesis as a method of nitrogen fixation was a landmark in the development of catalytic processes. From 1820 to 1900 many futile attempts were made to use platinum and various other substances to catalyze the reaction.

$$N_2 + 3H_2 \longrightarrow 2NH_3 \qquad (10.10)$$

$$-\Delta H_{500°C} = 109 \text{ kJ/mol}$$

Failure was in large part due to the fact that the thermodynamic concepts of equilibrium were still embryonic and the pressures and temperatures chosen for study were in many cases intrinsically unfavorable for reaction. At the beginning of the twentieth century, problems of gas equilibrium became a major challenge to physical chemists, and the system of nitrogen, hydrogen, and ammonia was a favorite topic. From 1904 to 1907 the quantitative studies of Ostwald, Nernst, and Haber on the decomposition and formation of ammonia, especially at elevated pressures, led to the first clear understanding of the equilibrium relationships in this system. Figure 10.6 (Bridger and Snowden, 1970, p. 126) shows the effect of pressure, temperature, and the presence of 10 percent inert gas, in contrast to a pure stoichiometric mixture, on the equilibrium conversion. Early workers used a variety of catalysts, including platinum foil, osmium, dispersed iron, and electrolytically deposited manganese, but for industrial purposes it was necessary to develop a rugged and active catalyst.

Mittasch, in the Badische Anilin und Soda Fabrik, engaged in a search for such a catalyst in which over 8000 compositions were tried. He gradually learned that traces of arsenic, phosphorus, and sulfur were strong catalyst poisons that had to be minimized, and he finally developed a promoted iron catalyst that, with the addition or substitution of other promoters, is the catalyst universally used today. A *singly promoted* catalyst contains alumina, which acts as a textural promoter. A *doubly promoted* catalyst contains K_2O in addition, and a *triply promoted* catalyst also contains CaO. A *quadruply promoted* catalyst contains Al_2O_3, K_2O, CaO, and MgO. Catalysts used today are of the multiply promoted type.

Typically the catalyst is made by fusion of magnetite ore plus promoter precursors into a melt (at 1500°C, for example), followed by cooling, casting, crushing, and sieving. Magnetite, Fe_3O_4, is easier to reduce than Fe_2O_3, and an optimum activity of the ultimate catalyst comes from use of a Fe^{2+}/Fe^{3+} ratio approximately equal to that in magnetite. Since finely divided iron is pyrophoric, the catalyst is reduced to the metallic form in the reactor. This must be done at as low a temperature and as low a partial pressure of water vapor (which is formed by reduction) as possible to minimize

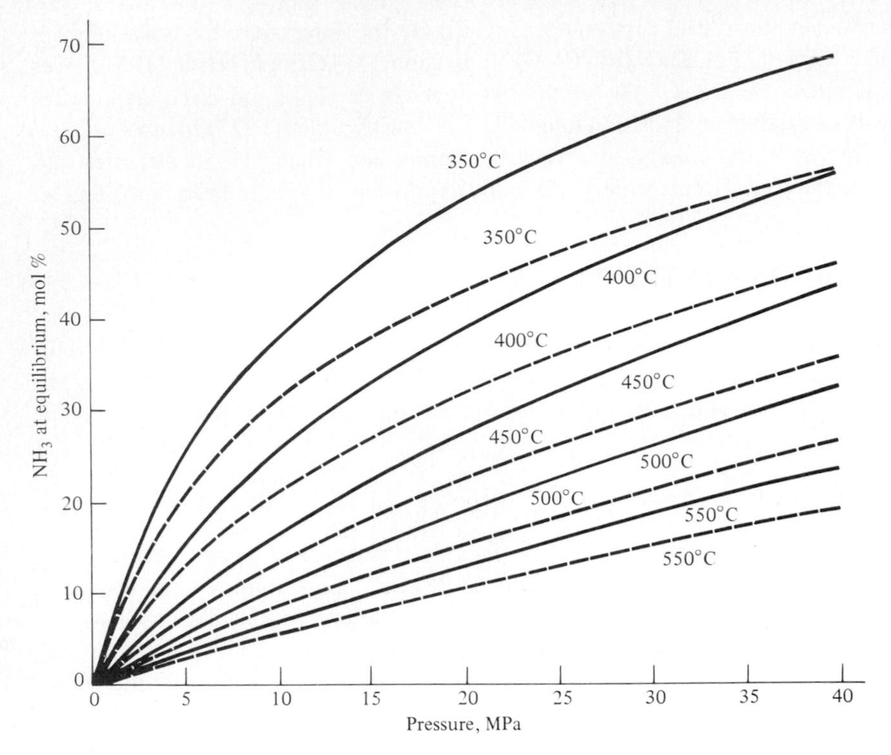

Figure 10.6 Effect of pressure, temperature, and inert gas on equilibrium ammonia concentration. Solid curves, inlet H/N = 3 : 1; dashed curves, inlet H/N = 3 : 1 with 7% CH_4 and 3% A. *(Bridger and Snowdon, 1970, p. 127.)*

crystal growth and maximize subsequent activity. This typically requires a few days. To shorten this time a prereduced catalyst is sometimes used. This has been substantially reduced by the catalyst manufacturer and then slightly reoxidized on the surface to make it nonpyrophoric during shipment.

Magnetite, Fe_3O_4, has a spinel structure $Fe^{II}[Fe_2^{III}O_4]$. During fusion some of the promoters dissolve in the magnetite; for example, Al^{3+} replaces Fe^{3+} and Mg^{2+} replaces Fe^{2+}. As the oxide is reduced, porosity is developed; dissolved Al_2O_3 and MgO come out of the structure and form a coating of very small crystallites on the iron crystals. These separate the iron crystals and act as a textural promoter. This structure may be seen in electron photomicrographs and is substantiated by comparison of total-area measurements (e.g., by the BET method) compared to iron-area measurements (by chemisorption) on the reduced catalyst. These show, for example, that although alumina typically amounts to only about 2 wt % of the total mass, it makes up one-third or more of the total surface. One of the functions of the promoters dissolved in the magnetite structure is to minimize sintering of the iron during reduction, as well as during the subsequent synthesis reaction. K_2O acts essentially as a structural promoter since it changes the chemisorption isobar of hydrogen and the kinetic order of the reaction. In the absence of alumina, K_2O does not prevent sintering, and the rate of reaction is, if anything, less than that of iron without any promoter. The mechanism of action of K_2O is still uncertain, but it may keep the iron surface free from NH, NH_2, or NH_3 groups, which otherwise might exert an inhibiting effect on the reaction rate at high pressure.

The action of CaO is complex. Its major action may be to impart resistance to sintering at elevated temperatures during the synthesis reaction. A variety of reactions can occur between different promoters, between promoters and impurities in the iron ore, and between these substances and traces of poisons in the feed stream. Since the promoters cover a large percentage of the iron surface in the reduced form of the catalyst, the maximum exposed iron surface per unit weight of catalyst is obtained with a relatively low promoter content. The optimum composition varies with reaction conditions, and the optimum for maximum activity may not be the optimum for long-term stability. Sometimes a reactor may be charged with the more active composition in low-temperature zones, and the more stable composition in the higher temperature zones. Krabetz and Huberich (1977) discuss promoter effects and interactions in some detail and discuss methods of manufacture.

Five commercial catalysts in recent usage (Guacci et al., 1977) have chemical compositions within the ranges listed in the second column of Table 10.4. That of two commercial prereduced catalysts are listed in the third column. The compositions of a large number of catalysts commercially available in 1964 to 1966 are given by Kravetz and Huberich (1977). Some of the minor substances present, such as silica, may be impurities in the magnetite ore or may be added deliberately, and they may have a significant effect. The activity of these catalysts depends not only on the reduction procedure and the elemental composition, but also to some extent on the particular ore used in manufacture. Details of the mixing, fusion, and cooling process can also be important.

Catalysts are used as crushed granules in different ranges varying from about 2 to

Table 10.4 Chemical composition of industrial ammonia-synthesis catalysts*

	Unreduced types, %	Prereduced types, %
Fe_2O_3	57.5-70.5	1.1-1.7
FeO	33.9-24.2	14.3-14.6
Fe	0-0.54	79.7-81.6
Al_2O_3	2.5-3.1	1.5-2.1
CaO	1.8-3.9	0.1-0.2
SiO_2	0.16-0.70	0.3-0.7
MgO	0.03-0.3	0.3-0.6
K_2O	0.44-0.65	0.1-0.5
Total promoters, %	5.5-7.9	3.7-4.1
Porosity, %	1.8-4.4	40-45

*Guacci et al., 1977. (Reprinted with permission from *Industrial and Engineering Chemistry, Process Design and Development.* Copyright by the American Chemical Society.)

10 mm or larger in size. The ammonia-synthesis reaction is somewhat diffusion-limited on the larger sizes. Outer layers of larger particles may also be intrinsically less active because of a degree of irreversible poisoning caused by water vapor escaping from the interior iron oxide during reduction. Smaller particle sizes permit a higher effectiveness factor (Chap. 11), but at the price of higher pressure drop per unit length and hence higher power consumption. The optimum particle size is in the range of 2 mm for use in a radial-flow or horizontal converter, 6 to 10 mm in a conventional axial-flow converter, and as high as 12 to 21 mm in other designs. Figure 4.10 shows commercial catalyst of these three size ranges. The surface area of these catalysts in the reduced form is typically 10 to 20 m^2/g. Much of this is promoter area; but even the surface area of iron alone, as determined by chemisorption, is not a very good indication of activity because of the complex effects of promoters.

In order to achieve appreciable conversion and a commercially acceptable rate, pressures in the range of 15 to 30 MPa and a minimum temperature of about 430 to 480°C are required. Extensive prepurification of the nitrogen and hydrogen is necessary to minimize poisoning. Water vapor and carbon oxides cause a temporary poisoning that, if the poisons are present in small concentration, can usually be reversed by their removal from the feed stream. However, a recrystallization may occur that lowers the effective area and hence activity of the catalyst. Most sulfur compounds cause severe and irreversible poisoning and must be minimized. In typical commercial operation a catalyst life of 5 to 8 years may be achieved before activity drops to a level at which the catalyst must be replaced.

10.5.1 Reactors

A variety of ammonia converters have been designed and used. Since the reaction is exothermic and the gas volume decreases on reaction, maximum conversion at equilibrium occurs at high pressure and low temperature. As with methanol synthesis the temperature profile through the reactor must steer a compromise between inadequate rate if the temperature is too low, and a thermodynamic limitation if the temperature is too high. The maximum catalyst temperature is limited to about 500°C to avoid significant decline in activity with time. The exit gas will typically contain about 12 to 14% NH_3. This is condensed, and the remaining gas is recycled and added to fresh gas. A very high degree of removal of NH_3 is not justified economically, so the gas mixture fed to the reactor will typically contain in the range of 4% NH_3.

In one of the more widely used designs the catalyst is held in several baskets in series between which *cold-shot* or *quench* cooling (a supply of additional synthesis gas at a lower temperature) is provided. This is rather similar to the methanol-synthesis reactor shown in Fig. 10.4. The exact design is dictated to a large extent by mechanical considerations set by operation at high pressure. Thus all inlet and exit lines are brought through the ends of the high-pressure vessel. Cold synthesis gas is passed between the inside of the pressure vessel and the catalyst containers to prevent decarburization (hydrogen embrittlement) that would occur at reaction temperature and pressure.

A radial-flow reactor may also be used. Because of the shorter path length through

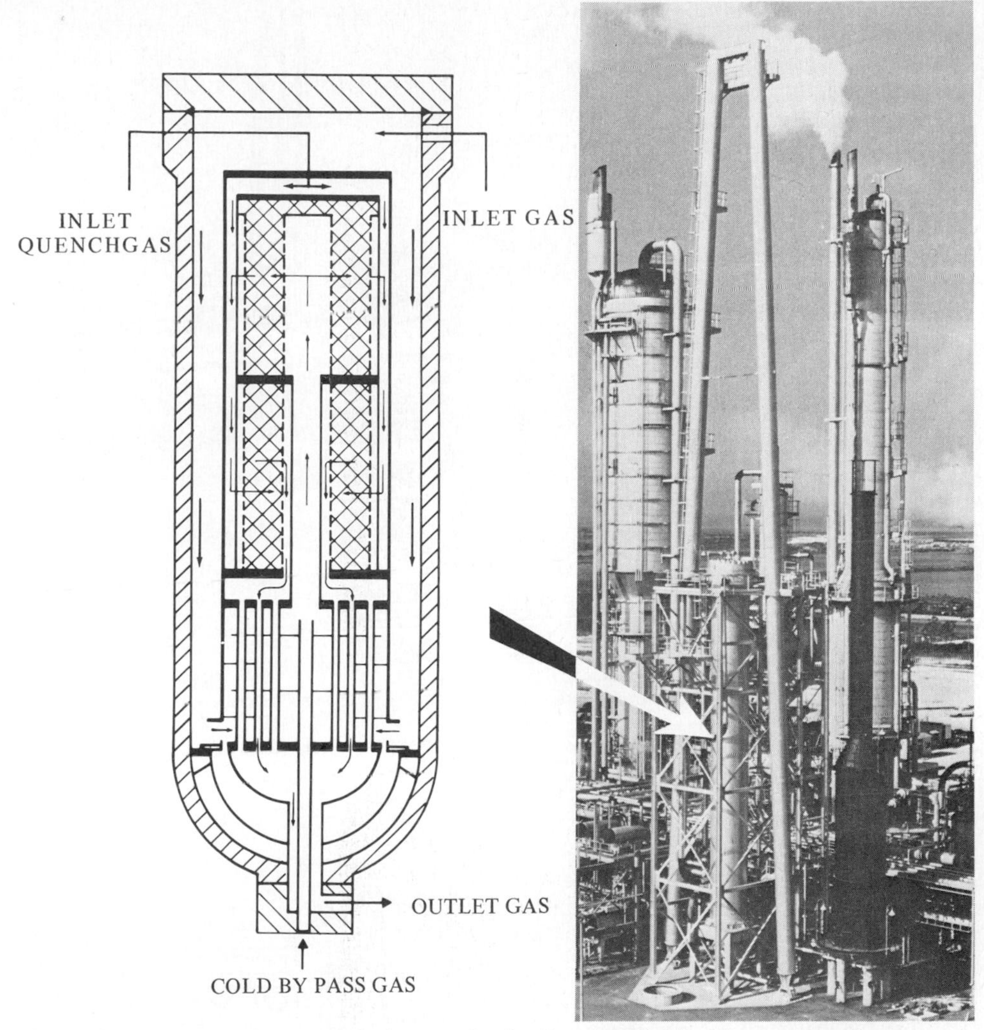

Figure 10.7 Schematic drawing and photograph of a Topsøe radial-flow ammonia-synthesis converter. (*Courtesy of Haldor Topsøe A/S.*)

the bed, smaller particle sizes, for example, 1.5 to 3 mm, may be used without excessive pressure drop. This results in a higher catalyst effectiveness factor and therefore a reduction in the total volume of catalyst required. Figure 10.7 shows a schematic diagram of the design of the Topsøe two-bed radial converter, which uses quench gases for interstage cooling. The accompanying photograph is of a reactor having a capacity of 1500 tons of ammonia per day. The accompanying overhead derrick is for installation and removal of the catalyst beds.

A horizontal converter may also be designed to achieve the same goals. Figure 10.8 shows a photograph of such a reactor, and Fig. 10.9 a schematic drawing of a unit with three beds of catalyst (Eschenbrenner and Wagner, 1972).

Figure 10.8 Photograph of a Kellogg horizontal ammonia converter. (*Courtesy of M. W. Kellogg, Inc.*)

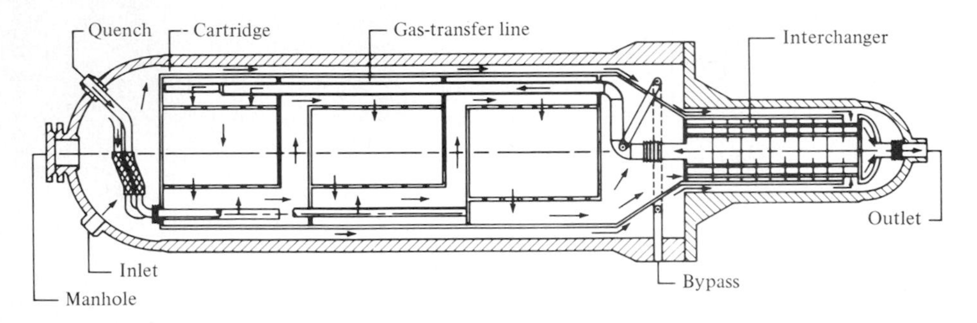

Figure 10.9 Schematic drawing of the converter shown in Fig. 10.8. (*Courtesy of M. W. Kellogg, Inc.*)

10.5.2 Kinetics

The detailed mechanism of ammonia synthesis on iron is still not clearly established, but the rate equation used almost always for correlation of data and prediction of performance of industrial reactors is based on the formulation of Temkin and Pyzhev. The original form was derived by assuming that the rate-limiting step is the chemisorption of nitrogen to form nitrogen atoms. These are assumed to be the main adsorbed species, and their surface concentration is determined by an equilibrium with hydrogen and ammonia in the gas phase. (This is not the equilibrium with the actual concentration of gaseous nitrogen, but instead with the partial pressure of nitrogen that would exist if N_2 were present in equilibrium with the gaseous hydrogen and ammonia present.) It was further assumed that the sites have a linear distribution of the heat of adsorption, and of the activation energy for adsorption and desorption of nitrogen. The resulting expression for the rate of reaction is:

$$r = k_1 P_{N_2} \left(\frac{P_{H_2}^3}{P_{NH_3}^2} \right)^{\alpha} - k_2 \left(\frac{P_{NH_3}^2}{P_{H_2}^3} \right)^{\beta} \tag{10.11}$$

The first term is for the rate of formation of ammonia, and the second for the rate of decomposition. α and β are both positive, and $\alpha + \beta = 1$.

The expression has been modified in various ways. Since operation is at high pressure, fugacities are usually substituted for pressures. Other derivations have used different assumptions concerning the energy distribution of the sites and the nature of the principal adsorbed species (for example, assuming NH to be present instead of N atoms), while retaining the two-site adsorption of nitrogen as the rate-limiting step (see, e.g., Boudart, 1962). The net effect is that in essence the form of the Temkin-Pyzhev equation is retained, but α then has a different mechanistic interpretation. The sum of α and β remains 1, but the numerical value of α may vary. Temkin proposed that $\alpha = 0.5$ on all iron catalysts.

Careful experimentation with commercial iron catalysts under practical conditions has led various researchers to the conclusion that the best correlation of the experimental data on various individual catalysts was obtained by use of a value of α in the range from 0.4 to 0.75 (Nielsen, 1968; Nielsen et al., 1964; Guacci et al., 1977). It is not unreasonable to expect that α would vary with catalyst composition, but comparisons of correlations of experimental data using values of α from 0.5 to 0.75 show little or no significant difference (Dyson and Simon, 1968; Guacci et al., 1977).

Equation (10.11) clearly cannot be applicable at zero NH_3 concentration. This is generally unimportant for industrial operations, since the feed to the converter contains some NH_3 from recycled gas. In the absence of NH_3, Temkin and coworkers have established that the expression is of the form

$$r = k p_{H_2}^{\alpha} \, p_{N_2}^{1-\alpha} \tag{10.12}$$

where $\alpha = 0.5$ (Nielsen, 1968).

Equation (10.11) is sometimes recast into the functional form:

$$r = k_2 \left[K^2 \cdot f_{N_2} \left(\frac{f_{H_2}^3}{f_{NH_3}^2} \right)^{\alpha} - \left(\frac{f_{NH_3}^2}{f_{H_2}^3} \right)^{1-\alpha} \right] \tag{10.13}$$

where K is the equilibrium constant for the reaction $1.5H_2 + 0.5N_2 \rightleftharpoons NH_3$ and k_2 is the reaction rate constant for the *reverse* reaction. (The latter is, of course, proportional to the constant for the forward reaction, k_1.) The value of k_2 seems to be independent of pressure if fugacities instead of pressures are used in Eq. (10.13). k_2 can be expressed in the Arrhenius form, $k_2 = Ae^{-E/RT}$, but the value of E depends on the value of α chosen. Nielsen et al. (1964) reported a value of $E = 177$ kJ/mol for their (triply-promoted) catalyst, taking $\alpha = 0.64$. Dyson and Simon recalculated a value of $E = 171$ kJ/mol from the same data with $\alpha = 0.50$. Bridger and Snowdon quote "best-fit" values of about 159 kJ/mol. Guacci et al., in their analysis of the activity of several commercial catalysts, also show how the comparison may vary depending upon the value of α chosen and the temperature.

An enormous literature exists on ammonia synthesis, which is not surprising in view of its central importance for the manufacture of fertilizer and explosives and the long-time study devoted to it. The book by Nielsen (1968) gives an extensive treatment of industrial catalysts. Composition, characterization, and detailed results of rate measurements under industrial conditions are presented, with extensive references. Much of the understanding of the mechanism of ammonia synthesis is due to P. H. Emmett, who surveys progress in a recent review (Emmett, 1975). Surveys of various aspects of the subject may also be found in the book by Vancini (1971), and in earlier reviews by Frankenburg (1955) and Bokhoven et al. (1955). Mittasch (1950) in a personal memoir describes the development of his early concepts of promoter action and multicomponent catalysts, which led to the first practicable catalyst composition. A pithy summary of ammonia-synthesis catalysts, focusing on promoters, poisons, kinetics, and thermodynamics is given by Bridger and Snowdon (1970). A four-part detailed treatise by Slack and James (1974 to 1977) covers ammonia from an industrial point of view, including the processes for preparation of the synthesis gas, reactor design, and economics. Recent developments are reviewed by Shannon (1978).

10.6 METHANATION

The reaction of low concentrations of CO in a mixture with H_2 to form CH_4 was developed as a gas-purification process in the 1950s. Commercial catalysts in essentially all cases are nickel on an alumina or other oxide support. Typical surface areas in the oxide form are 30 to 80 m^2/g. Iron is less active and more subject to carbon deposition. The gas to be purified typically contains 0.2 to 0.4% CO and contains a similar amount of CO_2. CO reacts more readily to form CH_4 than does CO_2, but the reaction rate of the latter becomes significant after the CO concentration has been reduced to 200 to 300 ppm. Some CO_2 may also be converted to CO by the reverse of the water-gas shift.

Representative reaction conditions with an inlet composition of 0.5% CO and 0.2% CO_2 are an inlet temperature of 315°C, an exit temperature of 365°C, and exit composition of 5 ppm of total carbon oxides for operation in an adiabatic reactor at 3 MPa (Campbell et al., 1970). With a lower inlet concentration of carbon oxides, the overall adiabatic temperature rise will be less. The inlet temperature is then typically slightly increased to restore adequate activity in the methanator. Under representative operating conditions the reaction is highly diffusion-limited. Rehmat and Randhava

(1970) report that for selective removal of CO with minimum reaction of CO_2, ruthenium is effective. A Raney nickel-type catalyst was almost as effective as ruthenium, but a supported nickel catalyst was much less satisfactory.

Studies on catalysts and processes for making synthetic CH_4 were pursued by the British Fuel Research Board from 1939 to 1945 and by the U.S. Bureau of Mines from 1951 on. In recent years there has been an upsurge in interest in these processes, spurred by the rapidly rising costs of natural gas and the possibility of replacing high-Btu "pipeline" gas by synthetic CH_4 from coal. A supported nickel catalyst is again preferred. With undiluted synthesis gas the reaction is highly exothermic. Temperature control in an adiabatic reactor requires use of a high recycle ratio of partly cooled product gas to fresh feed. Practicable pressures are in the range of 1 to 3 MPa. The ignition temperature is about 200°C under commercial conditions, and this minimum temperature must also be exceeded to avoid formation of nickel carbonyl. The maximum temperature is limited by an excessive rate of inactivation of the catalyst by sintering.

The CO content of the product gas must be less than about 0.1 percent so as not to constitute a health hazard. Taking representative inlet and exit temperatures as, say 290 and 450°C respectively, the CO in the mixed feed to the reactor, consisting of recycled gas plus fresh feed, cannot exceed about 3 percent for adiabatic operation. Alternately a "tube-wall" reactor may be used, in which the catalyst is applied directly to the heat-exchange surface. This allows better temperature control and the ability to avoid or minimize recycling, but a long-lived catalyst that will adhere is required (Haynes et al., 1970). With any type of reactor, operation must also be under conditions to avoid carbon deposition (see Sec. 10.1).

10.6.1 Rate Expressions

Rate expressions have been published by a number of authors. Those of Schoubye (1969) and Van Herwijnen et al. (1973) are for conditions that resemble the dilute concentrations as found in tail-end cleanup in ammonia-synthesis plants. Their reaction conditions were about 90 to 150°C, 0.1 to 1.5 MPa, and up to 20% CO in H_2 (Schoubye), and 170 to 210°C, 0.1 MPa, and less than 2% CO in H_2 (Van Herwijnen). Van Herwijnen used a Girdler G-65 catalyst in particle sizes of 0.03 to 0.4 mm, diluted with quartz, and concluded that diffusion limitations were not significant. Schoubye observed diffusion limitations on particles larger than 0.3 mm. Lee et al. (1970) studied compositions consisting of 2.4 to 10% CO, 2% CO_2, 13 to 34% H_2, and 53 to 82% CH_4, at 0.1 to 6.8 MPa and 275 to 480°C. These conditions are more representative of those for making synthetic methane, but diffusion effects may have been significant.

The rate expressions differ somewhat in form, and none allows for approach to equilibrium, although this may not be significant under industrial reaction conditions. It is difficult to obtain intrinsic kinetic data for this reaction by conventional packed-bed studies at industrial pressures and temperatures, and this is a case in which a recirculating reactor (Sec. 11.8.4) is particularly applicable. Lee's paper includes a summary of the expressions reported by many other investigators. Hausberger et al. (1975) give an approximate empirical procedure used for preliminary design with commerical nickel catalysts, and they compare it to a number of other equations in the literature. An additional complication in using any of these expressions comes from uncertainties about the extent to which direct methanation of CO_2 will occur. It is probably a good assumption that the water-gas shift reaction will be in essential equilibrium under almost all circumstances.

A detailed review of catalysts, possible reaction mechanisms, some of the published rate expressions, and industrial considerations are given by Mills and Steffgen (1973). A symposium edited by Seglin (1975) contains papers dealing with kinetics, thermodynamics, catalysts, and reactor considerations. A review by Greyson (1956) covers earlier developments.

REFERENCES

Allen, D., in A. V. Slack and G. R. James (eds.): *Ammonia*, part 2, Dekker, New York, 1974, p.3.

——, E. R. Gerhard, and M. R. Likins, Jr.: *Ind. Eng. Chem., Process Des. Dev.*, **14**, 256 (1975).

Andrew, S. P. S.: *Ind. Eng. Chem., Prod. Res. Dev.*, 8, 321 (1969).

Bhatta, K. S. M., and G. M. Dixon, *Ind. Eng. Chem., Prod. Res. Dev.*, 8, 324 (1969).

Bohlbro, H.: *An Investigation on the Kinetics of the Conversion of Carbon Monoxide with Water Vapour over Iron Oxide Based Catalysts*, Gjellerup, Copenhagen, 1966. Addendum; E. Mogensen, M. H. Jørgensen and K. Søndergaard: *Industrial Use of Shift Catalysts*, Gjellerup, Copenhagen, 1969.

Bokhaven, C., C. van Heerden, R. Westrik, and P. Zwietering in P. H. Emmett (ed.): *Catalysis*, vol. 3, Reinhold, New York, 1955, p. 265.

Boudart, M.: *Chem. Eng. Prog.*, **58**, 73 (1962).

Bridger, G. W., and G. C. Chinchen: *Catalyst Handbook (I.C.I.)*, Wolfe Scientific Books, London, 1970, chap. 5; Springer-Verlag, New York.

——, and C. B. Snowdon: *Catalyst Handbook (I.C.I.)*, Wolfe Scientific Books, London, 1970, p. 126; Springer-Verlag, New York.

Campbell, J. S.: *Ind. Eng. Chem., Process Des. Dev.*, 9, 588 (1970).

——, P. Craven, and P. W. Young: *Catalyst Handbook (I.C.I.)*, Wolfe Scientific Books, London, 1970, p. 97; Springer-Verlag, New York.

Cappelli, A., A. Collina and M. Dente: *Ind. Eng. Chem., Process Des. Dev.*, **11**, 184 (1972).

Dart, J.C.: *AIChE Symp. Ser. No. 143*, **70**, 5 (1974).

Denny, P. J., and D. A. Whan, in *Catalysis*, vol. 2, The Chemical Society (London), 1978, Chap. 3.

Dent, F. J., and J. W. Cobb: *J. Chem. Soc.*, **2**, 1903 (1929).

——, L. A. Moignard, A. H. Eastwood, W. H. Blackburn, and D. Hebden: *Trans. Inst. Gas Eng.*, 602 (1945).

Dry, M., T. Shingles, and L. Boshoff: *J. Catal.*, **25**, 99 (1972).

Dyson, D. C., and J. M. Simon, *Ind. Eng. Chem., Fundam.*, 7, 605 (1968).

Emmett, P. H., in E. Drauglis and R. I. Jaffee (eds.): *The Physical Basis for Heterogeneous Catalysis*, Plenum, New York, 1975, p. 3.

Eschenbrenner, G. P., and G. A. Wagner, III: *Chem. Eng. Prog.*, January 1972, p. 62.

Frankenburg, W. G., in P. H. Emmett (ed.): *Catalysis*, vol. 3, Reinhold, New York, 1955, p. 171.

Guacci, U., F. Traina, G. Buzzi Ferraris, and R. Barisone: *Ind. Eng. Chem., Process Des. Dev.*, **16**, 166 (1977).

Hausberger, A. L., C. B. Knight, and K. Atwood: *Adv. Chem. Ser. No. 146*, 1975, p. 47.

Haynes, W. P., J. J. Elliott, A. J. Youngblood, and A. J. Forney, *Am. Chem. Soc. Div. Petr. Chem. Prepr.*, **15** (4) A121 (1970).

Henrici-Olivé, G., and S. Olivé, *Angew. Chem. Int. Ed. Engl.*, 15 (3), 136 (1976).

Herman, R. G., K. Klier, G. W. Simmons, B. P. Finn, J. B. Bulko and T. P. Kobylinski, *Am. Chem. Soc. Div. Pet. Chem. Prepr.*, **23**(2), 595 (1978).

Kotera, Y., M. Oba, K. Ogawa, K. Shimomura, and H. Uchida in B. Delmon, P. A. Jacobs, and G. Poncelet, (eds.): *Preparation of Catalysts*, Elsevier, Amsterdam, 1976, p. 589.

Krabetz, R., and T. Huberich in A. V. Slack and G. R. James (eds.): *Ammonia*, part 3, Dekker, New York, 1977, p. 123.

LeBlanc, J. R., S. Madhavan, and R. E. Porter: *Encyclopedia of Chemical Technology*, vol. 2, 3d ed., Interscience, New York, 1978, p. 470.

Lee, A. H., H. L. Feldkirchner, and D. G. Tajbl: *Am. Chem. Soc. Div. Pet. Chem., Prepr.*, **15**(4) A93 (1970).

Madon, R. J., and H. Shaw: *Catal. Rev., Sci. Eng.*, **15**, 69 (1977).

Mills, G. A., and F. W. Steffgen: *Catal. Rev.*, 8, 159 (1973).

Mittasch, A.: *Adv. Catal.*, **2**, 82 (1950).

Natta, G., in P. H. Emmett (ed.): *Catalysis*, vol. 3, Reinhold, New York, 1955, p. 349.

Nicklin, T., and F. Farrington: U.S. Patent 3,847,836 (Nov. 12, 1974).

Nielsen, A.: *An Investigation on Promoted Iron Catalysts for the Synthesis of Ammonia*, 3d ed., Gjellerup, Copenhagen, 1968.

——, J. Kjaer, and B. Hansen, *J. Catal.*, **3**, 68 (1964).

Pichler, H.: *Adv. Catal.*, **4**, 271 (1952).

Pichler, H., and G. Krüger: *Herstellung flüssiger Kraftstoffe aus Kohle*, Gersbach & Sohn, München, 1973.

Podolski, W. F., and Y. G. Kim: *Ind. Eng. Chem., Process Des. Dev.*, **13**, 415 (1974).

Rehmat, A., and S. S. Randhava: *Ind. Eng. Chem., Prod. Res. Dev.*, **9**, 512 (1970).

Rostrup-Nielsen, J. R.: *Steam Reforming Catalysts*, Danish Technical Press, Copenhagen, 1975.

——, *Chem. Eng. Prog.*, September 1977, p. 87.

——, and P. B. Tøttrup: paper presented at symposium, *Science of Catalysis and its Application to Industry*, Sindri, India, February, 1979.

Schoubye, P.: *J. Catal.*, **14**, 238 (1969).

Seglin, L. (ed.): "Methanation of Synthesis Gas," *Adv. Chem. Ser. No. 146*, 1975.

Shah, Y. T., and A. J. Perrotta: *Ind. Eng. Chem. Prod. Res. Dev.*, **15**, 123 (1976).

Shannon, I. R., in *Catalysis*, vol. 2, The Chemical Society (London), 1978, Chap. 2, p. 28.

Slack, A. V., and G. R. James: *Ammonia*, Dekker, New York 1974–1977, four volumes.

Stiles, A. B., *AIChEJ.*, **23**, 362 (1977).

Storch, H. H., N. Golumbic, and R. B. Anderson: *The Fischer-Tropsch and Related Syntheses*, Wiley, New York, 1951.

Supp, E.: *Chemtech*, July 1973, p. 430.

Vancini, C. A., *Synthesis of Ammonia*, English translation, Macmillan and CRC Press, New York, 1971.

Van Herwijnen, T., H. Van Doesburg, and W. A. De Jong: *J. Catal.*, **28**, 391 (1973).

Vannice, M. A.: *Catal. Rev., Sci. Eng.*, **14**, 153 (1976).

Williams, R. J. J., and R. E. Cunningham: *Ind. Eng. Chem., Prod. Res. Dev.*, **13**, 49 (1974).

Young, P. W., and C. B. Clark: *Chem. Eng. Prog.*, **69** (5), 69 (1973).

CHAPTER

ELEVEN

EXPERIMENTAL METHODS

We are concerned here with carrying out studies on catalysts at laboratory or pilot plant scale. The ways in which this may be done will vary considerably with circumstances, such as scouting new catalysts, determining kinetics, proving out a practical catalyst, or studying new feedstocks. Generally information is needed on activity, selectivity, and catalyst life, but life tests are so expensive that they are generally reserved for the last stages of experimentation. In all these studies it is of the utmost importance that catalysts be evaluated, if at all possible, free of gradients of concentration and temperature within and between catalyst pellets in the reactor. This is not easy to do, since porous catalysts are usually desired for high area and therefore high activity per unit of reactor volume and heats of reaction are often substantial. In many cases significant gradients cannot be economically avoided in the industrial reactor, but in research and development it is important, if at all possible, to be able to determine the intrinsic behavior of a catalyst as observed in the absence of these gradients.

Consider a particle or pellet of solid catalyst in contact with a gas or liquid in which the reactants are present. The solid may be in a fixed bed or may be in motion with respect to the fluid. The latter occurs in a fluidized bed with a gas, or in a stirred vessel of liquid, sometimes termed a *slurry reactor.* The reactants are transferred to the active surface of the catalyst, reaction takes place, and the products are transferred back to the main body of the ambient fluid. If the catalyst is porous, the reactants must diffuse first from the fluid to the outside surface of the pellet and then through minute and irregularly shaped pores to the interior. Chemical potential decreases in the direction of diffusion through the porous structure, so the catalyst surface in the interior of the pellet is in contact with a fluid of lower reactant concentration and higher product concentration than the external or ambient fluid. The internal surface is not as "effective" as it would be if it were all exposed to contact with the external fluid.

Measurements that the experimenter uses for interpretation of data, measurements of composition and temperature, for example, are almost invariably those of the bulk of the fluid. However the observed course of the reaction is the sum of the events occurring throughout the catalyst, being determined by the conditions actually existing at each point on the internal surface of the catalyst. When gradients of concentration or temperature are significant, a "falsification of the kinetics" occurs. This is in the sense that the rate and selectivity of the reaction change with measured concentration and temperature in a different manner than they would in the absence of such gradients. In most cases selectivity is affected adversely.

The term *intrinsic kinetics* refers to the behavior of the reaction in the absence of concentration or temperature gradients. The adjective *apparent* or *effective* refers to that which is actually observed. When the difference is significant, the terms *mass- (or heat-) transfer limitation*, *diffusion limitation*, or *mass- (or heat-) transfer regime* are frequently used, but this phraseology is subject to misinterpretation. Mass- and heat-transfer effects may interact with each other such that effects may be observed when the potential difference between the bulk fluid and the reaction sites amounts to but a few percent of the overall decrease in potential. It is to be emphasized that these gradients *always* exist during reaction. In experimental work the objective is to reduce them to a minimum consistent with other objectives.

It is difficult to avoid a significant degree of coupling of physical phenomena with chemical reaction, particularly as reactors are scaled up for industrial processing, where high reaction rates are desired. Scientists or engineers engaged in research or development need to be able to conduct studies free of these physical transport limitations if possible in order to interpret their results correctly. They must know how to design such experiments properly, be aware of warning signs to look for in the data, and have some knowledge of what can be done if these effects become significant. Engineers concerned with development, design, and operation of reactors need to be aware of what changes in conversion and selectivity may occur as they change scale or alter operating parameters.

This is a large and complicated subject that has received extensive study, and the purpose of this chapter is to introduce the topic. Quantitative methods of analyzing these effects are considered in detail elsewhere (Satterfield, 1970).

11.1 COMMERCIAL REACTORS

The ultimate goal is to design and predict the performance of industrial-scale reactors using solid catalysts. This is the task of the chemical reactor engineer, but some understanding of the kinds of reactors most commonly used and of the kinds of information needed for scaleup helps in guiding laboratory work. The basic types are listed in Table 11.1. We will be concerned here primarily with the case of a gas phase in contact with a solid catalyst held in a fixed-bed reactor.

11.1.1 Adiabatic Reactor

The adiabatic reactor is a packed bed of catalyst without internals for transferring heat. The diameter is usually sufficiently great that heat transferred through the out-

Table 11.1 Basic types of commercial catalytic reactors

Gas-phase reactant	Gas- and liquid-phase reactants
Adiabatic packed bed	Trickle bed; flooded bed
Multitube with heat exchange	–
Fluidized bed	Slurry reactor

side walls is small relative to that evolved or absorbed by reaction. The fluid moves through the reactor in nearly plug flow, and the temperature rise (or drop) for a simple reaction is about in proportion to the percent conversion. Such a reactor may be used for either an exothermic or endothermic reaction. However, most applications involve exothermic reactions, and the following discussion is concerned with this type. An adiabatic reactor is the cheapest kind to build, and usually the first choice if practicable. Reacting gases are fed to the bed at the *ignition temperature*, that at which the rate becomes economically rapid.

A representative design is shown in Fig. 11.1. A set of baffles or perhaps a simple horizontal plate below the opening may be installed to provide more uniform flow of vapors onto the bed. A layer of dense, nonporous, inert material such as alumina or a fused ceramic larger in size than the catalyst particles is usually placed on top of the catalyst, to catch scale and impurities and to assist in flow distribution. The catalyst is typically supported on a grid, perhaps with screens and/or a layer of inert material between the grid and the catalyst packing. Spent catalyst is usually removed by vacuuming.

In an alternate design, product may be removed at the bottom of the vessel through a collector such as a spider, or through a stack of horizontal plates separated to provide a slotted opening. This is then usually surrounded by inert packing larger than the sizes of the openings. The inert material may be graded in size–coarse to fine in the top of the reactor and fine to coarse in the bottom.

Several beds may be used in series, and often they are all held inside a single shell for mechanical or structural reasons, especially for operation under pressure. Gas from the exit of one bed may be removed, cooled, and fed to a second bed (Fig. 8.5) or cooling coils may be installed in sections between beds. (Fig. 10.4.) Cooling coils are seldom embedded within a catalyst bed because this may cause irregular flow through the packing and undesirable temperature gradients. Alternately, reacting gases may be cooled between beds by injection of fresh, cool reactant, *quench gas*, which is mixed with the reaction products to lower the gas temperature before it enters the next bed (Figs. 10.7 and 10.9), or quench gas may be introduced into a single bed of catalyst by specially designed distributors (Fig. 10.5). The use of quench gas is also termed *cold-shot* cooling. With several adiabatic beds in series, the individual beds are often varied in depth, in order to achieve an optimum temperature at the exit of each. This is usually determined by either (1) the temperature at which the reaction rate drops to too low a level because of approaching equilibrium or (2) the temperature at which side reactions become significant or at which catalyst life may be adversely affected.

To make the temperature change through a reactor sufficiently small that an

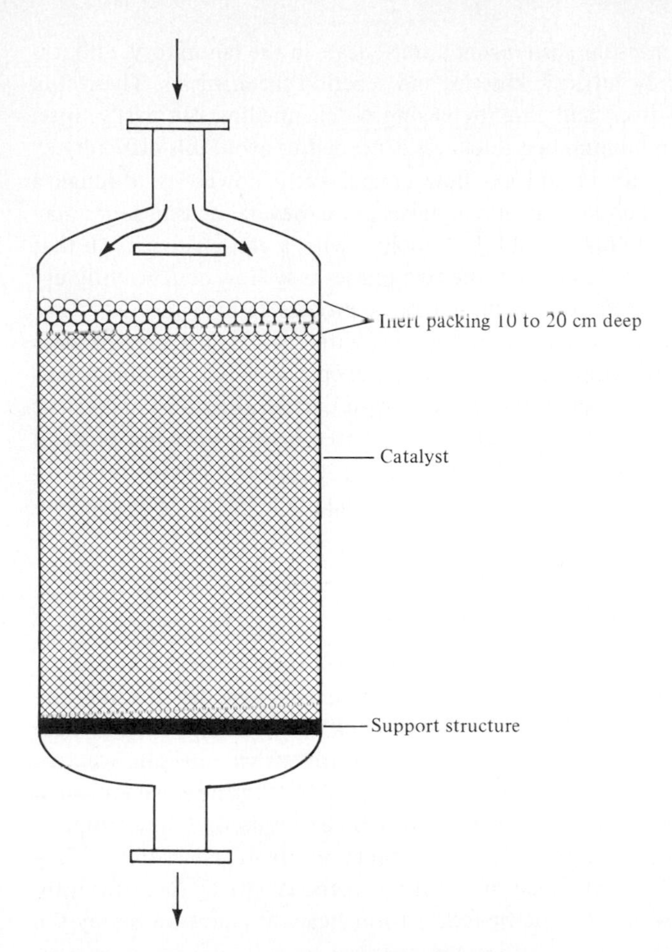

Figure 11.1 Representative design of an adiabatic reactor.

adiabatic reactor can be used: (1) a portion of the product may be recycled and mixed with fresh feed, (2) an excess of one reactant may be utilized, or (3) an inert gas may be added to the feed. Any of these increases the heat capacity relative to the quantity of heat released. Adiabatic reactors are generally most practical for large-scale, relatively slow reactions not involving large heat effects.

A group of highly exothermic reactions occurs extremely rapidly at high temperature, with reaction times on the order of 0.01 s or less. Examples are NH_3 oxidation to NO, synthesis of HCN from NH_3, CH_4, and air, and conversion of CH_3OH to HCHO on a silver catalyst. The catalyst beds are very thin, frequently 1 cm or less (see Fig. 8.10). The reactions are highly mass transfer-controlled, so the concentration of reactants and products at the outside catalyst surface is much different than that in the bulk. The catalyst-surface temperature greatly exceeds the bulk-gas temperature, especially at the inlet. Radiation heat transfer from the top and bottom of the bed may be important, although the reactor as a whole is usually nearly adiabatic. It is difficult to simulate these reaction systems on a small scale in the laboratory, and it is exceedingly difficult to study intrinsic kinetics and reaction mechanisms. These thin beds have a low pressure drop, and thus bypassing or channelling can easily arise, causing lower yields. Some minimum bed thickness is needed to avoid this difficulty.

In a *trickle-bed reactor* liquid and gas flow concurrently downward through a packed bed while reaction, essentially adiabatic, takes place. Several beds in series may be utilized with interstage cooling or cold-shot cooling with a gas, analogous to that used in gas-solid adiabatic beds. Alternately the two phases may flow concurrently upward, with the liquid being the continuous phase, a process termed *flooded flow.* Countercurrent operation is not usually used because, with the catalyst particle sizes of interest, the hydrodynamic flow patterns usually become irregular and unpredictable at gas and liquid flow rates below those usable with parallel flow. The equivalent of the multitube reactor with external heat exchange (see below) does not seem to have been used with trickle-bed reactors. To distribute both gas and liquid uniformly and reliably to several thousand tubes in parallel would require an intricate and expensive design.

11.1.2 Multitube Reactor with Heat Exchange

If the adiabatic temperature rise is so great that poor selectivity is encountered, or there is the possibility of a runaway reaction or unacceptable catalyst deactivation, the multitube reactor with external cooling is usually the next choice. The reactions here are usually more rapid and more exothermic than those for which adiabatic reactors are suitable. The heat-exchange fluid may be a thermally stable organic substance such as a mixture of diphenyl ethers or of terphenyls, which can be used up to a temperature of about 370 to 430°C. A molten salt, for example, a eutectic consisting of 53% KNO_3, 40% $NaNO_2$, and 7% $NaNO_3$, can be used up to about 540°C. At sufficiently low reaction temperatures water may be boiled to form steam at a pressure set by the desired temperature to be achieved.

The tubes are typically 2.5 to 5 cm in diameter and from 4 to 7 m long. The smaller the tube diameter, the more closely may isothermal operation be approached, but for a given capacity the number of tubes increases inversely with the square of the internal diameter. Use of tubes much smaller than 2.5 cm would require an unacceptably large number. In a representative design, 7000 to 10,000 tubes 2.5 cm in diameter may be manifolded in parallel in a single vessel. Two or more reactors of such design are typically required for a chemical process to produce a few hundred million pounds per year of product, a representative capacity for a commodity-type chemical. With tubes larger than about 5 cm in diameter, departures from isothermality usually become too great to be acceptable. A closer approach to isothermality can be achieved by diluting the catalyst with inert material; but additional reactor volume is then required, and in practice this may not be economically justified.

With either adiabatic or multitube packed beds, catalyst pellets are typically 1.5 to 6 mm in diameter. Smaller pellets usually cause excessive pressure drop. Larger pellets may lead to diffusion limitations, and if sizes larger than about 6 mm diameter are used, they usually are formed as rings, of about equal height and outside diameter, to minimize diffusion problems.

11.1.3 Fluidized-Bed Reactor

Here the reacting gas is passed up through a bed of finely divided solid catalyst, which is thus highly agitated. A particular advantage of this reactor type is the excellent uniformity of temperature. This is achievable throughout the bed because of the motion of the solid and the good heat exchange between solid and gas. The ease of adding and removing solid is an additional advantage. Hence a fluid-bed reactor is of value for a very exothermic reaction that cannot be adequately controlled with a multitube reactor, or when catalyst must be removed and replaced frequently. For a partial oxidation reaction this kind of reactor also permits a method of readily introducing air and reactant at different locations in the reactor. This avoids formation of an explosive composition that could exist if they were introduced together. A fluid-bed reactor may be less expensive to construct than a multitube reactor of the same capacity, and heat exchange may be simpler than with the adiabatic multibed reactor. However the hydrodynamics of fluidized beds are complex, scaleup procedures are still relatively empirical, solids-separation equipment must be provided, and the catalyst must be attrition resistant.

With representative catalyst particle sizes, and at atmospheric or slightly higher pressures, the maximum linear velocity usable is about 60 cm/s. Much higher velocities cause elutriation of solids from the bed. Cyclone separators are usually installed internally to return the fines to the bed, and these will become overloaded at excessive gas velocities. The minimum bed height to accommodate these and other internal structures, such as heat-exchange surfaces, is typically about 3 m, so it is difficult to use a fluid-bed reactor for reaction times of a few seconds or less.[1] A minimum height may also be needed to be able to introduce different feed gases at different locations.

A *transport line reactor* has also been used, in which a solid catalyst and a gas are caused to rise upwards concurrently at very high velocities, of the order of 6 m/s. In modern catalytic cracking processes, this is termed a *riser cracker.*

11.1.4 Slurry Reactor

The reaction of a liquid is often carried out by suspending a solid catalyst in a finely divided form in the liquid. This is often termed a *slurry reactor.* If a gas is to be reacted with the liquid, it may be introduced through a distributor in the bottom of the vessel or it may be dispersed into the liquid by a mechanical agitator. This also acts to keep the solid suspended. Elucidation of mass-transfer effects in multiphase reactors such as trickle beds and slurry reactors is an important subject, but is beyond the scope of the present treatment.

11.1.5 Contact Time

To determine the true average residence time of a fluid in a reactor often requires information not readily available. In particular, expansion or contraction of a vapor may

[1] Fluid beds are used for various kinds of solids processing, as for combustion, drying, etc., in which case particle sizes may be larger and shorter gas contact times may be utilized. The overall design problem is different.

occur, caused by a change in number of moles on reaction, or temperature and pressure gradients, or both. The void fraction of a packed or fluidized bed may also be unknown. To sidestep these difficulties a *contact time* or *superficial contact time* consisting of the reactor volume (unpacked) divided by the volumetric flow rate is often quoted. The latter may be calculated at the inlet or reactor conditions, or at standard temperature and pressure (abbreviated STP or NTP) and is usually based on the volume of entering reactant. The reciprocal of this is the *space velocity*, which has units of reciprocal time. In some cases the space velocity is given in terms of the volumetric feed rate of a liquid, even though it may be vaporized and mixed with other reactants before entering the catalyst bed. This is termed the *liquid hourly space velocity*, or LHSV.

11.2 REACTION REGIMES

Visualize a porous solid catalyst pellet in contact with a fluid reactant, and consider how the rate of the reaction will change as the temperature is increased. As pointed out by Wicke (1957), three different catalytic reaction regimes may be observed. These are shown diagrammatically on the Arrhenius-type diagram, Fig. 11.2, and on Fig. 11.3. At sufficiently low temperatures the rate of the reaction will be so low that the potential required to provide the diffusion flux is insignificant and intrinsic kinetics will be observed (regime A).

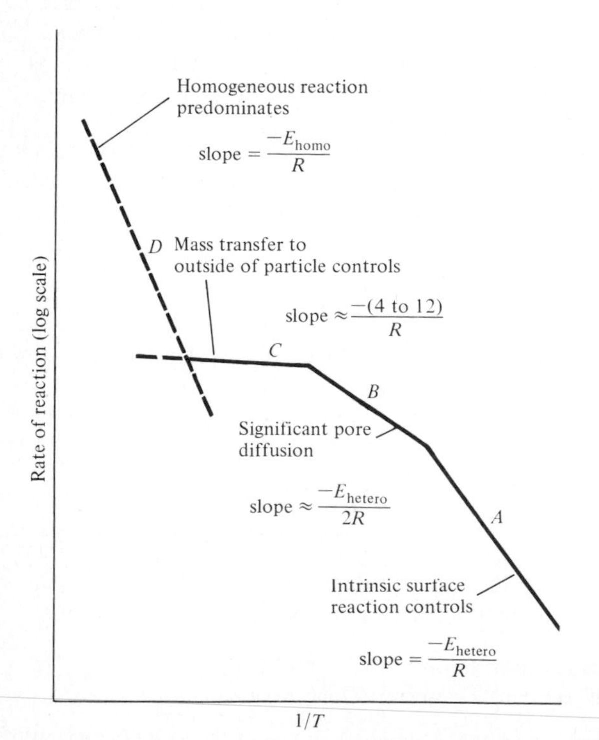

Figure 11.2 Possible kinetic regimes in a gas-phase reaction occurring on a porous solid catalyst.

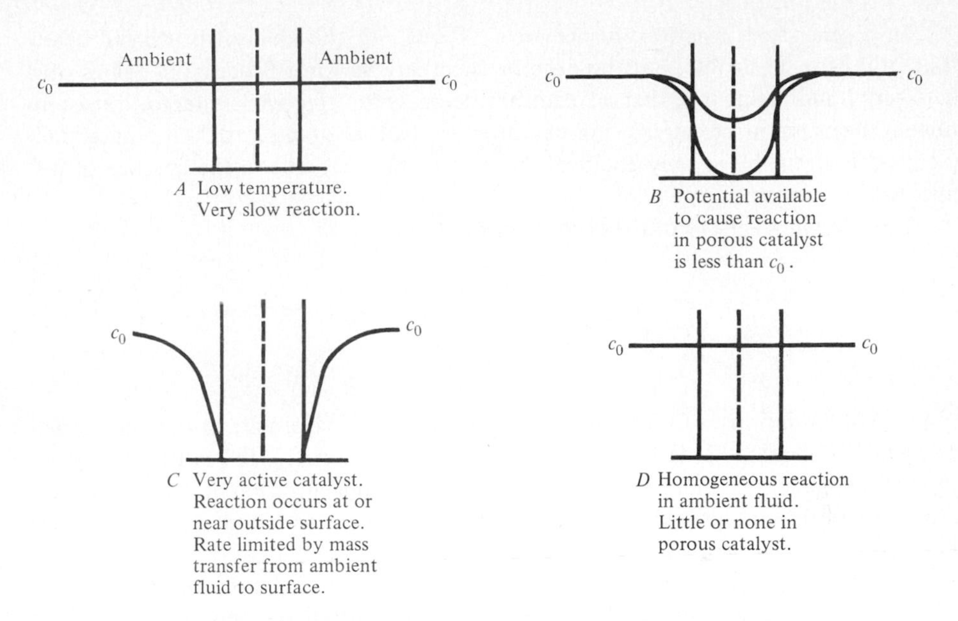

Figure 11.3 Concentration profiles in a porous catalyst under different reaction regimes.

With increased temperature, the rate of diffusion per unit potential difference (the diffusion coefficient) increases but slowly, whereas the intrinsic rate constant increases exponentially. Thus an increasing fraction of the total available potential is required for diffusion, leaving less to drive the chemical reaction. A significant concentration gradient of the reactant then develops through the pellet (Fig. 11.3). (Concentration gradients within the catalyst pores usually become significant before those in the ambient fluid.) This second regime, B, in which pore diffusion is significant, is sometimes termed the *internal-diffusion regime.* The apparent activation energy as calculated from an Arrhenius plot will be the arithmetic average of that for the intrinsic reaction and that for diffusion, provided that the reaction is simple, temperature gradients are negligible, and the intrinsic kinetics can be expressed by a power-law relationship. In gas-phase reactions the effect of temperature on diffusion rates is equivalent to an activation energy of the order of only 4 to 12 kJ/mol. This is small compared to that of most heterogeneous reactions, in which case the observed activation energy will be little more than one-half the intrinsic value. The apparent order of the reaction will shift toward first order; for example, an intrinsic second-order reaction will appear to be three-halves order. The reasons for this behavior are developed elsewhere (Satterfield, 1970).

In complex reactions the selectivity toward an intermediate product will very likely be affected by a change from the intrinsic-kinetics regime to the internal-diffusion regime. The degree of diffusion limitation in the latter is characterized by the *effectiveness factor* η, defined as the ratio of the observed rate of reaction to that which would occur in the absence of diffusion effects within the pores of the catalyst. (Sometimes an effectiveness factor is defined relative to the absence of all diffusion effects, both internal and external.)

In a series-type reaction, for example, A → B → C, the yield of B will fall below that otherwise attainable; but the drop in selectivity to form B occurs at values of η between 1 and about 0.3, that is, a further decrease in effectiveness factor causes no further decrease in selectivity. At effectiveness factors of $\eta < \sim 0.3$, the maximum yield of B attainable is only about 50 percent of that possible in the absence of diffusional limitations.

For parallel reactions of the type

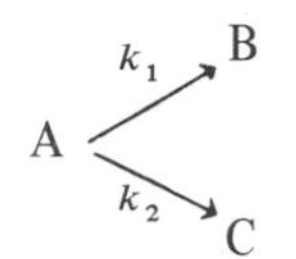

diffusional limitations do not affect selectivity if both reactions are of the same order. However if reaction (2) is of a higher order than reaction (1), the ratio B/C will increase with decreased effectiveness factor since as the concentration of A drops toward the center of the catalyst pellet, the rate of reaction (2) will fall more rapidly than that of reaction (1).

As temperature is further increased, reactant is effectively consumed before it can penetrate very far into the pellet. The concentration difference between the bulk of the fluid and the outside of the catalyst pellet then becomes significant. The internal effectiveness factor continues to drop. In this reaction regime, C, sometimes termed the *external-diffusion regime*, the concentration of reactant at the outside surface of the catalyst pellet approaches zero. The rate-limiting process is one of mass transfer from the ambient fluid and shows the same characteristics as bulk diffusion. The apparent activation energy is then about 4 to 12 kJ/mol for gases, 10 to 20 kJ/mol in liquid hydrocarbons, and 8 to 10 kJ/mol in aqueous systems. In this regime all reactions appear to be first order regardless of their intrinsic kinetics, since mass transfer is a first-order process. All catalysts will appear to have the same activity and selectivity, as determined by the relative rates of bulk diffusion of reactants, intermediates, and products, rather than by the intrinsic characteristics of the reaction.

In regime C the reaction rate is sometimes said to be limited by mass transfer to the catalyst surface, or is sometimes termed a *film-diffusion-limited* process, but the terminology of a rate-limiting process may be confusing. When two processes occur in series, the two rates must be equal under steady-state conditions. The rate-limiting process is the one that consumes the major portion of the chemical potential available. In regime B, diffusion through the pore structure occurs simultaneously with reaction. Although reactions in this regime are sometimes described loosely as being *limited by pore diffusion*, the process is not controlled by a single process as in regime C, since diffusion and reaction occur simultaneously rather than in series. In the external-diffusion regime, if the reaction is highly exothermic, as it usually is, a substantial temperature difference will exist between the outside surface and the bulk fluid, and various instability phenomena may be encountered.

If the same reaction can occur homogeneously as well as catalyzed heterogeneously, the effective activation energy for the homogeneous path is almost invariably greater than that for the heterogeneous route. Since the two competing processes occur in parallel, whichever is faster is the one that will be observed. Homogeneous reaction may predominate over catalytic reaction even at low temperature, depending upon the system, but it is shown on the left in Fig. 11.2. This emphasizes that it will play an increasing role at higher temperatures and that the possibility of contributions from homogeneous reaction must be considered in analyzing the results of a seemingly "catalytic" reaction.

The above picture is somewhat simplified. It represents the intrinsic reaction rate by an Arrhenius expression, whereas more complex kinetics are frequently encountered. Indeed, in a few cases, such as the catalytic hydrogenation of ethylene, the intrinsic rate may exhibit a maximum with increase in temperature (Sec. 3.3.3). Thus a decrease in apparent activation energy with increased temperature does not necessarily indicate the onset of the internal-diffusion regime. It may be solely a reflection of the intrinsic kinetics (Chap. 3). The temperature of the gas and solid are taken to be the same, whereas with highly exothermic (or endothermic) reactions significant temperature gradients between the two may occur. With an exothermic reaction, instability effects may develop from the consequences of coupling between temperature and concentration gradients. Nevertheless, the above broad outline describes the transitions from one regime to another as temperature is changed, and has been clearly demonstrated in many experimental studies. (Figure 11.2 represents the separate regimes by intersecting straight lines and omits representation of transition regions.)

The order in which the three catalytic regimes will be encountered with increased temperature will be as shown in Fig. 11.2, except perhaps in a few highly complex situations. The relative location of the three lines with respect to one another depends on several factors. In the internal-diffusion regime B the rate, but not the apparent activation energy, will be increased by reducing the size of the catalyst particle or by altering the pore structure so as to increase the diffusivity. The line will shift upward, its slope remaining the same. With nonporous solids, regime B will be eliminated and transition will occur from intrinsic kinetics directly to the external diffusion regime. Here the rate is a strong function of the linear velocity of the ambient fluid, whereas in the internal regime it is independent of it, bulk concentration being held constant. The relative importance of homogeneous reaction depends upon the ratio of bulk-gas volume to catalytic surface, as well as the relative rate constants and the appropriate rate expressions.

There are two general methods of determining whether heat- and mass-transfer gradients are causing significant effects. The most reliable is by experiment, to determine the effects of particle size, temperature, and agitation or fluid velocity. The second is by calculation. The accuracy of the calculation method is limited by the accuracy with which diffusivities and other physical properties can be predicted, by a knowledge of the basic kinetics, and by the complexity of the reaction. The transition temperatures between regimes will vary widely with different reactions and different catalysts, as will be seen.

11.2.1 Experimental Methods

Some examples of experimental studies to determine the reaction regime are shown in Figs. 11.4 to 11.9. Figure 11.4 (Satterfield and Cortez, 1970) shows the percent oxidation of dilute hexene in air on passage through a single nonporous platinum gauze, at

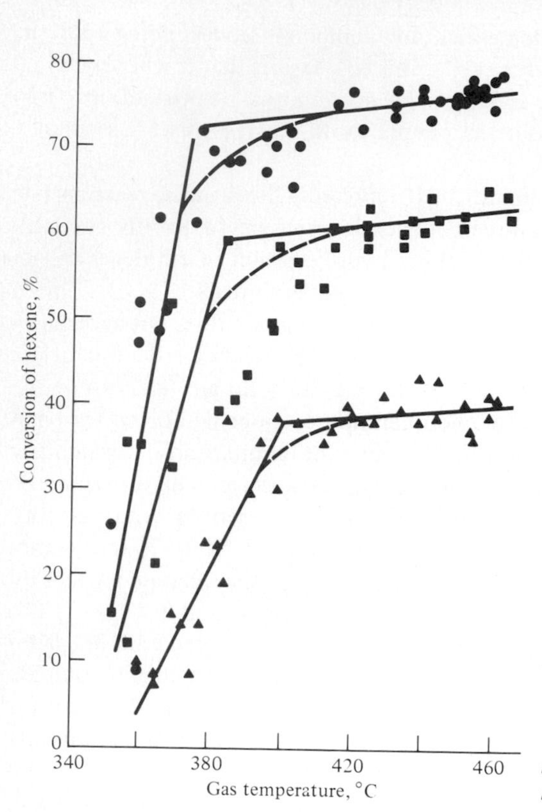

Figure 11.4 Effect of temperature on combustion of hexene; 52-mesh platinum gauze, 0.133% hexene in air. $V_0 = 165$ (●), 250 (■), and 450 (▲) cm^3 (at STP)/s. (*Satterfield and Cortez, 1970.*) (*Reprinted with permission from Industrial and Engineering Chemistry, Fundamentals. Copyright by the American Chemical Society.*)

each of three velocities. The rate increased markedly with increased temperature up to about 380 to 400°C, but it was relatively insensitive to temperature above this point. This suggests that the external-diffusion regime was then encountered. This conclusion was further supported by showing that the observed rates at high temperatures agreed closely with those predicted from mass-transfer correlations for a similar geometry but in nonreacting systems. The scatter in the data is not atypical. One seldom encounters a sharp kink in the transition between two regimes, and the use of two straight lines, as here, emphasizes the transition region. In the film-diffusion regime the conversion drops with increased linear velocity since the rate of mass transfer increases with velocity, but to a power less than 1.

Changing mass velocity alone in a packed bed is not a significant test for the external-diffusion regime, since contact time is also changed. It is necessary to change velocity and bed depth in proportion, in order to keep contact time constant. An example is seen in the results of a study of the water-gas shift reaction on 9.4-mm-diameter iron oxide pellets as reported by Hulburt and Srini Vasan (1961). The reverse reaction was negligible under these conditions. As shown in Figure 11.5, it was necessary in this case to increase the mass velocity to a value of 0.034 $kg/m^2 \cdot s$ (25 $lb/ft^2 \cdot h$) before bulk mass-transfer resistance was eliminated.

Figure 11.6 shows data on SO_2 oxidation to SO_3 on a commercial catalyst over a

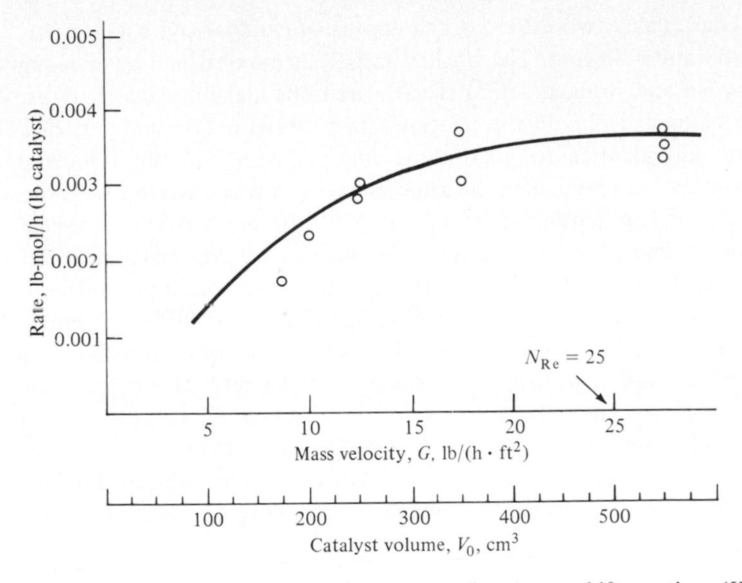

Figure 11.5 Effect of mass velocity on rate of water-gas shift reaction. (*Hulbert and Srini Vasan, 1961.*)

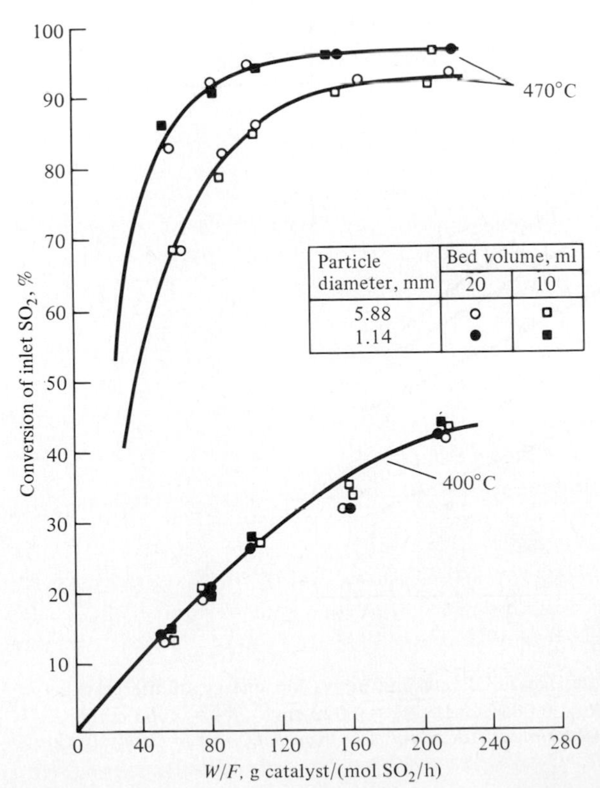

Figure 11.6 Effect of particle size and linear velocity on conversion. Oxidation of SO_2 to SO_3. (*Adapted from Dowden and Bridger, 1957.*) (*Reprinted with permission from Advances in Catalysis. Copyright by Academic Press.*)

range of linear velocities, using two different bed depths of catalyst and catalyst particle diameters of 5.88 and 1.14 mm. The smaller particles were obtained by crushing the largest size (Dowden and Bridger, 1957). These variations make it possible to test for both the pore-diffusion and film-diffusion regimes. (Studies with 2.36-mm particles, not shown, gave essentially identical results to those obtained with 1.14-mm particles.) There is no effect of mass velocity, since the same conversion was obtained with the same value of the ratio of bed depth to flow rate at two different bed depths. Hence the external-diffusion regime is not encountered. At 400°C the conversion was the same for all particle sizes, hence internal-diffusion gradients were unimportant here even with the largest particles. At 470°C, however, significant pore-diffusion limitations were encountered with the largest particles. The leveling off of conversion with increased contact time at 470°C utilizing the smaller particles reflects the approach of equilibrium.

Figure 11.7 from the studies of Weisz and Prater (1954) on cracking of cumene shows an increase in apparent activation energy as particle size was reduced, holding the temperature range constant. This is in agreement with theory for reaction in the internal-diffusion regime.

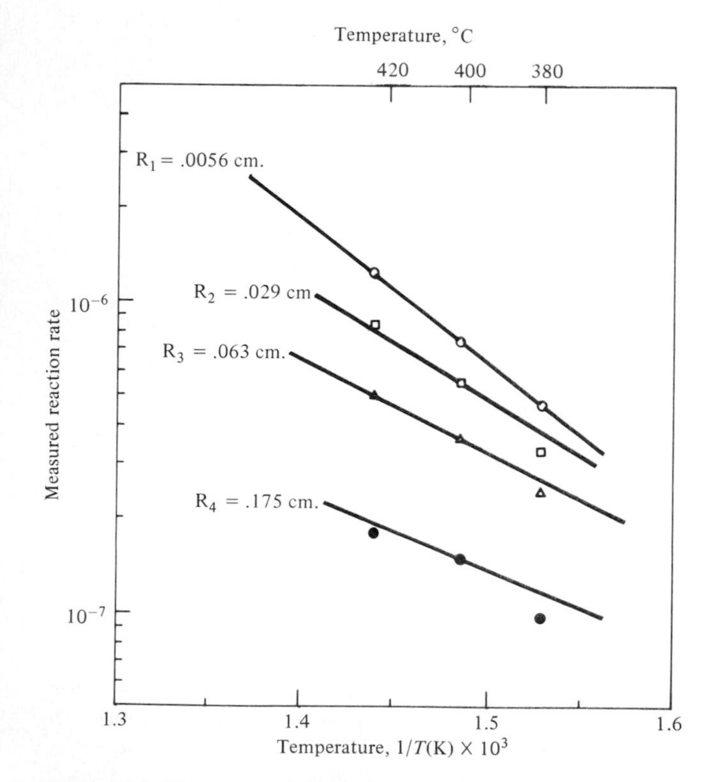

Figure 11.7 Effect of diffusion on the measured rate and activation energy of the cracking of cumene on $SiO_2 \cdot Al_2O_3$ catalyst. ○, R_1 = 0.0056 cm; □, R_2 = 0.029 cm; △, R_3 = 0.063 cm; ●, R_4 = 0.175 cm. (*Weisz and Prater, 1954.*) (*Reprinted with permission from Advances in Catalysis. Copyright by Academic Press.*)

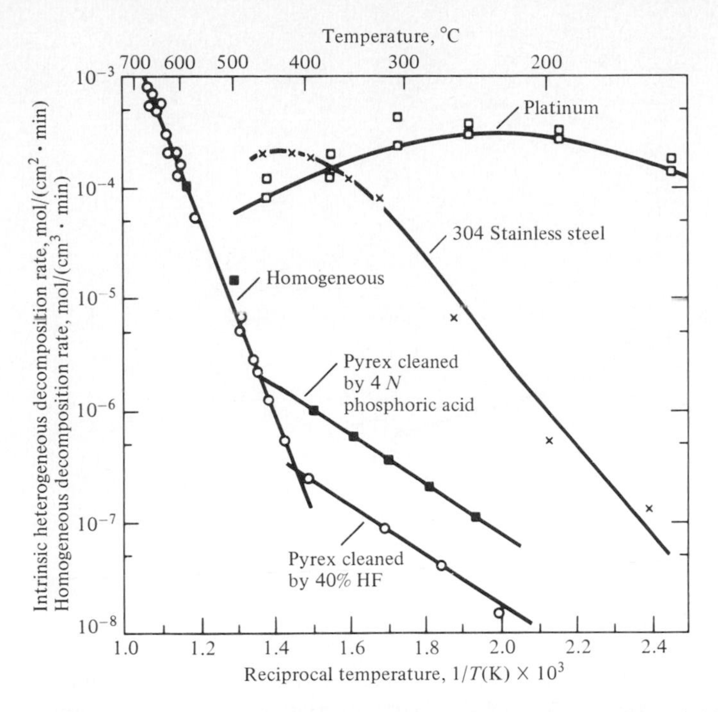

Figure 11.8 Intrinsic decomposition rate of H_2O_2 vapor. (*Satterfield and Yeung, 1963.*) (*Reprinted with permission from Industrial and Engineering Chemistry, Fundamentals. Copyright by the American Chemical Society.*)

Figure 11.8 shows data on the decomposition of hydrogen peroxide vapor on (1) Pyrex glass cleaned with either hydrofluoric or phosphoric acid, (2) on stainless steel, and (3) on platinum, all surfaces being nonporous. With the glass surfaces a marked shift from heterogeneous to homogeneous reaction as the dominating mechanism is seen to occur with increased temperature since the observed rate at the lower temperatures is markedly affected by the method used for cleaning the glass, but at the higher temperatures the observed rate is independent of it. The temperature at which this transition would occur depends on the surface/volume ratio and surface reactivity. For the glass surfaces here it was about 420 to 460°C. For a more active surface it would be higher. A maximum in the catalyzed rate with increased temperature is observed with stainless steel and especially with platinum. Over a moderate temperature range the seemingly flat curve might, in the absence of other information, be erroneously interpreted as indicating the film-diffusion regime. These results illustrate the hazards of trying to reach conclusions solely from the effect of temperature on apparent activation energy.

Figure 11.9 illustrates the decrease in selectivity observed for a reaction of the type $A \rightarrow B \rightarrow C$ as particle size is increased to where the internal-diffusion regime is encountered (Voge and Morgan, 1972). The reaction is the dehydrogenation of butene to butadiene in the presence of steam at 620°C on a porous commercial iron oxide (Shell 205) catalyst. The butadiene may react further to form carbon dioxide and

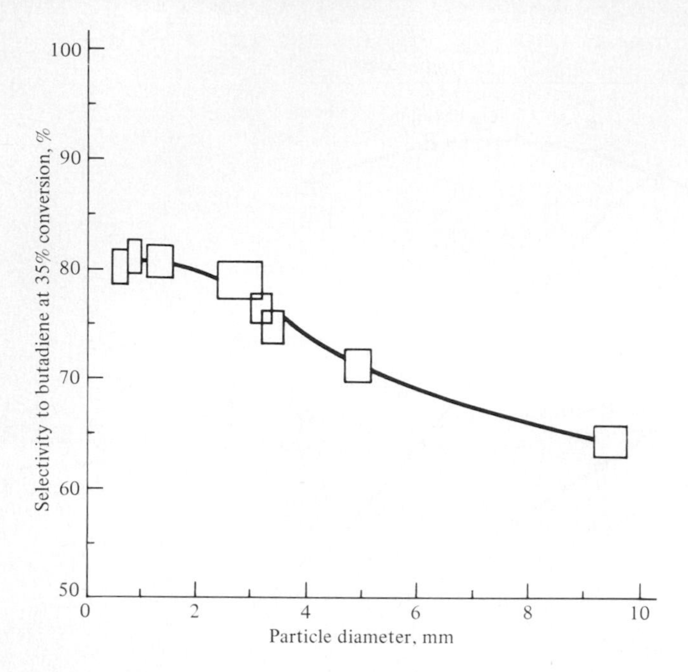

Figure 11.9 Dehydrogenation of butene to butadiene. Selectivity at 35 percent conversion as a function of particle size. (*Voge and Morgan, 1972.*) (*Reprinted with permission from Industrial and Engineering Chemistry, Process Design and Development. Copyright by the American Chemical Society.*)

cracked products. In accordance with theory the selectivity to butadiene achieved was less with the larger particles.

A technique suggested by Koros and Nowak (1967) may also be useful, especially for research studies. Pellets are made from a powdered catalyst that is mixed with inert powder of the same size. Measurements are made with pellets made up of two or more ratios of catalyst to inert. If the ratio of the rate constants (or better, turnover number) for the two kinds of pellets is the same as the ratio of amounts of catalyst in the two pellets, diffusional effects are deemed to be negligible. The inert material must be shown to be truly inert; ideally it should be nonporous and have the same deformable characteristics as the catalyst. The diffusional characteristics of the two pellets must be the same; this may be difficult to achieve, but is possibly attainable by pelletizing the two pellets in the same fashion.

Coke deposits occur often in catalytic reactions under reducing conditions, and the distribution of coke through a catalyst pellet is affected by diffusion limitations. This may be readily observed by sectioning catalyst pellets after use. Murakami et al. (1968) studied, both theoretically and experimentally, parallel- and series-type reactions in which the reactant was converted to product or coke. The disproportionation of toluene on an alumina-boria catalyst to form either xylene and benzene or coke and benzene was chosen as representative of parallel-type kinetics. The dehydrogenation of an alcohol on an alkaline-alumina catalyst to form an aldehyde that then formed coke

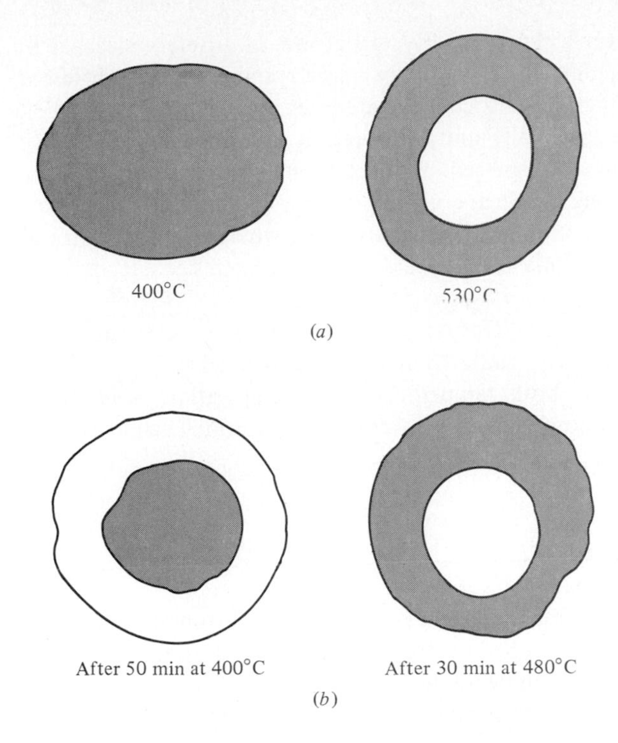

Figure 11.10 Effect of time and nature of reaction on distribution of coke deposits in a catalyst pellet. (*a*) Cross section of catalyst pellet after disproportionation of toluene for 10 min. (*b*) Cross section of catalyst pellet after dehydrogenation of *n*-butyl alcohol. (*Murakami et al., 1968.*) (*Reprinted with permission from Industrial and Engineering Chemistry, Fundamentals. Copyright by the American Chemical Society.*)

was taken as representative of a series-type reaction. With the series-type reaction, under a moderate degree of internal-diffusion limitation, coke deposition occurs preferentially in the interior of the pellet. With a high degree of diffusion limitation, the coke deposit appears preferentially at the outside with either type of kinetics. The drawings in Fig. 11.10 illustrate the photographs they published of the cross section of catalyst pellets after the reaction had proceeded for the time interval and at the temperature specified. With the toluene disproportionation reaction there were no diffusion limitations at 400°C but substantial limitations at 530°C. At the higher temperature the toluene essentially all reacted within the outer portion of the pellets (see also Fig. 11.3), so coke appears as an annular ring. With the series-type reaction, a moderate degree of diffusion limitation was encountered at 400°C, but a more severe degree occurred at 480°C. At 400°C coke appears only in the center of the pellet where the aldehyde concentration has built up to a level that permits significant further reaction to form coke. Under the more strongly diffusion-limiting conditions that occur at higher temperature (480°C), the entire series of reactions goes to completion near the outside pellet surface. In this case coke appears only in an annular ring because no appreciable aldehyde precursor survives to reach the center.

Some additional precautions need to be borne in mind in making experiments of this type. Studying the effect of particle size assumes the particle composition and pore structure to be uniform. The most useful tests usually proceed by crushing a given catalyst to produce two or more smaller particle sizes for study. Studying two or more pellet sizes as made in the laboratory or plant may be misleading since they may have different pore structures. Studying crystals grown to different sizes, as of zeolites, may likewise be misleading since they may have crystal defects to different degrees. If the active catalyst ingredient is concentrated near the outside of a pellet, the effectiveness factor may be essentially unity, whereas it might not be if the active catalyst ingredient were uniformly dispersed. With catalysts having some very fine pores, as with those incorporating zeolites, crushing the catalyst may reveal nothing about possible diffusion limitations in the fine-pore structure when the particle size of the crushed material is still orders of magnitude greater than the pore size.

It is important to note that as conditions are altered to cause a shift from the intrinsic kinetic region into the internal-diffusion regime, the most sensitive indicator is often a change in reaction selectivity rather than a change in activity. To determine activity requires measurement of both temperature and composition, whereas to determine selectivity requires a knowledge only of composition, which can usually be determined more accurately.

11.3 THEORETICAL CRITERIA

A number of criteria have been published that make it possible to estimate whether or not a single catalyst particle or packed-bed reactor is being studied under essentially ideal circumstances. These are based on estimated values of pore diffusivity, thermal conductivity, axial dispersion, heat- and mass-transfer coefficients, nature of the kinetics, and other parameters. The accuracy with which such criteria can be applied varies widely from case to case. The most reliable conclusions are usually reached from careful experimentation as discussed above rather than from theoretical calculations. This is especially true the more complex the reaction mixture. However, calculations may help to buttress conclusions derived from observations and provide a method of reaching conclusions when experimentation is impractical. It is especially difficult to estimate bulk heat- and mass-transfer coefficients and axial dispersion at the relatively low flow rates characteristically encountered in many laboratory reactors.

Some of the more useful criteria for application to laboratory and pilot plant-scale studies are summarized below. A larger number are discussed by Mears (1971) and by Butt and Weekman (1974). Three kinds of gradients of either concentration or temperature, or both, may exist:

I. Intraparticle (inside catalyst particle)
II. Interphase (between fluid and particle)
III. Reactor gradients
    A. Radial gradients
    B. Axial dispersion

These gradients may interact with each other in complex ways. Most of the following criteria assume each can be treated separately, which is usually valid if the criteria indeed are met. To obtain reliable guidance from a criterion that treats two or more kinds of gradients simultaneously usually requires that the values of several parameters be known with a degree of precision that is seldom available.

The degree of departure from ideality is usually expressed as an effectiveness factor, η. This is defined here as the ratio of the actual reaction rate to that which would occur if all of the surface throughout the inside of the catalyst particle were exposed to reactant and product of the same concentration and temperature as that existing in the bulk fluid. For the ideal situation, η approaches unity. Under isothermal conditions, η decreases in value as diffusion limitations become significant, but with a highly exothermic reaction η may exceed unity. This occurs when the catalyst temperature exceeds that of the bulk gas, and the increase in reaction rate by increase in temperature more than outweighs the decrease in rate caused by lower concentration.

These criteria utilize various dimensionless groups of properties, which are summarized in Table 11.2.

Table 11.2 Selected dimensionless groups

Name	See the following:	Proportional to:
Reynolds number,* Re $= \dfrac{LG}{\mu}$	Sec. 11.6	$\dfrac{\text{Inertial force}^{\dagger}}{\text{Viscous force}}$
Prandtl number, Pr $= \dfrac{c_p \mu}{k_f}$	Secs. 11.6.1, 11.6.2	$\dfrac{\text{Kinematic viscosity}}{\text{Thermal diffusivity}}$
		$\dfrac{\text{Momentum viscosity}^{\ddagger}}{\text{Thermal diffusivity}}$
Schmidt number, Sc $= \dfrac{\mu}{\rho D}$	Sec. 11.6	$\dfrac{\text{Kinematic viscosity}}{\text{Molecular diffusivity}}$
Lewis number, Le $= \dfrac{\text{Pr}}{\text{Sc}}$		$\dfrac{\text{Thermal diffusivity}}{\text{Molecular diffusivity}}$
Biot number, Bi $= \dfrac{h r_m}{\lambda}$		$\dfrac{\text{Midplane thermal internal resistance}}{\text{Surface film resistance}}$
Nusselt number,* Nu $=$ Pe $\cdot$ St $= \dfrac{hL}{k}$	Sec. 11.6	$\dfrac{\text{Total heat transfer}}{\text{Conductive heat transfer}}$
Sherwood number,* Sh $= \dfrac{k_c L}{D}$ (Nusselt number for mass transfer)	Sec. 11.6	$\dfrac{\text{Mass diffusivity}}{\text{Molecular diffusivity}}$
Stanton number, St $= \dfrac{h}{c_p G} = \dfrac{\text{Nu}}{\text{Re} \cdot \text{Pr}}$		$\dfrac{\text{Heat actually transferred}}{\text{Thermal capacity of fluid}}$
Peclet number,*,§ Pe $= \dfrac{L\bar{U}}{D_a}$	Secs. 11.3.4, 11.6	$\dfrac{\text{Bulk mass transport}}{\text{Diffusive mass transport}}$

Table 11.2 Selected dimensionless groups (*Continued*)

Name	See the following:	Proportional to:
Damköhler group II,[¶] $\Phi_s = \dfrac{r_p^2(-r)}{c_s D_{\mathrm{eff}}}$	Eq. (11.1)	$\dfrac{\text{Chemical reaction rate}}{\text{Molecular diffusion rate}}$
Damköhler group IV, $\Phi_s \lvert\beta\rvert$	Eq. (11.5)	$\dfrac{\text{Heat liberated}}{\text{Conductive heat transfer}}$
Arrhenius group, $\gamma = \dfrac{E}{RT_s}$	Eq. (11.4)	$\dfrac{\text{Activation energy}}{\text{Potential energy of fluid}}$
Heat generation function, or *thermicity*, $\beta = \dfrac{c_s(-\Delta H) D_{\mathrm{eff}}}{\lambda T_s}$	Eq. (11.3)	

*L is a characteristic length, usually taken as d_p for beds of particles.

†Inertial forces tending to disrupt laminar motion are proportional to $\rho V^2 L^2$. Internal (viscous) forces tending to stabilize flow are proportional to μVL.

‡Kinematic viscosity, μ/ρ, is a diffusivity for momentum or for velocity. If Pr = 1, heat and momentum diffuse through fluid at the same rate.

§In this definition D_a is the effective dispersion coefficient, which may include mixing by turbulence and eddies, as well as molecular diffusivity. In its original definition the term Peclet number was reserved for the case in which D is the molecular diffusivity. Then Pe = Re · Sc. When D is a dispersion coefficient, the above dimensionless group is sometimes termed the *Bodenstein number based on particle diameter*, especially in the European literature. A *Peclet number for heat transfer* is usually defined as Re · Pr.

¶This should not be confused with the Thiele modulus, frequently designated by ϕ_s, which is defined differently.

11.3.1 Intraparticle

A. Isothermal system The basic parameter is a form of the Thiele modulus suggested by Wagner in 1943. This consists of four quantities that are known or can be estimated. Symbols are defined in Table 11.11, pages 367–368.

$$\Phi_s = \frac{r_p^2(-r)}{D_{\mathrm{eff}} c_s} \tag{11.1}$$

The smaller the value of Φ_s, the smaller the concentration gradient through the pellet and the closer the effectiveness factor η approaches unity. Typically a criterion specifies conditions under which $\eta \geqslant 0.95$. This is determined by the intrinsic order of the reaction as well as the value of Φ_s. For an irreversible reaction of a single reactant whose kinetics can be represented by a power-law relationship, $\eta \geqslant 0.95$ if

$$\begin{aligned} \Phi_s &< 6 && \text{zero-order reaction} \\ &< 1 && \text{first-order reaction} \\ &< 0.3 && \text{second-order reaction} \end{aligned} \tag{11.2}$$

If the reaction is strongly inhibited by product adsorption, considerably lower values of Φ_s need be achieved for $\eta \geqslant 0.95$. A change in volume upon reaction affects the effective diffusivity if the bulk-diffusion regime is encountered inside catalyst pores (see below). This is usually of secondary importance, and becomes negligible for a dilute reactant. For nonspherical shapes, taking $r_p = 3V_p/S_p$ usually introduces little error. Methods of handling more complex systems, e.g., reversible reactions, and complex kinetics are summarized elsewhere (Satterfield, 1970). In a mixture of two or more reactants, Φ_s may be estimated based on the key reactant—the one which diffuses into the porous catalyst least rapidly. For example, in a stoichiometric gas mixture of hydrogen and a reactant A, A is usually the key reactant since hydrogen has a higher diffusivity than any other molecule. In a mixture of A and a great excess of B, where A and B have about the same diffusion coefficient, A will be the key reactant since the concentration gradient for B greatly exceeds that for A.

B. Nonisothermal systems Two new dimensionless groups are now utilized, β and γ.

$$\beta = \frac{c_s(-\Delta H)\, D_{\text{eff}}}{\lambda T_s} \tag{11.3}$$

ΔH is the enthalpy change on reaction, and λ is the thermal conductivity of the porous catalyst. For an exothermic reaction β is a positive number and for most reactions is ± 0.1 or less. Even for a highly exothermic reaction, β seldom exceeds 0.2. The value of β is also useful to know since it equals $\Delta T_{\text{max}}/T_s$, the maximum possible temperature difference between the outside pellet surface and the pellet interior, relative to the surface temperature, that occurs under highly diffusion-limited circumstances.

$$\gamma = \frac{E}{RT_s} \tag{11.4}$$

γ is sometimes termed the *Arrhenius group*. E is the intrinsic activation energy, R the gas constant, and T_s the temperature at the outside pellet surface. Values of γ typically will vary in the range between 10 and 40.

The observed rate will deviate from the rate under isothermal conditions by less than 5 percent (Anderson, 1963) if

$$\Phi_s |\beta| = \frac{|\Delta H|\,(-r)\, r_p^2}{\lambda T_s} < \frac{1}{\gamma} \tag{11.5}$$

This criterion is valid whether or not diffusional limitations exist in the pellet.

If both temperature and concentration gradients occur simultaneously, the criterion for $0.95 < \eta < 1.05$ is

$$\Phi_s < \frac{1}{|n - \gamma\beta|} \tag{11.6}$$

This applies to either endothermic or exothermic reactions, and to power-law kinetics, where n is the order of reaction but does not equal zero. When $|n - \gamma\beta|$ is close to or equals zero (a rather unusual case), more detailed examination is necessary. An asymptotic analysis by Peterson (1965) gives a value of 13 for the right-hand side of Eq. (11.6) when n and $\gamma\beta$ each equal 1.0.

A criterion for isothermal operation of a catalyst particle was developed by Peterson, by an asymptotic solution for first-order kinetics, as follows:

$$|\gamma\beta| < 0.3 \tag{11.7}$$

11.3.2 Interphase Transport

For vapor-phase systems, as temperature differences become important, the temperature difference between the bulk gas and the outside pellet surface is usually much greater than the temperature difference between the pellet surface and the pellet center. On the other hand, concentration differences between the outside pellet surface and the interior normally become severe before the difference between the concentration in bulk gas and that at pellet surface becomes appreciable.

In order for the observed rate not to deviate by more than 5 percent because of a temperature difference between catalyst particle and the bulk fluid (Mears, 1971),

$$\frac{(-\Delta H)(-r)\, r_p}{hT} < 0.15\, \frac{RT_b}{E} \tag{11.8}$$

Here h is the heat-transfer coefficient and T_b is the bulk temperature of the fluid. If the criterion holds, $T \approx T_b$. The criterion is valid whether or not diffusion limitations exist in the particle. Equation (11.8) is similar in form to Eq. (11.5), with h replacing λ/r_p.

Under *isothermal* conditions, a concentration difference between bulk fluid and outside pellet surface will be insignificant if

$$\frac{(-r)\, r_p}{c_b k_c} < \frac{0.15}{n} \tag{11.9}$$

Here c_b is the bulk-fluid concentration, and k_c is the mass-transfer coefficient between gas and particle. Mears points out that comparison of Eqs. (11.9) and (11.8) for typical cases demonstrates that temperature gradients become the source of deviations from ideality long before concentration gradients do so.

11.3.3 Reactor Gradients

A packed-bed laboratory reactor is usually operated as an integral reactor, i.e., with substantial conversion from entrance to exit. Temperature and concentration may vary from point to point both radially and axially. Hence analysis of the integrated data obtained from such a system to indicate transport limitations is particularly complex. The most difficult requirement to achieve is isothermality in both the radial and axial directions. If this is attained, it is seldom that radial concentration gradients (external to catalyst particles) will be significant.

Mears develops the following criterion for the observed reaction rate not to deviate more than 5 percent from the isothermal case, assuming plug flow and assuming that there are no concentration or temperature gradients inside particles or between particles and bulk fluid.

$$\frac{|-\Delta H|(-r)r_R^2}{\lambda T_w} < 0.4 \frac{RT_w}{E} \tag{11.10}$$

Here r_R is the reactor radius, λ is the effective thermal conductivity across the bed, and T_w is the wall temperature in degrees kelvin. The expression is seen to be similar in form to Eq. (11.5). Strictly the above applies only to the point in the bed at which the highest temperature is reached, which is typically in the region of 20 to 40 percent conversion. An averaged value through the bed will usually be less, so the above criterion must be used cautiously if the rate of reaction varies greatly from entrance to exit. The above criterion also assumes that the heat-transfer resistance at the inside wall of the reactor is negligible relative to that through the bed radially. This is valid, for example, for $r_R/r_p > 50$ to 100, but for small laboratory reactors the right-hand side should be decreased by a factor of about 2, for example, for Re $< \sim 100$ and $0.05 < r_p/r_R < 0.2$. This makes the criterion twice as stringent. The Reynolds number, Re, characterizes the fluid flow.

The criterion of Eq. (11.10) is very sensitive to reactor radius. This emphasizes the importance of using small-diameter laboratory reactors to obtain isothermality. The catalyst may also be diluted by mixing with inert material of about the same particle size. The rate of reaction, $-r$, in this equation is based on packed reactor volume and includes inert diluent if any. Thus diluting a catalyst 9:1 with inert material would reduce this term by a factor of 10.

11.3.4 Axial Dispersion

Deviations from plug flow may be caused by one or more of several effects. The void fraction of a packed bed next to the wall is higher than in the center. Because of the lower resistance at the wall, the linear velocity next to the wall is greater. This contribution of wall flow to the total flow may be significant for low ratios of r_R/r_p, for example, 10 or less. The extent to which this may affect reactor performance, however, depends on several other factors such as the value of Re, the bed length, and the conversion. At low Reynolds numbers molecular diffusion may cause significant axial dispersion. Regardless of the cause, axial dispersion is usually represented by a Peclet number, $\text{Pe} = \bar{u}d_p/D_a$, where D_a is the effective axial dispersion and $\bar{u}$ the mean fluid velocity.

Usually axial dispersion is of negligible importance except for very short beds and high conversions at low flow rates. Since this can usually be overcome by lengthening the bed, Mears presents the following criterion for the minimum reactor length necessary to avoid a significant dispersion effect.

$$\frac{L}{d_p} > \frac{20n}{\text{Pe}} \ln \frac{c_i}{c_f} \tag{11.11}$$

Here n is the order of the reaction, and c_f/c_i is the fractional conversion. The Peclet number is about 2 for gases at Reynolds numbers above about 2 and if the ratio r_R/r_p is sufficiently great that wall bypassing is not serious. It is difficult to reduce the latter to a more quantitative statement, and various suggestions appear in the literature. For Re $\gtrsim$ 2, values of r_R/r_p above about 10 seem to suffice. If these conditions do not apply, Pe will have a lower value, but to a degree that is difficult to predict. Values of r_R/r_p considerably less than 10 can still be used satisfactorily if the percent conversion sought is not very high.

With the use of larger particles such that r_R/r_p is, say 2 or 3, a useful diagnostic test is to fill the interstices between the particles with a finer inert material since this will eliminate or minimize wall bypassing. The question of possible wall bypassing often arises with the use of a tubular reactor equipped with a thermocouple well, in which the catalyst is packed in the annulus around the well. Here, a useful diagnostic test may be to remove the thermocouple well for a few runs, or possibly replace it with a single fine-wire thermocouple (but be careful that the thermocouple itself does not have catalytic activity).

Almost all the information on the effect of the Reynolds number on the Peclet number has been obtained from nonreacting, isothermal systems. At low values of Re, convection effects in laboratory reactors may be a far more important source of deviation from plug flow than is generally recognized; for example, if an exothermic reaction in a downflow reactor is being studied, convection effects may be suspected, especially in a short, squat type of configuration. It is difficult to predict such effects, but experimentally, if a fairly coarse catalyst is being studied and dispersion by convection is suspected, filling the voids with a finer inert material may be an effective diagnostic procedure. If dispersion by convection is indeed significant, this will reduce convection and likely increase conversion. Comparing upflow and downflow configurations may also be useful.

11.4 EFFECTIVE DIFFUSIVITY

To utilize the criteria in Sec. 11.3, it is necessary to determine or estimate the effective diffusivity of a porous catalyst and the thermal conductivity of a porous catalyst or bed of particles. Procedures for calculating values to the maximum degree of certainty possible are too extensive to be treated here. Instead a summary is presented below which should permit values to be estimated generally to within an order of magnitude or less, which may suffice for many cases of interest.

An effective diffusivity may be determined by passing two gases past opposite faces of a catalyst pellet and measuring the flux of one gas into the other. This is sometimes termed the *Wicke-Kallenbach experiment*, and data have been published on a large number of catalysts as well as on other porous materials. The effective diffusivity is defined by

$$N = D_{\text{eff}} \frac{dc}{dx} \tag{11.12}$$

(See Table 11.11, pages 367-368, for nomenclature.) This is *Fick's first law*.

Measurements are made with a pair of inert gases such as hydrogen and nitrogen or helium and nitrogen, and usually at atmospheric pressure and room temperature. For application to a catalytic reaction it is necessary to be able to extrapolate data to the composition, pressure, and temperature of reaction, or to predict D_{eff} without a diffusion measurement and with only knowledge of certain physical characteristics of the catalyst such as the pore-size distribution.

11.4.1 Bulk Diffusion

A large amount of data are available on diffusion coefficients in binary gas mixtures, D_{12}, and coefficients not known can be closely estimated from theoretical expressions based on kinetic theory. D_{12} is inversely proportional to pressure up to 20 atm (2 MPa) or more, so values are frequently quoted in the form of the product $D_{12}P$ (where P is in atmospheres and D_{12} in square centimeters per second). D_{12} increases with temperature, being proportional to T^n, where n is generally between 1.5 and 2. A selection of representative values is given in Table 11.3. For moderately small mole-

Table 11.3 Diffusion coefficients for binary gas systems (experimental values of $D_{12}P$, where D_{12} is in cm^2/s and P is in atm)

Gas pair	T, (K)	$D_{12}P$	Gas pair	T, (K)	$D_{12}P$
Air-ammonia	273	0.198	Ethane-methane	293	0.163
-benzene	298	0.0962	-propane	293	0.0850
-carbon dioxide	273	0.136	Helium-argon	273	0.641
	1000	1.32	-benezene	298	0.384
-chlorine	273	0.124	-ethanol	298	0.494
-diphenyl	491	0.160	-hydrogen	293	1.64
-ethanol	298	0.132	Hydrogen		
-iodine	298	0.0834	-ammonia	298	0.783
-methanol	298	0.162	-benzene	273	0.317
-mercury	614	0.473	-ethanol	340	0.578
-naphthalene	298	0.0611	-ethylene	298	0.602
-oxygen	273	0.175	-methane	288	0.694
-sulfur dioxide	273	0.122	-nitrogen	293	0.760
-toluene	298	0.0844	-oxygen	273	0.697
-water	298	0.260	-propane	300	0.450
	1273	3.253	Nitrogen		
Argon-neon	293	0.329	-ammonia	298	0.230
Carbon dioxide-benzene	318	0.0715	-ethylene	298	0.163
-ethanol	273	0.0693	-iodine	273	0.070
-hydrogen	273	0.550	-oxygen	273	0.181
-methane	273	0.153	Oxygen-ammonia	293	0.253
-methanol	299	0.105	-benzene	296	0.0939
-nitrogen	298	0.167	-carbon tetrachloride	298	0.071
-propane	298	0.0863	-ethylene	293	0.182
Carbon monoxide			Water-hydrogen	307.2	1.020
-ethylene	273	0.151	-helium	307	0.902
-hydrogen	273	0.651	-methane	307.6	0.292
-nitrogen	288	0.192	-ethylene	307.7	0.204
-oxygen	273	0.185	-nitrogen	307.5	0.256
Dichlorodifluoromethane			-oxygen	352	0.352
-ethanol	298	0.0475	-carbon dioxide	307.4	0.198
-water	298	0.105			

cules of the types encountered in most catalytic reactions, $D_{12}P$ is in the vicinity of 0.1 $cm^2/s \cdot atm$ at ambient temperature except for pairs with hydrogen, in which the diffusivity is somewhat higher. Methods of estimation and data on a large number of systems are given by Reid et al. (1977) and Satterfield (1970).

11.4.2 Bulk Diffusion in Porous Catalysts

Pore diffusion may occur by bulk or Knudsen diffusion. If the pores are large and gas relatively dense (or if the pores are filled with liquid), the process is that of *bulk*, or ordinary, *diffusion.* If the pores were an array of cylinders parallel to the diffusion path, the diffusion flux per unit total cross section of the porous solid would be the fraction θ of the flux under similar conditions with no solid present. However, the length of the tortuous diffusion path in real pores is greater than the distance along a straight line in the mean direction of diffusion. Moreover, the channels through which diffusion occurs are of irregular shape and of varying cross section; constructions offer resistances that are not offset by the enlargements. Both of these factors cause the flux to be less than would be possible in a uniform pore of the same length and mean radius. We may thus express a bulk diffusion coefficient per unit cross section of porous mass, $D_{12,\text{eff}}$, as

$$D_{12,\text{eff}} = \frac{D_{12}\theta}{\tau} \tag{11.13}$$

τ is a tortuosity factor which allows for both the varying direction of diffusion and varying pore cross section. It is essentially an adjustable parameter. For diffusion through a randomly oriented system of cylindrical pores, τ equals 3. Experimental measurements on a variety of commercial catalysts not subjected to excessive sintering conditions yield in almost all cases values of τ in the range of 2 to 7. The higher values are generally encountered with materials having lower void fractions. In the absence of other information it is recommended that a value of $\tau = 4$ be taken for estimation purposes. Values of θ vary from about 0.3 to 0.7. In the absence of other information a value of $\theta = 0.5$ for estimation purposes is recommended.

11.4.3 Knudsen Diffusion

If the gas density is low or if the pores are quite small, or both, the molecules collide with the pore wall much more frequently than with each other. This is known as *Knudsen diffusion.* The molecules hitting the wall are momentarily adsorbed and then given off in random directions (diffusively reflected). The gas flux is reduced by the wall "resistance." This causes a delay because of both the diffuse reflection and the finite time the molecule is adsorbed. Knudsen diffusion is not observed in liquids. Kinetic theory provides the following relations for Knudsen diffusion in gases *in a straight cylindrical pore:*

$$N = \frac{D_K}{x_0}(c_1 - c_2) = \frac{D_K}{RT}\frac{(p_1 - p_2)}{x_0} = \frac{2r_e\bar{u}}{3RT}\frac{(p_1 - p_2)}{x_0} \tag{11.14}$$

$$= \frac{2r_e}{3RT}\left(\frac{8RT}{\pi M}\right)^{1/2}\frac{(p_1 - p_2)}{x_0} \tag{11.15}$$

By substitution of Eq. (11.15) into Eq. (11.14),

$$D_K = 9700\, r_e \sqrt{\frac{T}{M}} \tag{11.16}$$

In Eq. (11.16) r_e is the pore radius in centimeters, T the temperature in degrees kelvin, and M the molecular weight. The symbols refer to a single component. Since molecular collisions are negligible, flow and diffusion are synonymous, and each component of a mixture behaves as though it alone were present.

The internal geometries of consolidated porous solids are but poorly understood, and an empirical factor must be introduced to make the theory useful. The cylindrical pore of radius r_e has a volume/surface ratio of $r_e/2$. We may logically define the mean pore radius as

$$r_e = \frac{2V_g}{S_g} = \frac{2\theta}{S_g \rho_p} \tag{11.17}$$

where S_g, in square centimeters per gram, is the total surface and ρ_p, in grams per cubic centimeter, is the pellet density. With this substitution, the Knudsen diffusion coefficient *for a porous solid* becomes

$$D_{K,\text{eff}} = \frac{D_K \theta}{\tau_m} = \frac{8\theta^2}{3\tau_m S_g \rho_p} \sqrt{\frac{2RT}{\pi M}} = 19{,}400 \frac{\theta^2}{\tau_m S_g \rho_p} \sqrt{\frac{T}{M}} \tag{11.18}$$

As in Eq. (11.14), the void fraction θ has been introduced so that the flux N given by $D_{K,\text{eff}}$ will be based on the total cross section of porous solid, not just the pore cross section. As with bulk diffusion, the factor τ_m allows for both the tortuous path and the effect of the varying cross section of individual pores. The subscript m reminds us that τ_m is the value of the tortuousity factor obtained when D_K is calculated from a mean pore radius, defined in Eq. (11.17).

If a significantly broad range of pore sizes exists, the proper average pore radius to use in Eq. (11.16) is given by Eq. (11.19) rather than Eq. (11.17), provided that the flux is completely in the Knudsen range.

$$\bar{r} = \frac{\int_{V_1}^{V_2} r\, dV}{V_2 - V_1} \tag{11.19}$$

dV is the volume of pores having radii between r and $r + dr$.

11.4.4 The Transition Region

Bulk diffusion occurs when the collisions of molecules with the pore wall are unimportant compared to molecular collisions in the free space of the pore. Knudsen diffusion occurs when this condition is reversed. For a given system at specified conditions of pressure, temperature, and concentration there is a range of pore sizes where both types of collisions are important. This is the *transition region*. However, as pressure is increased, for example, the change from Knudsen to bulk diffusion does not occur suddenly when the mean free path of the gas molecules becomes equal to the pore radius; rather there is a gradual change in the relative contributions from the two mechanisms. Unfortunately the diffusion process in many high-area porous catalysts under reaction conditions is in the transition regime. For equimolar counterdiffusion, which occurs when there is no change in number of moles on reaction,

$$\frac{1}{D_{\mathrm{eff}}} = \frac{1}{D_{K,\mathrm{eff}}} + \frac{1}{D_{12,\mathrm{eff}}} \tag{11.20}$$

In effect two resistances in series exist. The flux is limited by molecules colliding with the wall or with each other, or both. Whether Knudsen or ordinary diffusion predominates depends on the ratio D_{12}/D_K and not solely on pore size or pressure. D_{12} varies inversely with pressure and does not depend on pore size; D_K is proportional to pore diameter and independent of pressure. D_K is proportional to $T^{1/2}$, D_{12} to $T^{3/2 \text{ to } 2}$. In the above formulation the value of $D_{12,\mathrm{eff}}$ is affected if there is a change in number of moles on reaction since there is then a net molar flux in or out of the catalyst. This is a minor effect if the ratio of molar fluxes is not too great (for example, within the range of roughly 3 to $\frac{1}{3}$), or if much of the gas present does not take part in the reaction. The greater portion of the flux that occurs in the Knudsen and transition region, the less important is this effect on D_{eff}.

Figure 11.11 illustrates the effect of pore size at fixed pressure on the diffusion flux for the binary system hydrogen-nitrogen. The ratio of the fluxes of the two species is inversely proportional to the square root of the ratio of their molecular weights not only in the Knudsen and transition regime, but also in the bulk regime

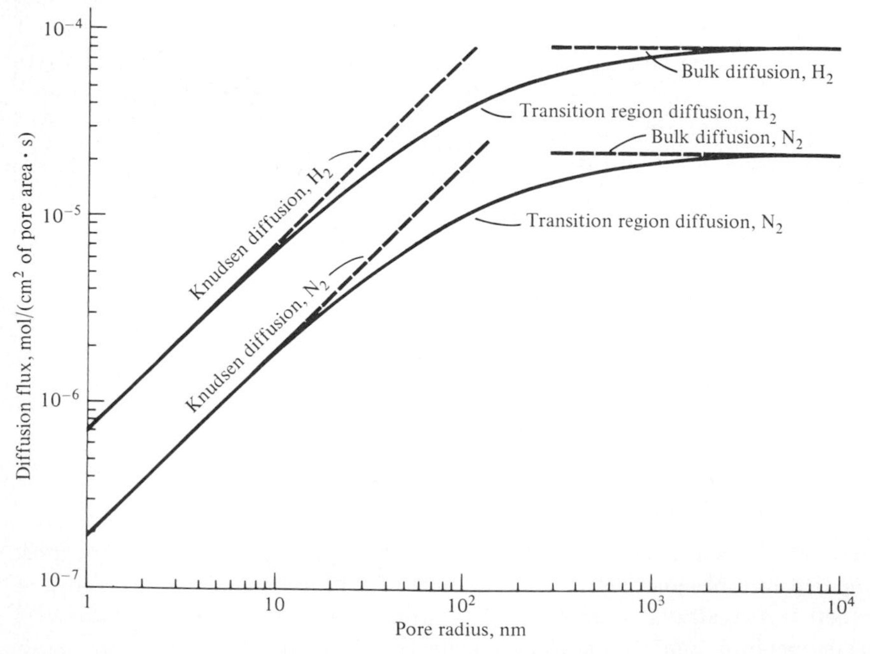

Figure 11.11 Counterdiffusing flux of hydrogen and nitrogen as a function of pore size; atmospheric pressure, $T = 298$ K.

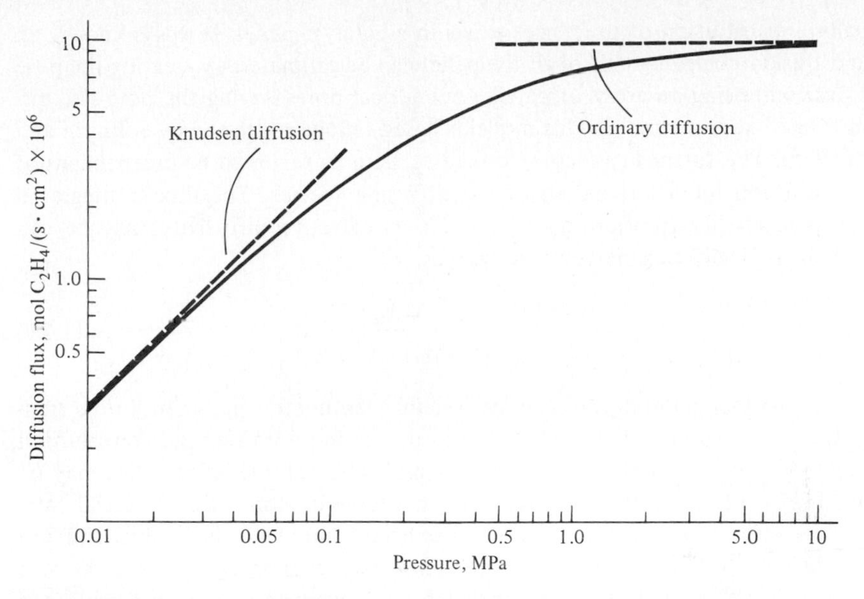

Figure 11.12 Steady-state diffusion flux of ethylene in presence of hydrogen. T = 298 K, θ = 0.4, r_e = 57 nm, P = 1 atm, D_{12} = 0.602 cm²/s.

when the pores are of the order of a micrometer in size. The range of pore sizes for which diffusion will occur in the transition region moves in the direction of smaller pores with increased pressure. Figure 11.12 shows the calculated effect of pressure on the flux through a porous plug in which all pores are taken to be 57 nm in radius. To illustrate the effect of composition, Table 11.4 gives the transition region limits for some binary gas pairs at 300°C and atmospheric pressure. The transition region is arbitrarily defined as that in which the diffusion flux is 10 percent or more below that predicted by the Knudsen or bulk diffusion equation, whichever is less.

Table 11.4 Transition region* for selected binary gas pairs; 300°C, atmospheric pressure

Gas pair	Limiting pore radius, nm	
	Lower	Upper
Hydrogen-carbon monoxide	24.5	2630
Hydrogen-benzene	13.7	1810
Hydrogen-pentadecene	1.9	318
Pentadecene-pentadecane	0.8	53.5
Air-naphthalene	7.6	635

*Diffusion flux 10 percent or more below that predicted for Knudsen or bulk diffusion, whichever is less.

Usually a distribution of pore sizes exists in a catalyst pellet. If this is known, an effective diffusion coefficient through the pellet can be estimated by treating thc pore structure as comprising an array of parallel cylindrical pores having the pore-size distribution found experimentally. This model is based on one suggested by Johnson and Stewart (1965). The tortuosity factor, termed τ_p here, is taken to be independent of pore size, diffusion mode, or nature of the diffusing species. The flux is integrated across the pore-size distribution, and from this an effective diffusivity may be calculated. Mathematically this may be expressed as

$$D_{\text{eff}} = \frac{\rho_p}{\tau_p} \sum_{r=0}^{r=\infty} \frac{\Delta V_g}{(1/D_{1m}) + (1/D_K)} \tag{11.21}$$

where ΔV_g is the incremental pore volume in cubic centimeters per gram. In the transition or Knudsen region, it is found that with decreasing pore size the contribution to the flux and hence D_{eff} per unit of incremental pore volume is less. This may be seen from Figure 11.11. From measurements on a large number of commercial catalysts the tortuosity factor τ_p for this model has been found to be between 3 and 7 in almost all cases, provided that the catalyst has not been excessively sintered. As with bulk diffusion, a value of 4 is recommended for use when no other information is available.

The average pore radius $r_e = 2V_g/S_g = 2\theta/\rho_p S_g$. Taking as representative values $\theta = 0.5$ cm^3 pores per cubic centimeter of pellet and $\rho_p = 2$ g/cm^3, $r_e \approx 1/(2S_g)$. For a catalyst with $S_g = 50$ m^2/g, $r_e = 10$ nm. Thus Fig. 11.11 might imply that for high-area catalysts ($S_g > 50$ m^2/g) the flux would be completely in the Knudsen regime, but this is incorrect. In most industrial catalysts a pore-size distribution exists, and the flux at atmospheric pressure in the larger pores may be in the transition or bulk mode.

Calculations of D_{eff} were made for 11 pelleted commercial catalysts of known pore-size distribution with $S_g > 40$ m^2/g using Eq. (11.18) and a value of $\tau_m = 4$, and were compared to D_{eff} determined from measurements at ambient temperature and pressure. The experimental and calculated values of D_{eff} are given in Table 11.5, and the nature of the catalysts is described in Table 11.6. For the one very low area catalyst (G-58) the flux was predominantly in the bulk regime. For the others, it was in the Knudsen and transition regime. For this group of catalysts, τ_p in the parallel-path model varied between extremes of 2.8 and 7.3. The calculated values from Eq. (11.18) fall below the experimental values by a factor varying from nearly 1 to 10 or more. Thus this expression should not be used even for high-area catalysts except for a very few cases, such as some high-area gelled catalysts, where the pore size distribution is narrow and diffusion is essentially completely in the Knudsen regime.

In the above cases 50 percent or more of the surface area was in pores with $r < 10$ nm, but by the parallel-path model they contributed less than 20 percent of the total flux at most. At higher pressure the dominating regime moves towards bulk diffusion, and the micropore region will contribute to an increasing degree. This emphasizes the importance of knowing the micropore-size distribution for applications at substantial pressure. Figure 11.13 gives the helium flux through nitrogen for five commercial catalysts as a function of pressure from 0.1 to 6.5 MPa. The shapes of the

Table 11.5 Effective diffusivities for selected commercial catalysts and supports*

Designation	Nominal size, mm	Surface area, m^2/g	Total void fraction	$D_{eff} \times 10^3$, cm^2/s†	$r_e = 2V_g/S_g$, nm	$D_{eff} \times 10^3$ from Eq. (11.18)
T-126	4.8 × 3.2	197	0.384	29.3	2.9	3.30
T-1258	4.8 × 3.2	302	0.478	33.1	2.36	3.39
T-826	4.8 × 3.2	232	0.389	37.7	2.14	2.45
T-314	4.8 × 3.2	142	0.488	20.0	4.15	6.00
T-310	4.8 × 3.2	154	0.410	16.6	3.43	2.78
G-39	4.8 × 4.8	190	0.354	17.5	2.24	2.32
G-58	4.8 × 4.8	6.4	0.389	87.0	54.3	62.4
T-126	6.3 × 6.3	165	0.527	38.8	4.9	7.66
BASF	5 × 5	87.3	0.500	11.8	4.1	6.05
Harshaw	6.3 × 6.3	44	0.489	13.3	9.1	13.1
Haldor Topsøe	6.3 × 6.3	143	0.433	15.8‡	2.58	3.29

*Satterfield and Cadle, 1968a.

†Diffusivity of hydrogen through nitrogen, measured at room temperature and atmospheric pressure, average of five sets of samples.

‡Diffusivity of helium through nitrogen, multiplied by $\sqrt{4/2}$, average of two sets of samples.

curves are as anticipated from theory; the flux increases rapidly with pressure at low temperature and finally approaches a constant value at the highest pressure, where diffusion is almost completely by the bulk mechanism.

Measurements on a group of 17 commercial catalysts at pressures from 0.1 to 6.5 MPa (Satterfield and Cadle, 1968a) and on another group of 12 commercial catalysts (Brown et al., 1969) at pressures from 0.1 to 2.0 MPa showed for this wide

Table 11.6 Catalysts and supports in Table 11.5 and Fig. 11.13

Catalyst	Description
T-126	Activated γ-alumina
T-1258	Activated γ-alumina
T-826	3% CoO, 10% MoO_3, and 3% NiO on alumina
T-314	About 8–10% Ni and Cr in the form of oxides on an activated alumina
T-310	About 10–12% Ni as the oxide on an activated alumina
G-39	A cobalt-molybdenum catalyst, used for simultaneous hydrodesulfurization of sulfur compounds and hydrogenation of olefins
G-58	Palladium/alumina catalyst, for selective hydrogenation of acetylene in ethylene
G-52	Approximately 33 wt % Ni on a refractory oxide support, prereduced. Used for oxygen removal from hydrogen and inert gas streams
BASF	A methanol-synthesis catalyst, prereduced
Harshaw	A methanol-synthesis catalyst, prereduced
Haldor Topsøe	A methanol-synthesis catalyst, prereduced

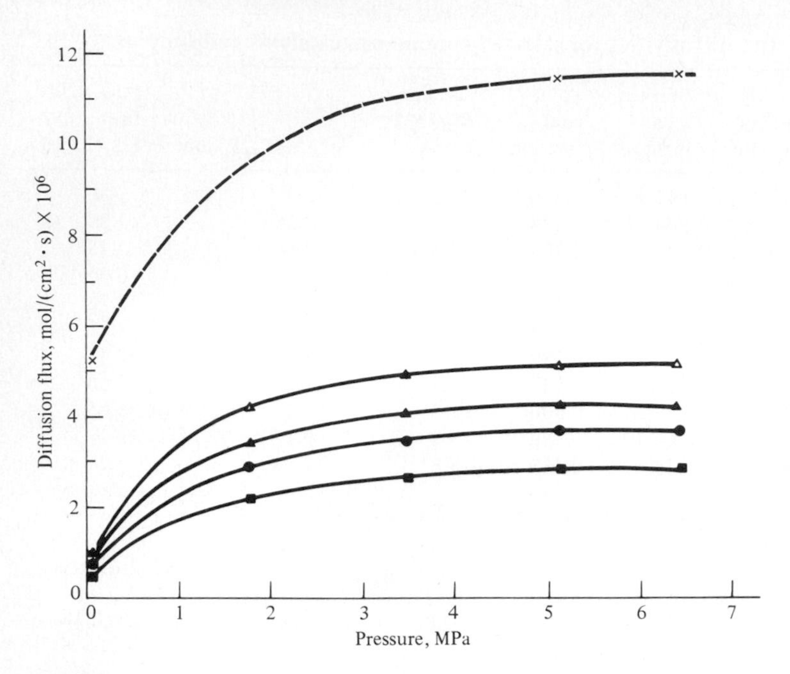

Figure 11.13 Helium flux through nitrogen as a function of pressure. x, G-58, Pd/Al_2O_3; △, G-52, Ni-oxide support; ▲, Haldor-Topsøe $ZnO\text{-}Cr_2O_3$; ●, Harshaw $ZnO\text{-}Cr_2O_3$; ■, BASF $ZnO\text{-}Cr_2O_3$. (*Satterfield and Cadle, 1968b.*) (*Reprinted with permission from Industrial and Engineering Chemistry, Fundamentals. Copyright by the American Chemical Society.*)

variety of catalysts that τ_m as calculated from the parallel-path pore model was essentially independent of pressure. Changing the pressure to this extent allows the entire diffusion regime from Knudsen to bulk to be studied. It leads to the useful conclusion that τ may indeed be regarded essentially as a characteristic of the porous structure that is not significantly affected by the diffusion regime.

Catalysts made by a gelling process, such as some silica-alumina and chromia-alumina catalysts, generally have a high area and fine pores with a fairly narrow pore-size distribution. Hence the flux is usually completely in the Knudsen regime. Some representative values of D_{eff} for these types of catalysts are given in Table 11.7. For homogeneous bead catalysts the diffusivity is approximately inversely proportional to surface area. A clay cracking catalyst has a broader pore-size distribution, and this can also be produced in a gelled catalyst by incorporation of a finely divided solid material such as α-alumina into it. Such compositions have an effective diffusivity severalfold greater than homogeneous gel structures of the same surface area.

11.4.5 Recommended Procedures

In summary, for accurate prediction of D_{eff} it is necessary to know the pore-size distribution of the catalyst pellet. Application of the parallel-path model with a τ_m of 4 should predict D_{eff} within a factor of 2 to 4 for most cases. If only information on

Table 11.7 Diffusion in gel-derived and related catalysts

Catalyst	θ	S_g, m^2/g	$D_{eff} \times 10^3$,* cm^2/s	Reference
A. Silica-alumina "homogeneous" beads				
1. High-density beads, three samples	0.33–0.37	22–381	17.5–0.6	Weisz and Schwartz (1962)
2. Low-density beads, 35 samples	0.42–0.61	46–450	42–3.0	Weisz and Schwartz (1962)
B. Chromia-alumina and chromia-alumina-molybdena beads, two samples	0.53–0.63	155–162	20–15	Weisz and Schwartz (1962)
C. Silica-magnesia beads, one sample	0.34	442	0.45	Wesiz and Schwartz (1962)
D. Commercial pelleted clay cracking catalysts, in regenerated state after commercial use, two samples	0.26–0.30	65–105	42–49	Weisz and Schwartz (1962)
E. Silica-alumina "homogeneous" beads. Laboratory preparations, three samples		270–400	6–9	Weisz and Goodwin (1963)
F. Commercial molecular-sieve bead catalyst containing 4.5 or 10% zeolite X, 30–40% inert α-alumina, dispersed in silica-alumina matrix, average of 15 particles				Cramer et al. (1967)
1. Fresh catalyst			27–28	
2. After 3 h at 700°C in air			15	
3. Fresh catalyst after 24 h at 650°C in 0.2 MPa of steam		101–121	28–29	
4. After 5 h at 650°C in 0.8 MPa of steam		57–73	36–40	

*D_{eff} calculated as diffusion flux of hydrogen through nitrogen at room temperature and atmospheric pressure, divided by concentration gradient (cm^2/s).

surface area is available, and one is dealing with a high-area catalyst under conditions where Knudsen diffusion is dominating, a value of D_{eff} calculated from Eq. (11.18) with a value of $\tau_m = 4$ will generally be conservative (i.e., the calculated value will tend to be lower than the true value). Use of this equation, however, is not recommended in general, since there may be considerable variation in the pore-size distribution and hence diffusivity among catalysts of approximately the same surface area. If bulk diffusion is clearly the dominating regime, use of Eq. (11.13) with a value of $\tau = 4$ is recommended. Note however that if a considerable portion of the diffusion in fact occurs in the transition range, assumption of bulk diffusion will lead to an unrealistically high value of D_{eff}. For more detailed treatment see Satterfield (1970, Chap. 1). Diffusion coefficients in zeolites are very low and are discussed in Chap. 7.

In some cases a catalyst has a bimodal pore-size distribution comprising a group of pores in the vicinity of 1 μm in size and another group of the order of 10 nm or less. Application of the above procedures to the larger pores will indicate whether diffusion *in and out* of the pellet as a whole is a rate-limiting process, but it is still possible for molecules to become trapped or to diffuse very slowly in the very smallest pores or dead-end pores. Thus some "diffusion-limited" type of behavior can occur even when calculations as outlined above indicate that no diffusion limitation occurs in a gross sense through the pellet.

Surface diffusion may contribute significantly to the total flux if one or more of the reactants is a vapor not much above its condensation temperature. Then a physically adsorbed multilayer may form. Under such conditions condensation of reactant (or product) in fine pores may also occur (Chap. 5), possibly significantly blocking diffusion. This may also lead to unusual apparent kinetic behavior as reaction conditions are changed, stemming from hysteresis effects in condensation and evaporation from pores. Surface diffusion may also contribute to the total flux in a few other cases at higher temperatures, encountered primarily with high-area catalysts and possibly at high pressures. Little information is available to indicate how important this mechanism is in general in catalysis.

Multicomponent diffusion in gases, liquids, and solids is treated in a monograph by Cussler (1976). A monograph by Jackson (1977) treats gaseous diffusion in porous media from a fundamental approach, emphasizing the modeling of simultaneous diffusion and reaction in porous catalyst pellets. Aris (1975) has published a detailed and definitive treatment of the mathematical theory of diffusion and reaction in porous and permeable substances.

11.5 THERMAL CONDUCTIVITY OF POROUS CATALYSTS

Relatively little information has been published on the thermal conductivity of porous solids of the types used in catalysis. The principal studies appear to be those of Sehr (1958), Mischke and Smith (1962), Masamune and Smith (1963a), and Harriott et al. (1975).

Table 11.8 shows thermal conductivities for several porous oxide catalysts as reported by Sehr at a mean temperature of about 90°C in air at atmospheric pressure. Mischke and Smith measured the thermal conductivity of a series of alumina catalyst pellets in the presence of helium or air at pressures from vacuum to atmospheric. The catalysts were prepared by pelletizing alumina powder under different pressures. Masamune and Smith reported similar studies in the presence of air, carbon dioxide, or helium on a series of metal catalysts prepared by compacting microporous silver powder. Representative results from these two studies are presented in Table 11.9, plus data reported by other workers, usually incidental to other studies. θ_{macro} is the fraction of the porous pellet occupied by pores above about 10 or 12 nm in radius. θ_{micro} is the fraction occupied by smaller pores. $(1 - \theta_{macro} - \theta_{micro})$ is the solid fraction present.

Harriott and coworkers measured thermal conductivities by a non-steady-state method over the temperature range from ambient temperature to about 140°C. Spherical pellets enclosing air in the pores were compacted to various degrees from powders of silica or silica-alumina either as such or after impregnation to deposit

Table 11.8 Thermal conductivities of some porous catalysts in air at 90°C, atmospheric pressure*

Catalyst	$\lambda_{particle}$, W/m · K	λ_{powder}, W/m · K	Density, g/cm^3	
			Particle	Powder
Nickel-tungsten	0.47	0.31	1.83	1.48
Cobalt-molybdenum dehydrogenation catalysts	0.35	0.21	1.63†	1.56†
	0.24	0.14	1.54‡	1.09‡
Chromia-alumina reforming catalyst	0.29	0.18	1.4	1.06
Silica-alumina cracking catalyst	0.36	0.18	1.25	0.82
Platinum-alumina reforming catalyst	0.22	0.13	1.15	0.88
Activated carbon	0.27	0.17	0.65	0.52

*Sehr, 1958.
†3.6% CoO and 7.1% MoO_3 on α-alumina, 180 m^2/g.
‡3.4% CoO and 11.3% MoO_3 on β-alumina, 128 m^2/g.

nickel. In all cases impregnation was carried out before compacting. The thermal conductivities in Table 11.10 (Harriott et al., 1975) were read from figures and show data points for the two extreme values of θ_{macro} studied. In all cases thermal conductivity decreased with increasing θ_{macro}, as expected. The values range from 0.13 to 0.24 W/m · K.

Except for a pellet of relatively dense silver, the spread of values of thermal conductivity is remarkably small for the wide range of catalysts studied by several investigators. It does not vary greatly with major differences in void fraction and pore-size distribution, brought about by varying the pelletizing pressure (void fractions were not reported by Sehr). The study of Harriott shows that the presence of nickel on the support, even in large amounts, has very little effect on the effective conductivity. (The results for nickel impregnated on silica-alumina are slightly high relative to the others, but the reasons for this are not clear.)

The thermal conductivity under vacuum of compacted silver powder is not much different than that of compacted alumina. This emphasizes that this property is dependent primarily upon geometrical considerations rather than the thermal conductivity of the solid as such. This is further brought out by a summary published by Masamune and Smith (1963b) of studies of the thermal conductivity of beds of spherical particles of a wide range of metallic and nonmetallic substances (silicon carbide, glass beads, quartz, steel balls, steel shot, lead, glass, and magnesium oxide). The ratio of thermal conductivity of the porous bed of solid under vacuum to that of the nonporous solid itself varied from about 0.05 to 0.001, although the void fractions of the porous beds were all within the range of 0.27 to 0.45. There was no significant correlation between conductivity of the solid and the conductivity of the bed. For sintered copper-tin alloy particles at atmospheric pressure the ratio varied from about 0.02 to 0.07 at void fractions of 0.33 to 0.4, to about 0.13 at a void fraction of 0.2.

The last two groups of studies in Table 11.9 are representative of the various

Table 11.9 Thermal conductivities of selected porous materials

Substance	Fluid in pores	Temperature, °C	Pellet Density g/cm^3	θ_{macro}	θ_{micro}	λ, (W/m · K)		Reference
						100 kPa	Vacuum	
Alumina (boehmite) pellets	Air	50	1.12 0.67	0.134 0.450	0.409 0.275	0.22 0.13	0.16 0.07	Mischke and Smith (1962)
Pellets of silver powder	Air	34	2.96 1.35	0.144 0.61	0.574 0.261	0.71 0.17	0.06 0.09	Masamune and Smith (1963a)
Cu on MgO pellets*	Air (?)	25–170	0.7–1.20*	–	–	0.08–0.17	–	Cunningham et al. (1965)
Pt on alumina pellets, 0.05 wt %	Air	–	1.34	0.35	0.15	0.15	–	Miller and Deans (1967)
Pt on alumina pellets	H_2	68	0.57	0.56	0.23	0.26	–	Maymo and Smith (1966)
Bed of stainless steel shot, 71 μm in diameter	Air	42	5.77 (bed density)	0.264	–	0.26	0.02	Masamune and Smith (1963b)
Bed of glass beads 29, 80, 200 or 470 μm in diameter	Air	42	1.50 (bed density)	0.38	–	0.18	0.05	Masamune and Smith (1963b)

*Carbon deposits present to various degrees.

Table 11.10 Thermal conductivity of nickel-impregnated silica and silica-alumina*

Composition	Range of θ_{macro}	Range of thermal conductivity, W/m · K
1. Pure silica (particle diameter ~ 2-3 μm)	0.16–0.5	0.17–0.13
2. Porous silica–25% alumina (particle diameter ~ 70 μm)	0.12–0.45	0.20–0.13
3. Coprecipitated 62% Ni on silica	0.35–0.43	0.21–0.16
4. 62% Ni impregnated on silica	0.2–0.4	0.21–0.13
5. 5% Ni impregnated on silica	0.16–0.32	0.18–0.16
6. 0.5% Ni impregnated on silica	0.2–0.33	0.17–0.16
7. 0.5% Ni impregnated on silica–25% alumina	0.16–0.33	0.24–0.22

*Harriott et al., 1975.

investigations with beds of spherical particles. For these beds conductivity under vacuum is dependent primarily on the area of contact between particles, and would be expected to be greater for rough than for smooth particle surfaces. As a bed of powder is compacted, the thermal conductivity will increase as the contact area between particles is increased. This may also be affected by size, shape, particle-size distribution, and any possible subsequent sintering.

Even the most dense porous silver pellet in Table 11.9 exhibited a thermal conductivity that is much less than that of solid silver (which is about 400 W/m · K). However, if porous metals were prepared by a sintering process, as by the usual powder metallurgy methods, one anticipates that the ratio of thermal conductivity of the porous metal to that for pure solid would be substantially greater than that achieved by pressing. This is indicated by the results with copper-tin alloy particles.

Both the solid and fluid phases are continuous in porous catalysts, so the thermal conductivity may be modeled as two conducting paths in parallel, with transfer of heat between the two. The latter contribution is particularly complicated since the pore-size distribution of many catalysts is such that at atmospheric pressure diffusion is in the transition region between the Knudsen and bulk modes. Here the thermal conductivity of a gas varies significantly with pore size or pressure. The prediction of a model will vary with the geometry postulated for the two paths, but it is evident that the effective conductivity will be determined primarily by that of the phase with the greater thermal conductivity. The greater the differences between the thermal conductivities of the two phases, the greater the divergence between the predictions of different models. Krupiczka (1967) gives several examples of models, with particular reference to beds of granular materials, and gives an extensive compilation of experimental data.

Some feeling for magnitude is helpful at this point. Reaction conditions will usually be at atmospheric pressure or higher and at higher temperatures than those at which measurements have been made. The thermal conductivity of a gas is almost independent of pressure when the mean free path is substantially less than the pore size. At room temperature the thermal conductivity of air is about 0.03 W/m · K, that of hydrogen is about 0.18, and that of a wide range of organic vapors, polar and nonpolar, varies from about 0.01 to 0.03. Omitting hydrogen (and helium), these are an order of magnitude less than that of the usual porous catalysts under vacuum. (Thermal conductivities of simple organic liquids are usually 10 to 100 times greater than that of the vapor at the same temperature. Typical values for nonpolar liquids at room temperature are in the range of 0.08 to 0.20 W/m · K and are 2 or 3 times greater than this for highly polar substances.)

Over the temperature range of about 60 to 200°C, the thermal conductivity of a gas typically increases about linearly with temperature and doubles over about a 200°C temperature range. Data in Tables 11.8, 11.9, and 11.10 are for catalyst pores filled with air at atmospheric pressure, with one exception. The thermal conductivity of a liquid or vapor under reaction conditions will usually be greater than that of air at room temperature, so the values reported for effective thermal conductivity of porous catalysts in air usually represent the minimum values to be expected during reaction. Temperature gradients calculated by their use would thus represent probable maximum values.

The values of thermal conductivity for packed beds given above are in the presence of stagnant gas. The flow of gas, as in a reactor, increases the effective bed conductivity. The effect is minor for typical laboratory conditions. The radial thermal conductivity is generally expressed by an equation of the form

$$\lambda = A + B \cdot \mathrm{Re} \tag{11.22}$$

where A is the bed conductivity at zero flow rate. Typically it requires a value of Re in the neighborhood of 150 to 200 to double the bed conductivity over the value for stagnant conditions. The Reynolds number in laboratory reactors seldom exceeds about 20.

11.6 BULK-MASS TRANSFER

Data on rates of mass transfer between a gas stream and a solid have been obtained by a variety of means, e.g., by observing the rate of volatilization of a low-volatile solid or the rate of evaporation of a liquid from a porous solid. Rates of mass transfer are commonly expressed in terms of a mass-transfer coefficient k_c defined by

$$N = k_c(c_0 - c_s) \tag{11.23}$$

where N is the diffusion flux of the constituent in question (mol/s · cm^2), c_s is the concentration at the surface, and c_0 is the concentration in the ambient fluid. Alternately the potential may be expressed in terms of the partial pressure of the diffusing substance, in which case it may be convenient to define a coefficient k_G by

$$N = k_G(p_0 - p_s) \tag{11.24}$$

where $k_G = k_c/RT$.

Dimensional considerations suggest representing the mass-transfer coefficient in the form of the Sherwood number, $Sh = k_c d_p / D_{lm}$. This in turn is a function of the Schmidt number, Sc, and the Reynolds number, Re. Alternately the mass-transfer coefficient may be represented in terms of the group j_D suggested by Chilton and Colburn in 1934 as the basis for an analogy between mass transfer and heat transfer

$$j_D = \frac{\text{Sh}}{\text{Re} \cdot \text{Sc}^{1/3}} = \frac{k_c \rho}{G} \text{Sc}^{2/3} = \frac{k_G P}{G_M} \text{Sc}^{2/3} \tag{11.25}$$

j_D in turn is expressed as a function of Re, which characterizes the flow condition. Much more data are available on heat transfer than on mass transfer, so analogies between the two are useful for checking and extending mass-transfer data.

The Schmidt number and the Reynolds number are defined as

$$\text{Sc} = \frac{\mu}{\rho D_{1m}} \qquad \text{Re} = \frac{d_p G}{\mu} \tag{11.26}$$

Here μ and ρ are the viscosity and density of the fluid, D_{1m} is the molecular diffusion coefficient for the diffusing species in the fluid, and G is the mass velocity of the fluid in grams per second per square centimeter of total or superficial bed cross section normal to mean flow. The Schmidt number for gas mixtures seldom falls outside the range of 0.5 to 3. For dilute aqueous liquids representative values of Sc are in the range of 400 to 2500; for organic molecules in benzene, a range of 300 to 500 is representative. The length d_p appearing in Re is taken to be the particle diameter in the case of spheres; for cylinders d_p represents the diameter of a sphere having the same surface.

In laboratory work Re will typically range from about 10 to 20 down to as low as 0.1 or less. Petrovic and Thodos (1968) present new data and recalculated various earlier studies by Thodos and coworkers to correct previously published data for axial mixing. They recommended the following correlation, for $3 < \text{Re} < 2000$.

$$\epsilon j_D = \frac{0.357}{\text{Re}^{0.359}} \tag{11.27}$$

This is for $1.8 \text{ mm} < d_p < 9.4 \text{ mm}$.

For lower flow rates and smaller particles, mass-transfer coefficients may be very low, but the situation is confused because of the difficulties of avoiding axial dispersion and end effects in experimental work and the lack of reliable theory for prediction. More data have been reported on heat transfer than on mass transfer, and as a very rough guide the results of several workers may be approximately represented by

$$\text{Sh} \approx \text{Nu} \approx 0.07 \text{Re}_p \tag{11.28}$$

for data obtained at $0.1 < \text{Re}_p < 10$ (Kunii and Suzuki, 1967; Cybulski et al., 1975; Wakao and Tanisho, 1974). Here $\text{Nu} = h d_p / k_f$, and is the heat-transfer analogy to the Sherwood number, Sh. These and other results agree moderately well with a theory put forward by Nelson and Galloway (1975), although the basis for the theory is debatable.

It is likely that at these very low flow rates local mass-transfer coefficients may vary substantially from point to point (Harriott, 1974). A low mass-transfer coefficient in one zone of a packed bed is not necessarily compensated for by a higher coefficient elsewhere, so the effect of mass transfer may be more important than would be estimated from an averaged coefficient. Especially with some degree of particle-size distribution and hence distribution of voidage, it might be expected that much of the flow would channel around small blocks of fine particles, within which mass transfer is largely by molecular diffusion. Kunii and Suzuki (1967) develop a model for heat and mass transfer between particles and fluid in a packed bed based on this concept. Martin (1978) was able to match the experimental features of data on heat transfer by a model of a packed bed with nonuniform distribution of the void fraction, and consequently nonuniform fluid velocity. Next to the wall the void fraction is much larger than the average, especially at low Reynolds numbers. In Martin's analysis Nu is presented in terms of a Peclet number for heat transfer, defined as $\text{Pe} = \text{Re} \cdot \text{Pr} = C_p G d_p / k_f$.

In summary, qualitative tests to determine whether external mass transfer is a significant resistance by varying, e.g., linear velocity, may give anomalous results at values of Re below about 1 to 10. Also, external mass-transfer coefficients may be so low that external resistance becomes more important than internal resistance. This is the reverse of the usual pattern observed at higher flow rates and with larger particles.

11.6.1 Heat Transfer

Heat and mass are transferred between solid and fluid by similar mechanisms, and data on heat transfer in fixed beds are correlated in the same way as data on mass transfer. Thus

$$j_H = \frac{h}{c_p G} \text{Pr}^{2/3} \tag{11.29}$$

$$h = \frac{q}{T_s - T_0} \tag{11.30}$$

where

$$\text{Pr} = \frac{c_p \mu}{k_f} \tag{11.31}$$

Here h is the heat-transfer coefficient, q is the heat flux (per unit pellet outside surface area), c_p is the heat capacity per unit mass of fluid, Pr is the Prandtl number, T_s is the pellet surface temperature, T_0 is the fluid stream temperature, and k_f is the thermal conductivity of the fluid. j_H is approximately equal to j_D, and this forms the basis for estimating mass-transfer coefficients from heat-transfer data.

11.6.2 Temperature Difference between Solid and Fluid

The relationship between the degree of bulk mass-transfer control of a reaction and temperature difference between pellet outside surface and fluid may be easily derived for steady-state conditions. The rate of mass transfer of a reacting species from fluid to solid multiplied by the heat of reaction per mole of diffusing species must equal the rate of heat transfer from solid back to fluid. Hence

$$k_c(c_0 - c_s)(-\Delta H) = h(T_s - T_0) \tag{11.32}$$

Substituting the expressions for the Prandtl and Schmidt numbers, and for the j_D and j_H functions,

$$(T_s - T_0) = \frac{j_D}{j_H}\left(\frac{\text{Pr}}{\text{Sc}}\right)^{2/3} \frac{(-\Delta H)}{\rho c_p}(c_0 - c_s) \tag{11.33}$$

The extent to which the reaction is bulk mass transfer-controlled, f, may logically be defined as the ratio $(c_0 - c_s)/c_0$, whence Eq. (11.33) becomes

$$(T_s - T_0) = \frac{j_D}{j_H}\left(\frac{\text{Pr}}{\text{Sc}}\right)^{2/3} \left[\frac{(-\Delta H)c_0}{\rho c_p}\right] f \tag{11.34}$$

The temperature difference is seen to be directly proportional to the heat of reaction per mole of diffusing component and to the fractional drop in concentration between bulk fluid and solid. The product $(-\Delta H)c_0$ is the heat that would be released by complete reaction of 1 cm^3 of reactant mixture. The product ρc_p is the volumetric heat capacity of the gas. The quotient $(-\Delta H)c_0/\rho c_p$ represents the temperature rise that would be calculated for complete adiabatic reaction of the fluid mixture.

For many simple gas mixtures, the ratio Pr/Sc is in the vicinity of unity, and j_D/j_H may be taken as 1. For a completely mass transfer-controlled gas-phase reaction ($f = 1$) corresponding to the above circumstances, the temperature difference between gas phase and solid would thus be approximately equal to the calculated adiabatic temperature rise for complete reaction of the fluid. If fluid properties did not change through the reactor, in theory the solid would be at the same temperature, namely the adiabatic reaction temperature, throughout the bed. In fact such a reaction would probably proceed at high temperature, and radiative heat transfer might be important. A mass transfer-controlled reaction at the high velocities at which industrial reactors can be operated results in such a short reaction zone that heat transfer by radiation from the bottom and top may result in substantial temperature gradients. For a number of partial oxidation reactions in air the ratio Pr/Sc is actually somewhat less than unity, but the ratio could conceivably exceed unity for some reaction system. In this case it would be possible for the catalyst surface temperature to exceed the adiabatic reaction temperature.

Equation (11.34) emphasizes the fact that, if the heat of reaction is large, mass-transfer limitations may be small, yet heat transfer can still cause significant effects. Consider, for example, a case in which the calculated adiabatic temperature rise for the reaction is 500°C and the mass-transfer rate at a point in a reactor results in only a 4 percent concentration difference between bulk gas and catalyst, that is, $f = 0.04$. Taking $(j_D/j_H)(\text{Pr/Sc})^{2/3}$ as unity, the temperature difference would be 20°C, sufficient to cause a marked increase in the observed rate of reaction over that which would occur if the catalyst were indeed at the bulk-vapor temperature. With sufficiently active catalysts a region of unstable catalyst temperatures exists, as in ammonia oxidation on a platinum gauze (Sec. 8.10). Upon increasing the temperature of the reacting gas, the temperature of the catalyst will suddenly jump to a new, higher level. The situation is analogous to the ignition of a fuel such as carbon in a stream of air.

One of the useful applications of a recirculating differential reactor (Sec. 11.8.4) is to highly exothermic reactions such as the above. An increase in the linear gas velocity through a catalyst bed increases the heat-transfer and mass-transfer coefficients proportionately. If the concentration difference between fluid and outside surface is small to begin with, increased velocity has little effect on surface concentration. Therefore there is little increase in rate of reaction and rate of heat generation. However, increased fluid velocity increases the heat-transfer coefficient and thus reduces the temperature of the catalyst particle down closer to that of the fluid.

11.7 EXAMPLES OF USE OF CRITERIA

Example 11.1 Estimation of D_{eff} Estimate D_{eff} for the diffusion of thiophene in hydrogen at 660 K and 30 atm (3 MPa) in a catalyst having a BET surface of 180 m^2/g, a void volume of 40 percent, a pellet density of 1.40 g/cm^3, and exhibiting a narrow pore-size distribution. D_{12} is 0.052 cm^2/s. Substituting in Eq. (11.13), we obtain

$$D_{12,\text{eff}} = \frac{D_{12}\theta}{\tau} = \frac{0.052 \times 0.40}{\tau} = \frac{0.0208}{\tau}\ \text{cm}^2/\text{s}$$

Substituting in Eq. (11.18), we have

$$D_{K,\text{eff}} = \frac{19400 \times 0.4^2}{\tau_m \times 1{,}800{,}000 \times 1.40}\sqrt{\frac{660}{84}} = \frac{0.00344}{\tau_m}\ \text{cm}^2/\text{s}$$

Taking $\tau = 4$, $D_{12,\text{eff}} = 0.0052$ and $D_{K,\text{eff}} = 0.00172$ cm^2/s, Knudsen diffusion may be expected to predominate, since $D_{12,\text{eff}}$ is so much larger than $D_{K,\text{eff}}$. Applying Eq. (11.20),

$$\frac{1}{D_{\text{eff}}} = \frac{1}{D_{K,\text{eff}}} + \frac{1}{D_{12,\text{eff}}} = 774$$

$$D_{\text{eff}} = 0.0013\ \text{cm}^2/\text{s}$$

Example 11.2 Use of Φ_s Archibald et al. (1952) studied the rate of catalytic cracking of a West Texas gas oil at 550 and 630°C and atmospheric pressure by passing the vaporized feed through a packed bed containing a silica-alumina cracking catalyst of each of several sizes ranging from 8 to 14 mesh to 35 to 48 mesh (See Table 5.3). They report that at 630°C the apparent catalyst activity was inversely proportional to catalyst particle size. This implies that the catalyst is operating at a relatively low effectiveness factor and Φ_s should be considerably greater than unity. We can check this by suitable calculations based on their run on 8 to 14 mesh catalyst. The average particle radius may be taken as 0.088 cm.

They report 50 percent conversion at a liquid hourly space velocity (LHSV) of 60 cm^3 of liquid per cubic centimeter of reactor volume per hour. The liquid density is 0.869, and its average molecular weight is 255. The effective density of the packed bed was about 0.7 g catalyst per cubic centimeter of reactor volume. The molecular weight of the products was about 70.

Pore-structure characteristics for a commercial homogeneous silicia-alumina are as follows: average pore radius = 2.8 nm, catalyst particle density $\rho_p = 0.95$, $\theta = 0.46$, $S_g = 338\ m^2/g$. Take $\tau = 4$. From Eq. (11.18),

$$D_{\text{eff}} = 19{,}400 \frac{\theta^2}{\tau S_g \rho_p} \sqrt{\frac{T}{M}}$$

$$= \frac{(19{,}400)(0.46)^2}{(4)(338 \times 10^4)(0.95)} \sqrt{\frac{903}{255}} = 6.0 \times 10^{-4}\ \text{cm}^2/\text{s}$$

The rate of reaction is 50 percent of the rate of feed,

$$-r = \frac{(60)(0.869)}{255} \left(\frac{1}{3600}\right) \left(\frac{1}{0.7}\right) (0.95)(0.5)$$

$$= 3.86 \times 10^{-5}\ \text{mol/s} \cdot \text{cm}^3 \text{ pellet volume}$$

For the average concentration of reactant outside the pellets, an arithmetic average of inlet and exit concentrations is sufficiently precise for this example. This corresponds to conditions at 25 percent conversion.

Each mole of gas oil produces 255/70 = 3.64 moles of products. Per 100 moles entering, at 25 percent conversion there remain 75 moles of gas oil and (25)(3.64) = 91 moles of products, for a total of 166 moles.

The average reactant concentration is therefore

$$\frac{1}{22{,}400} \left(\frac{273}{903}\right) \left(\frac{75}{166}\right) = 0.61 \times 10^{-5}\ \text{mol/cm}^3$$

The average catalyst particle radius is 0.088 cm; whence, from Eq. (11.1),

$$\Phi_s = \frac{(0.088)^2}{6 \times 10^{-4}} (3.86 \times 10^{-5}) \times \frac{1}{0.61 \times 10^{-5}} = 81$$

Regardless of the kinetics, this reaction is highly diffusion-limited.

Example 11.3 Use of Φ_s Weisz (1957) reports measurements of the rate of burnoff in air of carbonaceous deposits formed on a silica-alumina cracking catalyst. Oxygen consumption rates are reported for two particle sizes, 0.20-cm beads and 0.01-cm powder, at temperatures of 460°C and higher. At 460°C, the rate per unit mass of catalyst was the same for both particle sizes, demonstrating that the effectiveness factor for the 0.20-cm beads at 460°C was nearly unity. Significant differences between the rates on the two particle sizes began to appear at a temperature of about 475°C. The corresponding rate of oxygen consumption was about 4×10^{-7} mol/s · cm^3. The diffusivity of catalyst samples, using hydrogen and nitrogen on opposing faces, was measured and found to be 6.2×10^{-3} cm^2/s for hydrogen at 20°C. Temperature gradients within the pellet are unimportant in this reaction. Compare the rate of reaction at the temperature at which diffusion begins to be appreciable with that expected from the critical value of Φ_s.

Carbon combustion is about first order in oxygen, and, in the presence of excess air, inhibition by products is probably negligible. The critical value of the parameter Φ_s will be about 1.0. Diffusion is assumed to be by the Knudsen mode.

$$D_{\text{eff}} = 6.2 \times 10^{-3} \sqrt{\frac{748}{293} \times \frac{2}{32}} = 2.48 \times 10^{-3} \text{ cm}^2/\text{s}$$

$$c_s = \frac{0.21}{22{,}400} \times \frac{273}{748} = 3.42 \times 10^{-6} \text{ mol/cm}^3$$

Calculate the value of the maximum reaction rate in the 0.20-cm particles above which diffusion will become appreciable. From Eqs. (11.1) and (11.2),

$$\frac{r_p^2(-r)}{D_{\text{eff}} c_s} = \Phi_s = 1$$

$$\frac{(0.20)^2}{2.48 \times 10^{-3}} \times \frac{1}{(3.42 \times 10^{-6})} \times (-r) = 1$$

$$-r \approx 2.12 \times 10^{-7} \text{ mol/s} \cdot \text{cm}^3$$

This compares closely with the value of about 4×10^{-7} found experimentally.

Example 11.4 Nonisothermality Consider the data on catalystic cracking of gas oil on a silica-alumina catalyst as analyzed in Example 11.2, in which isothermality was assumed. Test the validity of this assumption by use of the criterion of Eq. (11.5).

The endothermic heat of reaction varies with degree of reaction because of secondary processes, but the maximum is about 167 kJ/mol. Assume the thermal conductivity of silica-alumina to be that given by Sehr (Table 11.8), namely 0.36 W/m · K. The effective diffusivity was estimated in the earlier example to be about 6.0×10^{-4} cm^2/s. Consider conditions at the entrance to the bed, with a reaction temperature of 630°C:

$$\beta = \frac{c_s(-\Delta H) D_{\text{eff}}}{\lambda T_s} = \frac{\left(\frac{1}{22{,}400} \times \frac{273}{903}\right)(-167 \times 10^3)(8.0 \times 10^{-4}) \times 10^2}{(0.36)(903)}$$

$$= -0.00055$$

Since $\beta = \Delta T_{\max}/T_s$, the maximum temperature difference that could exist between particle surface and interior is about 0.5°C, and this will occur only under highly diffusion-limiting conditions.

The maximum activation energy for this reaction is about 167 kJ/mol. Hence, $\gamma = 167{,}000/(8.32)(903) = 22$. Applying the criterion of Eq. (11.5),

$$\Phi_s|\beta| < \frac{1}{\gamma}$$

$$(81)(0.00055) < \tfrac{1}{22}$$

$$0.04 < 0.05$$

The criterion is just barely met.

Example 11.5 Nonisothermality Consider the dehydrogenation of cyclohexane over a platinum-alumina reforming catalyst at 2.5 MPa and 450°C. A large excess of hydrogen is used to prevent carbon formation on the catalyst, so a 4:1 hydrogen/hydrocarbon ratio will be considered. Prater (1958) gives the following data:

$$\Delta H \text{ of reaction} = +220 \text{ kJ/mol}$$

$$D_{\text{eff}} = 16 \times 10^{-3} \text{ cm/s}^2 \text{ for cyclohexane}$$

(assume this is completely Knudsen diffusion)

$$\text{Thermal conductivity} = 0.22 \text{ W/m} \cdot \text{K}$$

At 450°C, the reaction will be effectively irreversible. The value of β is

$$\frac{\left(\dfrac{1}{22{,}400} \times \dfrac{273}{723} \times \dfrac{2.5}{0.5}\right)(-220 \times 10^3)(16 \times 10^{-3}) \times 10^2}{(0.22)(723)} = -0.18$$

The reaction is highly endothermic. Under highly diffusion-limiting cases, the temperature in the center of a catalyst pellet could be as much as (0.18)(450 + 273) = 130°C below that at the outside surface.

11.8 EXPERIMENTAL LABORATORY REACTORS

A variety of reactors may be utilized at the laboratory or pilot plant scale depending upon the particular purpose at hand. A gaseous reaction catalyzed by a solid is the most common, and fast, highly exothermic reactions are some of the most difficult to study effectively. In the following we will consider four types of reactors that are representative of those that may be utilized in a research and development program leading to the final design of a fixed bed, adiabatic or multitubular, reactor involving an exothermic reaction.

11.8.1 A Scouting Laboratory Reactor

The purpose of this reactor is to provide a means of characterizing intrinsic catalyst activity with a minimum of interference from extraneous effects caused by concentration and temperature gradients. This is to ensure that a promising catalyst composition is not overlooked because it is tested under conditions such that the data are falsified by these extraneous effects.

This reactor is typically a straight or U tube about 6 mm in inside diameter, held at nearly constant temperature in a fluidized sand bath or tube furnace. In the latter case a straight reactor tube may be inserted in a metal block to minimize temperature gradients. The block in turn is held in the furnace. A fluidized sand bath generally gives a more uniform temperature along the reactor wall than does a tube furnace, but it may be more time-consuming to replace catalyst samples, which is of importance if this is to be done frequently. The narrow reactor diameter maximizes the degree of approach to isothermality. Thermocouples may be embedded into the outside portion of the reactor wall, or, with a fluidized sand bath, the temperature of the bath may usually be taken as a close approximation of the wall temperature. This requires, however, that the bath be well fluidized and that the thermocouple be placed close to the reactor tube. A straight-tube reactor should always be arranged vertically rather than horizontally to avoid bypassing that could occur with settling of catalyst. Downflow is generally preferred in order to avoid blowing catalyst out of the reactor.

The quantity of catalyst required is small, of the order of a few cubic centimeters, so a large number of catalyst preparations can be rapidly screened, the temperature and gas composition being varied to search for an optimum. Since small quantities of reactants are utilized, this scale of operation is desirable for studies with model compounds that may be expensive. Catalyst particles should in general not be finer than about 400 μm, in order to avoid excessive pressure drop and bed plugging, and not coarser than about 2 mm, in order to minimize the possibility of diffusion limitations and significant bypassing. Particles as small as 100 to 200 μm may be satisfactory if a narrow size distribution is used and the Reynolds number is not so low ($\mathrm{Re} < \sim 1$) that poor heat and mass transfer occur. If the particles dust, break up readily, are sticky, or are too fine, they may be spread out on a coarse inert support (for example, 8 to 14 mesh quartz) that will act to keep the bed open. In any case it is preferable to have as narrow a particle-size distribution as possible in order to minimize packing and channelling. It is desirable, especially with fine powder, to increase gas flow gradually in bringing the reactor on stream. This minimizes the possibility of forming an impenetrable plug. Studies should also be made with the reactor emptied of catalyst in order to determine if reactor wall contributes significant catalytic activity or if homogeneous reaction is important.

It is not unusual, especially with an amorphous catalyst, to observe a drop in performance from acceptable to unacceptable levels after a period of as long as 50 to 100 h. Catalysts that survive this long will usually have reached a relatively stable structure and will exhibit only slow further change, except for that caused by poisoning, fouling, etc.

If an inert diluent is used, a nonporous substance such as glass, quartz, or α-alumina is preferred. Blank runs should be made to ensure that the diluent is truly inert, not only to the primary reactant but also to intermediates that may be formed. Thus intermediates as well as reactants should be tested. High-area substances, such as silica gel, that are inert by themselves can nevertheless sometimes significantly affect performance. By adsorbing impurities or coke precursors, they can cause performance to be altered over that observed in their absence.

With the above precautions, the reactor should operate in essentially plug flow and essentially isothermally, particularly if a fluidized sand bath is used. If there is reason to suspect significant departure from these ideal conditions, measurements may be made in two reactors varying in diameter by 50 percent or more and with two or more particle sizes.

Studies are sometimes done with a *micropulse* reactor, which utilizes a small quantity of catalyst held in a fixed bed at specified temperature. An inert gas (e.g., helium) or one of the reactants (e.g., hydrogen) is fed continuously through the bed. From time to time a short pulse of another reactant is injected into the feed stream, by use of a hypodermic needle through a rubber septum, for example. The exit gas may be run directly into a gas chromatograph or other system for on-line analysis. Such apparatus is simple to set up and use, and avoids the necessity for precise metering and feed pumps for a liquid reactant. Hypodermic injection is practicable to pressures up to about 1 MPa.

Such apparatus is useful in oxidation catalysis for revealing the change in catalyst activity and selectivity as its oxidation state changes, as upon successive pulsing of a reactant without intermediate reoxidation of the catalyst. The effects of poisoning on catalyst behavior may also be readily studied. This should, however, be regarded as a secondary experimental procedure. The catalyst is not at equilibrium with the gas, and adsorption-desorption effects may be much different than in a steady-state reactor. Kinetic data are sometimes reported from micropulse reactor studies, but the non-steady-state nature of the process introduces uncertainties. Reliable rate data are almost impossible to obtain because the instantaneous concentrations are unknown. This type of reactor is reviewed by Langer et al. (1969) and Choudhary and Doraiswamy (1971).

11.8.2 A Catalyst-Optimization Reactor

The best catalyst candidates from the screening reactor above must now be prepared and tested in a form that would be satisfactory commercially, e.g., as pellets or extrudates of sufficient size that an acceptably low pressure drop can be obtained in the final reactor. Considerable experimentation with carriers, binders, and methods of preparation may be required at this stage. The reactor will now necessarily be of larger diameter, probably about the same as that expected to be ultimately used commercially if a multitube reactor is chosen. This typically would be about 2.5 cm. Generally, the laboratory unit will be from 30 to 100 cm long; it will be heated in a tube furnace or with electrical windings and have a thermocouple well (e.g., about 3 mm OD) down the center to obtain axial temperature profiles.

The question of concentration and temperature gradients within catalyst pellets and between the pellet outside surface and bulk fluid should again be reexamined both experimentally and theoretically. It may well develop that marked temperature gradients cannot be avoided commercially, and suitable mathematical simulation becomes appropriate at this point to indicate the optimum in tube diameter, flow rates, and other variables (Sec. 11.9). In order to avoid effectiveness factors much below unity and, consequently, poorer selectivity, it may be desirable at this point to consider other catalyst formulations. These may have a more open pore structure, larger pores, or a bimodal pore structure. The possibility of decreasing particle size somewhat or the use of rings instead of pellets should also be examined.

If the reaction is ultimately to be run at substantially atmospheric pressure, the laboratory reactor will almost invariably be operated at atmospheric pressure. However the commercial reactor will usually be operated at a pressure sufficient to overcome the pressure drop through downstream processing equipment. The reactor pressure will typically be in the neighborhood of from 200 kPa to as much as 400 kPa. This may affect the kinetics, but, possibly more important, it may make the heat-transfer problem more difficult. With a first-order reaction, an increase in pressure of from 100 to 400 kPa quadruples the reaction rate and hence the rate of heat release, but it has only a small effect on the heat-transfer coefficient. Hence temperature gradients will be exacerbated. A reactor of the type considered makes it possible to explore this possibility.

If pellets as large as, say, 6 mm in diameter are being studied in a 25-mm-diameter tube with a thermowell, only about two pellets can be positioned side by side in the annulus, so some bypassing may occur. However, for a specified contact time the linear flow rate here will be considerably greater than in the laboratory reactor. With a correspondingly higher Reynolds number (for example, 10 to 100) axial dispersion may be much less important (see Sec. 11.3.4).

The possibility of a fluidized-bed reactor may arise at this point, and scouting work might be done with a bed of 2.5 to 5 cm diameter. Wall effects may be significant in small-diameter, fluidized-bed reactors. The performance of a laboratory reactor of this size may be somewhat better than that achievable in a plant-scale reactor, because gas-solid contacting usually becomes worse in larger vessels and at higher superficial velocities.

11.8.3 The Prototype Reactor

If a multitube reactor appears to be that desired for the ultimate commercial reactor, a full-scale experiment on a jacketed and packed tube of the final proposed size is generally required. This is to prove out the activity and selectivity that can be expected to be obtained in the full-scale reactor. Typically the tube is 2.5 to 5 cm in diameter and 3 to 6 m long and is cooled with the heat-transfer medium to be ultimately used. Although expensive, this stage of experimentation may be needed for several possible reasons, including the following:

1. The much higher linear flow rates and possibly different tube diameter than that used in the catalyst-optimization reactor may lead to somewhat different temperature profiles along the tube, with consequent effect on performance. Although these can be estimated from a reactor model, correlations of the thermal conductivity of packed beds may be insufficiently accurate. In particular, it may be desirable to use relatively large particles for low pressure drop, resulting in a tube

diameter/particle diameter ratio of, say, 3 to 6. There is no inherent reason why this needs to be avoided, but the effective bed thermal conductivity for these systems can be predicted much less accurately than for more conventional cases in which a continuum model applies.

2. The pressure drop through the reactor may significantly affect either the kinetics or the economics. This needs to be established.
3. The major heat-transfer resistance is inside the packed tube. Typically it is divided approximately equally between resistance at the inside wall and that through the bed. It is important to establish the actual temperature profile along the tube, which is also determined in part by the reactant flow rate and characteristics of the heat-transfer medium. These should be similar to that achievable in the commercial reactor.
4. Reproducibility of pressure drop through the tube may be poor on repacking, especially at low tube diameter/particle diameter ratios. This needs to be checked. The heat-transfer coefficient at the interior wall varies with particle shape, and small changes in catalyst size and shape can significantly affect the manner of packing.

In the design of the commercial reactor several new problems require attention. These include means of securing even flow of coolant around tubes and even distribution of gas into all tubes. The resistance to flow must be the same through each tube in order to avoid different residence times in different parallel tubes, which could reduce overall performance. The pressure drop is often measured across each tube under standardized conditions after it is substantially loaded, and the amount of catalyst present is adjusted if necessary so that no packed tube deviates beyond a specified degree from the average.

11.8.4 Gradientless Reactors

For careful kinetic studies it is desirable to work with a *differential reactor*, one in which the gas and solid composition, pressure, and temperature are essentially uniform throughout the reactor. It is sometimes possible to achieve this in a packed-bed reactor of the type described in Sec. 11.8.1, but often flow rates are so low that possible heat- and mass-transfer gradients raise doubts about the validity of the results obtained. The term *differential reactor* means a reactor in which the conversion is limited to not more than a few percent. It does *not* mean a particular piece of apparatus. A laboratory fixed-bed reactor may possibly be operated in a differential mode, but it is difficult to do so over a very wide range of operating conditions.

A variety of gradientless reactors have been designed and studied, and a considerable amount of performance information and results of specific studies have been reported. In each case a small amount of catalyst is used, and the relative velocity of the fluid with respect to the solid is increased over that achieved in an ordinary laboratory fixed-bed reactor. This improves the heat- and mass-transfer characteristics external to catalyst particles. Two representative designs are shown in Figs. 11.14 and 11.15. In the Carberry (Notre Dame) reactor pellets are held in mesh baskets and

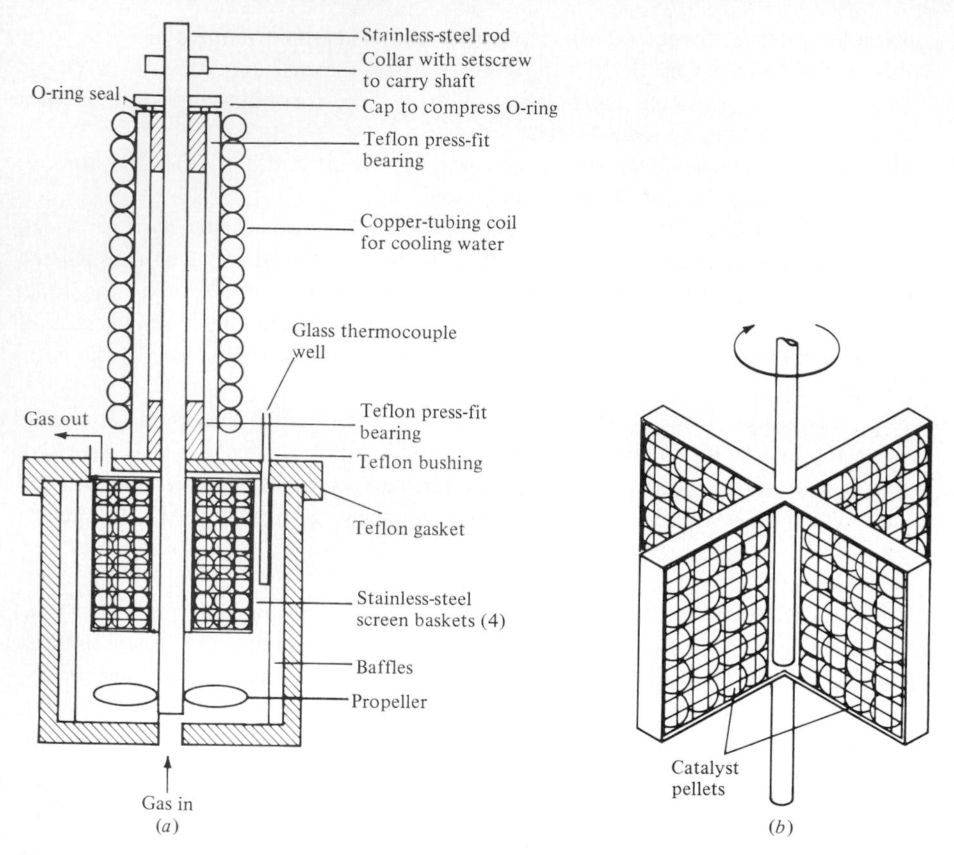

Figure 11.14 (*a*) The Notre Dame gradientless reactor for gas-solid and gas-liquid-solid studies. (*b*) The catalyst disposition. (*Carberry, p. 406, 1976.*)

revolved. In the Berty (Union Carbide) reactor, the pellets are stationary and fluid is recirculated. In the latter a magnetic coupling is used to avoid sealing and contaminant problems. This particular design can be used at high pressure and has a low gas volume to minimize homogeneous reaction. The flow rate through the bed is calculated by measuring the pressure drop through the bed.

The particular advantages of these reactors are:

1. They provide a means of obtaining differential kinetic data over a range of experimental conditions often not achievable by other reactors. The fresh feed rate can be varied over a wide range to simulate local conditions from the entrance to the exit of an integral reactor.
2. Commercial reactors are much longer than laboratory reactors. Hence for a given contact time the linear velocity in the commercial reactor will be much greater than in a conventional laboratory unit. If bulk heat- and mass-transfer limitations exist industrially, the gradientless reactor provides a means of simulating commercial conditions in the laboratory.
3. In a packed-bed reactor a poison is often adsorbed preferentially on the upstream

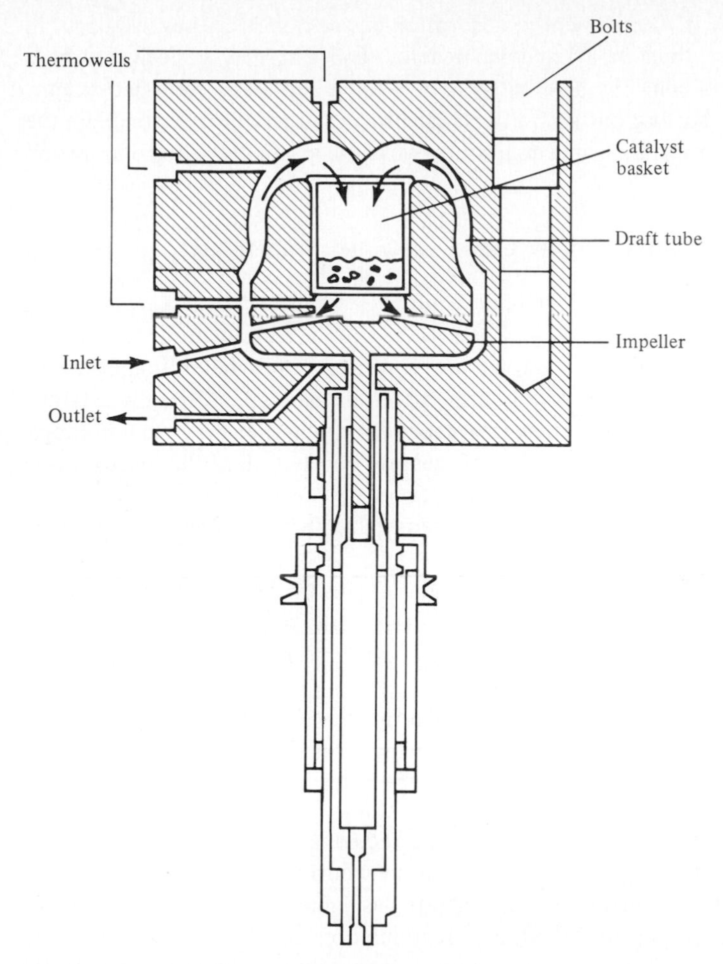

Figure 11.15 Berty gradientless reactor.

portions of the bed. A band of poisoned catalyst moves downward with time, increasing in depth in a chromatographic effect while the remainder of the bed remains active. The results may thus be difficult to interpret. Gradientless reactors are an effective means of studying poisoning since a thin bed of material can be used without concern for bypassing. The role of certain intermediate products, which may be formed and then disappear before the gas leaves the reactor, can be studied by deliberate addition.

4. The composition of a catalyst depends on the environment, for example, the value of x in a metal oxide catalyst MO_x. The relative amounts of reactants, intermediates, and products adsorbed on the catalyst are also affected. This in turn can affect activity and selectivity. In a packed-bed reactor under steady-state conditions, the catalyst assumes a steady-state composition both in the bulk and on the surface. This, however, may vary substantially from inlet to exit. In a fluid-bed reactor the solid moves in a complex pattern with respect to the gas and in many cases is not

in equilibrium with it. Consequently, adsorption-desorption processes will occur in a different manner than in a fixed-bed reactor, and this may conceivably be a reason for different behavior of a catalyst in a fixed-bed and a fluid-bed reactor mode. One way of sorting out such effects is with a recirculating reactor. Here the catalyst composition will be in equilibrium with the particular gas composition present.

Some precautions to consider in the use of gradientless reactors are:

1. The ratio of void volume to packed-bed volume is much higher in a recirculating reactor than in a conventional packed bed. Data from a recirculating reactor may be misleading if homogeneous reactions are significant or if heterohomogeneous processes occur. Sometimes reactive intermediates are formed, as in some catalytic partial oxidation reactions, and their fate depends on whether they react homogeneously or catalytically. The possibility of reaction on the wall of the reactor or on mechanical bearings, etc., should be considered.
2. Some ranges of temperature and pressure achievable in packed-bed reactors may not be obtainable in a recirculating reactor.
3. The possibility of contamination from sealing mechanisms should be considered.
4. It is assumed that the gas composition is uniform, but this needs to be checked, e.g., by pulse tracer experiments. Operating conditions must be established that provide an adequate recycle rate relative to feed rate in the internal recirculation reactor. In the revolving basket reactor, design must ensure that bypassing of feed gas to the exit does not occur. Also, the bulk mass-transfer characteristics vary substantially between pellets close to the axis and those far from the center.

For development of a commercial process, a recirculating reactor can reveal important information, but it is evident that a reactor configuration of the type to be used industrially must be constructed and operated as an integral reactor to prove out the reaction. Various types of laboratory recirculation reactors are reviewed by Doraiswamy and Tajbl (1974) and by Carberry (1969). Livbjerg and Villadsen (1971) describe an internal recirculating reactor for studying the SO_2 oxidation and give references to previous apparatus of this type. The operating characteristics of the Berty reactor are described by Berty (1974). He generally recommends a lower pressure limit of about 300 kPa in order to maintain good recirculation rates, but a set of conditions for suitable operation at lower pressures has been published by Kuchcinski and Squires (1976). Gradientless reactors in general are discussed by Bernard et al. (1972) and Bennett et al. (1972). The various types of laboratory reactors are also critically compared and contrasted by Weekman (1974) and Sunderland (1976).

11.9 REACTOR MODELING: AN EXAMPLE

During the course of laboratory experimentation a point will arise at which a tentative design of a commercial reactor is needed to estimate performance and costs. For a multitube reactor typical variables are tube diameter, contact time (tube length), and wall temperature. For a highly exothermic reaction it is impossible to hold the reaction temperature substantially constant without the use of tubes of impracticably small diameter. A hot zone, at a temperature considerably above that of the tube wall and the coolant, will appear at some distance down from the inlet. An optimum design aims at keeping the temperature here within bounds to maximize selectivity and minimize catalyst aging, and at the same time using tubes of as large diameter as feasible to lower costs. With a reliable kinetic expression in hand and information on physical and transport properties, the reactor performance can be modeled by computer simulation and the effects of changing variables can be quickly ascertained.

The following illustrates such reactor modeling for a reaction of industrial importance, the partial oxidation of *o*-xylene in air to phthalic anhydride. This in turn may be further oxidized to carbon monoxide and carbon dioxide. Some *o*-xylene may also be directly oxidized to these byproducts without the intermediate formation of phthalic anhydride. These steps are shown in Figure 11.16.

This reaction was modeled using the kinetic data and other information published by Froment (1967) for a vanadium oxide type of catalyst. The inlet mole fraction of xylene is taken as 0.93 percent. Since a large excess of air is used, each of the three reactions may be expressed as pseudo-first order. The heat of reaction (1) is 1285 kJ/mol; that of reaction (3) is 4564 kJ/mol. The activation energies for the reactions are E_1 = 113 kJ/mol, E_2 = 131 kJ/mol, and E_3 = 120 kJ/mol. A minimum temperature is needed to achieve an economic rate of reaction, but it is evident that increased temperature increases the undesired reactions relative to the desired ones. Reaction (3) releases much more heat than reaction (1), so an increased temperature, which will increase reaction (2) relative to reaction (1) will also increase the heat load and therefore the heat-transfer problem. In a reaction of this type it is often desirable to achieve nearly complete consumption of reactant if this can be achieved at acceptable selectivity. This avoids the cost of recovery and recycle of reactant.

For the case here the superficial mass velocity will be held constant at 4684 kg/$m^2 \cdot h$, and the consequences of changing the effect of wall temperature, taken to be the same as the inlet temperature, and the tube diameter will be explored. The heat-transfer resistance is all lumped at the wall; that is, at any distance downstream the bed temperature is taken to be uniform radially, but it is higher than the wall. The heat-transfer coefficient is taken to be 96.2 $W/m^2 \cdot s \cdot K$.

Representative results of the simulation are given in Figs. 11.17*a* through 11.17*f*, which show plots of Y, the fractional conversion of *o*-xylene to any type of product;

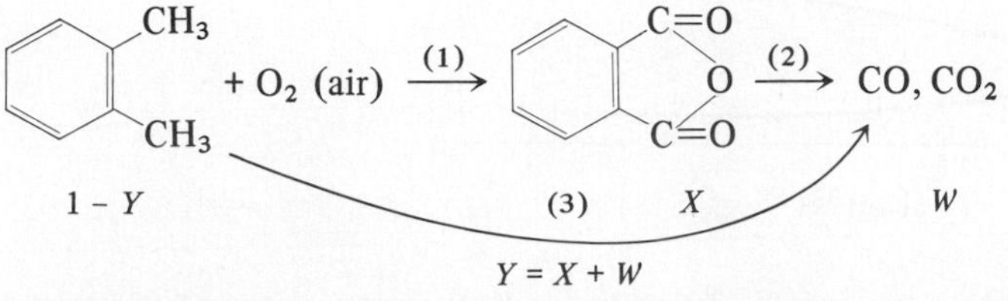

CH_3, CH_3 + O_2 (air) $\xrightarrow{(1)}$ C=O, O, C=O $\xrightarrow{(2)}$ CO, CO_2

$1 - Y$ (3) X W

$Y = X + W$

Figure 11.16 Reaction rate model for partial oxidation of *o*-xylene to phthalic anhydride.

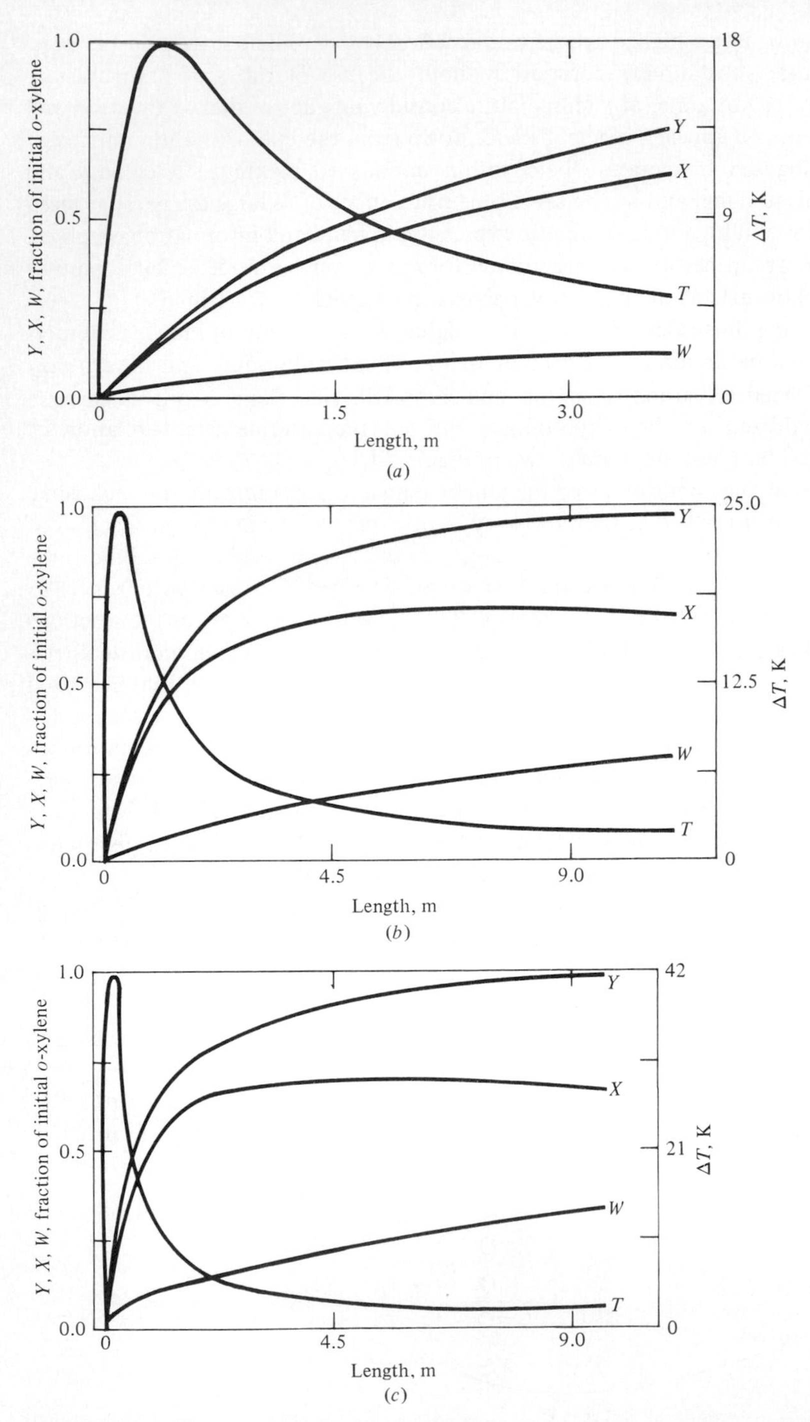

Figure 11.17 Reactor simulation (see text for definitions of X, Y, and W). (*a*) d_t = 0.025 m, wall T = 625 K. (*b*) d_t = 0.025 m, wall T = 630 K. (*c*) d_t = 0.025 m, wall T = 635 K. (*d*) d_t = 0.025 m, wall T = 637 K. (*e*) d_t = 0.030 m, wall T = 630 K. (*f*) d_t = 0.035 m, wall T = 630 K.

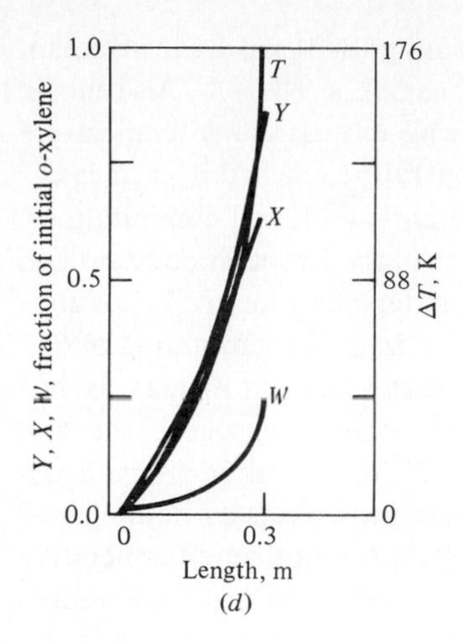
1.0
0.5
0.0
176
88
0
T
Y
X
W
0
0.3
Y, X, W, fraction of initial o-xylene
ΔT, K
Length, m
(d)

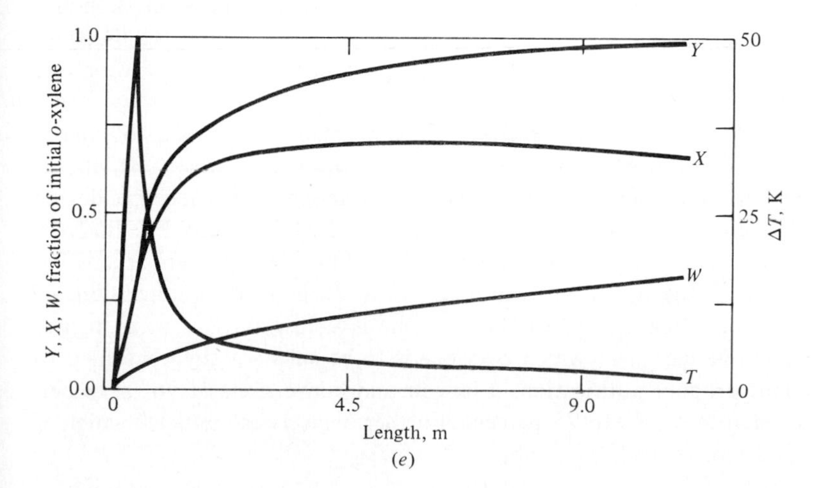
1.0
0.5
0.0
50
25
0
Y
X
W
T
0
4.5
9.0
Y, X, W, fraction of initial o-xylene
ΔT, K
Length, m
(e)

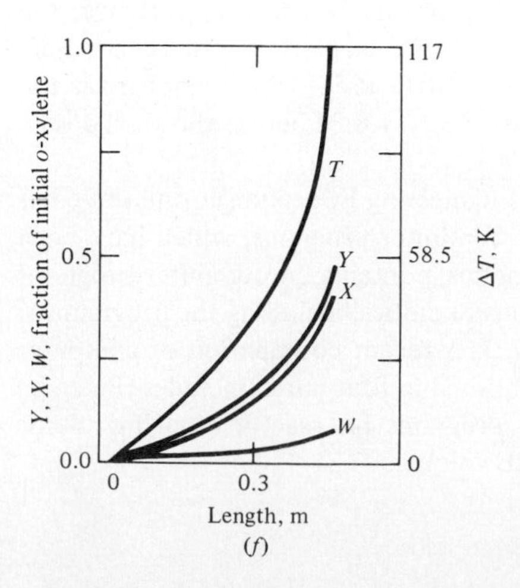
1.0
0.5
0.0
117
58.5
0
T
Y
X
W
0
0.3
Y, X, W, fraction of initial o-xylene
ΔT, K
Length, m
(f)

X, the fraction of the feed that is converted to phthalic anhydride; and W, the fraction of the feed that is converted to byproducts (left vertical axis). $X + W = Y$. Also shown is the temperature rise, in degrees kelvin, above the wall (equals inlet) temperature (right vertical axis), scaled relative to the maximum temperature reached in each case.

A variety of comparisons can be made from these and additional computations. Figures 11.17*a, b, c*, and *d* show the effects of holding the tube diameter constant and systematically increasing wall temperature. The maximum temperature of the hot zone increases with wall temperature. An instability is reached in the neighborhood of 635 to 637 K at which a slight increase in temperature causes a marked increase in bed temperature. It was assumed here that if the bed temperature exceeded the wall temperature by 200 K this would be unacceptable, and the computer program was terminated at this point. Figures 11.17*b*, *e*, and *f* show the effects of holding wall temperature constant and increasing tube diameter. Again, the hot-zone temperature is increased, and between a tube diameter of 0.030 and 0.035 m the hot zone reaches an unacceptable level. It is clear that an operating region of *parametric sensitivity* exists, and this will need to be explored carefully by more detailed simulation and experimentation for a suitable and reliable design. These calculations also illustrate why only a fairly narrow range of tube diameters are practicable for highly exothermic reactions.

Calculations such as those shown in Figure 11.17*a–f* can also be used to explore other design alternatives. For example, selectivities at 8 m (a reasonable tube length) or at, say, 75 percent conversion can be compared. This might be a practicable degree of conversion if unconverted *o*-xylene were to be recovered for reuse. For this particular case the selectivity (X/Y) is not markedly affected by the maximum acceptable temperature reached in the hot zone. It is about 83 to 84 percent at conversions of 73 to 75 percent even when the ΔT of the hot zone is as low as 18°C or as high as 50°C. In an alternative design in which *o*-xylene is to be substantially consumed to simplify subsequent product purification, it is seen that conversions of 96 to 98 percent result in selectivities of 70 to 75 percent. Further analysis can indicate which of these two designs is economically preferable.

Some of the simplifying assumptions in the above analysis should be noted. Radial temperature gradients have been neglected and, probably of more importance, the reaction in fact is more complex than the above scheme suggests, as discussed in Sec. 8.6.1. Any simulation can be only as good as the data used, which emphasizes the importance of studies in a prototype reactor (see Sec. 11.8.3) before the final design is established.

The CACHE (Computer Aids for Chemical Engineering Education) Committee has compiled, edited, and tested a large number of computer programs, which have been published in seven volumes. Volume 2, on kinetics, contains 24 computer programs for various problems in reactor design and chemical kinetics, including the program for the above analysis (Satterfield and Jones, 1972). A recent compilation of computer programs of interest to chemical engineers, published in four parts, includes (Peterson et al., 1978, part II) brief descriptions of 26 programs for reactor modeling. Some of these, but not all, are published in the CACHE volume.

Table 11.11 Nomenclature for Chap. 11*

Symbol	Definition
c	Concentration (mol/cm^3); c_0 or c_b for concentration in the bulk-gas phase; c_s for concentration at outside particle surface; c_g for local gas-phase concentration; dc/dx for concentration gradient
c_p	Heat capacity (kJ/kg · K)
d	Diameter of pore (cm)
d_p	Diameter of particle (cm)
D	Coefficient of molecular diffusion (molecular diffusivity) (cm^2/s). D_a for effective dispersion, as in Pe.
D_{12}	Diffusion coefficient for bulk diffusion of species 1 in a mixture of species 1 and 2 (cm^2/s); D_{1m} for diffusion of species 1 in a multicomponent mixture.
D_K	Knudsen diffusion coefficient for a straight cylindrical pore (cm^2/s)
D_{eff}	Effective diffusion coefficient for a porous solid, based on *total* cross section normal to direction of diffusion; equal to ratio of flux to concentration gradient, $$\frac{\text{mol/s}\cdot\text{cm}^2\text{ of solid}}{(\text{mol/cm}^3\text{ fluid})\,(1/\text{cm of solid})}=\frac{\text{cm}^3\text{ fluid}}{\text{cm solid}\cdot\text{s}}=\frac{\text{cm}^2}{\text{s}}$$
$D_{12,\,\text{eff}}$	Effective diffusion coefficient for a porous solid under conditions of bulk or ordinary diffusion, based on *total* cross section of porous solid normal to direction of diffusion (cm^2/s); $D_{12,\,\text{eff}} = D_{12}\theta/\tau$
$D_{K1,\,\text{eff}}$	Effective Knudsen diffusion coefficient for species 1 in a porous solid (cm^2/s); $D_{K1,\,\text{eff}} = D_{K1}\theta/\tau$
E	Activation energy for reaction (kJ/mol)
G	Mass velocity of fluid (g/s · cm^3 of total or superficial bed cross section normal to mean flow); G_M for molal velocity, $= G/M$
h	Heat-transfer coefficient (W/m^2 · K)
ΔH	Enthalpy change on reaction (kJ/mol)
j_D	$(k_c\rho/G)\,\text{Sc}^{2/3} = (k_g P/G_M)\,\text{Sc}^{2/3}$
j_H	$\dfrac{h}{c_p G}\,\text{Pr}^{2/3}$
k	Reaction rate constant; k_s for intrinsic first-order reaction-rate constant for surface reaction (cm/s)
k_c	Mass-transfer coefficient (cm/s)
k_f	Thermal conductivity of fluid, (J · m/s · m^2 · K) = (W/m · K)
k_G	Mass-transfer coefficient (mol/s · cm^2 · atm); $k_G = k_c/RT$
L	Length (cm). In dimensionless groups (Table 11.2) L is a characteristic length, usually taken as d_p for packed beds of particles.
M	Molecular weight
N	Diffusion flux (mol/s · cm^2)
P	Partial or total pressure (in atm in Chap. 11)
r	Reaction rate. As used in Chap. 11, r is the *observed* rate of reaction (mol/s · cm^3 of catalyst pellet).
r_e	Equivalent radius of pore (cm); $= 2V_g/S_g$
r_m	Distance from midpoint to surface (cm)
r_p	Radius of particle (cm)
r_R	Radius of cylindrical reactor (cm)
R	Gas constant; = 8.314 kJ/mol · K or 82.057 cm^3 · atm/mol · K
S_g	Total surface area per unit mass of porous solid (cm^2/g or m^2/g) (as measured, for example, by the BET method)

Table 11.11 Nomenclature for Chap. 11* (*Continued*)

Symbol	Definition
S_v	Total area of catalyst per unit volume of catalyst pellet (cm^2/cm^3) = $\rho_p S_g$
T	Temperature (degrees kelvin or centigrade); T_s at outside surface of particles; T_0 or T_b for bulk-gas phase; T_w at the reactor wall
U	Fluid velocity (cm/s)
V_c	Catalyst particle volume (cm^3)
V_g	Pore volume per unit mass of catalyst (cm^3/g); $V_g = \theta/\rho_p = (1/\rho_p) - (1/\rho_t)$
x	path length for diffusion (cm)
Greek	
β	Heat generation function = $c_s(-\Delta H)\, D_{\text{eff}}/\lambda T_s$
γ	Exponent in Arrhenius reaction rate expression = E/RT_s
η	Effectiveness factor; equals ratio of actual rate of reaction in a porous catalyst to that which would occur if the pellet interior were all exposed to reactants at the same concentration and temperature as that existing at the outside surface of the pellet
θ	Porosity (void fraction) of catalyst (fraction of gross volume of catalyst pellet that is pore space) = $V_g \rho_p$ (cm^3 pores/cm^3 pellet)
λ	Thermal conductivity of packed bed, of a porous structure, or of a solid, watts/(m^2 of total cross section) (K°/m) = W/m · K
μ	Viscosity (N · s/m^2)
ρ	Density of fluid (g/cm^3)
ρ_p	Particle density (g/cm^3 of particle volume); ρ_t for true density of solid material in porous catalyst (g/cm^3); $\rho_p = \rho_t(1 - \theta)$
τ	Tortuosity factor, an empirical factor to correct for "tortuosity" and for nonuniformity of pore cross section; $\tau = D\theta/D_{\text{eff}}$; τ_m for tortuosity factor obtained assuming completely Knudsen diffusion and a mean pore radius defined by Eq. (11.17) to calculate D_K; τ_p for tortuosity factor calculated by parallel-path pore model, Eq. (11.21).

*Some symbols used only once are defined at point of use and are not included in this list.

REFERENCES

Anderson, J. B.: *Chem. Eng. Sci.*, **18**, 147 (1963).
Archibald, R. C., N. C. May, and B. S. Greensfelder: *Ind. Eng. Chem.*, **44**, 1811 (1952).
Aris, R.: *The Mathematical Theory of Diffusion and Reaction in Permeable Catalysts*, Oxford, New York, 1975, 2 vols.
Bennett, C. O., M. B. Cutlip, and C. C. Yang: *Chem. Eng. Sci.*, **27**, 2255 (1972).
Bernard, J. R., B. L. Villemin, and S. J. Teichner: *Can. J. Chem. Eng.*, **50**, 431 (1972).
Berty, J. M.: *Chem. Eng. Prog.*, 70(5), 78 (1974). *Catal. Rev. Sci. Eng.*, **20**, 75 (1979).
Brown, L. F., H. W. Haynes, and W. H. Monague: *J. Catal.*, **14**, 220 (1969).
Butt, J. B. and V. W. Weekman, Jr.: *AIChE Symp. Ser.*, **70**(143) 27 (1974).
Carberry, J. J.: *Catal. Rev.*, **3**, 61 (1969).
——, *Chemical and Catalytic Reaction Engineering*, McGraw-Hill, New York, 1976.
Choudhary, V. R., and L. K. Doraiswamy: *Ind. Eng. Chem., Prod. Res. Dev.*, **10**, 219 (1971).
Cramer, R. H., A. F. Houser, and K. I. Jagel: U.S. Patent 3,312,615 (Apr. 4, 1967) (to Mobil Oil Co.).
Cunningham, R. A., J. J. Carberry, and J. M. Smith: *AIChE J.*, **11**, 636 (1965).
Cussler, E. L.: *Multicomponent Diffusion*, Elsevier, Amsterdam, 1976.

Cybulski, A., M. J. Van Dalen, J. W. Verkerk, and P. J. Van Den Berg: *Chem. Eng. Sci.*, **30,** 1015 (1975).
Doraiswamy, L. K., and D. G. Tajbl: *Catal. Rev. Sci. Eng.*, **10,** 177 (1974).
Dowden, D. A., and G. W. Bridger: *Adv. Catal.*, **9,** 669 (1957).
Froment, G.: *Ind. Eng. Chem.*, **59**(2), 23 (1967).
Harriott, P.: *Chem. Eng. Sci.*, **29,** 1309 (1974).
——, *Chem. Eng. J.*, **10,** 65 (1975). [See also C. S. Sharma, P. Harriott, and R. Hughes, *Chem. Eng. J.*, **10,** 73 (1975).]
Hulburt, H. H., and C. D. Srini Vasan: *AIChE J.*, **7,** 143 (1961).
Jackson, R., *Transport in Porous Catalysts*, Elsevier, Amsterdam, 1977.
Johnson, M. F. L., and W. E. Stewart, *J. Catal.*, **4,** 248 (1965).
Koros, R. M., and E. J. Nowak: *Chem. Eng. Sci.*, **22,** 470 (1967).
Krupiczka, R.: *Int. Chem. Eng.*, **7,** 122 (1967).
Kuchcinski, G. R., and R. G. Squires: *J. Catal.*, **41,** 486 (1976).
Kunii, D., and M. Suzuki: *Int. J. Heat Mass Transfer*, **10,** 845 (1967).
Langer, S. H., J. Y. Yurchak, and J. E. Patton: *Ind. Eng. Chem.*, **61**(4), 11 (1969).
Livbjerg, H., and J. Villadsen: *Chem. Eng. Sci.*, **26,** 1495 (1971).
Martin, H.: *Chem. Eng. Sci.*, **33,** 913 (1978).
Masamune, S., and J. M. Smith: *J. Chem. Eng. Data*, **8,** 54 (1963a).
——, and ——: *Ind. Eng. Chem. Fundam.*, **2,** 137 (1963b).
Maymo, J. A., and J. M. Smith: *AIChE J.*, **12,** 845 (1966).
Mears, D: *Ind Eng. Chem., Process Des. Dev.*, **10,** 541 (1971).
Miller, F. W., and H. A. Deans: *AIChE J.*, **13,** 45 (1967).
Mischke, R. A., and J. M. Smith: *Ind. Eng. Chem. Fundam.*, **1,** 288 (1962).
Murakami, Y., T. Kobayashi, T. Hattori, and M. Masuda: *Ind. Eng. Chem. Fundam.*, **7,** 599 (1968).
Nelson, P. A., and T. R. Galloway: *Chem. Eng. Sci.*, **30,** 1 (1975).
Peterson, E. E.: *Chemical Reaction Analysis*, Prentice-Hall, Englewood Cliffs, N.J., 1965.
Peterson, J. N., C-C. Chen and L. B. Evans: *Chem. Eng.*, **85,** July 3, 1978, p. 69.
Petrovic, L. J., and G. Thodos: *Ind. Eng. Chem. Fundam.*, **7,** 274 (1968).
Prater, C. D.: *Chem. Eng. Sci.*, **8,** 284 (1958).
Reid, R. C., J. M. Prausnitz, and T. K. Sherwood: *The Properties of Gases and Liquids*, 3d ed., McGraw-Hill, New York, 1977.
Satterfield, C. N.: *Mass Transfer in Heterogeneous Catalysis*, M.I.T., Cambridge, Mass., 1970. (Reprint edition available from M.I.T. Department of Chemical Engineering.)
——, and P. J. Cadle: *Ind. Eng. Chem., Process Des. Dev.*, **7,** 256 (1968a).
——, and ——: *Ind. Eng. Chem. Fundam.*, **7,** 202 (1968b).
——, and D. H. Cortez: *Ind. Eng. Chem. Fundam.*, **9,** 613 (1970).
——, and R. L. Jones in M. Reilly, (ed.): *Computer Programs for Chemical Engineering*, vol. 2, Sterling Swift Pub. Co., P.O. Box 188, Manchaca, Texas, 78652, 1972, p. 180.
——, and R. S. C. Yeung: *Ind. Eng. Chem. Fundam.*, **2,** 257 (1963).
Sehr, R. A.: *Chem. Eng. Sci.*, **9,** 145 (1958).
Sunderland, P.: *Trans. Inst. Chem. Eng.*, **54,** 135 (1976).
Voge, H. H., and C. Z. Morgan: *Ind. Eng. Chem., Process Des. Dev.*, **11,** 454 (1972).
Wagner, C.: *Z. Phys. Chem.*, **A193,** 1 (1943).
Wakao, N., and S. Tanisho: *Chem. Eng. Sci.*, **29,** 1991 (1974).
Weekman, V. W., Jr.: *AIChE J.*, **20,** 833 (1974).
Weisz, P. B.: *Z. Phys. Chem. Neue Folge*, **11,** 1 (1957).
——, and R. D. Goodwin: *J. Catal.*, **2,** 397 (1963).
——, and C. D. Prater: *Adv. Catal.*, **6,** 144 (1954).
——, and A. B. Schwartz: *J. Catal.*, **1,** 399 (1962).
Wicke, E.: *Chem.-Ing.-Tech.*, **29,** 305 (1957).

APPENDIX

A

THE LITERATURE

A.1 GENERAL TREATMENTS OF HETEROGENEOUS CATALYSIS IN ONE VOLUME

Books in Section A.1 are listed chronologically. The earlier publications are primarily of historical interest.

Sabatier, P.: *Catalysis in Organic Chemistry*, Van Nostrand, New York, 1918.
(A pioneering book, translated into English by E. Emmet Reid, 1923.)

Rideal, E. K., and H. S. Taylor: *Catalysis in Theory and Practice*, Macmillan, New York, 1919.
(A famous pioneering book.)

Ellis, C.: *Hydrogenation of Organic Substances, including Fats and Fuels*, 3d ed., Van Nostrand, New York, 1930.
(986 pp.; a detailed collection of information.)

Ipatieff, V. N.: *Catalytic Reactions at High Pressures and Temperatures*, Macmillan, New York, 1936.
(Primarily a detailed review of the author's pioneering work in this area.)

Schwab, Georg-Maria: *Catalysis*, translated into English with additions by H. S. Taylor and R. Spence, Van Nostrand, New York, 1937.
(357 pp.; a landmark book of its time.)

Berkman, S., J. C. Morrell, and G. Egloff: *Catalysis: Inorganic and Organic*, Reinhold, New York, 1940.
(1130 pp.; detailed treatment and extensive bibliography.)

Griffith, R. H., and J. D. F. Marsh: *Contact Catalysis*, 3d ed., Oxford, New York, 1957.

Germain, J. E.: *Catalyse Heterogene*, Dunod, Paris, 1959.
(230 pp.; a good coverage in one volume. Emphasis is primarily on electronic theory of catalysis.)

Bond, G. C.: *Catalysis by Metals*, Academic, New York, 1962.
(519 pp.; detailed treatment and numerous references. The first chapters are of broad interest.)

Ashmore, P. G.: *Catalysis and Inhibition of Chemical Reactions*, Butterworth, London, 1963.
(374 pp.; treats homogeneous, enzyme, and polymerization catalysis as well as heterogeneous catalysis. Some discussion also of chain reactions.)

Prettre, M.: *Catalysis and Catalysts*, Dover, New York, 1963.
(Paperback, 88 pp.; translated into English by D. Antin. A good brief introduction, but now somewhat out-of-date.)

Balandin, A. A., et al.: *Catalysis and Chemical Kinetics*, Academic, New York, 1964.
(Reviews of Russian and Polish work on catalysis; in English.)

Emmett, P. H., P. Sabatier, and E. E. Reid: *Catalysis Then and Now*, Franklin Publishing Company, Englewood, N.J., 1965.
(A reprint of the English translation by Reid of Sabatier's pioneering book *Catalysis in Organic Chemistry*, first published in French in 1918. It is accompanied by a "Survey of Advances in Catalysis" by Emmett, which focuses especially on reactions of hydrocarbons.)

Thomas, J. M., and W. J. Thomas: *Introduction to the Principles of Heterogeneous Catalysis*, Academic New York, 1967.
(544 pp.; comprehensive coverage with considerable emphasis on mathematical aspects. Extensive references.)

Balandin, A. A., et al.: *Scientific Selection of Catalysts*, 1966. English translation, Keter Pub. House, Israel, 1968.
(A collection of papers by Russian authors presenting theoretical concepts.)

Rideal, E. K.: *Concepts in Catalysis*, Academic, New York, 1968.
(A personal account by a famous investigator, but without references.)

Thomson, S. J., and G. Webb: *Heterogeneous Catalysis*, Wiley, New York, 1968.
(197 pp.; somewhat selective treatment.)

Catalyst Handbook, Wolfe Scientific Books, 10 Earlham Street, London WC2, 1970; Springer-Verlag, New York.
(231 pp.; individual chapters written by members of the staff of the Agricultural Division of Imperial Chemical Industries. Gives considerable practical information about catalytic desulfurization, steam reforming of hydrocarbons to form synthesis gas, the water-gas shift reaction, and ammonia synthesis, as well as other topics. The first three reactions are typically operated together for the manufacture of hydrogen.)

Krylov, O. V.: *Catalysis by Non Metals, Rules for Catalyst Selection*, English translation, Academic, New York, 1970.
(282 pp.; emphasizes solid-state properties of catalysts and presents correlations for a number of reactions on simple oxide catalysts. Omits complex oxides or systems with promoter or carrier effects.)

Thomas, C. L.: *Catalytic Processes and Proven Catalysts*, Academic, New York, 1970.
(284 pp.; information on specific industrial catalysts used to carry out industrial reactions and typical operating conditions. Focus is primarily on those involved in hydrocarbon and fuels processing.)

Schlosser, E.-G.: *Heterogene Katalyse*, Verlag Chemie, Weinheim, Germany, 1972.
(177 pp.; an overall treatment with emphasis on theoretical aspects.)

Bond, G. C.: *Heterogeneous Catalysis*, Oxford, New York, 1974.
(120 pp.; an elementary survey with considerable attention to applications.)

Anderson, J. R.: *Structure of Metallic Catalysts*, Academic, New York, 1975.
(469 pp.; an overall, balanced account from a scientific point of view focusing on experimental methods as well as interpretation of results. Extensive references.)

Delmon, B., P. A. Jacobs, and G. Poncelet (eds.): *Preparation of Catalysts*, Elsevier, Amsterdam, 1976.
(706 pp.; proceedings of a symposium, including discussion.)

Szabo', Z. G., and D. Kallo' (eds.): *Contact Catalysis*, English translation, Elsevier, Amsterdam, 1976.
(Two volumes, 540 and 480 pp.; written by a group of authors under the guidance of the Catalysis Club of the Hungarian Academy of Sciences. The first volume is primarily on theoretical aspects; the second is primarily on preparation and various physical methods of characterization. Detailed treatment and extensive references.)

A.2 ENCYCLOPEDIAS AND SERIAL PUBLICATIONS

Advances in Catalysis, a series of annual reviews published by Academic Press. Volume 1 in 1948. Volume 17 (1967) has cumulative author and subject indices to previous volumes. Volumes subsequent to Volume 17 list contents of all previous volumes.

Catalysis. The Chemical Society (London) publishes various series of *Specialist Periodical Reports.* A series on *Catalysis*, devoted to reviews of recent literature, was initiated by the publication of Volume 1 in 1977. Volume 2 appeared in 1978. Topics covered include homogeneous as well as heterogeneous catalysis with emphasis on fundamental research but with some attention to applied and industrial catalysis.

Catalysis, edited by P. H. Emmett, Reinhold, New York.

Volume 1: *Fundamental Principles*, 1954.
Volume 2: *Fundamental Principles*, 1955.
Volume 3: *Hydrogenation and Dehydrogenation*, 1955.
Volume 4: *Hydrocarbon Synthesis, Hydrogenation and Cyclization*, 1956.
Volume 5: *Hydrogenation, Oxo-Synthesis, Hydrocracking, Hydrodesulfurization, Hydrogen Isotope Exchange, and Related Catalytic Reactions*, 1957.
Volume 6: *Hydrocarbon Catalysis*, 1958.
Volume 7: *Oxidation, Hydration, Dehydration and Cracking Catalysts*, 1960.

Catalysis Reviews, a series of reviews published quarterly by Marcel Dekker, New York, Volume 1 in 1968; fairly similar in contents to *Advances in Catalysis*. Beginning with Volume 9 (1974) the title was changed to *Catalysis Reviews, Science and Engineering*, edited by H. Heinemann and J. J. Carberry, and the scope expanded to include more applications research and chemical reaction engineering.

Comprehensive Chemical Kinetics, edited by C. H. Bamford and C. F. H. Tipper, Elsevier, New York, 1978. Volume 20, *Complex Catalytic Processes*. Solid-catalyzed reactions with emphasis on kinetics and mechanism. Chapter 1 deals with hydrogenation, Chapter 2 with selective oxidation reactions, covering primarily the literature during 1970 to 1976. Chapter 3 deals with a variety of elimination, addition, and substitution reactions on solid acid-base catalysts including dehydration, hydration, and hydrolysis.

Handbuch der Katalyse, edited by G.-M. Schwab, Springer-Verlag, Berlin. Seven volumes, published in 1940 to 1943 except for Volume 5 *Heterogene Katalyse II* published in 1957. *Heterogene Katalyse I* and *III* were published as Volumes 4 and 6.

A.3 ADSORPTION AND RELATED TOPICS

Clark, A.: *The Theory of Adsorption and Catalysis*, Academic, New York, 1970.
(418 pp.; the first half is a comprehensive detailed treatment of adsorption. The discussion of catalysis in the second half is, as the title states, largely theoretical.)

deBoer, J. H.: *The Dynamical Character of Adsorption*, Oxford, New York, 1953.

Gregg, S. J., and K. S. Sing: *Adsorption, Surface Area and Porosity*, Academic, New York, 1967.
(369 pp.; comprehensive treatment of physical adsorption and its application to characterization of porous solids.)

Hayward, D. O., B. M. W. Trapnell: *Chemisorption*, 2d ed., Butterworth, London, 1964.
(Selective treatment; velocities of adsorption and desorption, adsorption isotherms, heat and mechanism of adsorption, mobility of adsorbed layers, catalytic specificity, and mechanisms of catalytic reactions.)

Linsen, B. G. (ed.): *Physical and Chemical Aspects of Adsorbents and Catalysts*, Academic, New York, 1970.
(650 pp.; a "Festschrift" in honor of J. H. deBoer. Comprises eleven survey papers primarily on adsorption phenomena, structure, and properties of catalyst supports. Has individual chapters on alumina, porous silica, active magnesia, hydrous zirconia.)

Parfitt, G. D., and K. S. W. Sing (eds.): *Characterization of Powder Surfaces*, Academic, New York, 1976.
(464 pp.; subtitled *with Special Reference to Pigments and Fillers*. Nine chapters by individual authors including two on physical and chemical surface characterization and individual chapters on carbon blacks, silicas, and clays.)

A.4 HYDROGENATION

Augustine, R. L.: *Catalytic Hydrogenation*, Marcel Dekker, New York, 1965.
(188 pp.; subtitled *Techniques and Applications in Organic Synthesis*. Extensive references.)

Freifelder, Morris: *Practical Catalytic Hydrogenation, Techniques and Applications*, Wiley, New York, 1971.
(663 pp.; directed primarily to the chemist concerned with laboratory work. The author was a hydrogenation specialist at Abbott Laboratories for many years. Extensive references.)

——: *Catalytic Hydrogenation in Organic Synthesis: Procedures and Commentary*, Wiley, New York, 1978.
(191 pp.; detailed specific descriptions of laboratory procedures.)

Kieboom, A. P. G., and F. van Rantwijk: *Hydrogenation and Hydrogenolysis in Synthetic Organic Chemistry*, Delft Univ. Press, The Netherlands, 1977.
(157 pp.; a treatment from both a mechanistic and preparative point of view, directed to organic chemists concerned with synthesis.)

Rylander, P. N.: *Catalytic Hydrogenation over Platinum Metals*, Academic, New York, 1967.
(550 pp.; extensive references. The author is with Engelhard Industries, a principal commercial supplier of platinum-group catalysts.)

Sokol'skii, D. V.: *Hydrogenation in Solutions*, 1962. English translation, Daniel Davey & Co., New York, 1964.
(532 pp.; useful for its treatment of earlier Russian literature.)

A.5 SPECIAL TOPICS

Anderson, R. B. (ed.): *Experimental Methods in Catalytic Research*, Academic, New York, 1968.
(498 pp.; individual chapters on methods of measuring surface area and pore structure, surface potentials, acidity, field ion emission microscopy, chemisorption, low-energy electron diffraction, semiconductivity, spectra of adsorbed species, magnetic methods, electron spin resonance.) R. B. Anderson and P. T. Dawson (eds.): *Preparation and Examination of Practical Catalysts*, Academic, New York, 1976, vol. 2, 300 pp.; *Characterization of Surfaces and Adsorbed Species*, vol. 3, 1976, 344 pp.

Bohlbro, H.: *An Investigation on the Kinetics of the Conversion of Carbon Monoxide with Water Vapor over Iron Oxide Based Catalysts*, Gjellerup, Copenhagen, 1966.
(135 pp.; detailed treatment from the Haldor-Topsøe laboratories of catalyst properties and kinetics of the reaction.) Also, *Addendum* by E. Mogensen, M. H. Jørgensen, and K. Sondergaard, 1969; 55 pp.)

Cusumano, J. A., R. A. Dalla Betta, and R. B. Levy: *Catalysis in Coal Conversion*, Academic, New York, 1978.

Delmon, B., and G. Jannes (eds.): *Catalysis: Heterogeneous and Homogeneous*, Elsevier, New York, 1975.
(547 pp.; proceedings of a symposium held in Brussels, Oct. 23-25, 1974, including discussion. Largely fundamental and theoretical studies.)

Drauglis, E., and R. I. Jaffee (eds.): *The Physical Basis for Heterogeneous Catalysis*, Plenum, New York, 1975.
(596 pp.; proceedings of a symposium, including discussion. Focus of attention on application of modern theoretical and experimental surface physics methods to catalysis.)

Hucknall, D. J.: *Selective Oxidation of Hydrocarbons*, Academic, New York, 1974.
(212 pp.; catalytic oxidation of ethylene, propylene, C_4 and C_5 hydrocarbons. Does not treat aromatics.)

International Union of Pure and Applied Chemistry, Information Bulletin. *Appendices on Provisional Nomenclature, Symbols, Units and Standards, Number 39. Definitions, Terminology and Symbols in Colloid and Surface Chemistry*. Part II: *Heterogeneous Catalysis*. Published in *Adv. Catal.*, **26**, 351 (1977).

Kuczynski, G. C. (ed.): *Sintering and Catalysis*, Plenum, New York, 1975.
(38 papers, 512 pp.; proceedings of a symposium held at the University of Notre Dame, May 26-28, 1975.)

Nielsen, A.: *An Investigation on Promoted Iron Catalysts for the Synthesis of Ammonia*, 3d ed., Gjellerup, Copenhagen, 1968.
(264 pp.; a basic reference on this subject, from the Haldor-Topsøe laboratories. Much material on reactor design as well as catalysts.)

Ozaki, A.: *Isotopic Studies of Heterogeneous Catalysis*, Academic, New York, 1977.
(239 pp.)

Rabo, J. A. (ed.): "Zeolite Chemistry and Catalysis," *ACS Monogr. No. 171*, 1976.
(796 pp.; a comprehensive treatment including structure, chemistry, and technology.)

Rostrup-Nielsen, J. R.: *Steam Reforming Catalysts*, Danish Technical Press Inc., Copenhagen, 1975.
(240 pp.; a detailed treatment from the Haldor-Topsøe laboratories with emphasis on development of catalyst and process for naphtha reforming.)

Rylander, P. N.: *Organic Syntheses with Noble Metal Catalysts*, Academic, New York, 1973.
(331 pp.; extensive references.)

Satterfield, C. N.: *Mass Transfer in Heterogeneous Catalysis*, M.I.T., Cambridge, Mass., 1970.
(Paperback edition available from Department of Chemical Engineering, M.I.T., Cambridge, Mass.)

Storch, H. H., N. Golumbic, and R. B. Anderson: *The Fischer-Tropsch and Related Syntheses*, Wiley, New York, 1951.

Tamaru, K.: *Dynamic Heterogeneous Catalysis*, Academic, New York, 1978.
(140 pp.; a monograph focusing on adsorption studies, especially under dynamic conditions, and their use to interpret the mechanism of catalytic reactions.)

Vancini, C. A., *Synthesis of Ammonia*, English translation by L. Pirt, Macmillan and CRC Press, New York, 1971.
(A condensed version of original in Italian, published in 1961. Extensive references.)

Vol'kenshtein, F. F.: *The Electronic Theory of Catalysis on Semi-Conductors*, English translation, Pergamon, New York, 1963.
(169 pp.; a condensed version of the Russian original.)

Weisser, O., and S. Landa: *Sulphide Catalysts, Their Properties and Applications*, English translation, Pergamon, New York, 1973.
(506 pp.; exhaustive references, includes a numerical patent index.)

A guide to the literature on zeolites is given in Sec. 7.7.7, and to that on catalytic oxidation in Sec. 8.14.

A.6 INTERNATIONAL CONGRESSES ON CATALYSIS

These have been held at 4-year intervals, with emphasis on theory and fundamentals. Proceedings have been published as follows:

Year	Location	Publication
1956	Philadelphia	Published as Volume 9 of *Advances in Catalysis.*
1960	Paris	*Actes du Deuxieme Congres International de Catalyse, Paris, 1960*, Editions Technip, Paris, 1961. (2 volumes.) (*Proceedings of the International Congress on Catalysis, Paris, 1960,* 2811 pp.)
1964	Amsterdam	*Proceedings of the Third International Congress on Catalysis, Amsterdam, July 20-25, 1964*. Edited by W. M. H. Sachtler, G. C. A. Schuit, and P. Zweitering. Interscience Division, Wiley, New York, 1965. (2 volumes, 1445 pp.)
1968	Moscow	*The Fourth International Congress on Catalysis, Moscow, USSR, June 23-29, 1968*. Reprints of the papers, in English (6 volumes, paperback) were compiled for the U.S. Catalysis Society by J. W. Hightower, Rice University, Houston. Also published by Akadémiai Kiadó, Budapest, 1972.
1972	Miami Beach, Fla.	*Proceedings of the Fifth International Congress on Catalysis, Miami Beach, Florida, August 20-26, 1972*. Edited by J. W. Hightower, American Elsevier, New York, 1973. (2 volumes, 1483 pp.)
1976	London	*Proceedings of the Sixth International Congress on Catalysis, London, July 12-16, 1976*, The Chemical Society, London, 1977. (2 volumes, 1133 pp.)
1980	Tokyo	Seventh International Congress on Catalysis, Tokyo, June 30-July 4, 1980.

A.7 PETROLEUM AND HYDROCARBON PROCESSING

Germain, J. E.: *Catalytic Conversion of Hydrocarbons*, Academic, New York, 1969.
(322 pp.; emphasis on mechanisms of reactions. Numerous references.)

Hahn, A. V. G.: *The Petrochemical Industry: Market and Economics*, McGraw-Hill, New York, 1970.
(620 pp.)

Hobson, G. D., and W. Pohl (eds.): *Modern Petroleum Technology*, 4th ed., Wiley, New York, 1973.
(996 pp.)

Waddams, A. L.: *Chemicals from Petroleum*, 3d ed., Wiley, New York, 1973.
(326 pp.; paperback)

The following two multivolume works contain a number of reviews on catalytic processes:

Brooks, B. T., C. E. Boord, S. S. Kurtz, and L. Schmerling (eds.): *The Chemistry of Petroleum Hydrocarbons*, Reinhold, 1954-55. (3 volumes.)

McKetta, J. J. (ed.): *Advances in Petroleum Chemistry and Refining*, Wiley-Interscience, New York, 1958-1965. (10 volumes.)

Some of the Proceedings of the World Petroleum Congresses have useful reviews on catalytic processing, for example, Volume 4 of *Eighth World Petroleum Congress*, 1971, and earlier proceedings.

A.8 JOURNALS

Papers reporting on original research on catalysis appear in a great variety of journals. Of particular interest are: the *Journal of Catalysis*, first volume in 1962, and *Kinetika i Kataliz*. The latter is available in English translation as *Kinetics and Catalysis*. *Industrial and Engineering Chemistry, Product Research and Development* has a catalyst section which publishes papers on catalysts and catalytic reactions of industrial interest. See also the preprints of the Division of Petroleum Chemistry and Division of Fuel Chemistry of the American Chemical Society. Some of these are eventually published elsewhere, but many useful papers appear only in the preprints.

A.9 GENERAL

The catalysts and reaction conditions used in the industrial manufacture of the more important organic chemicals are described in a series of papers by Austin (1974) and in books by Waddams (1973) and Tedder et al. (1975). Innes (1955) gives a classification of heterogeneous catalytic reactions in general as of the early 1950s. Considerable information on catalysts and processes is also available in the *Kirk-Othmer Encyclopedia of Chemical Technology* (1963-1970) and the *Encyclopedia of Chemical Processing and Design* edited by McKetta (1976).

Austin, G. T.: *Chem. Eng.*, Jan. 21, 1974, p. 127; Feb. 18, 1974, p. 125; March 18, 1974, p. 87; April 15, 1974, p. 86; April 29, 1974, p. 143, May 27, 1974, p. 101; June 24, 1974, p. 149; July 22, 1974, p. 107; Aug. 5, 1974, p. 96.

Innes, W. B., in P. H. Emmett (ed.): *Catalysis*, vol. 2, Reinhold, New York, 1955, p. 1.

Kirk-Othmer Encyclopedia of Chemical Technology, 2d ed., Wiley-Interscience, New York, 1963-1970 (22 volumes); 3d ed., 1978-.

McKetta, J. J. (exec. ed.): *Encyclopedia of Chemical Processing and Design*, Marcel Dekker, New York, 1976-.

Tedder, J. M., A. Nechvatal, and A. H. Jubb: *Basic Organic Chemistry*. Part 5: *Industrial Products*, Wiley, New York, 1975. (Paperback, 646 pp.)

Waddams, A. L.: *Chemicals from Petroleum*, 3d ed., Wiley, New York, 1973. (Paperback, 326 pp.)

Flowsheets and some process information on petrochemical processes available for license are published every 2 years (odd-numbered years) in the November issue of *Hydrocarbon Processing* (for example, November 1973, 1975, and 1977). Similar information on refining processes is published every 2 years (even-numbered years) in the September issue (for example, September 1974, 1976 and 1978). In both cases noncatalytic and separation processes are included in addition to catalytic processes.

APPENDIX

B

PROBLEMS

1. Taylor and Liang [*J. Am. Chem. Soc.*, **69**, 1306 (1947)] have reported a number of experiments in which hydrogen was adsorbed on zinc oxide. In one typical experiment, the solid was held at 0°C for the first 100 min. The temperature was then raised quickly to 111°C and held there until 300 min from the start of the run. At 300 min the temperature was raised to 154°C. At 1000 min, the temperature was dropped quickly to 111°C. At 1150 min the temperature was raised again to 154°C and held there until 1250 min. The pressure is presumed to be held constant throughout the experiment at 0.1 MPa. The amount of hydrogen which was found to be adsorbed by the solid is plotted as a function of time in Fig. B.1.

Present your interpretation of these results, together with the facts that support each part of your interpretation.

2. With respect to Prob. 1:

(*a*) The slope of the curve at point B (just before the temperature was raised to 154°C) is less than at point C (just after the temperature was raised). Is this reasonable?

(*b*) Is there any evidence that any of the zinc oxide was being reduced to zinc metal? Is there any evidence that it was *not* being reduced?

(*c*) Suppose that the temperature had been *held constant at 111°C* throughout the experiment and for an additional 2 or 3 days. Show in relationship to the present results how you would expect the amount of adsorbed hydrogen to change with time, noting any special features concerning the shape or location of the curve that you anticipate.

(*d*) Is it significant that the amount of hydrogen adsorbed between A and B seems to be the same as the amount desorbed between B and C?

3. A study has been made of the chemisorption of hydrogen onto a zeolite (a calcium aluminosilicate) which contained 0.5% platinum introduced by a cation-exchange process replacing some of the calcium. By this means the platinum was dispersed atomically.

The results of a typical run are shown in Fig. B.2, which plots the amount of hydrogen adsorbed (as atoms of H per atom of Pt present) as a function of time and at a constant pressure of 80 kPa. Adsorption of hydrogen on the support is negligible, and diffusion effects into and out of the zeolite are insignificant.

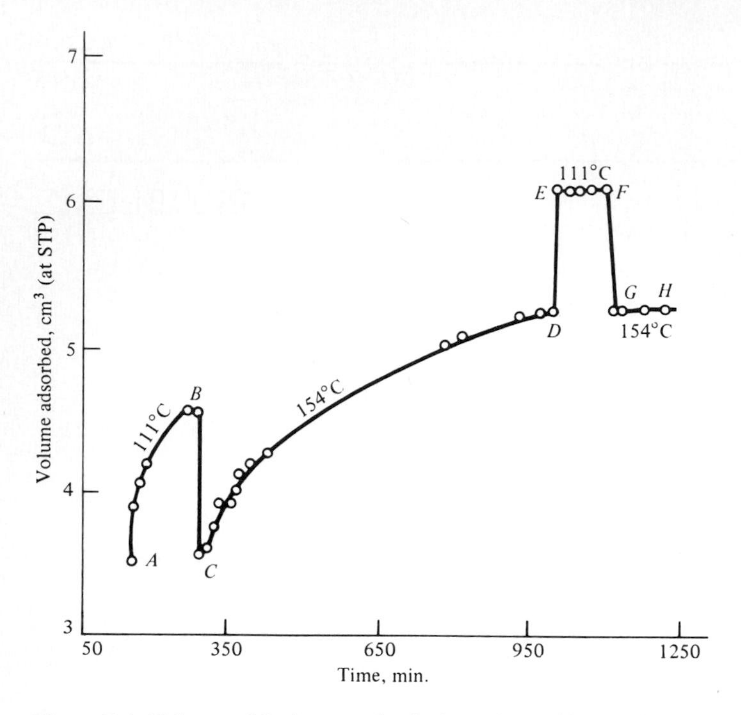

Figure B.1 Volume of hydrogen adsorbed on zinc oxide as a function of time and temperature. *(Taylor and Liang, 1947.) (Reprinted with permission from the Journal of the American Chemical Society. Copyright by the American Chemical Society.)*

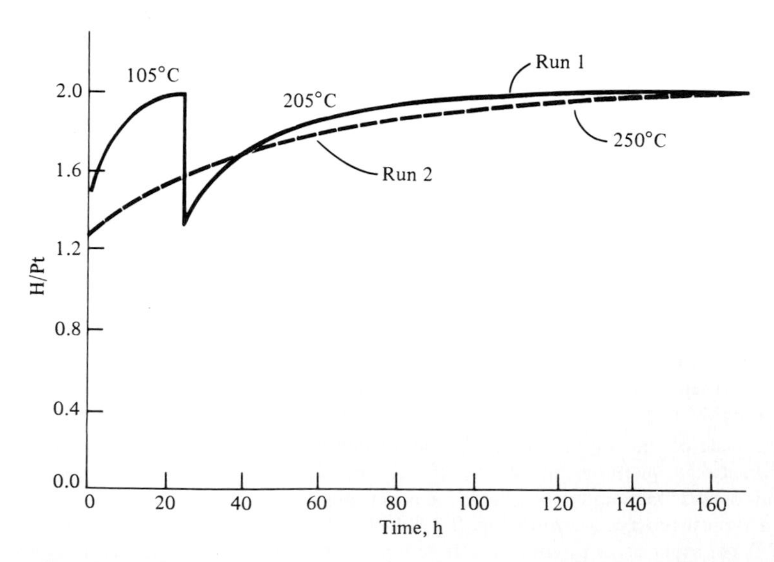

Figure B.2 Effect of time and temperature on ratio of H atoms adsorbed to Pt atoms exposed.

In the first run for the first 25 h the temperature was held constant at 105°C; it was then raised to 205°C. In the second run the temperature was held constant at 250°C throughout. A fresh catalyst is presumed to be used for each study.

(*a*) Interpret these observations.

(*b*) What can you speculate about the form(s) of the adsorbed hydrogen under these different conditions:

4. Wauquier and Jungers [*Bull. Soc. Chim. Fr.*, **10,** 1280 (1957)] have reported that when a mixture of *p*-xylene and tetralin is hydrogenated over a Raney nickel catalyst in a *batch* process, the rate of reaction (moles of tetralin plus *p*-xylene reacted per unit time) increases with extent of conversion. (See Sec. 3.5.) A typical set of data is reported in the table below:

Composition of mixture		Total rate of hydrogenation of mixture, arbitrary units
p-xylene $[A_1]$	tetralin $[A_2]$	
620	280	8.5
566	235	8.8
462	139	9.4
334	57	10.4
159	10	11.3

The hydrogen pressure was kept constant, and it is assumed that the extent of hydrogen adsorption on the catalyst is unaffected by the course of the reaction. Since the reaction occurs in the liquid phase, it is further assumed that all catalyst sites are occupied; thus the fraction of available sites occupied by species 1 is

$$\theta_1 = \frac{K_1[A_1]}{K_1[A_1] + K_2[A_2]}$$

where K_1 and K_2 are adsorption constants and $[A_1]$ and $[A_2]$ are concentrations of reactants in the liquid phase.

(*a*) If the organic species compete for one kind of site, show that the rate of reaction as defined above should follow the expression:

$$-r = \frac{k_1 K_1 [A_1] + k_2 K_2 [A_2]}{K_1 [A_1] + K_2 [A_2]}$$

(*b*) How would the rate change as hydrogenation was allowed to continue beyond the last datum reported in the table if the above equation continued to be obeyed? In fact, would you expect the basic assumptions and the above equation to be followed until hydrogenation was completed?

(*c*) What rate would you expect for hydrogenation of pure tetralin, at a concentration of 100 (arbitrary units)?

Note: The rate of hydrogenation of *p*-xylene alone was found to be zero order and equaled 12.9 (arbitrary units).

5. Data have been obtained on the rates of catalytic hydrogenation of mixtures of acetone (A) and cyclohexene (C) in the presence of various solvents and expressed in terms of a selectivity *S*, which is the ratio of the relative effective rate constants of the two reactions (same as the ratio of the rates under the same set of conditions). From arguments such as that used in Prob. 4, *S* is formulated as

$$S = \frac{k_C K_C}{k_A K_A}$$

Here k_i is the reaction rate constant for the surface reaction, and K_i is an adsorption constant (partition coefficient).

Explain the observation that for a series of *nonreactive* solvents it was found that S increased in the order

$$\text{Cyclohexane} < \text{benzene} < \text{1-octanol} < \text{isopropanol}$$

Assume that A and C compete for adsorption on one kind of site.

6. Several researchers have made careful studies of the kinetics of ethylene hydrogenation to ethane over a supported nickel catalyst, an irreversible reaction. It has been concluded that the best mechanism fitting the evidence was reaction occurring between an adsorbed ethylene molecule and a colliding gaseous hydrogen molecule as the rate-controlling step.

The best correlation of the various data on nickel is given by the following expression:

$$\text{Rate} = \frac{Ae^{+6050/T}(P_{H_2} P_{C_2H_4})}{1 + Be^{+11,100/T}(P_{C_2H_4})}$$

(*a*) Is the *form* of this equation consistent with the proposed mechanism? Show by suitable derivation.

(*b*) Is it reasonable for the algebraic sign to be positive in both exponential terms?

(*c*) In one study it was reported that the apparent activation energy (from an Arrhenius plot) for the rate of this reaction on a nickel catalyst was about 41.9 kJ/mol at 100°C and 26.8 kJ/mol at 150°C. Is this observation consistent with the above rate expression?

(*d*) This reaction has been extensively studied by many investigators on a variety of different metal catalysts. It is reliably reported that if a sufficiently wide temperature range is covered for a fixed gas composition, on some catalysts the rate goes through a maximum with increased temperature (over a significant temperature range), but on other catalysts the rate continues to increase with increased temperature without a drop.

(*i*) If the above rate expression applies, would a maximum indeed be observed?

(*ii*) Is this different kind of behavior with different catalysts consistent with the postulated mechanism?

7. A reaction of the following type is being studied.

$$A + B \longrightarrow C \xrightarrow{+B} D$$

A is a gas and B, C, and D are liquids. All are organic compounds.

A is bubbled up through a vessel containing B and a finely divided catalyst, and B is always present in great excess. C is an unsaturated compound, but D is quite stable.

If the pressure of A is increased, the catalyst life drops appreciably. Suggest an explanation.

8. Are there any cases in which the apparent activation energy (as determined from an Arrhenius plot) for a catalyzed reaction would *increase* with increased temperature under circumstances in which homogeneous reaction does not occur?

9. Consider the hydrogenation of an olefin to the corresponding paraffin. Assume that the mechanism involves the simultaneous reaction of an adsorbed olefin molecule with two (dissociated and adsorbed) hydrogen atoms. Formulate the rate of reaction in terms of the Langmuir-Hinshelwood model (one type of site) if the products are not significantly adsorbed.

10. The hydrogenation of ethylene to ethane catalyzed by copper has been studied in a differential reactor over the temperature range of 0 to 200°C. At 200°C the rate expression was found to be: $-r = kP_{H_2}P_{C_2H_4}$. At 0°C: $-r = k'P_{H_2}/P_{C_2H_4}$

Assuming that on this catalyst the Langmuir-Hinshelwood model on a single kind of site applies and there is no dissociation of hydrogen, explain the results. Is it reasonable to expect this kind of a shift in the rate expression with temperature?

11. Studies have been reported of the dehydrogenation of cyclohexane to benzene on a noble metal catalyst in the presence of hydrogen-deuterium mixtures at an elevated pressure. At temperatures below 250°C the only reaction observed is the formation of deuterated cyclohexanes $C_6H_{12-x}D_x$. At temperatures between 250 and 300°C some benzene is formed as well as deuterated cyclohexanes. Above 300°C no deuterated cyclohexanes are found.

(*a*) What can be inferred about how the mechanism of the dehydrogenation reaction changes with temperature?

(*b*) Consider conditions corresponding to 50 percent conversion in a differential reactor. How would the *form* of the kinetic expression be expected to change over the temperature range? Under all conditions the mole ratio of hydrogen (or deuterium) to hydrocarbon in the gas phase greatly exceeds the stoichiometric ratio.

12. Rideal quotes data on the decomposition of various alcohols on an alumina catalyst. The products are the corresponding olefin and water, for example,

$$CH_3CH_2CH_2OH \longrightarrow CH_3CH{=}CH_2 + H_2O$$

Water is much more strongly adsorbed than either the alcohol or the olefin under all conditions.

The apparent activation energy of the reaction (as determined by the slope of the rate versus $1/T$ on an Arrhenius plot) is substantially different if it is studied at a relatively high pressure versus a relatively low pressure, as shown in the following table. Assume that at each combination of pressure and temperature a moderate range of conversion was studied, for example, 20 to 40 percent, and that reverse reaction was negligible. Also assume that the Langmuir-Hinshelwood formulation of kinetics on a single kind of site applies.

	E, kJ/mol		
Alcohol	High pressure	Low pressure	Difference
n-C_3H_7OH	172	119	53
iso-C_3H_7OH	163	109	54
n-C_4H_9OH	184	117	67

(*a*) Show by suitable calculations what the difference between the two sets of values of *E* for each alcohol, which averages about 58 kJ/mol, represents.

(*b*) Is it reasonable for the difference between the high-pressure value of *E* and the low-pressure value to be essentially the same (within experimental error) for all three alcohols?

13. A screening program is being carried out for new catalysts for a hydrogenation reaction $A + H_2 \rightarrow C + D$ (all vapor phase). The two most active catalysts found will be termed M and P. All studies are done in a fixed catalyst bed with the same quantity of catalyst and at constant temperature and pressure. At high flow rates, M is more active than P (see definition below). At low flow rates, P is more active than M. Feed composition is the same under all conditions.

(*a*) Explain the results.

(*b*) Would you expect M to continue to be "more active" than P at a fixed higher temperature?

Notes: There are no mass- or heat-transfer limitations or diffusion effects. Consider the reaction to be irreversible and conversion to be considerably less than 100 percent under all conditions. There are no side reactions.

Activity is defined as

$$\frac{\text{Moles reacted}}{\text{(Unit time) (grams of catalyst)}}$$

14. An irreversible catalytic reaction of the type $A \rightarrow B$ is being studied in the laboratory. B is much more strongly adsorbed than A, and A is very weakly adsorbed. Suppose the reaction is studied over a range of temperatures in a reactor, keeping the percent conversion constant at 20 percent. An Arrhenius plot is made of the log of the rate [moles reacted/(minute) (gram of catalyst)] versus $1/T$.

(*a*) Qualitatively, what will be the shape of the curve?

(*b*) Repeat for the case in which B is not adsorbed:

(*i*) When A is relatively strongly adsorbed.

(*ii*) When A is relatively weakly adsorbed.

Assume that the Langmuir-Hinshelwood single-site model applies.

15. Two zinc catalysts are prepared for the synthesis of methanol by the reaction:

$$CO + 2H_2 \longrightarrow CH_3OH$$

Catalyst A is made by decomposing zinc acetate, $Zn(CH_3\overset{\overset{\displaystyle O}{\|}}{C}{-}O)_2$, in a high-temperature oven in a stream of dry nitrogen; catalyst B is made by exactly the same procedure, with zinc carbonate, $ZnCO_3$, as the starting material. In both cases the starting compound is a fine powder and is spread out in a thin layer in a container in a furnace, and in each case the residue left after decomposition is put into a reactor in contact with the same reaction mixture. Catalyst A is twice as active per unit weight as catalyst B initially, and moreover, it does not lose its activity as fast.

Account for the difference between the rate of deterioration of the two catalysts.

16. We are considering using nickel oxide in a reactor system in which it will be alternately oxidized and reduced in a cyclic process. The nickel oxide is prepared by precipitating nickel hydroxide by adding ammonium hydroxide to a solution of nickel carbonate followed by filtration, drying, calcination, and pelleting.

(*a*) In what ways might it be desirable to modify the preparation so as to increase the life of the material in this application?

(*b*) We would like to be able to increase the rate of reduction of the nickel oxide, especially during the early portions of the reaction. Do you have any suggestions?

17. A nickel catalyst is to be prepared by coprecipitation by NH_4OH of a mixture of $Ni(NO_3)_2$ and a second substance which will become a textural promoter in the final catalyst. Neither alumina nor silica is suitable. What can you suggest?

18. A liquid organic compound is hydrogenated in batches using a finely divided 5% palladium/carbon catalyst. The carbon by itself is inactive, but if we add to the vessel an amount of carbon support equal to 3 to 5 times the amount of the catalyst, then reaction is essentially completed in a substantially shorter time. Suggest an explanation.

19. We are searching for a hydrogenation catalyst for a specific feedstock available in the plant, and we are testing a large number of catalyst preparations by a simple batch lab test. The catalysts are all powders or finely ground materials, and the standard test consists of adding 0.1 g of catalyst to 50 ml of the feedstock in a stirred flask and measuring the rate of uptake of hydrogen over a specified length of time. Our objective is to find the most active catalyst to use in the plant. Results are reproducible with any one catalyst, but in many cases if the quantity of catalyst is doubled, the rate increases four- to six-fold. What may be happening?

20. A hydrogenation reaction of the type $A + H_2 \rightarrow B$ is being considered, to be carried out in the vapor phase in a fixed bed at about 300°C and atmospheric pressure. A is a highly purified organic chemical, and in order to avoid the extra costs of purification, we wish to make the B as pure as possible; that is, we desire very high selectivity to B. In the laboratory a catalyst consisting of platinum supported on alumina has been used that gives us over 99.5 percent conversion to B with only traces of other products. However, its life is short for reasons that are not clear. An experienced representative of a catalyst manufacturer suggests that the particular alumina we are using is slowly deteriorating and recommends that we use instead a *stabilized alumina* containing 0.5 to

2 wt % silica, a material that has been highly satisfactory as a support in other hydrogenation reactions. What is your reaction?

21. In one of the processes for converting butylene to butadiene a mixture of butylene and steam is passed over a calcium nickel phosphate catalyst, typically at about 660°C and slightly above atmospheric pressure. This catalyst is usually made in a well-stirred flow reactor by mixing a stream of aqueous ammonia and a second stream containing calcium and nickel salts plus an orthophosphate. Flow rates and concentrations are adjusted so the resultant mixture is at a pH of about 8, and a gel-like precipitate is formed, which is then filtered, washed, dried, pulverized, and pressed into pellets of desired size and shape.

Different batches of catalyst may vary in their properties as a result of variations in the procedures for making, drying, and pressing the catalyst into pellets. It has been reported that the best of such catalysts, in the form of uniform pellets, have a bulk density of from about 0.95 to 1.1 g/ml. More dense catalysts tend to be less active; less dense catalysts are usually the most active, but also have the lowest selectivity, i.e., produce more byproducts per mole of butylene converted. Suggest an explanation.

22. Two silver catalysts for a certain reaction that occurs under reducing conditions are being compared. Catalyst A is made by decomposing silver oxalate, $Ag_2C_2O_4$, in an oven in a stream of dry nitrogen. Catalyst B is made by exactly the same procedure, but starting with silver nitrate, $AgNO_3$. In each case a fine powder is spread out in a boat in the furnace. Catalyst A is initially less active per unit weight than catalyst B, but catalyst B deteriorates more rapidly than catalyst A. For example, after 20 h of reaction time the activity of B is only one-fifth that of A.

The test procedure is the same for the two catalysts, and the test reaction follows the same kinetic expression and shows the same apparent activation energy with each catalyst. Catalyst A looks black, while catalyst B looks silvery white.

Suggest an explanation.

23. The surface area and pore-size distribution is being measured by nitrogen adsorption of a porous solid with pores in the range of 10 to 30 nm diameter. The area is calculated by the standard BET method using nitrogen and also by the summation of the areas of the pores as determined from the pore-size distribution (Sec. 5.2).

(*a*) The nitrogen sorption curve shows a hysteresis. Would you expect the adsorption branch or the desorption branch to predict the higher area?

(*b*) The surface area calculated from the pore-size distribution using the *desorption* branch is greater than that calculated by the BET method. Assuming all measurements are accurate, how can this be so?

24. R. M. Barrer and E. Strachan [*Proc. Roy. Soc.* (London), Ser. A, **231**, 52 (1955)] reported measurements of the surface area per gram of pellets made by compressing finely divided carbon powder under different degrees of compression. Some of their data are as follows:

Void fraction of pellet	BET area, m^2/g
0.79	987
0.64	933
0.53	890
0.45	795
0.37	839

Accepting the validity of the BET method for determining surface area, what may cause a drop and then rise of surface area per gram with increasing degree of compression?

25. A pore-size distribution (PSD) is measured on a given high-area catalyst by mercury porosimetry on either whole pellets (5 mm) or pellets gently crushed to give somewhat smaller particles (for example, 0.8 mm). The PSD on whole pellets shows considerably smaller pores on the average than that on crushed pellets. What could be the explanation?

26. Some surface-area measurements by the BET method have been made of cuprous oxide films on copper metal. Each batch of material under study consisted of 25 copper disks, each 5 cm in diameter and 0.05 mm thick, separated from one another and stacked inside the sample container. The cuprous oxide film ranged from 100 to 300 nm in thickness and was prepared by successively oxidizing and reducing the copper disks several times. Krypton gas was used as the adsorbing vapor. The area covered by the spacers between the disks is negligible.

The following data were obtained for three different catalyst samples:

V, volume of gas adsorbed (STP), cm^3	P, equilibrium pressure in sample container, Pa
I	
0.0301	37.2
0.0349	51.2
0.0383	61.8
0.0426	76.4
0.0475	89.4
P_0, saturation vapor pressure of liquid krypton = 505 Pa	
II	
0.0133	37.8
0.0183	58.1
0.0198	75.8
0.0200	91.8
0.0205	108
P_0 = 384 Pa	
III	
0.0197	30.8
0.0309	48.5
0.0369	73.0
0.0375	90.2
0.0424	106
P_0 = 421 Pa	

The cross-sectional area of a krypton molecule may be taken to be 0.195 nm^2.

(*a*) What was the surface area of each sample, as calculated from the BET method?

(*b*) Comment on the results.

27. What is the surface area of each of the catalysts and supports for which a BET plot is given in Fig. 5.2?

28. If a batch of fine particles of nickel oxide, NiO, is reduced with hydrogen, it is found that the rate of reduction is at first slow, then increases to a maximum at about 50 percent reduction before dropping off again (Fig. B.3). It may take a very long time to reduce the NiO completely. Explain.

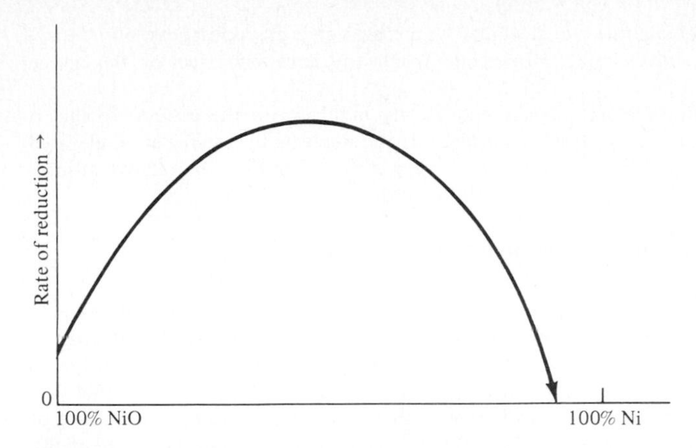

Figure B.3 Variation of reduction rate of NiO with solid composition.

29. Ethylene is produced by the thermal cracking, at approximately atmospheric pressure, of any of a variety of hydrocarbon feedstocks ranging from a field condensate (containing primarily ethane and propane) to naphthas and heavier hydrocarbons. The product mixture is then compressed to about 1.3 MPa for subsequent fractionation. Small amounts of acetylenes and various other unsaturated compounds are formed that must be essentially all removed before the ethylene can be used for chemical synthesis, a typical specification allowing a maximum of 5 ppm of unsaturated impurities. A commonly used method is selective hydrogenation, which is done by passing the gas mixture over a catalyst in a fixed bed. If we start with a C_2 to C_3 stream, this might be done on either the light-ends fraction following initial separation of the product from the thermal cracker, or after final fractionation. Typical gas compositions for the two cases are reported to be as follows:

Gas	Light-ends fraction	After fractionation
H_2	19.0%	–
CH_4	30.0%	0.25
C_2H_4	40.0%	98.5
C_2H_6	10.0%	1.0
C_2H_2	0.1%	0.25
C_3H_6	0.9%	–

In the light-ends fraction there typically are also minor amounts of sulfur compounds, of $CH_3C{\equiv}CH$, and of $H_2C{=}C{=}CH_2$. In the second type of operation it is necessary to add additional H_2, typically about twice the stoichiometric amount relative to C_2H_2.

(*a*) Compare and contrast the advantages of carrying out the hydrogenation reaction in the two different locations.

(*i*) Suggest possible catalysts and carriers.

(*ii*) How might the location of the hydrogenation reactor affect the choice of catalyst?

(*iii*) Assuming information were available on the apparent activation energies of suitable catalysts, would a low E or a high E be desired, or does it make any difference?

(*b*) With a naphtha feed considerable quantities of butadiene, benzene, toluene, etc., are produced, and it is desirable to recover these as byproducts. If the C_4+ fraction is removed from the cracker product, the remaining mixture is similar to the light-ends fraction above but the hydrogen content is much less, for example, 10 percent. Would this have any effect on the choice of a catalyst?

Notes: Assume that carbonaceous deposits may slowly build up on the catalyst so that it may be necessary to regenerate the catalyst by burning off the deposits by passing air at elevated temperature through the bed periodically (e.g., once every 2 to 8 months). The relative adsorptivities of the kind of hydrocarbons present here are in the order

$$\text{Acetylenes} > \text{diolefins and aromatics} > \text{olefins} > \text{saturates}$$

30. A mixture consisting of 1 mol % C_2H_2, 90 mol % C_2H_4, and 9 mol % H_2 is passed over a supported metal catalyst in an isothermal continuous-flow, fixed-bed reactor, keeping the quantity of catalyst constant. At high flow rates the ratio of moles of ethylene reacted to moles of acetylene reacted is quite low (a small fraction of unity). As the flow rate is decreased, however, we find that beyond a critical point the ratio increases substantially. In the high-flow-rate region the rate of reaction [defined as moles of H_2 disappeared/(second) (cubic centimeter of catalyst)] is essentially independent of flow rate. Below the critical flow rate, the rate of reaction *increases* with decrease in flow rate, at least over a considerable range of flows.

Interpret the results. There are no mass-transfer or diffusion limitations.

31. It has been reported that 2,2,4-trimethylpentane is converted to *p*-xylene with about 95 percent selectivity upon passage over a chromia catalyst supported on alumina, at about 500°C. If the chromia is supported on silica-alumina, the selectivity drops drastically and a wide variety of products is obtained. Explain these observations.

32. In a confined environment such as a nuclear submarine it is necessary to have a method of purifying the air of small amounts of organic impurities present as vapor or as finely divided aerosols. These may consist of smoke or aerosols from various sources including possible leaks in the high-pressure hydraulic fluid systems, cooking odors, human waste products, etc. Consider whether catalytic oxidation may be a feasible method.

(*a*) Compare and contrast this problem with that of catalytically oxidizing auto exhaust. Suggest one or more catalysts that should be suitable and any specific features of the catalyst-bed design that may be important.

(*b*) Suggest one or more methods other than catalytic oxidation for solving the problem, and compare their advantages and disadvantages relative to catalytic oxidation.

33. The catalytic partial oxidation reation of an organic compound A to form a desired product B is being studied using a fixed bed of a vanadium oxide catalyst:

$$A + O_2 \longrightarrow B$$

Substantial quantities of CO and CO_2 are also formed. Gas samples are removed at points 1 and 2, and at the exit, point 3, and analyzed, with results shown in Fig. B.4 (locations 1, 2, and 3 are not linear in scale).

(*a*) Explain the changes in selectivity observed (moles B formed/moles A consumed).

(*b*) Suggest a commercial reactor design that might maximize selectivity.

Note: There are no equilibrium limitations or diffusion limitations. The reactor operates isothermally, and adsorption of water, CO_2, or CO is not a significant factor.

34. The vinyl acetate–synthesis reaction (Sec. 8.7) is being studied in a laboratory fixed-bed reactor utilizing a high-area supported metal catalyst. The inlet feed stream is a vapor-phase mixture comprising about 80% C_2H_4, 7% O_2, and 13% CH_3COOH. The CH_3COOH does not react with the oxygen. Temperature is held constant at 150°C, and the *total* pressure is varied from a value somewhat below 0.1 MPa to a value considerably greater than 0.1 MPa, holding the inlet composition and feed rate constant. The rate of reaction (moles per unit of time) is found to vary with total pressure as shown in Fig. B.5.

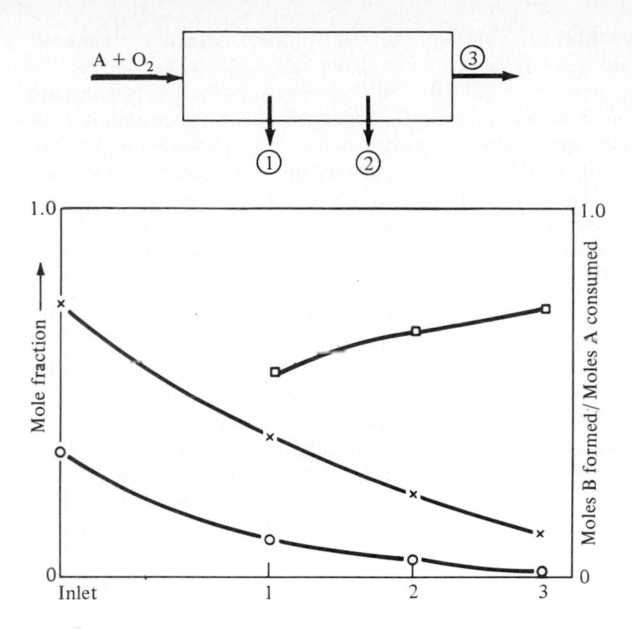

Figure B.4 Changes in gas composition and product selectivity with position in a partial oxidation reactor. x, mole fraction A; ○, mole fraction oxygen; □, moles B formed/moles A consumed.

At pressures below P_1 or above P_2 the rate is closely reproducible, but at pressures between P_1 and P_2 the rate depends on the way the pressure has been previously varied. If the run is started at a low pressure and the pressure is increased, the upper curve (*ABCDE*) is found; if the run is started at a high pressure and the pressure is lowered, the lower curve (*EDBA*) is found. Explain.

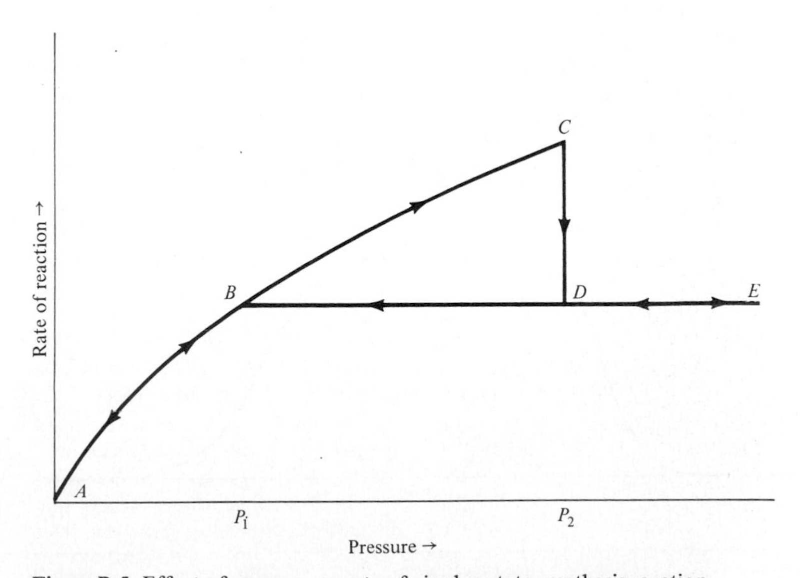

Figure B.5 Effect of pressure on rate of vinyl acetate-synthesis reaction.

35. A waste stream from a synthetic fiber plant consists essentially of dilute organic acids in water. A pilot plant has been set up to study the feasibility of oxidizing the acids to CO_2 and H_2O by heating the stream plus air under pressure to 200 to 250°C and passing the mixture upwards through a tall column packed with commercial FeO-Cr_2O_3 pelleted catalyst. The reactor is 15 cm in diameter and operates in the liquid phase, essentially isothermally and with excess air. Over the course of a few weeks the exit portion of the bed becomes choked up with deposits, and the reactor must be shut down because of excessive pressure drop. Analysis of the deposits shows them to contain iron and chromium in the form of a gelatinous precipitate. Stainless steel (18% Cr, 8% Ni, remainder Fe) is used for the equipment, but corrosion is negligible. What may be happening and what are some possible courses of action?

36. A catalyst manufacturer is studying the characteristics of a supported metal oxide catalyst that seems to offer promise for removal of NO from automobile exhaust by reduction to N_2. It is tested by flowing through the catalyst at 350°C a very dilute mixture of NO, CO, C_3H_6, and H_2O vapor in N_2, to which is added varying amounts of O_2, to simulate the exhaust gas produced under various operating conditions of the car. Figure B.6 shows a set of typical results. Up to time (1) the concentration of O_2 present in the inlet gas is only 0.5 mol % and a high percentage of the inlet NO is removed. At time (1) the concentration of O_2 in the feed gas is doubled to 1.0 mol %. At time (2) the concentration of O_2 is lowered back to 0.5 mol %. Elapsed time between (1) and (2), or (2) and (3), is 15 to 25 min. The gas residence time in the bed is a fraction of a second. What is noteworthy about this behavior? Explain the results.

37. A catalyst bed containing a hopcalite catalyst (an unsupported mixture of copper oxide and manganese oxide) is being evaluated as a means of removing traces of organic impurities from the air by catalytic oxidation to be used in a confined environment such as a space capsule or submarine. Some studies are being made with an experimental unit in the lab. The feed is a low concentration (less than 0.1 mol %) of a single, saturated, pure hydrocarbon (C_4H_{10}) in air, and the amount of reaction is being determined by removing a sample of the exit gas, determining its CO_2 content, and calculating percent conversion by a material balance based on the C_4H_{10} entering. (No CO is present.)

The operator has been examining data on CO_2 concentration versus time for the several runs made thus far and is concerned that the catalyst bed does not come up to full activity until a considerable number of hours have passed after the start of a run. A similar phenomenon occurs at each of several temperature levels studied. In each run the catalyst bed is brought up to the desired temperature, and then the hydrocarbon vapor is introduced. At the end of the run the reverse

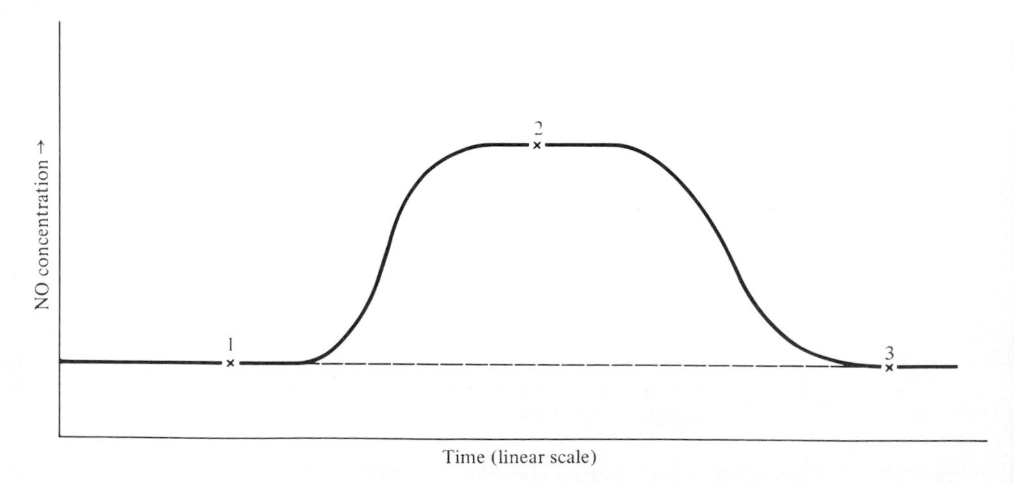

Figure B.6 Variation of exit NO concentration with time. Auto-exhaust catalyst.

procedure is adopted; that is, the hydrocarbon feed is shut off first, and then the bed is allowed to cool down with air being passed through it.

(*a*) What is your analysis?

(*b*) How can you check the validity of your hypothesis?

(*c*) What should we do about the situation?

38. We are searching for a catalyst to be used on an "oxygen sensor" that will determine the ratio of fuel to air being fed to an automobile engine and generate a signal that can be fed back through a microprocessor, which in turn causes appropriate adjustments to be made to the engine operation. The sensor is a thin wafer of zirconium oxide mounted in the exhaust line, immediately downstream from the engine; it is exposed to outside air on one side and to exhaust gases (which may contain some partially oxidized products) on the other. It generates a voltage that is a function of the difference in oxygen partial pressures on the two sides of the wafer. The automobile will operate on nonleaded gasoline, and the sensor should start to work just as soon as possible after the cold engine is started up.

Suggest a catalyst to be used on the exhaust-gas side.

39. Some scouting work is being done on a possible new process for partial oxidation of an organic substance in which a metal oxide is to be used as an "oxygen carrier." The oxide would be of a type that can exist in more than one valence form, and we visualize operating with two fluidized beds. In the first the organic substance would be contacted with the oxide in the highest oxidation state to form the desired product. The solid oxide leaving this reactor in a partly reduced state would pass to a second reactor, where it is reoxidized by contact with air. The operation would be controlled so as to prevent reduction to the metallic form. For economic operation the oxide must be capable of being recycled many times with little loss of activity.

(*a*) What might cause rapid loss in activity?

(*b*) How might the composition of the material be modified to increase its life?

(*c*) What are the possible advantages and disadvantages to such a process?

40. In the development of catalysts to oxidize automobile exhaust to CO_2 and H_2O, most investigators in the laboratory have used a mixture of CO, C_3H_6 (propylene), O_2, and N_2 to simulate the exhaust composition.

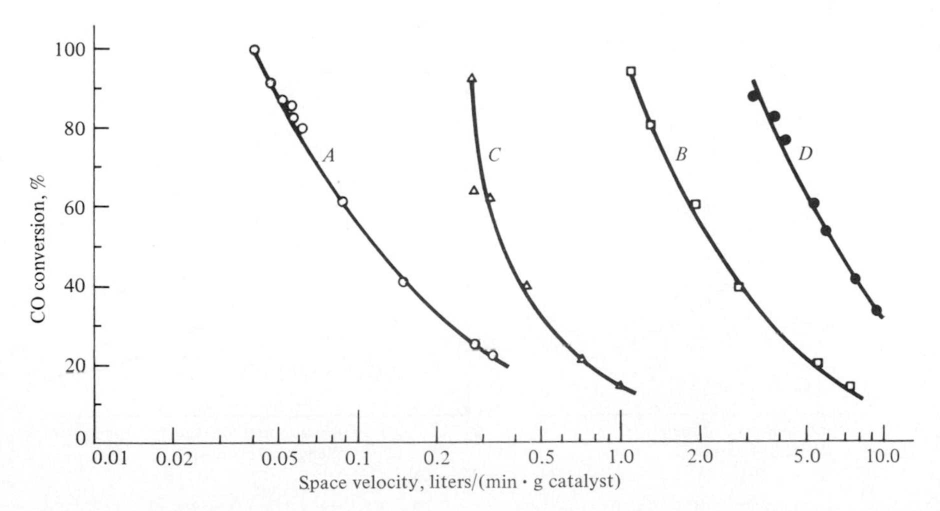

Figure B.7 Performance of an automobile-exhaust catalyst at 550°F. (*Voltz et al., 1973.*) (*Reprinted with permission from Industrial and Engineering Chemistry, Product Research and Development. Copyright by the American Chemical Society.*)

Voltz and coworkers [*Ind. Eng. Chem., Prod. Res. Dev.*, **12**, 294 (1973)] have performed studies with such a gas on a Pt/Al_2O_3 pelleted catalyst held in a fixed bed. Some of their results, at 550°F and atmospheric pressure, are shown in Fig. B.7.

(*a*) Qualitatively, what are the major features of the kinetics of this reaction in terms of enhancement or inhibition effects?

(*b*) If a rate expression were to be developed for this reaction, which are the *significant* terms that should appear in the numerator and the denominator? Why?

Notes: Assume that there are no concentration or temperature gradients in the reactor and that possible diffusion or mass-transfer effects may be neglected. *Space velocity* is the rate of feed of reactant, calculated at standard temperature and pressure, divided by quantity of catalyst.

	Feed composition*		
Curve	CO, %	O_2, %	C_3H_6, ppm
A	4.0	3.0	800
B	0.7	3.0	800
C	4.0	4.5	100
D	0.7	4.5	100

*Remainder is N_2.

41. The disproportionation of toluene into benzene and xylenes has been studied at 350°C and 200 kPa using a highly acidic, high-area catalyst (not a zeolite) in the form of pellets about 0.8 mm in size. Two reactors have been used: (1) a small continuous-flow reactor and (2) a pulse reactor. In this second reactor a continuous stream of helium was passed through the catalyst bed, and from time to time short pulses of toluene were injected into the stream so that alternating slugs of toluene vapor and helium contacted the catalyst. In both cases the toluene was completely vaporized before it reached the catalyst. In the first reactor the feed was pure toluene. The quantity of catalyst in the two reactors was about the same, and temperatures were the same; the flow pattern was essentially plug flow, so the contact time was about the same in both cases.

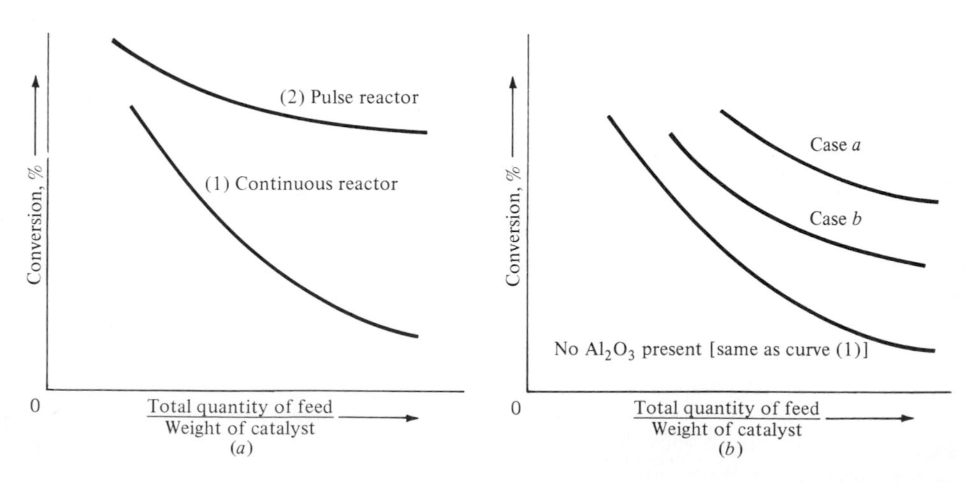

Figure B.8 Studies of the disproportionation of toluene. (*a*) Pulse-reactor studies. (*b*) Continuous-reactor studies.

Figure B.8*a* shows a plot of the percent conversion observed as a function of the ratio of total amount of toluene fed to the weight of catalyst in the reactor.

(*a*) Explain the difference between these two sets of results.

(*b*) The studies in the continuous reactor were repeated using a mixture of 10% catalyst and 90% γ-alumina in either of two forms: Case *a*, finely powdered catalyst and powdered alumina were intimately mixed and pelletized to the 0.8-mm size, and case *b*, 0.8-mm pellets of catalyst and of alumina were well mixed and placed in the reactor.

Figure B.8*b* shows the results, where weight of catalyst refers to the weight of active ingredient alone. The alumina by itself showed no catalytic activity whatsoever. Explain these results.

(*c*) Suggest further experiments that would assess the validity of your explanation to part (b).

42. With respect to Prob. 41

(*a*) Suppose the studies in part (*b*) were repeated except that a nonporous fine silica were to be used in place of alumina. Show qualitatively how you would expect the three curves in Fig. B.8*b* to appear under these circumstances.

(*b*) We are told that the beneficial effect of diluting the catalyst with alumina is less pronounced at higher temperatures. Does this seem plausible? The form of the alumina is not altered by the higher temperature.

43. Beuther and Larson [*Ind. Eng. Chem., Process Des. Dev.*, **4**, 177 (1965)] studied the hydrocracking of a "straight-run furnace oil" (containing 0.04 wt % sulfur, 18 ppm nitrogen compounds, 25 wt % aromatics) on a series of platinum catalysts supported on silica-alumina (25% alumina) in a reactor operated at 316°C, 5 MPa, and in the presence of a high partial pressure of hydrogen. The catalysts were prepared with various amounts of platinum (0.13 to 2.8 wt % platinum) and the total area of platinum exposed on each catalyst was determined (by chemisorption techniques) to vary from 0.05 to 0.6 m^2/g. The silica-alumina support had a surface area of about 350 to 400 m^2/g. Some of their results are shown in Figs. B.9 and B.10.

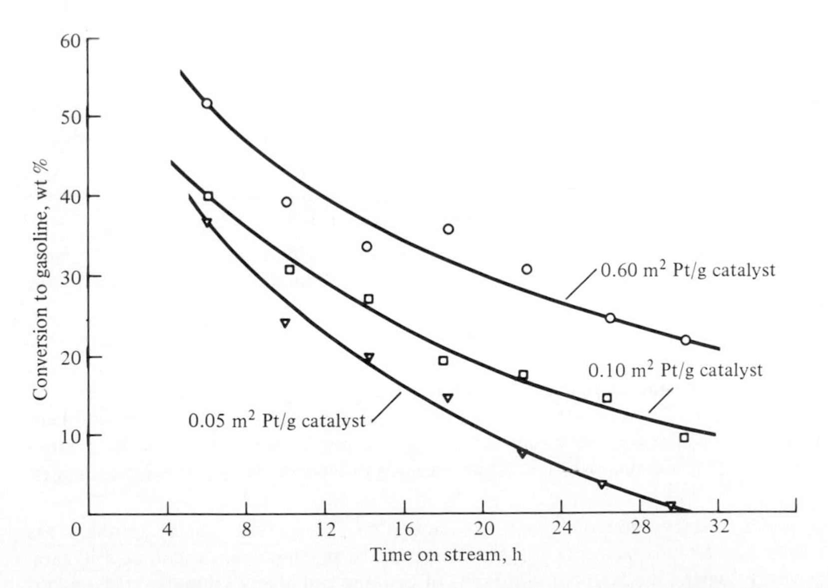

Figure B.9 Hydrocracking of furnace oil with a $Pt/SiO_2 \cdot Al_2O_3$ catalyst. Effect of metal area and time on aging rate. (*Beuther and Larson, 1965.*) (*Reprinted with permission from Industrial and Engineering Chemistry, Process Design and Development. Copyright by the American Chemical Society.*)

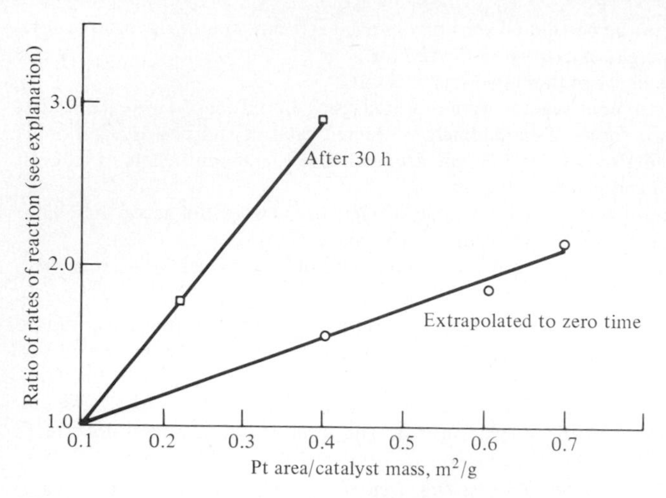

Figure B.10 Effect of Pt area on catalyst activity. (*Beuther and Larson, 1965.*) (*Reprinted with permission from Industrial and Engineering Chemistry, Process Design and Development. Copyright by the American Chemical Society.*)

Figure B.9 shows the percent conversion of furnace oil to gasoline as a function of time on stream, for three of the catalysts with different amounts of exposed platinum. Figure B.10 shows the ratio of the rate of reaction on catalysts having higher amounts of exposed platinum relative to that for the 0.10 m^2 platinum per gram of catalyst as a function of platinum area per gram of catalyst. These are shown for data extrapolated back to zero time and for data taken after 30 h on stream. (As an example, after 30 h the catalyst consisting of 0.40 m^2 platinum per gram of catalyst has an activity 2.9 times that of the 0.10 m^2 platinum per gram of catalyst.) On all these catalysts and under these reaction conditions the hydrocracking activity is lost fairly rapidly even in the absence of sulfur and nitrogen poisons. In the presence of poisons, activity declines more rapidly and the activity after a given time on stream is lower. Interpret these results in terms of what is happening on the catalyst surface.

44. Petroleum refiners find that they have poorer performance of their catalytic cracking reactors if the nitrogen content of the feed exceeds a few hundred parts per million. Why is this?

45. Consider a conventional fluidized-bed reactor in a petroleum refinery. Would one be more likely to find a higher concentration of multiring aromatic compounds in the gas oil *fed to* the reactor or in the *product from* the reactor having the same boiling-point range?

46. The isomerization of 1-butene has been studied on a series of mordenites that were prepared by previously treating a sodium mordenite with aqueous HCl of several different concentrations, followed by washing, drying, and calcination. By this means alumina was leached out and catalysts were prepared varying from 12 wt percent to 0.1 wt percent Al_2O_3. All other variables were kept constant, and it was found that the maximum activity was shown by a mordenite with an intermediate degree of removal of alumina. No significant amounts of chloride were left behind. Suggest an explanation.

47. A novel approach to conversion of coal to gasoline is to first convert coal to synthesis gas ($CO + H_2$) and then convert this to methanol. Methanol can be reacted over certain zeolite catalysts to form a product having the boiling-point range of gasoline and highly aromatic in character. The aromatics are predominantly benzene, toluene, and xylene, plus some trimethylbenzene. Under the same reaction conditions the aromatic products with hydrogen-mordenite (large-port type) are predominantly pentamethyl benzenes. With HY zeolite considerable hexamethylbenzene is observed, although it is not formed with the two other catalysts.

(*a*) Suggest a possible explanation.

(*b*) How would you test your hypothesis?

48. Crude petroleum contains varying amounts of metal porphyrins, especially of nickel and vanadium, which are found preferentially in the residual fraction from distillation. If such a residual fraction is hydrodesulfurized in a fixed-bed reactor, for some time on stream essentially all the metal is left behind on the catalyst. If individual catalyst particles are removed from the bed, the nickel is found to be distributed fairly uniformly through the particle but the vanadium is concentrated primarily in a thin layer on the outside of the particle.

Upon continued processing, metal content begins to appear in the exit stream. Would you expect the ratio of nickel to vanadium in the exit stream to be the same as or different from that in the feed stock?

49. In a manufacturing process, a pure dialkyl cyclohexane is being dehydrogenated to the corresponding aromatic derivative (the desired pure product) using a commercially available Pt/Al_2O_3 catalyst. Some undesirable isomerization is also occurring, which we wish to eliminate, but we want to continue using a Pt/Al_2O_3 catalyst. What can be done about the situation?

50. We have some conventional Pt/Al_2O_3 petroleum reforming catalyst with which we wish to hydrogenate some naphthalene derivatives, for example,

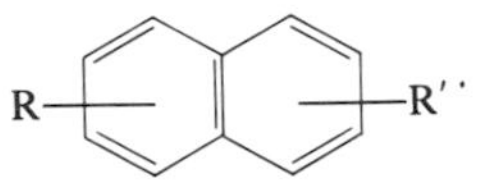

R—(naphthalene)—R′

Operating conditions will be about 350°C and 0.7 MPa.

(*a*) Do you forsee any difficulties in using this catalyst for this purpose?

(*b*) What might we do about it?

51. We wish to study the *thermal* cracking of a variety of individual hydrocarbons in a laboratory reactor consisting of a heated tube 2 mm in inside diameter and 1 m long. The tube is made of 18-8 stainless steel (18% Cr and 8% Ni), and we are obtaining an excessive amount of carbonaceous deposits on the inside walls. This does not occur to any substantial degree on Vycor glass (nearly pure silica), but we cannot use Vycor here. Can you suggest any treatment of the stainless steel that might reduce the amount of deposits?

52. A catalytic process for isomerization of normal hexane is being considered as part of a highly integrated petroleum refinery. It has been proposed to use a "standard" reforming catalyst for this purpose.

(*a*) Would you expect this catalyst to be suitable?

(*b*) What changes to the catalyst composition or modifications to the method of carrying out the reaction might be an improvement?

53. The isomerization of 1, 5-dimethylnaphthalene (DMN) is being studied on an acidic catalyst held in a fixed-bed laboratory reactor (the desired reaction is to form other dimethylnaphthalene isomers). Reaction conditions are 0.1 MPa total pressure and 300°C. For a series of experiments the catalyst bed volume has been held constant, the feed rate of DMN vapor has been held constant, and the partial pressure of DMN in the feed has been varied by changing the feed rate of nitrogen mixed with the DMN before the mixture enters the reactor. (Note carefully what has been varied.)

Figure B.11 shows how the percent conversion of DMN and the selectivity changed with partial pressure, when partial pressure was adjusted by the above procedure.

(*a*) Qualitatively, what kind of a kinetic expression is needed to give a conversion curve of this shape?

(*b*) Explain why the maximum selectivity occurs at the lowest partial pressure.

Note: The byproducts are predominantly substituted naphthalenes containing zero, one, three, or four methyl groups, presumably formed by disproportionation reactions. *Conversion* is the percent reactant disappeared. *Selectivity* is the ratio of moles of other dimethyl isomers formed to moles of 1, 5-DMN disappeared. Mass-transfer effects are insignificant.

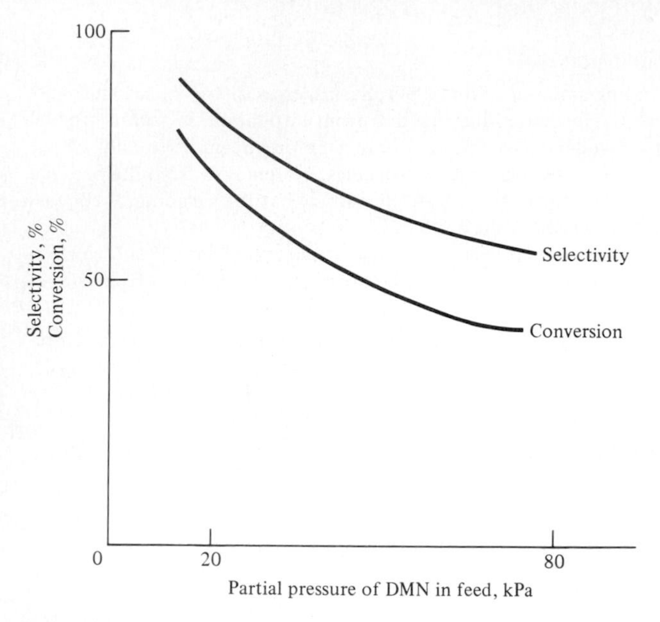

Figure B.11 Effect of partial pressure of dimethylnaphthalene on selectivity and conversion (Prob. 53).

54. In a refinery the product from a catalytic cracking reactor is separated into several fractions including a C_4 cut, a portion of which is polymerized with a "mounted" phosphoric acid catalyst to make *polymer gasoline.* This is primarily butylene dimer (C_8H_{16}), although it contains a considerable distribution of molecular weight. This stream is then hydrogenated with a commercial cobalt molybdenum catalyst, sulfided before use, to reduce the amount of unsaturation present and hence improve resistance to gum formation on storage. (Operating conditions are about 300°C, 3 MPa, and substantial excess H_2, in a fixed-bed reactor.)

It has been found, rather surprisingly, that under plant conditions the drop in catalyst activity with time on stream is considerably less than was found in careful laboratory studies with a small reactor using carefully purified octenes.

(*a*) Explain.

(*b*) If this catalyst were to be used for another type of hydrogenation reaction, what precautions are suggested?

(*c*) It has been reported that in another refinery, if the feed composition or operating conditions are slightly altered so that some of the "gasoline" is in the liquid phase in the reactor, the life of the catalyst is substantially increased. Explain.

55. Natta [in P. H. Emmett (ed.), *Catalysis*, vol. 3, p. 349] has reported the results of very careful studies of the methanol-synthesis reaction at temperatures of 330 to 390°C and pressures of 20 to 30 MPa on conventional ZnO/Cr_2O_3 catalyst. The expression for the rate of reaction is given in Sec. 10.4.3.

He specifically notes that the data are much better fitted with the denominator cubed than squared. He also reports the effect of temperature on the constants A, B, C, and D as shown in Fig. B.12. From adsorption studies with the separate species under nonreaction conditions he reports that the heats of adsorption vary in the order $CH_3OH > CO > H_2$. From adsorption studies at somewhat lower pressure in the absence of reaction, but the same temperature range, he reports that over short periods of time adsorption of CO is higher than that of H_2, but if equilibrium is

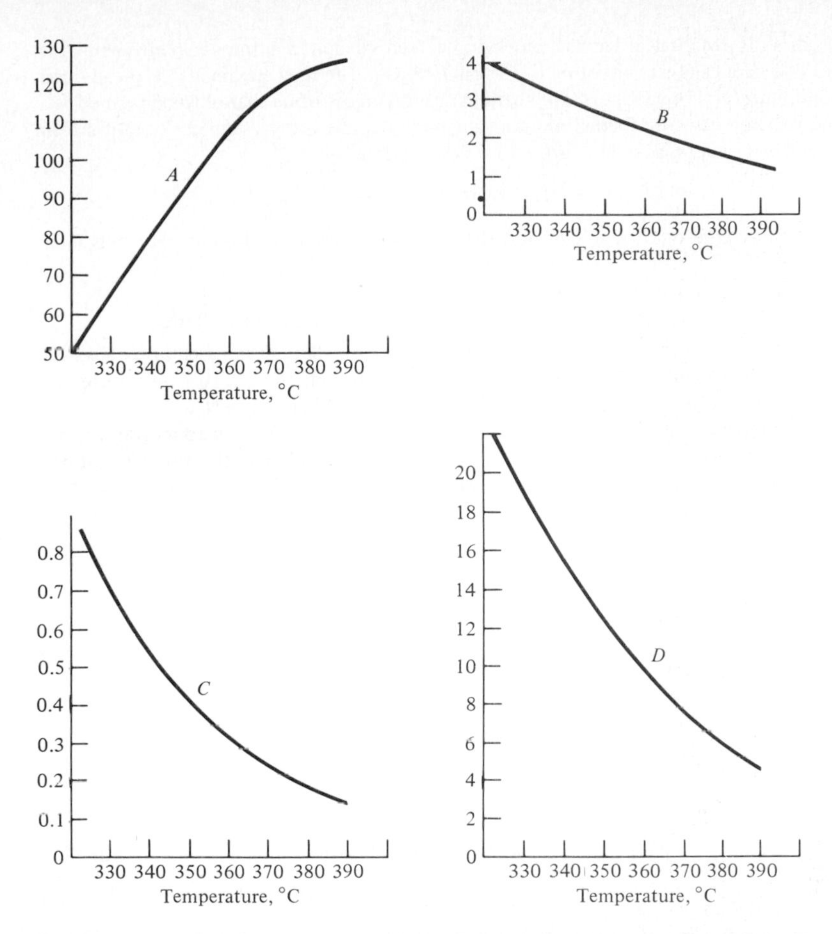

Figure B.12 Dependence on temperature of the constants A, B, C, and D in Eq. (10.9). (*Natta, 1955.*)

established the amounts of CO and H_2 adsorbed on this catalyst are about equal, for equal partial pressures.

(*a*) Derive Eq. (10.9) in terms of a Langmuir-Hinshelwood model.

(*b*) Are the results reported for the constants A, B, C, and D plausible within the framework of the Langmuir-Hinshelwood model?

(*c*) Are the adsorption studies and the kinetic studies consistent?

56. It has been reported that the apparent activation energy for MeOH synthesis on a conventional ZnO/Cr_2O_3 catalyst is about 115 to 125 kJ/mol at a low percentage conversion. The conditions are not stated. Is this consistent with Natta's equation if it is assumed that the dominating term in the denominator is that for CO adsorption? Representative values for the effect of temperature on the constants in the equation are:

$$\frac{1}{A^3} \alpha e^{+172/RT} \qquad \frac{B}{A} \alpha e^{+100/RT} \qquad \frac{C}{A} \alpha e^{+126/RT} \qquad \frac{D}{A} \alpha e^{+121/RT}$$

57. The water-gas shift reaction is carried out on the feed stream to ammonia-synthesis plants, utilizing a fixed bed of a catalyst consisting of Cu-ZnO/Al_2O_3. The feed stream also typically contains traces of ammonia, and the copper in the catalyst causes traces of methanol to be formed.

It is found in some cases that small amounts of methylamine are present in the exit stream from the above catalyst bed, presumably formed by the reaction:

$$CH_3OH + NH_3 \rightarrow CH_3NH_2 + H_2O$$

Methylamine can create odor problems and its formation is undesirable, but copper is indispensable in the catalyst.

(*a*) What might cause this reaction to occur?

(*b*) How might the catalyst composition be altered so as to minimize this reaction?

58. The rate of isomerization of cyclopropane to propylene on a silica-alumina catalyst has been studied in a vertical fixed bed. Temperatures were in the range of 50 to 110°C. Pressure was atmospheric, and the cyclopropane was present in the order of a few percent in N_2.

In each case a liquid was allowed to trickle down over the catalyst, flowing concurrently with the cyclopropane. Keeping all other variables constant, it was found that the rate of reaction dropped dramatically as the nature of the liquid was varied, as shown in the following table. In no case did any of the liquid react.

	Highly paraffinic (>99%) liquid, average MW ~400	Highly paraffinic (~97%) liquid, MW ~150	Highly aromatic liquid, MW ~150
Relative rate of reaction	1	0.3	0.02

Explain the results. The highly paraffinic liquid consisted mostly of cycloparaffins with alkyl side chains.

59. With respect to Prob. 58:

(*a*) If the same study were to be repeated with the two paraffinic liquids but at a higher temperature, would you expect the ratio of the two rates to remain at 0.3, to increase, or to decrease? Why?

(*b*) What would you expect to happen if glycerol were used as the liquid?

60. A researcher is studying the reaction

$$CH_2{=}CH{-}CH_2{-}CH_3 \longrightarrow H_2C{=}\overset{\displaystyle H}{\overset{|}{C}}{-}\overset{\displaystyle H}{\overset{|}{C}}{=}CH_2 + H_2O$$

by passing a stream of pure helium continuously through a bed containing coarse particles of a metal oxide. From time to time he injects a small pulse of 1-butene into the helium stream and measures the percent conversion by on-line gas chromatography. No gaseous oxygen is fed to the system at any time, and the system is isothermal. The quantity of 1-butene injected in each pulse is kept constant, and the interval between pulses is also kept constant. Fig. B.13 shows the percent conversion observed as a function of number of pulses.

The study is now repeated keeping all procedures exactly as before except that the interval between pulses is now doubled. (The size of the pulse is the same.) The new results are likewise shown in Fig. B.13.

What seems to be happening?

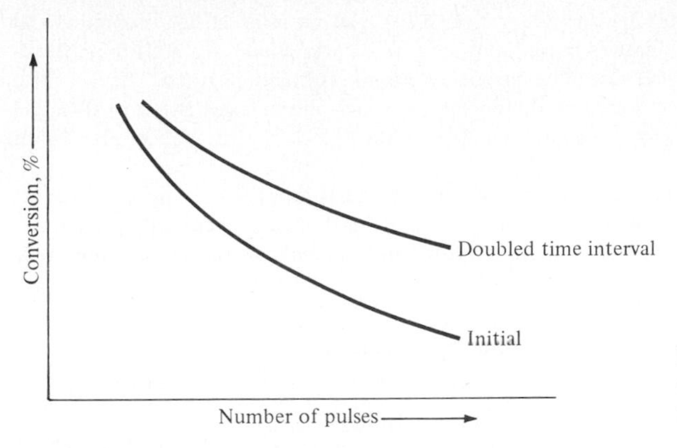

Figure B.13 Effect of pulsing interval on oxydehydrogenation of butylene.

61. Explain why the rate of reduction of particles of WO_3 by hydrogen gas is greatly accelerated if the WO_3 is coated with finely divided platinum.

62. An organic reactant in an inert solution is being used as a working fluid in a process through which it is recycled continuously. The reactant slowly decomposes to form high-molecular-weight material, which must be removed to prevent its concentration from rising to undesirable levels. A side stream is therefore withdrawn continuously and passed through a bed of alumina, which adsorbs this impurity. Recently it has been found necessary to raise the temperature of the adsorption bed to improve the degree of removal. The organic molecule has a paraffinic side chain, R, in the 1 position, but we are now finding in the working fluid some material with R in the 2 position, which is undesirable. What seems to be happening and what can be done about it?

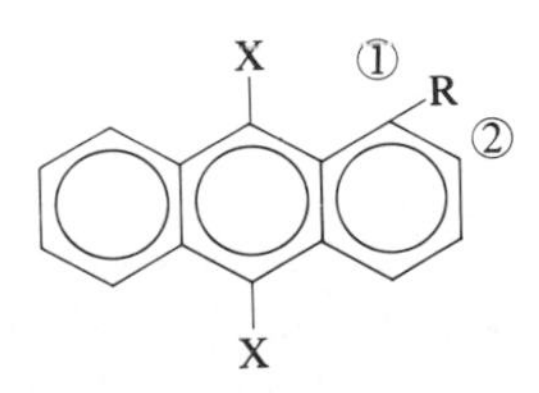

63. A plant adsorbs impurities continuously from an organic stream in a large vessel about 2 m in diameter and 6 m high packed with 8-mm alumina pellets. From time to time the bed is regenerated by a rather elaborate procedure; most of the impurities are first washed off with a hydrocarbon stream, and then the bed is treated with aqueous caustic, neutralized, steamed, etc., before being put back on stream.

After a number of regenerations the charge is intended to be dumped (by opening a discharge line in the bottom of the vessel) and replaced with fresh alumina.

The process was developed in a small pilot plant unit 15 cm in diameter × 1.3 m in height, and the plant unit duplicates the pilot plant cycle exactly with respect to cycle times, temperatures, etc. The performance of the commercial plant unit was found to be as predicted from the pilot plant unit, but when it first came time to empty the adsorption vessel, it was found that the alumina pellets in much of the vessel were cemented together and could only be removed by sending in workers with jackhammers. Cementing never occurred in the pilot plant unit.

(*a*) What seems to have happened?

(*b*) We wish to set up a program of studies with the pilot plant unit to indicate how to modify our commercial operation so cementing will not occur again. We wish to continue to use a fixed bed of alumina. What are your suggestions?

64. Phosgene is made by the reaction: $CO + Cl_2 \longrightarrow COCl_2$. A mixture of carbon monoxide and highly pure chlorine in the vapor phase is passed over activated carbon held in a multitube, fixed-bed reactor with cooling on the shell side. The carbon monoxide is highly purified. The last stage of the purifying process includes a water scrubbing step in which the exit gas is fed to silica gel driers, which reduce the water content to an average dew point of 0°C. The gas is then mixed with chlorine and fed to the reactor.

Over the course of several months the degree of conversion has dropped, and upon shutdown we find that some of the tubes are less than half full of catalyst, although all were fully packed initially. In laboratory work at the same reaction temperature this phenomenon was not encountered.

(*a*) What seems to have happened?

(*b*) What should we do about it?

65. Suggest one or more catalysts to convert ethylene to ethyl alcohol.

66. Suggest a solid catalyst to replace $AlCl_3$ in the alkylation of an aromatic compound with an olefin.

67. Suggest a catalyst composition for a coating to go on the walls of self-cleaning household ovens.

68. Suggest a catalyst for the anode of a fuel cell running on hydrogen and air (hydrogen is contacted at the anode). The electrolyte is KOH.

69. A liquid-phase catalytic reaction is being carried out continuously in either of two reactors; (1) a fixed bed containing pellets of catalyst or (2) a well-stirred vessel containing powdered catalyst. The feed stream contains tiny amounts of impurities, which cause the activity of the catalyst to decrease gradually. Reaction conditions (temperature, pressure) are the same in the two vessels, and there are no mass-transfer or heat-transfer gradients. The catalyst structure is identical in the two cases, both physically and chemically. For a given feed rate and given quantity of catalyst, the catalyst activity drops off much faster in the well-stirred vessel than in the fixed bed. What might cause this?

70. During a shutdown period a sample of catalyst has been removed from the top of a fixed-bed reactor used to catalyze a partial oxidation reaction utilizing air. Elemental analysis shows the presence of considerable quantities of P, Cl, and Fe. Where might these come from? (These are not present in the fresh catalyst.)

71. A partial oxidation gas-phase catalytic reaction has been studied in the laboratory using 3-mm-diameter pellets in a 12-mm-ID tubular reactor held vertically in a tube furnace. The bed length is about 20 cm. The results are somewhat different depending upon whether upflow or downflow is used.

(*a*) What might cause such differences?

(*b*) How might the experimental procedure be altered to obtain more reliable data?

72. We have been studying the feasibility of burning off coke from a catalyst made up of hard pellets about 0.8 mm in diameter and in length, by stirring them in a layer about 5 cm thick spread out on a hearth furnace in contact with air. Three runs have been made, each on a catalyst sample removed at a different time from the commercial reactor in which the coking occurs, and each being made at a different hearth furnace temperature. Typical data on the carbon content of the catalyst as a function of time at each of three temperatures are shown in Fig. B.14.

(*a*) Analyze these data in whatever fashion you believe will be most meaningful in order to use these results in the design of a larger production-scale furnace.

(*b*) Can you tell from these results whether all atoms of carbon on the catalyst are equally exposed to air or instead are present as small clumps, or is this question indeterminable from the data available?

(*c*) Compare the value of the reaction rate constant here with Fig. 9.4.

Notes: The total surface area of the catalyst is about 400 m^2/g. Assume that the three sets of data are representative of what would be obtained by further experimentation and that the reported temperatures are uniform in the furnace and are accurate. A large excess of air is present at all times, so the partial pressure of oxygen is about 21 kPa. Note that the three samples of catalyst

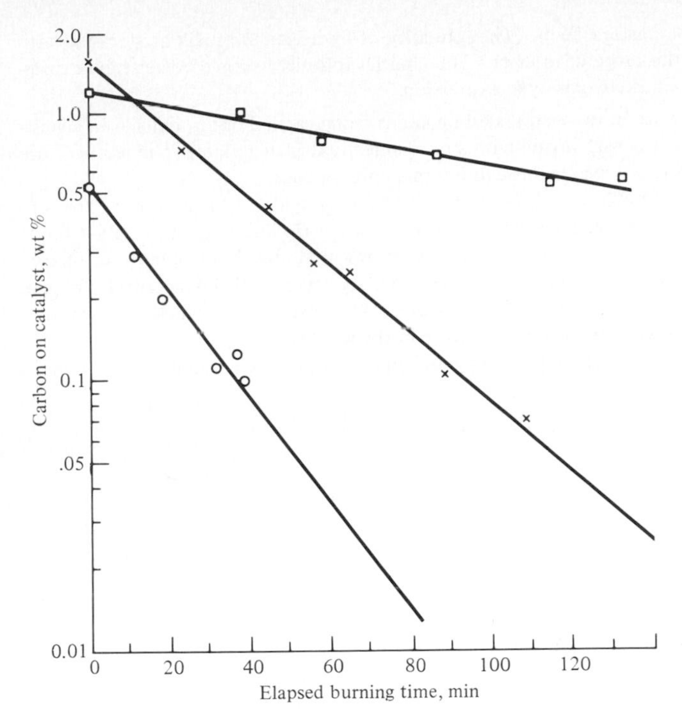

Figure B.14 Burnoff of carbon from a catalyst, □, 720 K; ×, 765 K; ○, 810 K.

have somewhat different initial concentrations of carbon. The catalyst is similar in type to silica-alumina and contains no heavy metals.

73. The kinetics of the SO_2 oxidation reaction are being studied in a laboratory fixed-bed reactor 3 cm in diameter and 40 cm long held essentially isothermal in a fluidized sand-bath heater and operated at industrial reaction temperatures. The catalyst is a conventional supported vanadium pentoxide (which is yellow) as 1.2-mm-diameter pellets, diluted about 4:1 with γ-alumina (which is white) of the same size. The two are well mixed initially.

When the reactor is opened up after several runs have been made, the alumina is yellow. What may have happened?

74. The hydrogenation of an organic compound A with a finely divided catalyst is being studied in a well-agitated flask held in a constant-temperature water bath in the laboratory using a 50:50 mixture of A plus water in which some sulfuric acid (solution is about 2 m) is dissolved for reasons that do not concern us here. The organic compound and the water are essentially immiscible in one another, but the agitation is excellent, one liquid is highly dispersed in the other, and the catalyst is well dispersed. The reaction is of the type $A \rightarrow B$, and side reactions are insignificant.

The rate of reaction is studied at atmospheric pressure and temperatures of 40 to 90°C, passing in hydrogen continuously and removing samples of the liquid from time to time for analysis. Under comparable conditions (e.g., same percent conversion) it is found that the rate increases with temperature, reaches a maximum at about 70 to 75°C, and then decreases. Results are reproducible.

What can cause this? If this reaction were to be scaled up and it is desired to run at about 100°C at a high rate, what do you suggest?

Notes: Visualize the organic compound as having properties like an alcohol or aldehyde, relatively nonviscous and of moderately low molecular weight (boiling point about 150°C). Assume there are no mass-transfer or heat-transfer effects. (The solubility of hydrogen *increases* moderately with increased temperature over the range of interest.) The kinetics follows a simple power-type expression rather than a Langmuir-Hinshelwood–type expression.

75. The reaction in Prob. 74 has now been scaled up and is being carried out in batches in a glass-lined 10,000-liter vessel, which is well stirred with a mechanically driven agitator. The temperature is well controlled by circulating city water through internal cooling coils.

The catalyst (a supported noble metal) is used over and over again until its activity drops to an unacceptable level. After a few months time, however, we are finding that the catalyst life is becoming shorter and shorter. Analysis of spent catalyst shows that significant quantities of iron are present, and this is causing the poisoning. Analysis of catalyst as supplied, reactant, hydrogen, acid, and distilled water charged to the reactor shows them all to be free of iron. No iron or iron-alloy surfaces come in contact with the reaction mixture in the reactor.

What might be happening, how could you test your hypothesis, and what could be done about it?

APPENDIX

C

SYMBOLS AND CONVERSION FACTORS FOR UNITS

SYMBOLS

meter	m
gram	g
second	s
temperature in kelvins	K
gram mole	mol
newton	$N = kg \cdot m/s^2$
joule	$J = N \cdot m$
watt	$W = J/s$
pascal	$Pa = N/m^2$

PREFIXES

10^{-9}	nano	n
10^{-6}	micro	μ
10^{-3}	milli	m
10^{-2}	centi	c
10^{-1}	deci	d
10^{3}	kilo	k
10^{6}	mega	M

CONVERSION FACTORS

Length

1 in	25.4 mm
1 Å (angstrom)	0.1 nm
1 μ (micron)	1 μm
1 mil	0.0254 mm

Volume

1 ft^3	0.0283 m^3
1 U.S. gal	3785.4 cm^3
1 barrel (for petroleum, 42 gal)	0.15899 m^3

Mass

1 oz	28.352 g
1 troy oz	31.1 g
1 ton (2000 lbm)	907.2 kg

Force

1 dyn	10^{-5} N
1 lbf	4.4482 N

Temperature Difference

1°F (°R)	$\frac{5}{9}$°C (K)

Energy

1 cal	4.1868 J
1 Btu	1.055 kJ
1 kW · h	3.6 MJ

Calorific Value

1 Btu/ft^3	37.259 kJ/m^3

Density

1 lb/ft^3	16.019 kg/m^3

Pressure

1 lbf/in^2	6.8948 kN/m^2 (or kPa)
1 lbf/ft^2	47.880 N/m^2 (or Pa)
1 standard atmosphere	101.3 kN/m^2 (or kPa)
1 bar	10^5 N/m^2 (or Pa)
1 mmHg (1 torr)	133.32 N/m^2 (or Pa)

Power

1 Btu/h	0.29307 W
1 hp	745.70 W

Viscosity, Dynamic

1 P (poise)	0.1 N · s/m^2 (or Pa · s)
1 lb/ft · s	1.4882 N · s/m^2 (or Pa · s)

Mass-Flux Density

1 lb/h · ft^2	1.3562 g/s · m^2

Heat-Flux Density

1 Btu/h · ft^2	3.1546 W/m^2
1 kcal/h · m^2	1.163 W/m^2

Heat-Transfer Coefficient

1 Btu/h · ft^2 · °F	5.6783 W/m^2 · K

Specific Enthalpy

1 Btu/lb	2.326 kJ/kg

Heat Capacity

1 Btu/lb · °F	4.1868 kJ/kg · K

Thermal Conductivity

1 Btu/h · ft · °F	1.7307 W/m · K
1 kcal/h · m · °C	1.163 W/m · K

INDEX